D0997133

DEEP-SEA SEDIMENTS

Physical and Mechanical Properties

MARINE SCIENCE

Coordinating Editor: Ronald J. Gibbs, Northwestern University

Volume 1—*Physics of Sound in Marine Sediments*
Edited by Loyd Hampton, 1974
An Office of Naval Research symposium
Consulting Editors: Alexander Malahoff and Donald Heinrichs
Department of the Navy

Volume 2—*Deep-Sea Sediments: Physical and Mechanical Properties*
Edited by Anton L. Inderbitzen, 1974
An Office of Naval Research symposium
Consulting Editors: Alexander Malahoff and Donald Heinrichs
Department of the Navy

In Preparation:

Natural Gases in Marine Sediments
Edited by Isaac R. Kaplan
An Office of Naval Research symposium
Consulting Editors: Alexander Malahoff and Donald Heinrichs
Department of the Navy

Suspended Solids in Water
Edited by Ronald J. Gibbs
An Office of Naval Research symposium
Consulting Editors: Alexander Malahoff and Donald Heinrichs
Department of the Navy

DEEP-SEA SEDIMENTS

Physical and Mechanical Properties

Edited by

Anton L. Inderbitzen

College of Marine Studies
University of Delaware
Lewes, Delaware

PLENUM PRESS • NEW YORK AND LONDON

Library of Congress Cataloging in Publication Data
Main entry under title:

Deep-sea sediments: physical and mechanical properties.

(Marine science, v. 2)
Proceedings of a symposium conducted by the Ocean Science and Technology Division of the Office of Naval Research at Airlie House, Va., Apr. 24-27, 1973.
Includes bibliographies.
1. Marine sediments — Testing — Congresses. I. Inderbitzen, Anton L., ed. II. United States. Office of Naval Research. Ocean Science and Technology Division.
GC380.D43 551.4'6083 74-7140
ISBN 0-306-35502-7

Proceedings of a symposium conducted by the Ocean Science and Technology Division of the Office of Naval Research on Physical and Engineering Properties of Deep-Sea Sediments held in Airlie, Virginia, April 24-27, 1973

A Division of Plenum Publishing Corporation
227 West 17th Street, New York, N.Y. 10011

United Kingdom edition published by Plenum Press, London
A Division of Plenum Publishing Company, Ltd.
4a Lower John Street, London, W1R 3PD, England

Printed in the United States of America

PREFACE

As part of its continuing program to stimulate superior basic research in the marine environment, the Office of Naval Research, Ocean Science and Technology Division, sponsored a series of closed seminar-workshops in 1972-1973. Each seminar focused upon one research area of marine geology which is relatively new and in need of a critical evaluation and accelerated support. The subjects areas chosen for the seminars were:

1. natural gases in marine sediments and their mode of distribution,
2. nephelometry and the optical properties of ocean waters,
3. physical and engineering properties of deep-sea sediments, and
4. physics of sound in marine sediments.

The objectives of each seminar-workshop were to bring into sharper focus the state-of-the-science within each subject area, to effect some degree of coordination among the investigators working within each of these areas and to provide the Ocean Science and Technology Division guidance for national program support.

This volume contains most of the papers presented at the seminar on the physical and engineering properties of deep-sea sediments. The seminar was held at Airlie House, Airlie, Virginia on April 24-27, 1973 and was organized and chaired by A. Inderbitzen. The attendees were invited from among the leading investigators in this field from both the engineering and scientific disciplines. Each attendee was requested to prepare a paper within his area of speciality.

The purpose of the papers was to present a fairly complete picture of the state of our present knowledge. Three workshops were also held during the seminar. Their purpose was to define the most serious problems presently facing this field of study, discuss possible solutions and recommend courses of action. All participants at the seminar contributed to the open give and take of the workshops and in the formulation of the recommendations as presented in

the workshop summaries at the end of this volume.

Each of the three workshops addressed a different facet of this area of research and its problems. At the first workshop, the basic phenomena that we are studying were discussed in an attempt to try to determine the primary research goals upon which effort should be focused. The second workshop discussed the limitations of our present instrumentation and possible means of improving and expanding the research tools. A promising attempt was made in the third workshop to standardize units, symbols, nomenclature and some test techniques to be used in studying and reporting on the mechanical properties of marine sediments. It was hoped, by the participants, that these units, terms and techniques would be adopted by workers in this field of research and become standards through general acceptance and use. Results of all three workshops are presented at the end of this volume.

The entire seminar-workshop was kept as informal as possible and open to continuous dialogue between scientists and engineers working in the field. Research work within this area of ocean science is being conducted by both scientists and engineers which often leads to a dichotomy in goals and techniques. It was the sincere hope of ONR that one by-product of this seminar-workshop would be a better understanding between engineers and scientists of each other's research goals, immediate problems, and approaches to solving these problems. Through a mutual understanding, greater progress will be made towards all of our research goals in studying the physical and mechanical properties of deep-sea sediments.

I believe the seminar-workshop was successful in reaching its goals of summarizing the present state-of-the-science, recommending some focus to the direction of future research efforts and providing a means for open dialogue between fellow research workers. After reading these papers and the workshop summaries, I hope you agree.

I would especially like to thank Mses. R. Foner, A. LeCates, and D. Fuhrman for their invaluable aid in editing and preparing the manuscripts for final typing and Ms. R. Stegner for typing the manuscripts. Without their help, I doubt if this book would have been published.

Anton L. Inderbitzen

University of Delaware
January, 1974

CONTENTS

PREDICTION OF DEEP-SEA SEDIMENT PROPERTIES: STATE-OF-THE-ART

EDWIN L. HAMILTON

Naval Undersea Center

ABSTRACT

The accuracy of predictions of sediment properties is dependent on relationships between various physical properties and/or their expectable statistical variability. In shallow water, sediment types, and consequently their properties, may vary greatly over short distances. In the deep sea, sediment types are fewer, and the main types occupy vast areas. Within the term "deep-sea sediments" must be included the turbidites which form large abyssal plains, rises, and fans in the deep sea.

Prediction of sediment surface properties in deep-sea sediments falls into two categories: (1) a sediment sample is available, or (2) location, only, is provided. Given a sediment surface sample in which water content, grain size, and grain density can be measured, porosity and density can be computed or measured. Given only a dried sample, grain size data provide predictive relations with the other properties. Having derived one or more of these properties, velocity of compressional waves can be predicted within about 1 to 2% because of its relations with grain size and porosity. Given certain restrictive conditions, reflection coefficients and bottom losses at normal incidence can be computed. All of these properties can be corrected to in situ conditions. With less confidence, values of compressional wave attenuation can be approximated because of relations with grain size and porosity, and values of elastic properties (e.g., shear-wave velocity) can be computed. The values of attenuation and the elastic properties require additional confirmation by future in situ measurements. There is apparently no usable relationship between soil mechanics shear strength and velocity, but there may be a relation between dynamic rigidity and attenuation of compressional waves.

If sediment samples are not available and location, only, is provided, the geologist or geophysicist must use available information to predict, in sequence, the physiographic province, sedimentary environment, and sediment type. After the sediment type is predicted, available tables can be entered to determine averaged laboratory properties which can then be corrected to in situ values.

Although surface properties may be reasonably predicted, there is a lack of data to predict many property gradients with depth in the sediment body. At present, compressional velocity vs. depth in sediments can be reasonably predicted because of sonobuoy measurements, but more measurements are needed. The attenuation of compressional waves with depth is not known, but is speculated to decrease with overburden pressures. A small amount of data is available on consolidation, which, if supplemented, will allow prediction of the variations of density and porosity with depth. The least known properties in deep-sea sediments are the velocity and attenuation of shear waves.

Tables of sediment properties previously published by the writer are supplemented by additional measurements from 12 different sources, and new regression equations and diagrams illustrate some interrelationships.

INTRODUCTION

This report will present previously unpublished results of measurements of sediment physical properties by the writer (since 1970), and will discuss the present state-of-the-art in predicting some properties. Those properties considered are indicated in table and text headings. They are mostly acoustic and related physical properties. Previous reports by the writer contain numerous references to the pertinent literature, and no attempt is made to compile an exhaustive bibliography.

Aside from use in some theoretical studies, the only utility in knowing the laboratory properties of sea-floor sediments is to be able to predict them for the in situ condition. This subject was discussed in some detail in a previous report (Hamilton, 1971b). Some aspects of correcting laboratory to in situ values will be noted in connection with specific physical properties.

After the 1971 report concerning prediction of in situ properties, the prediction of two other properties have been studied. Prediction of attenuation of compressional waves was reported by Hamilton (1972). Prediction of the vertical gradients of the velocities of compressional waves (previously unpublished) will be briefly discussed below. A brief discussion of prediction of interval velocities of sediment layers and thicknesses will be included.

TABLE 1a

CONTINENTAL TERRACE (SHELF AND SLOPE) ENVIRONMENT;
Average Sediment Size Analyses and Bulk Grain Densities

Sediment Type	No. Samples	Mean Grain Dia. mm	Mean Grain Dia. ϕ	Sand, %	Silt, %	Clay, %	Bulk Grain Density g/cm^3 (kg/m$^3\cdot 10^{-3}$)
Sand							
Coarse	2	0.5285	0.92	100.0	0.0	0.0	2.710
Fine	18	0.1638	2.61	92.4	4.2	3.4	2.708
Very fine	6	0.0915	3.45	84.2	10.1	5.7	2.693
Silty sand	14	0.0679	3.88	64.0	23.1	12.9	2.704
Sandy silt	17	0.0308	5.02	26.1	60.7	13.2	2.668
Silt	12	0.0213	5.55	6.3	80.6	13.1	2.645
Sand-silt-clay	18	0.0183	5.77	33.3	40.2	26.5	2.705
Clayey silt	54	0.0074	7.07	5.9	60.6	33.5	2.656
Silty clay	19	0.0027	8.52	4.8	41.2	54.0	2.701

Prediction of in situ sediment-surface properties falls into two categories: (1) a sediment sample is available, or (2) location, only, is provided. Given a saturated, "least disturbed", sediment-surface sample, many physical properties can be measured and computed. Given only a dried sample, grain-size data provide predictive relations with other properties. Having derived one or more properties, many others can be approximated by entry into diagrams or equations showing interrelationships between properties. Laboratory values can be corrected to in situ conditions, if necessary.

If sediment samples are not available and location, only, is provided, the geologist or geophysicist must use available information to predict, in sequence, the physiographic province, sedimentary environment, and sediment type. After the sediment type is predicted, available tables can be entered to determine average laboratory properties which can then be corrected to in situ values.

Although many sediment-surface properties may be reasonably predicted, there is lack of data to predict many property gradients with depth in the sediment body. Among the least-known sediment properties are the velocity and attenuation of shear waves. Additional measurements and study are required to confirm predicted values of elastic properties and the compressional-wave attenuation (especially at frequencies below 1kHz).

METHODS AND RESULTS OF RECENT MEASUREMENTS

Methods

The methods used in the laboratory and in situ measurements were as described and discussed in previous reports (Hamilton, 1970b; Hamilton et al., 1970). Sediment nomenclature in the tables and discussions follows Shepard (1954), except that, within the sand sizes, the various grades of sand follow the Wentworth scale.

As in previous reports, the sediments are divided into three general environments: (1) the continental terrace (shelf and slope), (2) the abyssal hill environment, and (3) the abyssal plain environment.

Most of the new measurements of velocity were made in the laboratory at 200 kHz as described in Hamilton (1970b). One set of velocity measurements was made in situ, by diving, as described in Hamilton et al. (1970). Samples were obtained by gravity corers, box corers, and by diver-held tubes.

Although this volume is concerned with deep-sea sediments, three sets of measurements on the continental shelf (Santa Barbara, San Diego Trough, and the Asiatic shelf) are included with previous measurements for completeness, and because layers in deep-sea turbidites can be composed of similar materials. New deep-sea sediment samples were from the Pacific Ocean (Gulf of Alaska, Central and South Pacific, and the Carnegie Ridge area), the Indian Ocean (Java Trench, south of the trench, and in the Bay of Bengal), the Bering Sea and Mediterranean Sea. As in previous reports, all reported samples are from the 0 to 30 cm interval below the sea floor for reasons noted in Hamilton (1970b).

Results

The various measured and computed properties involved in this study are listed in Tables 1 to 6. No measurements of cohesion or consolidation were made by the writer. However, cohesion and associated properties were measured in the cores from one expedition to the South Pacific and Indian Oceans by Anderson (1971). In previous reports, the tables included samples from the Bering and Okhotsk Seas (listed with abyssal plain sediments) where the surficial sediments are usually diatomaceous. The siliceous tests of diatoms cause the bulk grain density to be lower and sound velocity higher than in abyssal-hills or abyssal-plains, non-siliceous sediments of the same porosity. An additional suite of cores from the Bering Sea allowed separate listing of this sediment type in the tables, and separate notation in the figures. Calcareous ooze samples from the

Central Pacific and Carnegie Ridge allowed separation of calcareous material in the tables.

The averaged results (and standard error of the mean) of all these measurements, corrected where necessary to 23°C and 1 atmosphere pressure, are listed in Tables 1-6. New figures are presented for most of the data. Regression equations (Appendix) were computed for some illustrated data. These tables and figures are revised from Hamilton (1970b, 1970d, 1971a) and include all old and new data to date (June 1973). Some previously published figures are included for completeness and for the convenience of the reader.

DISCUSSION AND CONCLUSIONS

The empirical relationships between measured and computed physical properties, and some environmental differences, were discussed in recent reports with many references to earlier work (Hamilton 1970 a-d, 1971b). Basic, theoretical aspects of sound velocity, elasticity, and related properties were discussed in three

TABLE 1b

CONTINENTAL TERRACE (SHELF AND SLOPE) ENVIRONMENT;
Sediment Densities, Porosities, Sound Velocities, and Velocity Ratios

Sediment Type	Density, g/cm^3		Porosity, %		Velocity, m/sec		Velocity Ratio	
	Avg.	SE	Avg.	SE	Avg.	SE	Avg.	SE
Sand								
Coarse	2.034	___	38.6	___	1836	___	1.201	___
Fine	1.957	0.023	44.8	1.36	1753	11	1.147	0.007
Very fine	1.866	0.035	49.8	1.69	1697	32	1.111	0.021
Silty sand	1.806	0.026	53.8	1.60	1668	11	1.091	0.007
Sandy silt	1.787	0.044	52.5	2.44	1664	13	1.088	0.008
Silt	1.767	0.037	54.2	2.06	1623	8	1.062	0.005
Sand-silt-clay	1.583	0.029	67.2	1.59	1580	9	1.033	0.006
Clayey silt	1.469	0.017	72.6	0.91	1546	4	1.011	0.003
Silty clay	1.421	0.015	75.9	0.82	1520	3	0.994	0.002

Notes.

Laboratory values: 23°C, 1 atm; density: saturated bulk density; porosity: salt free; velocity ratio: velocity in sediment/velocity in sea water at 23°C, 1 atm, and salinity of sediment pore water.
SE: Standard error of the mean.

other reports (Hamilton, 1971a, 1972; Hamilton et al., 1970). This report will be confined to the present state-of-the-art in predicting some sediment properties, but of necessity, some material will be repeated. For extended discussions and fuller references, the reader is referred to the cited papers.

A report in preparation for the proceedings of another ONR symposium will review some basic aspects of the sediment elastic system and applications of sediment property data to geoacoustic models of the sea floor (Hamilton, in press).

Density-Porosity Relationships

General. The equation linking density, porosity, pore-water density, and bulk density of mineral solids in a gas-free system is:

$$\rho_{sat} = n\rho_w + (1 - n)\rho_s \tag{1}$$

where ρ_{sat} = saturated bulk density

n = fractional porosity (volume or voids/total volume)

ρ_w = density of pore water

ρ_s = bulk density of mineral solids.

When sea water is evaporated from sediments during laboratory measurements, dried salts remain with the dried mineral residues. A "salt correction" should be made to eliminate the false increment to the weight of dried minerals; otherwise, porosity, water content, and bulk grain density values are incorrect. Methods of making a salt correction were detailed by Hamilton (1971b). All values in the tables have been so corrected.

Density of pore-water. In computations involving pore-water density, it can be assumed that pore-water and bottom-water salinities are approximately the same. Values for the laboratory density of sea water can be obtained from Sigma-T tables (e.g., NAVOCEANO, 1966). For almost all deep-sea sediments, a laboratory value at 23°C of 1024 kg/m^3 (1.024 g/cm^3) will be within 2 kg/m^3 (0.002 g/cm^3) of any other density at reasonable "room temperatures." This value is recommended for laboratory computations. In situ values of water density can be computed from NAVOCEANO tables (1966); such values (when rounded off) would vary little in deep water from those given in Hamilton (1971b) for the Central Pacific.

Density of mineral solids. The bulk density of mineral solids in sediments varies widely because the mineral species present de-

TABLE 2a

ABYSSAL PLAIN AND ABYSSAL HILL ENVIRONMENTS;
Average Sediment Size Analyses and Bulk Grain Densities

Environment Sediment Type	No. Samples	Mean Grain Dia. mm	 ϕ	Sand, %	Silt, %	Clay, %	Bulk Grain Density g/cm^3 $(kg/m^3 \cdot 10^{-3})$
Abyssal Plain							
Sandy silt	1	0.0170	5.88	19.4	65.0	15.6	2.461
Silt	3	0.0092	6.77	3.2	78.0	18.8	2.606
Sand-silt-clay	2	0.0208	5.59	35.2	33.3	31.5	2.653
Clayey silt	21	0.0056	7.49	5.0	55.4	39.6	2.636
Silty clay	36	0.0021	8.91	2.7	36.0	61.3	2.638
Clay	5	0.0014	9.51	0.0	21.7	78.3	2.672
Bering Sea and Okhotsk Sea (Diatomaceous)							
Silt	1	0.0179	5.80	6.5	76.3	17.2	2.474
Clayey silt	5	0.0049	7.68	8.1	49.1	42.8	2.466
Silty clay	23	0.0024	8.71	3.0	37.4	59.6	2.454
Abyssal Hill							
Deep-sea ("red") clay							
Clayey silt	8	0.0051	7.62	3.8	58.7	37.5	2.611
Silty clay	50	0.0023	8.78	2.6	31.5	65.9	2.696
Clay	14	0.0014	9.52	0.4	20.9	78.7	2.751
Calcareous ooze							
Sand-silt-clay	5	0.0146	6.10	27.3	42.8	29.9	2.609
Silt	1	0.0169	5.89	16.3	75.6	8.1	2.625
Clayey silt	15	0.0069	7.17	3.4	60.7	35.9	2.678
Silty clay	4	0.0056	7.48	3.9	39.9	56.2	2.683

pend on mineralogy and nearness of source areas for terrigeneous components, on pelagic particles deposited from the water, and on diagenetic changes in mineralogy in the sea floor. The geographic variation in pelagic organisms such as diatoms and Radiolaria (silica) and Foraminifera (calcium carbonate) have marked effects on grain density. An average value for grain densities in diatomaceous sediments of the Bering and Okhotsk Seas (Table 2a) is 2.46 g/cm^3 ($gm/cm^3 = kg/m^3 \cdot 10^{-3}$). In the open Pacific to the south, the deep-sea clays have average grain density values between 2.61 and 2.75 g/cm^3 (avg. 2.69 g/cm^3).

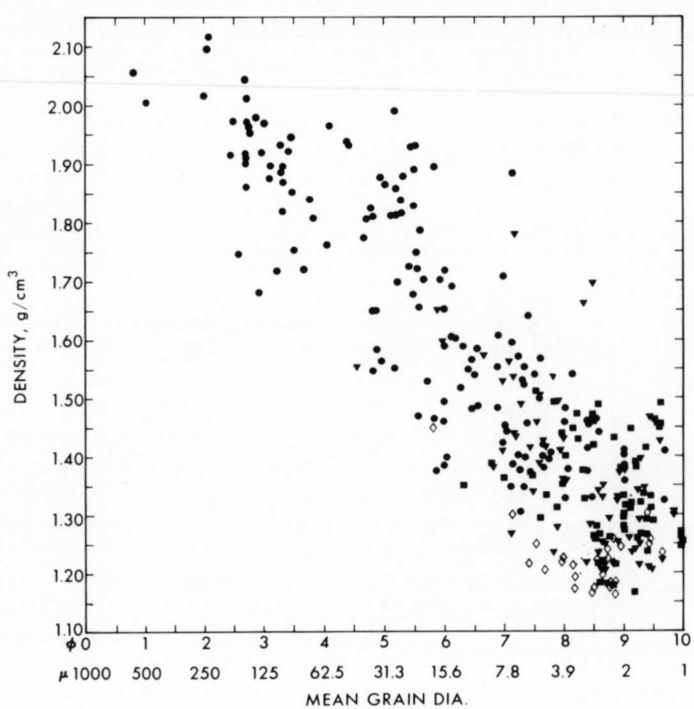

Figure 1. Mean diameter of mineral grains versus density. Round dots are continental terrace (shelf and slope) samples; squares are abyssal hill samples; triangles are abyssal plain samples; open diamonds are diatomaceous samples from the Bering and Okhotsk Seas.

The averages (in g/cm^3) of all samples in each of the three environments (not including diatomaceous and calcareous sediments) are: terrace-2.686, abyssal-hill "red" clay-2.694, abyssal plain (mostly fine-grained)-2.636. The overall average of the above is 2.672 g/cm^3. Keller and Bennett (1970) report an average for terrigeneous materials of 2.67 g/cm^3, and for the Pacific, 2.71 g/cm^3. Cernock (1970), for the Gulf of Mexico, reports 2.637 g/cm^3. Akal (1972) reports a general value of 2.66 g/cm^3. In soil mechanics computations a value of about 2.65 g/cm^3 is used for sands and silts when the value is unknown (e.g., Wu, 1966). Thus, there is enough information at hand to predict, with confidence, grain densities for general sediment types.

The conclusion is that the following grain densities be predicted and used in computations when no data is available.

Sediment Type	Avg. Bulk Density of Minerals (g/cm^3)
Terrigeneous	2.67
Deep-sea (red clay)	2.70
Calcareous ooze	2.71
Diatomaceous ooze	2.45

Saturated bulk density (or unit weight). The relationships of saturated bulk density to porosity (Equation 1) are not illustrated, but are indicated for these data in regression equations in the Appendix. Previous illustrations and discussions indicate the small errors for most sediments when either property is used as an index to the other (Hamilton et al., 1956, 1970b).

In the two deep-water environments (in the upper 30 cm), the least saturated bulk density was 1.16 g/cm^3 from the Okhotsk Sea, and the highest was 1.65 g/cm^3 in a silty layer in Japan Basin turbidites. Variations of density with depth were discussed by Hamilton (1959, 1964). In predicting density without any sediment data, one can enter the tables for the appropriate environment and use the data for the dominant sediment type. Silty clay is the most common type in both deep-water environments. Given mean grain size, M_z, a value of density can be derived by entering the diagram (Fig. 1), or regression equation (Appendix), related M_z to density.

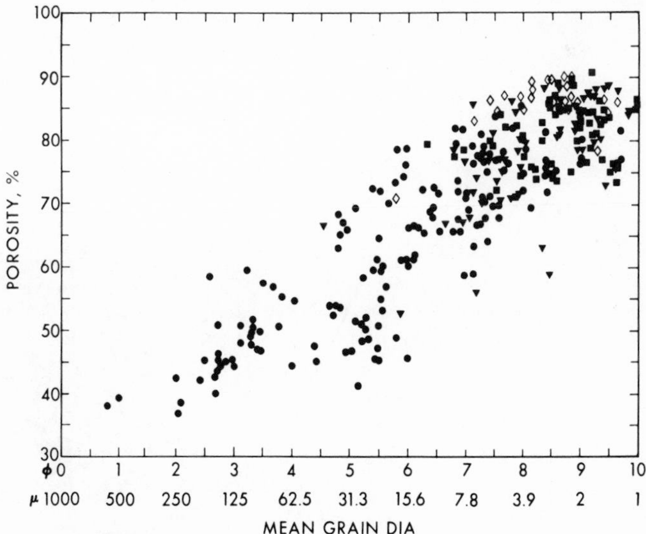

Figure 2. Mean diameter of mineral grains versus porosity, all environments: symbols as in Figure 1.

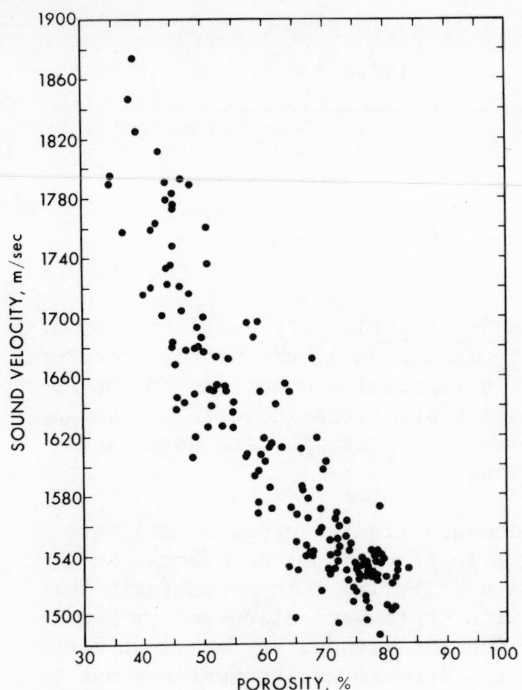

Figure 3. Porosity versus sound velocity, continental terrace

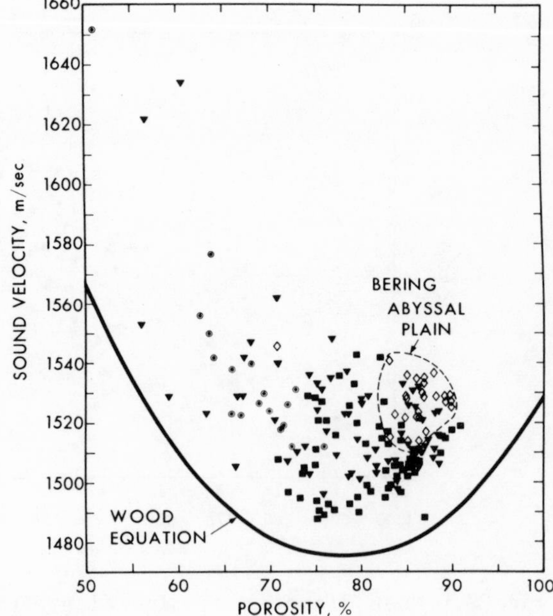

Figure 4. Porosity versus sound velocity, abyssal hill and abyssal plain environments; symbols as in Figure 1 except circled dots are calcareous samples.

Laboratory values of density can usually be used as in situ values, but the small correction can be easily made by computing saturated bulk density with equation (1), using in situ density of sea water. The increment to density for most high-porosity sediments varies with water depth, but is a maximum of 0.03 g/cm^3 to 6000 m water depth.

Porosity. The amount of pore space in a sediment is the result of a number of complex, interrelated factors; most important are the mineral sizes, shapes, and distributions, mineralogy, sediment structure, and packing of solid grains. This subject has been previously discussed with many references (Hamilton, 1970b). The interrelated effects of the above factors usually result in a general decrease in porosity with increasing grain size (Fig. 2). There is much scatter in the data because of the factors cited above.

The marked effect of mineralogy and environmental control in porosity-density can be seen in the tables and figures: the diatomaceous sediments of the Okhotsk and Bering Seas have significantly higher porosities and lower densities than do similar sediments of the same grain size. Silty clay in diatomaceous ooze has average densities of 1.214 g/cm^3 and porosities averaging 86.8%, whereas this sediment type in abyssal hills and other plains has densities around 1.33 g/cm^3 and porosities around 81%.

In predicting porosity or density, given the other property (or deriving it by using mean grain size), one can enter density vs. porosity equations or diagrams, but it is usually better procedure to assume values of grain density and pore-water density (laboratory or in situ) and compute the missing property with equation (1).

Compressional Wave (Sound) Velocity

General. In this section, empirical relationships between sound velocity and other physical properties will be discussed. Some of these empirical relationships are important in predicting sound velocity, but it should be emphasized that wave velocities are elastic properties of the sediment mass. Properties such as porosity and grain size affect sound velocity only in the effects they have on elasticity of the sediment (discussed at length in Hamilton, 1971a).

Sound velocity and porosity-density relationships. The relationships between sound velocity and porosity have received much attention in the literature because porosity is an easily measured or computed property likely to yield predictable relations with sound velocity. This is because porosity is the volume of water-filled pore space in a unit volume of sediment, and compressional-wave speed is largely determined by the compressibility of pore water, especially in high-porosity silt-clays. Many studies have empha-

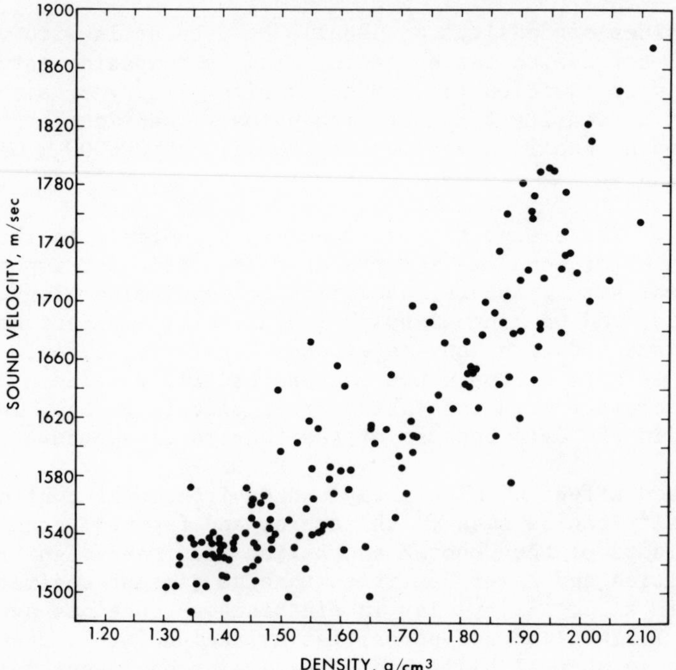

Figure 5. Density versus sound velocity, continental terrace

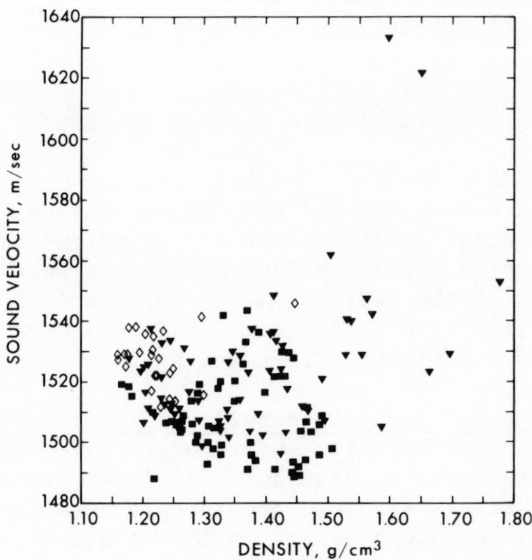

Figure 6. Density versus sound velocity, abyssal hill and abyssal plain environments; symbols as in Figure 1

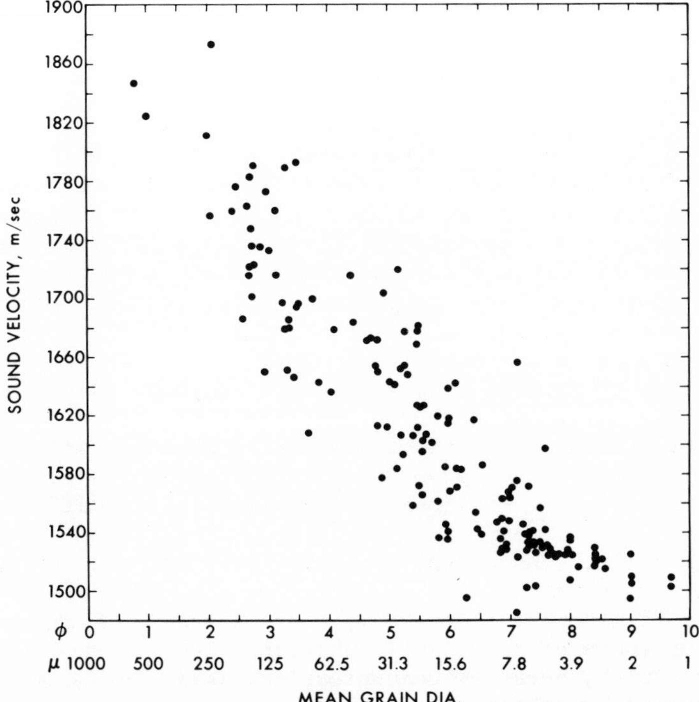

Figure 7. Mean diameter of mineral grains versus sound velocity, continental terrace

sized the relationships between sound velocity and porosity over the full range of porosity (Hamilton, 1956, 1965, 1970b; Hamilton et al., 1956; Sutton et al., 1957; Laughton, 1957; Nafe and Drake, 1957, 1963; Horn et al., 1968, 1969; Schreiber, 1968; McCann and McCann, 1969; Kermabon et al., 1969; Cernock, 1970; Buchan et al., 1972; McCann, 1972; Akal, 1972; Smith, personal communication). These studies have included sediments for all of the world's major oceans. The latest data of the writer is illustrated in Figures 3 and 4. The relationships between sound velocity and density (Figs. 5 and 6) are similar to those for sound velocity and porosity because of the linear relationship between porosity and density.

Sound velocity and grain-size relationships. Grain-size analyses in the laboratory usually include percentages of sand, silt, and clay, mean and median diameters of mineral grains, and other statistical parameters. The relationships between grain-size and velocity (Figs. 7, 8, & 9) are in accord with previous studies (Hamilton et al., 1956; Hamilton, 1970b; Sutton et al., 1957; Shumway, 1960; Horn et al., 1968; Schreiber, 1968). Empirically, mean grain size and percent clay size (Fig. 9), or percent sand and silt, are important indices to velocity. This is important because size analyses can be made on wet or dry material; and, frequently, size analyses are all the data available in published reports.

<u>Discussion of velocity indices</u>. The information currently available indicates that the higher-porosity silt-clays in the deep basins of the world's oceans have velocities within 1 to 2% of each other at any given porosity above about 65 to 70% (excluding special types such as diatom and calcareous ooze). It is difficult to compare velocity measurements when all have not been corrected to a common temperature, or where temperature is not reported for velocity measurements. Variations in "room temperature" can easily cause velocity variations on the order of 20 to 30 m/sec (about 1 to 2%). Variations are much greater if measurements are made in sediment soon after coring or removal from a refrigerator. Temperature measurements should always be made with velocity measurements because temperature variations can cause velocity changes which obscure, and can be greater than, environmental or lithological differences.

If abyssal plain and abyssal hill measurements are lumped together, velocity, at a porosity of 80%, from the Mediterranean (Horn et al., 1967; Kermabon et al., 1969), North Atlantic, Caribbean, and Gulf of Mexico (Horn et al., 1968; Schreiber, 1968; Cernock, 1970; McCann, 1972; Smith, personal communication; Akal, 1972), and Pacific and Indian Oceans (Hamilton, this report) averages about 1500±25m/sec. The lumping of data from various environments and unknown temperatures of measurement is not advised. Yet the results indicate the small velocity variations in high-porosity sediments of the world's oceans.

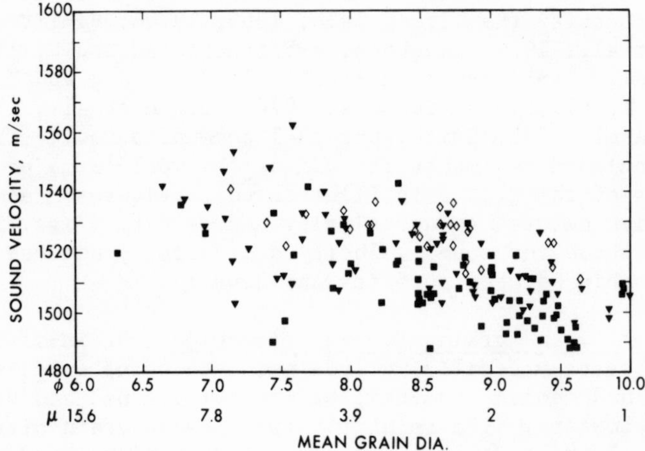

Figure 8. Mean diameter of mineral grains versus sound velocity, abyssal hill and abyssal plain environments; symbols as in Figure 1

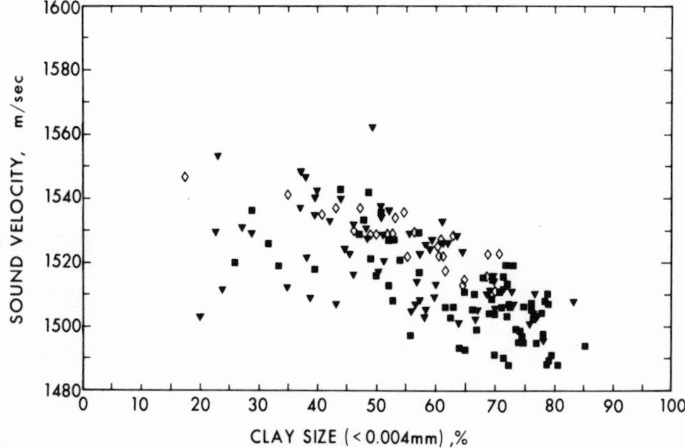

Figure 9. Percent clay size versus sound velocity, abyssal hill and abyssal plain environments; symbols as in Figure 1

As discussed in previous reports, general curves covering the full range of porosity, density, or grain size, wherein data from all environments and sediment types is lumped, should be abandoned in favor of specific curves for particular environments and/or geographic areas or sediment types. In other words, enough data is at hand to stop lumping and start splitting. Examples of this are illustrated in the velocity vs. porosity and density diagrams (Figs. 3-6). These figures and the tables indicate that at porosities around 80% abyssal-hill silt-clays have lower velocities than do abyssal plain or terrace sediments. The diatomaceous sediments of the Bering Abyssal Plain have significantly higher velocities at higher porosities and lower densities than either abyssal hill or other terrigenous abyssal plain sediments (Figs. 4 & 6).

A resumé of velocity vs. mean or median grain size data indicates that in the various ocean basins, deep-sea sediments of the same mean grain size are apt to have about the same velocities. At a mean grain size of 9.5 phi, the range of velocities from the Pacific and Indian Oceans and adjacent areas (Hamilton, this report), the Gulf of Mexico (Cernock, 1970), the Atlantic, Caribbean, and Mediterranean (Horn et al., 1968, 1969; Schreiber, 1968; Smith, personal communication) is about 1% (about 1495 to about 1510 m/sec). This is remarkably close considering the lack of temperature control and geographic range. However, as previously discussed in the cases of porosity and density vs. velocity, such lumping should be discouraged. An example, again, is the siliceous sediment of the Bering and Okhotsk Seas. At any given grain size between 8 and 10 phi, these diatomaceous sediments have higher velocities than do the other deep-water sediments (Fig. 8).

TABLE 2b

ABYSSAL PLAIN AND ABYSSAL HILL ENVIRONMENTS
Sediment Densities, Porosities, Sound Velocities, and Velocity Ratios

Environment Sediment Type	Density, g/cm^3 Avg.	Density SE	Porosity, % Avg.	Porosity SE	Velocity, m/sec Avg.	Velocity SE	Velocity Ratio Avg.	Velocity Ratio SE
Abyssal Plain								
Sandy silt	1.652	___	56.6	___	1622	___	1.061	___
Silt	1.604	___	63.6	___	1563	___	1.022	___
Sand-silt-clay	1.564	___	66.9	___	1536	___	1.004	___
Clayey silt	1.418	0.021	76.2	1.23	1528	3	0.999	0.002
Silty clay	1.323	0.021	81.7	1.21	1517	2	0.991	0.001
Clay	1.357	0.045	79.8	2.69	1504	3	0.982	0.002
Bering Sea and Okhotsk Sea (Diatomaceous)								
Silt	1.447	___	70.8	___	1546	___	1.011	___
Clayey silt	1.228	0.019	85.8	0.86	1534	2	1.003	0.001
Silty clay	1.214	0.008	86.8	0.43	1525	2	0.997	0.001
Abyssal Hill								
Deep-sea ("red") clay								
Clayey silt	1.352	0.029	79.8	1.78	1527	3	0.997	0.002
Silty clay	1.335	0.013	81.4	0.71	1508	2	0.986	0.001
Clay	1.343	0.025	81.1	1.28	1499	2	0.980	0.001
Calcareous ooze								
Sand-silt-clay	1.400	0.013	76.3	0.90	1581	8	1.034	0.005
Silt	1.725	___	56.2	___	1565	___	1.023	___
Clayey silt	1.573	0.020	66.8	1.22	1537	5	1.005	0.003
Silty clay	1.483	0.029	72.3	1.61	1524	7	0.996	0.005

Notes.

Laboratory values: 23°C, 1 atm; density: saturated bulk density; porosity: salt free; velocity ratio: velocity in sediment/ velocity in sea water at 23°C, 1 atm, and salinity of sediment pore water. SE: Standard error of the mean.

The status of prediction, given good samples, temperature control, and adequate statistical information, is exemplified by measurements by the writer on 17 samples of deep-sea "red" clay taken in the Central Pacific by the Kennecott Exploration Co., using box corers. Publication of tables and diagrams with regression equations took place when measurements were made (Hamilton, 1970b). Entry into the regression equation (mean grain size vs. velocity) using the average mean grain size of the 17 samples of 9.2 phi yielded a velocity of 1502 m/sec. The average velocity in the 17 samples at 23°C was 1507 m/sec.

TABLE 3

CONTINENTAL TERRACE (SHELF AND SLOPE) ENVIRONMENT
Average Sediment Impedances, Density (Velocity)^2, Reflection Coefficients, and Bottom Losses

Sediment Type	ρ_2V_2	$\rho_2(V_2)^2$	R	BL
Sand				
Coarse	3.7344	6.8564	0.4098	7.7
Fine	3.4302	6.0125	0.3739	8.5
Very fine	3.1662	5.3725	0.3389	9.4
Silty Sand	3.0129	5.0265	0.3168	10.0
Sandy silt	2.9732	4.9468	0.3108	10.1
Silt	2.8675	4.6534	0.2944	10.6
Sand-silt-clay	2.5008	3.9508	0.2307	12.7
Clayey silt	2.2715	3.5124	0.1847	14.7
Silty clay	2.1596	3.2822	0.1602	15.9

Notes.

Laboratory values: 23°C, 1 atmosphere

ρ_2V_2 = sediment impedance, g/cm^{2}sec x 10^5

$\rho_2(V_2)^2$ = sediment density X (velocity)^2, g/cmsec^2, or dynes or dynes/cm^2 x 10^{10}

R = Rayleigh reflection coefficient at normal incidence = $\frac{\rho_2V_2 - \rho_1V_1}{\rho_2V_2 + \rho_1V_1}$

BL = -20 log R, bottom loss, db

ρ_1, V_1: sea-water density, velocity; ρ_2, V_2: sediment density, velocity.

Another example of "prediction" is provided by in situ velocities (corrected to 23°C) measured in England by McCann and McCann (1969) in fine sand, very fine sand, and silt (4 stations). Entry into the regression equation for velocity vs. mean grain size for the continental terrace (this report), using the published mean grain sizes of McCann and McCann, in three cases yields velocities less than 1% different from those measured, and in one case yields a velocity 4.5% different. Average for the four cases is 1.6%.

There are three general ways to predict in situ sound velocity at the sediment surface.

1. Correct the laboratory velocity from 1 atmosphere pressure and temperature of measurement to the in situ temperature and pressure using tables for the speed of sound in sea water.

2. Multiply the laboratory velocity ratio (sediment velocity/sea water velocity at 1 atmosphere, temperature of sediment, and bottom water salinity) by the bottom-water velocity.

3. In the absence of sediment data, enter a table (e.g., Table 2b) and select a velocity or ratio for the particular environment and most common sediment type. Then correct to in situ as in (1) or (2) above.

The ratio method, (2), is the easiest to apply because the ratio remains the same in the laboratory or in situ, and all one needs for in situ computations is a curve of sound velocity vs. depth in the water mass. These methods were discussed at length (with a numerical example) in a special report concerned with prediction of in situ properties (Hamilton, 1971b).

Impedance

The characteristic acoustic impedance of a medium is the product of density, ρ, and velocity, V_p (impedance = ρV_p, g/cm^2 sec). Impedance is an important property of any material. The amount of energy reflected (or lost) when sound passes from one medium into another of different impedance is largely determined by impedance difference, or "mismatch" (e.g., Kinsler and Frey, 1962). In the field of marine geophysics, echo-sounding and continuous-reflection-profiling records indicate the travel-time of sound between impedance mismatches.

Average impedances were computed for the sediments of this study (Tables 3 and 4), using the averaged, measured values of sediment density and velocity in Tables 1b and 2b. Density and porosity vs. impedance were illustrated in an earlier report (Hamilton et al.,

1956) and more fully discussed in a recent report (Hamilton, 1970d). The figures and regression equations in the latter report adequately describe the writer's data to date.

Laboratory impedances require correction to in situ values. The methods of correcting laboratory density and velocity to in situ values have already been discussed. In situ impedance is merely the product of the corrected values.

TABLE 4

ABYSSAL PLAIN AND ABYSSAL HILL ENVIRONMENTS
Average Sediment Impedances, Density (Velocity)^2, Reflection Coefficients, and Bottom Losses*

Environments Sediment Type	$\rho_2 V_2$	$\rho_2 (V_2)^2$	R	BL
Abyssal Plain				
Sandy silt	2.6795	4.3462	0.2623	11.6
Silt	2.5071	3.9185	0.2311	12.7
Sand-silt-clay	2.4023	3.6899	0.2107	13.5
Clayey silt	2.1663	3.3094	0.1608	15.9
Silty clay	2.0067	3.0438	0.1234	18.2
Clay	2.0404	3.0679	0.1315	17.6
Bering Sea and Okhotsk Sea (Diatomaceous)				
Silt	2.2371	3.4585	0.1763	15.1
Clayey silt	1.8840	2.8904	0.0920	20.7
Silty clay	1.8514	2.8233	0.0833	21.6
Abyssal Hill				
Deep-sea ("red") clay				
Clayey silt	2.0638	3.1504	0.1371	17.3
Silty clay	2.0134	3.0367	0.1250	18.1
Clay	2.0130	3.0173	0.1249	18.1
Calcareous ooze				
Sand-silt-clay	2.2137	3.5003	0.1714	15.3
Silt	2.6996	4.2249	0.2658	11.5
Clayey silt	2.4175	3.7155	0.2138	13.4
Silty clay	2.2598	3.4435	0.1813	14.8

* See notes under Table 3.

Rayleigh Reflection Coefficients and Bottom Losses at Normal Incidence

The computations of reflection coefficients and bottom losses, herein discussed, are a simple, straightforward procedure, given accurate values of density and velocity for sediment and water. Comparisons of such computations with actual measurements at sea (Hamilton, 1970d) by Breslau (1965, 1967) and Fry and Raitt (1961), and the measurements and computations by Hastrup (1970, p. 183-184, Fig. 5) demonstrate that the method is valid and yields realistic predicted values for acoustic bottom losses (db) at the water-sediment interface, given certain restricted conditions.

The whole subject of reflection, refraction, and energy losses of sound incident on the sea floor is too complex for simple statements and is not within the scope of this report. For selected references and fuller discussion see Hamilton (1970d). The reader is cautioned against attempted use of Rayleigh reflection coefficients and bottom losses except under very restricted conditions of bottom sediment layering, sound energy levels, and frequency. In general, the Rayleigh fluid/fluid model is valid only when, for various reasons, any second, or other, layers in the sea floor cannot reflect sound which interferes with that reflected from the water-sediment interface (see Cole, 1965, for discussion).

More sophisticated models of reflectivity and bottom loss involve layers and varying layer properties. A linear viscoelastic model which includes shear moduli has been successful in predicting bottom losses (Bucker, 1964; Bucker et al., 1965). A simpler layered model without sediment rigidity has been successfully used by Cole (1965). Morris (1970) has reported on a successful bottom-reflection-loss model with a velocity gradient. A figure from her report is reproduced herein as Figure 10. It serves to illustrate the state-of-the-art in reconciling experiment with theory (or predicting) in cases of bottom loss vs. grazing angle when the acoustician is given realistic geoacoustic models of the layered sea floor. Another example is provided by Hastrup (1970) in the Mediterranean.

<u>The sea floor as a liquid model</u>. The simplest reflection model involves a simple harmonic, plane wave incident on a plane boundary between two fluids across which there is a change in velocity and density. Several recent textbooks include the derivations of the appropriate equations for this model (Ewing et al., 1957; Officer, 1958; Kinsler and Frey, 1962). The Rayleigh reflection coefficient for this model expresses the ratio of the amplitudes, or pressures, of a reflected wave to that of the incident wave. At normal incidence, the pressure reflection coefficient, R, and bottom loss, BL, (in decibels, db), are expressed by the equations under Table 3.

Reflection coefficients and bottom losses at normal incidence

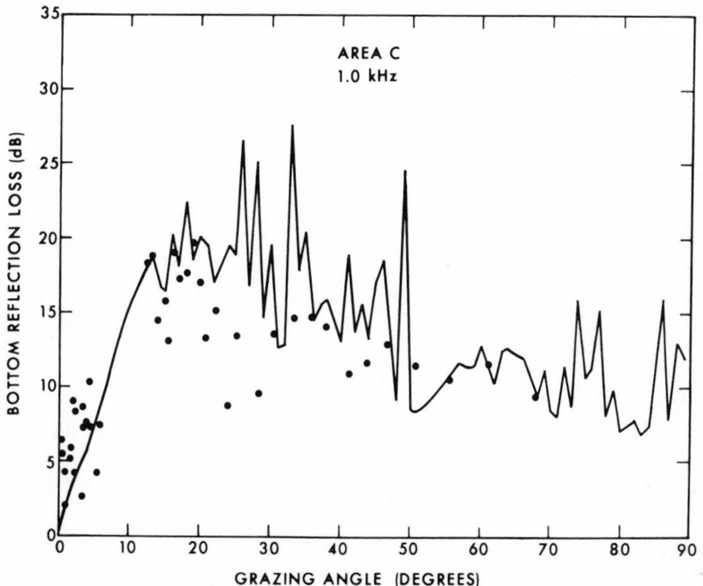

Figure 10. Comparison of theoretical (________) and measured (•) bottom reflection loss versus grazing angle at 1 kHz in Area C (figure 10 in Morris, 1970)

were illustrated by Hamilton et al. (1956) and are the subject of a separate paper (Hamilton, 1970d). For the present report, the values of reflection coefficients and bottom losses (Table 3 and 4) were computed using the average sediment density and velocity values in Tables 1b and 2b, plus appropriate values of water density, velocity, and salinity at 23°C. As discussed in the 1970d report, laboratory values of reflection coefficients and bottom losses at normal incidence are so close to corrected, in situ, values, that laboratory values can be used for in situ values in generalized studies. Figures 11 and 12 (from Hamilton, 1970d) illustrate the empirical relationships between density and Rayleigh reflection coefficients and bottom losses at normal incidence. Regression equations and other figures are included in the cited report.

Sound Velocity vs. Shear Strength (Cohesion)

Plots of sound velocity vs. shear strength (cohesion) in modern, surficial sea-floor sediments (Horn et al., 1968; Schreiber, 1968; Hamilton, 1970b) show that there is no relationship that would allow the use of cohesion as an empirical index property for sound velocity (and vice versa). In studies of sound velocity vs. shear strength (cohesion), the tacit assumption is that cohesion is a usable measure of dynamic rigidity, and that sound velocity increases with increas-

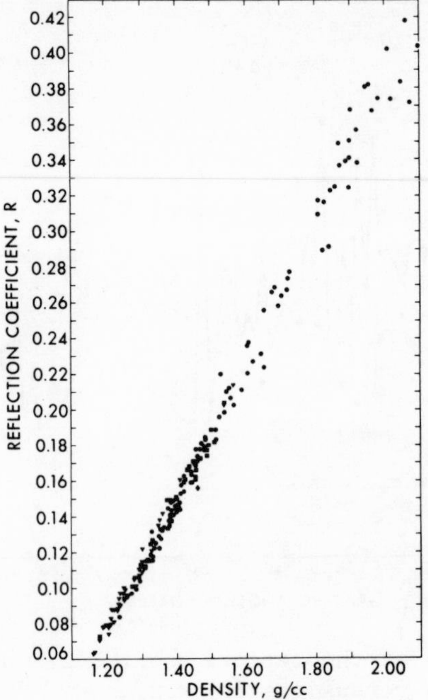

Figure 11. Density versus Rayleigh reflection coefficient at normal incidence, all environments (figure from Hamilton, 1970d)

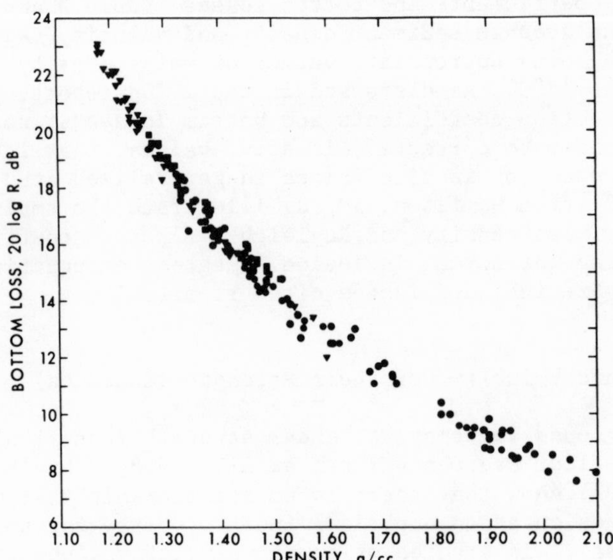

Figure 12. Density versus bottom loss at normal incidence, all environments (figure from Hamilton, 1970d)

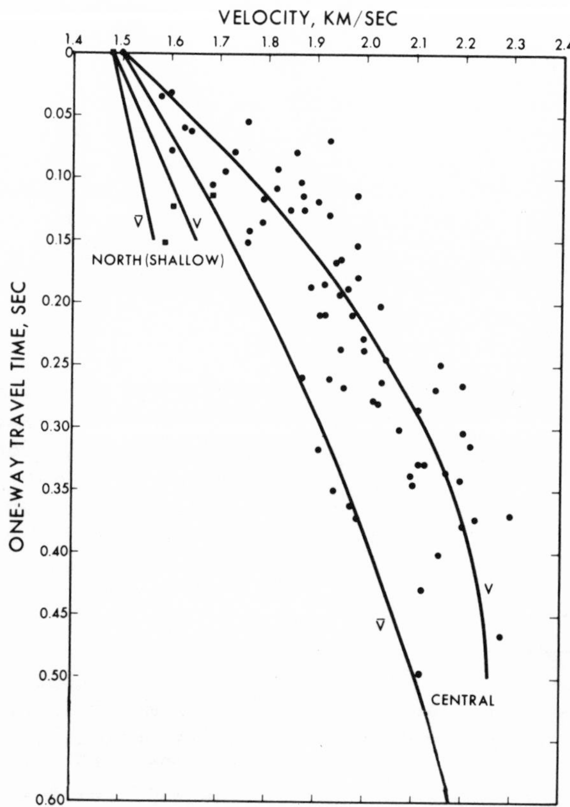

Figure 13. Instantaneous and mean velocity versus one-way travel time as determined from sonobuoy measurements of mean velocity in the first sedimentary layer in the Bay of Bengal (from Hamilton et al., 1973). Velocity at the sediment surface (t = 0) was determined from sediment laboratory measurements corrected to in situ.

ing rigidity (which is true). The lack of any usable correlation between sound velocity and cohesion is apparently caused by the fact that static tests of cohesion cannot be equated with dynamic rigidity, and by the expectable, small effect of the rigidity (shear) modulus on the compressional velocity in present-day, high porosity silt-clays. Usually dynamic rigidity is about 4 orders of magnitude greater than cohesion. This subject was discussed and illustrated in a previous report (Hamilton, 1970b).

Velocity Gradients and Layer Thicknesses

General. Reflection profiling has become an important tool in geologic, geophysical, and engineering studies. Reflection records indicate sound travel time between impedance mismatches within the sediment or rock layers of the sea floor. To derive the true thicknesses of these layers, it is necessary to measure or predict the interval or mean layer velocity, or to use a measured or predicted sediment surface velocity and a velocity gradient in the sediment body. True thicknesses of layers is a critically important requirement in studies of rates of sedimentation and accumulation of sediments, total volumes of sediments, in engineering site surveys, and other studies. Most of the following is from a report in preparation which contains a fuller discussion (Hamilton et al., in press).

The techniques of measuring sediment interval velocities at sea with expendable sonobuoys, and subsequent data reduction, were developed largely at the Lamont-Doherty Geological Observatory (Le Pichon et al., 1968; Houtz et al., 1968). Sonobuoy measurements have now been reported for many areas (Houtz et al., 1968; M. Ewing et al., 1969; J. Ewing et al., 1970; Houtz et al., 1970; Ludwig et al., 1971; Hamilton et al., in press). Velocity-gradient data for the sea floor are usually produced in the form of linear or non-linear curves based on plots of instantaneous and mean velocity vs. one-way travel time in the sediment or rock layer (e.g., Fig. 13).

Velocity gradients. Velocity gradients are usually expressed as an increase in velocity per linear increase in depth, m/sec/m, or sec^{-1}. In the upper levels of deep-water marine sediments these gradients are normally positive, and usually between 0.5 and 2.0 sec^{-1} (Ewing and Nafe, 1963; Houtz et al., 1968; this report). However, most velocity gradients are non-linear if followed to sufficient depths within the sediment body (e.g., Fig. 13; Houtz et al., 1968, 1970).

When the velocity gradient, a, is linear, the instantaneous velocity, V, at depth, h, is (Houtz and Ewing, 1963):

$$V = V_0 + ah \qquad (2)$$

At any depth within sediment layers, an average linear gradient, a, can be determined from the parabolic equations for V and $\bar{V}$ (mean or interval velocity in the sediment layer) vs. t (Houtz et al., 1968, equation 3) by:

$$a = (V - V_0)/h \qquad (3)$$

where v = instantaneous velocity at time t

V_0 = velocity at sediment surface (t = 0)

h = layer thickness at time t = $\bar{V}t$.

In most sediment sections, the linear velocity gradient decreases with increasing depth, or travel time. The average linear velocity gradient was computed with equation (3) at increments of 0.1 sec (from 0 to 0.5 sec) for each of 13 areas of mostly turbidite deposition: 4 from Lamont-Doherty investigations (Atlantic, Gulf of Mexico, Aleutian Trench, and Bering Sea-thin), and nine from Hamilton et al. (in press). The values at each 0.1 sec interval were averaged and plotted in Figure 14. These average gradients decreased from about 1.31 sec^{-1} at t = 0, to 0.77 sec^{-1} at t = 0.5 sec. As discussed in the next section, such average values can be used to compute a predicted true sediment thickness in many areas where no interval velocity data are available.

Thickness computations. There are three usual alternatives when computing true sediment thicknesses for an area where no interval velocities have been measured: (1) use an equation or curve for mean velocity vs. travel time from a similar area, (2) use a predicted linear gradient and a predicted V_0 (discussed below), or (3) assume an interval velocity. Of these three alternatives, the last one is most commonly used, and is the least accurate. These alternatives are briefly discussed, with examples, in the next section.

There is now sufficient, published data to show that most areas of turbidites have reasonably close velocity gradients in the upper, unlithified layers. For example, at a one-way travel time of t = 0.2 sec, the computed thickness of a layer, using the Atlantic and Gulf of Mexico equations of Houtz et al., (1968), and those for the Cen-

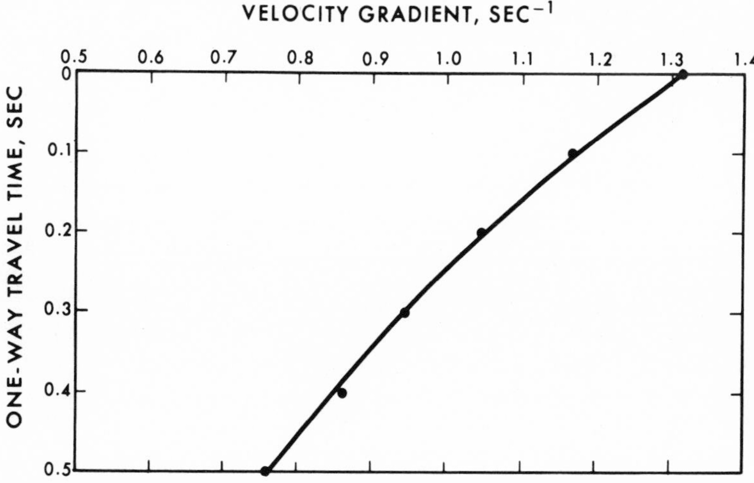

Figure 14. Linear gradient of velocity versus one-way travel time. Gradient at each increment of 0.1 sec is an average of the gradients in 13 areas of mostly turbidite deposition (from Hamilton et al., 1973).

tral Bengal Fan and Kamchatka Basin (Hamilton et al., in press) are respectively, 347 m, 341 m, 351 m, and 343 m. These figures have a variation of less than 3%. Thus, if one is computing sediment layer thicknesses for an area of turbidites where no measurements have been made, the use of equations for the most similar area will probably yield reasonable results. If the sediment type is calcareous ooze, the equations for the Pacific Equatorial Zone (Houtz et al., 1970) are recommended.

Given a linear gradient, a, the sediment surface velocity, V_0, and one-way travel time, t, the thickness of a layer can be computed (Houtz and Ewing, 1963) by:

$$h = V_0(e^{at} - 1)/a \quad (4)$$

where e = the base of natural logarithms

This is a very useful equation because V_0 can be closely estimated (Hamilton, 1971b). One-way travel time in a layer can be measured from a reflection record; and, as discussed in the preceeding section and illustrated in Figure 14, the velocity gradient can usually be reasonably estimated.

The third, most popular and least accurate, method for computing layer thicknesses is to measure travel time from a reflection record and assume an interval velocity. There is a tendency in marine geophysical literature to assume a value for the interval velocity in unlithified deep-sea sediments which is usually too high. Common assumptions are 2000 m/sec (which is convenient in computations), or for the Pacific, 2150 m/sec. Now that interval velocity and sediment-surface velocity information is available, this type of assumption should be abandoned in favor of actual measurements or more realistic estimates and assumptions.

DSDP Hole 118 (Leg 12) in the Biscay Abyssal Plain is a good example to illustrate the alternative methods of computing thicknesses. Reflection records over the site showed 0.24 sec (one-way) of turbidites over a reflector which was correlated with sandstone at a drilled depth of 400 m (Laughton et al., 1972). The Atlantic mean velocity equation of Houtz et al. (1968) would have predicted the reflector at 406 m. Using equation (4) with a predicted V_0 of 1530 m/sec, and a linear velocity gradient of 1.0 sec^{-1} (from Fig. 14) yields a computed thickness of 415 m. Assuming 2000 m/sec for the interval velocity results in a computed thickness of 480 m.

Attenuation of Compressional (Sound) Waves

General. A recent report on compressional-wave attenuation in marine sediments (Hamilton, 1972) contained the following: (1) a report on in situ measurements off San Diego of sound velocity, attenuation, and associated physical properties, (2) a literature review of attenuation in saturated sediments, (3) a discussion of relationships between frequency, velocity, and attenuation, (4) the causes of attenuation in saturated sediments, (5) a discussion of elastic and viscoelastic models which can be applied to marine sediments, and (6) a method to predict attenuation in marine sediments, given frequency and porosity or mean grain size. This section will contain only a resumé of information necessary to predict attenuation. The reader should see the cited paper for extended discussions of the other matters noted above.

The relationships between attenuation and frequency were expressed in the form:

$$\alpha = kf^n \tag{5}$$

where α = attenuation of compressional waves in db/m

k = a constant

f = frequency in kHz

n = the exponent of frequency.

It was concluded in Hamilton (1972) that velocity dispersion is negligible or absent in marine sediments from a few Hz to the MHz range, and that attenuation in db/unit length (e.g., db/m) is approximately dependent on the first power of frequency; that is, in equation (5), n is close to one.

If n in equation (5) is taken as one, the only variable in the equation for various sediments is the constant k. This constant is useful in relating attenuation to other sediment properties such as mean grain size and porosity. The relationships between k and common physical properties give an insight into causes of attenuation, and allow prediction of attenuation.

Assuming that linear attenuation is dependent on the first power of frequency, values of k can be easily computed by dividing attenuation by frequency. This was done for all measurements by the writer and for those in the literature in natural saturated sediments. These values of k were then plotted versus mean grain size and porosity (Figs. 15 and 16). Calcareous sediments, not included in this study, may require separate diagrams.

Average values of the best data were used to establish regression equations which appear in the figure captions in Hamilton (1972). The individual measurements which were averaged are shown in Figures 15 and 16 to better illustrate the trends and scatter of the data.

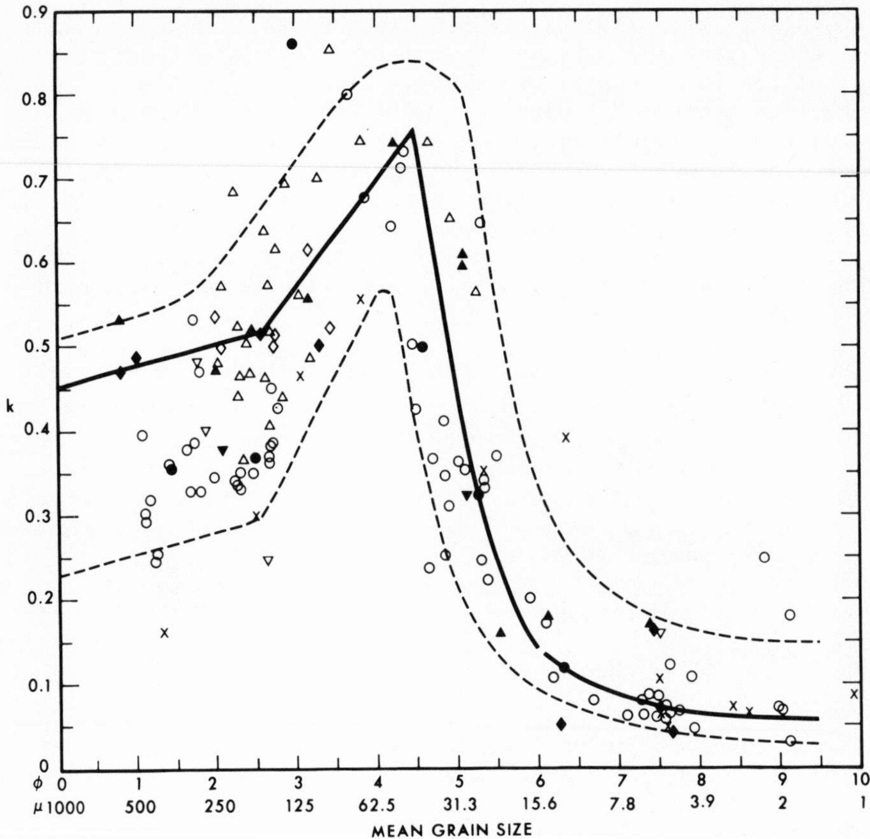

Figure 15. Mean diameter of mineral grains versus k (in $\alpha = kf^1$) in natural, saturated sediments (from Hamilton, 1972); see cited report for meaning of symbols and regression equations for solid lines. The area between dashed lines forms an envelope into which it is predicted most data should fall.

These regression equations are strictly empirical and are recommended only within the limiting values shown. The values of k so obtained are approximations, but it is predicted that most future measurements of attenuation will result in k values which fall within or near the indicated "envelopes." In predicting attenuation, one can use the central (heavy line) values as "most probably", and the upper and lower dashed lines as indicating "probably maximum" and "probably minimum."

Attenuation and dynamic rigidity. Figures 19 and 20, in the next section (on elasticity), revised from Hamilton (1971a, 1972), illustrate the probable variations of computed values of dynamic rigidity with grain size and porosity. Comparison of these figures with Figures 15 and 16 indicate that attenuation and values of rigidity respond in the same way to variations in mean grain size and porosity. This correlation and its causes were first noted in Hamilton (1970a) and are discussed in Hamilton (1972).

Prediction of Compressional-Wave Attenuation. Estimation or prediction of compressional-wave attenuation in saturated sediments is of considerable importance in geophysics and underwater acoustics. The relationships between attenuation (as expressed by the constant k) and grain size and porosity (Figs. 15 and 16) afford a very simple method for prediction of approximate values of attenuation at most frequencies of interest in geophysics and underwater acoustics in all major sediment types (with the possible exception of calcareous sediments).

The method for estimating or predicting approximate values of attenuation (given mean grain size, or porosity) is as follows: determine or predict mean grain size or porosity of the sediment, enter the mean grain size or porosity versus k diagram (Figs. 15 and 16) and determine a value of k (regression equations for these data are in the figure captions in Hamilton, 1972), and insert this value

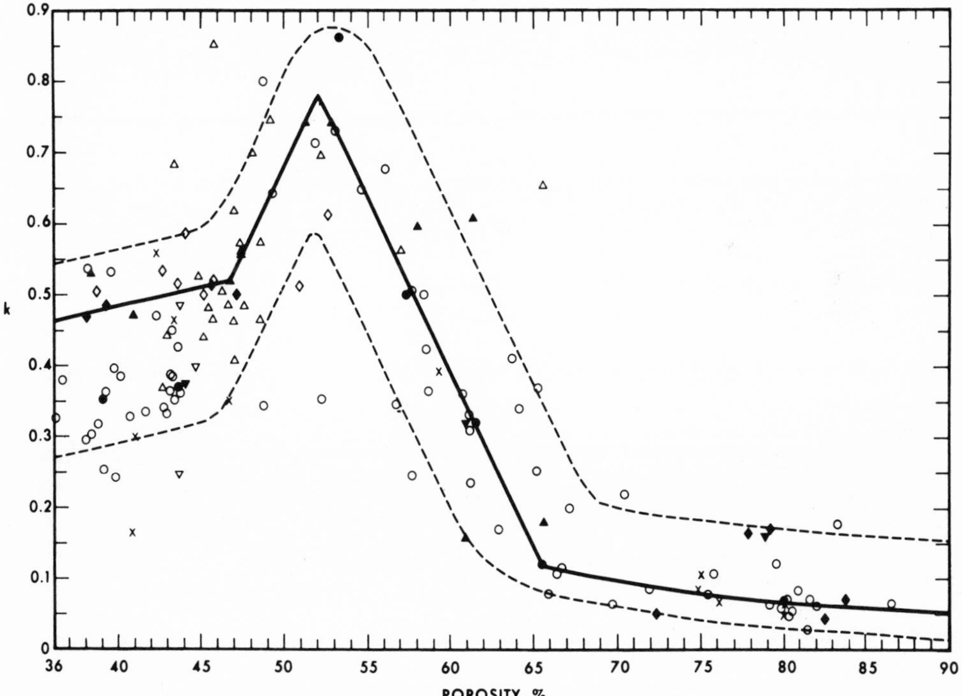

Figure 16. Porosity versus k (in $\alpha = kf^1$) in natural, saturated sediments (from Hamilton, 1972); remarks same as in Figure 15

of k in equation (5). The resulting equation can then be used to compute attenuation in db/m at any desired frequency (kHz).

Elastic Properties

The subjects of elastic and viscoelastic models for water-saturated porous media, and measurements and computation of elastic constants in marine sediments have been discussed in four recent reports (Hamilton, 1971a, b, 1972; Hamilton et al., 1970). The favored model for marine sediments, and concomitant equations, is a case of linear viscoelasticity (Ferry, 1961) which was discussed at some length by Hamilton (1972), and will be reviewed for another symposium (Hamilton, in press). A conclusion of the cited studies was that the equations of Hookean elasticity can be used to compute unmeasured elastic constants in water-saturated sediments. To compute such elastic constants, density and any two other elastic constants are required. In the case of marine sediments, the density and compressional-wave velocity can be easily measured or predicted. The third constant selected for use in computing the other constants was the bulk modulus (or incompressibility) of the sediment water-mineral system. This constant was selected because it appears possible to compute a valid bulk modulus from its components. The theoretical basis of this computation follows that of Gassmann (1951). See Hamilton (1971a) for an extended discussion.

The components of the computed system bulk modulus are sediment porosity, the bulk modulus of pore water, an aggregate bulk modulus of mineral grains, and a bulk modulus of the sediment structure (or frame) formed by the mineral grains. Good values for the bulk modulus of distilled and sea water, and most of the common minerals of sediments have been established in recent years. This leaves only a value for the frame bulk modulus, needed to compute a bulk modulus for the water-mineral system (following Gassmann, 1951).

A contribution of Hamilton (1971a: Fig. 2) was the derivation of a relationship between sediment porosity and the dynamic frame bulk modulus. Using this relationship, the frame bulk modulus was derived for each sample and used with the bulk moduli of pore water and minerals to compute the system bulk modulus. The method is fully discussed by Hamilton (1971a, b). The volumetric contributions to the system bulk modulus of minerals, pore water, and the frame, are strongly dependent on porosity. Therefore it is to be expected that the plots of the system bulk modulus vs. porosity show good correlation (Fig. 17). Regression equations for these data (and other environments) are in the Appendix.

The computed bulk modulus, and measured density and compressional-wave velocity were then used to compute the other elastic constants. The equations used are:

$$\text{Compressibility, } \beta = \frac{1}{\kappa} \tag{6}$$

$$\text{Lame's constant, } \lambda = \frac{3\kappa - \rho V_p^2}{2} \tag{7}$$

$$\text{Poisson's ratio, } \sigma = \frac{3\kappa - \rho V_p^2}{3\kappa + \rho V_p^2} \tag{8}$$

$$\text{Rigidity (Shear) Modulus, } \mu = (\rho V_p^2 - \kappa)3/4 \tag{9}$$

$$\text{Velocity of shear wave, } V_s = (\mu/\rho)^{1/2}. \tag{10}$$

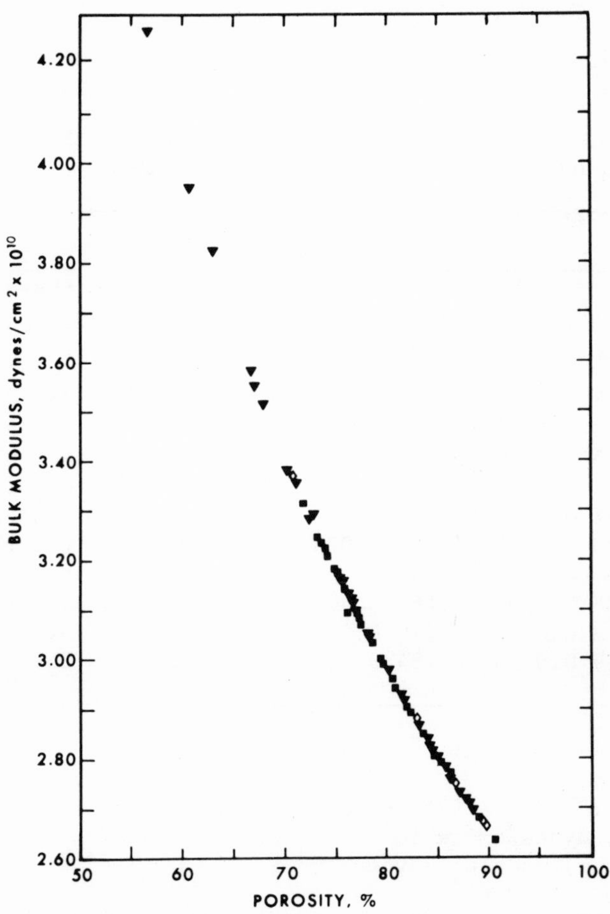

Figure 17. Porosity versus computed bulk modulus, abyssal hill and abyssal plain environments

The measured values of density and velocity are in Tables 1b and 2b. These values were used to compute density X (velocity)^2 in Tables 3 and 4. Computed values of the bulk modulus, Poisson's ratio, the rigidity (shear) modulus, and shear-wave velocity are listed in Tables 5 and 6 according to sediment type in each environment. All values are referred to 23°C and 1 atmosphere pressure.

Computations of elastic constants are strongly dependent on accurately measured values of density, porosity, and compressional velocity. If ρV_p^2 is less than the computed bulk modulus κ (which is strongly dependent on porosity), there is no rigidity or shear wave (equations 9 and 10), and Poisson's ratio is 0.50. In the computed elastic constants listed in Tables 5 and 6, $\kappa > \rho V_p^2$ in some principle sediment types. These are: 3 out of 18 cases in fine sand, 4 out of 73 in continental terrace silt-clays, 2 out of 62 in abyssal plains silt-clays, and 2 out of 72 in abyssal hills silt-

TABLE 5

CONTINENTAL TERRACE (SHELF AND SLOPE) ENVIRONMENT
Computed Elastic Constants in Sediments

Sediment Type	κ Avg.	κ SE	σ Avg.	σ SE	μ Avg.	μ SE	V_s Avg.	V_s SE	No. Samples
Sand									
Coarse	6.6859		0.491		0.1289		250		2
Fine	5.5063	0.1638	0.466	0.005	0.3713	0.0509	417	37	15
Very fine	5.0243	0.3479	0.456	0.010	0.4501	0.1228	472	62	5
Silty sand	4.5017	0.1327	0.459	0.006	0.3716	0.0452	447	27	13
Sandy silt	4.4487	0.2137	0.469	0.007	0.2745	0.0613	363	47	13
Silt	4.3320	0.1631	0.484	0.003	0.1324	0.0187	270	27	9
Sand-silt-clay	3.5903	0.0907	0.463	0.003	0.2784	0.0223	412	17	18
Clayey silt	3.3173	0.0450	0.476	0.002	0.1687	0.0135	324	12	50
Silty clay	3.1459	0.0353	0.484	0.002	0.1026	0.0101	263	12	19

Notes.

Laboratory values: 23°C, 1 atmosphere pressure.
κ = bulk modulus, dynes/cm^2 x 10^{10}.
μ = rigidity (shear) modulus, dynes/cm^2 x 10^{10}.
σ = Poisson's Ratio.
V_s = velocity of shear wave, m/sec.
SE = Standard error of the mean.

clays. These cases were not included in the averages shown in Tables 5 and 6. For this reason, and because of rounding off to a lower number of decimal places, the average values in the tables usually cannot be exactly interrelated with the elastic equations.

The bulk moduli of the deep-water sediments are plotted against

TABLE 6

ABYSSAL PLAIN AND ABYSSAL HILL ENVIRONMENTS
Computed Elastic Constants in Sediments*

Environment / Sediment Type	κ		σ		μ		V_s		No. Samples
	Avg.	SE	Avg.	SE	Avg.	SE	Avg.	SE	
Abyssal Plain									
Sandy silt	4.2572	—	0.492	—	0.0668	—	201	—	1
Silt	3.5798	—	0.484	—	0.1291	—	254	—	2
Sand-silt-clay	3.5670	—	0.488	—	0.0898	—	228	—	2
Clayey silt	3.1465	0.0479	0.480	0.002	0.1286	0.0126	292	16	21
Silty clay	2.8963	0.0391	0.487	0.001	0.0798	0.0078	238	11	34
Clay	3.0108	0.1048	0.493	0.001	0.0421	0.0079	173	20	5
Bering Sea and Okhotsk Sea (Diatomaceous)									
Silt	3.3610	—	0.489	—	0.0731	—	225	—	1
Clayey silt	2.7969	0.0222	0.488	0.004	0.0711	0.0247	224	41	5
Silty clay	2.7381	0.0191	0.488	0.002	0.0648	0.0079	225	12	22
Abyssal Hill									
Deep-sea ("red") clay									
Clayey silt	2.9955	0.0625	0.481	0.003	0.1165	0.0170	287	21	8
Silty clay	2.9455	0.0267	0.487	0.001	0.0759	0.0060	229	9	48
Clay	2.9474	0.0467	0.491	0.001	0.0512	0.0038	194	7	14
Calcareous ooze									
Sand-silt-clay	3.1370	0.0381	0.458	0.005	0.2730	0.0280	439	24	5
Clayey silt	3.5587	0.0529	0.488	0.001	0.0877	0.0077	234	10	14
Silty clay	3.3139	0.0779	0.485	0.002	0.0978	0.0137	255	17	4

* See notes under Table 5.

ρV_p^2 in Figure 18, together with a line $\rho V_p^2 = \kappa$. If these are true values of κ, the divergence of the data points from the line indicate the presence (and values) of rigidity ($\rho V_p^2 = \kappa + 4/3\mu$). The fact that almost all data points in Figure 4 (porosity vs. velocity) are well above Wood's equation for a suspension also points to the presence of appreciable rigidity.

The data summarized and discussed in Hamilton (1971a), and in Figure 4 and 18, indicate that almost all marine sediments have enough rigidity to allow transmission of shear waves. Some exceptions may be in sediments of bays and estuaries or near deltas. Other exceptions may be a few localities where fine sediments are deposited at a fast rate, lack appreciable structural rigidity, and for practical purposes are little more than suspensions.

The relationships between dynamic rigidity and mean grain size and porosity are illustrated in Figure 19 and 20. The probable causes of these interesting relationships were discussed in an earlier report (Hamilton, 1971a). The values for elastic constants

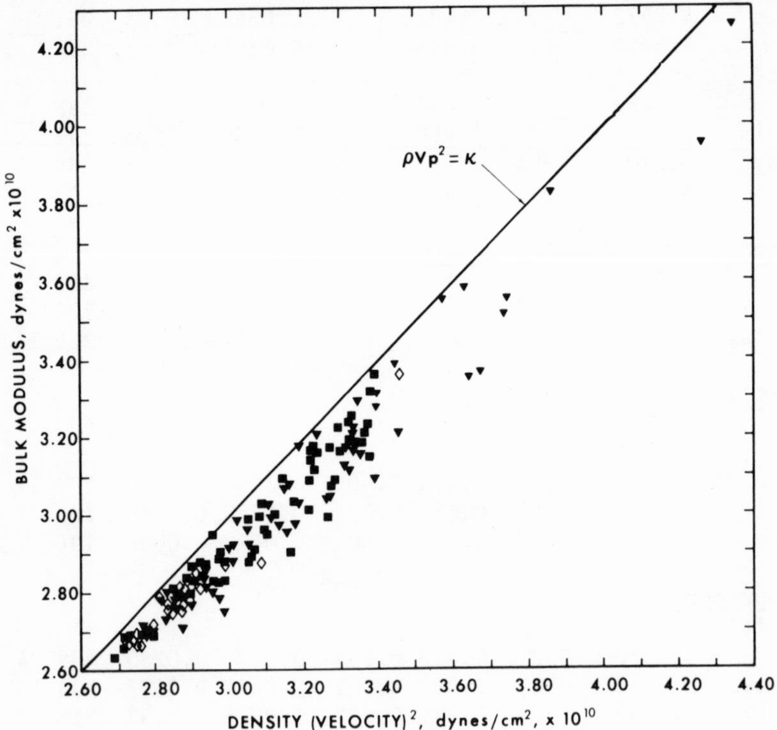

Figure 18. Density X (compressional velocity)^2 versus computed bulk modulus, abyssal hill and abyssal plain environments; $\rho V_p^2 = \kappa$ indicates relationships if the samples had no rigidity.

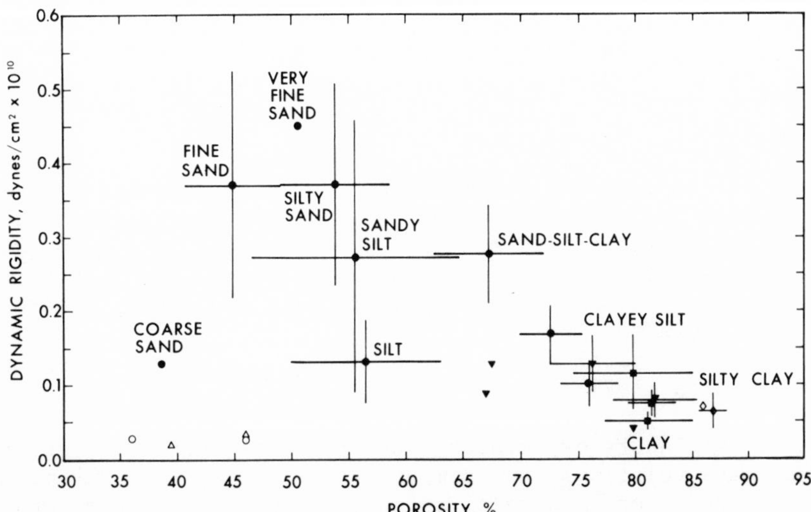

Figure 19. Porosity versus computed dynamic rigidity; symbols as in Figure 1; crosses (through mean values) are 3 times standard error of the mean; lower left: St. Peter's sand (circles) and Ottowa sand (triangles). This figure revised from Hamilton (1971a).

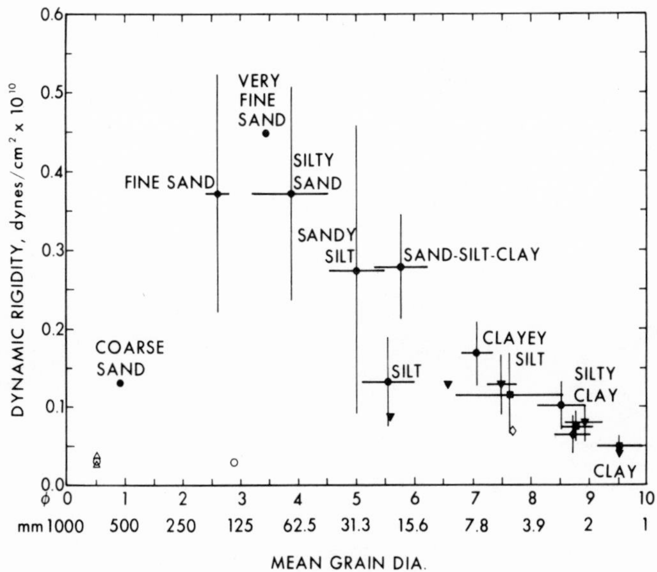

Figure 20. Mean diameter of mineral grains versus computed dynamic rigidity; remarks same as for Figure 19.

at 23°C and 1 atmosphere in the tables and scatter diagrams are usable for basic studies and interrelationships, but cannot be used as in situ values because all three of the constants, (ρ, V_p, and κ) used in the computations require correction from laboratory to in situ conditions.

Prior to computing in situ elastic constants, density and compressional-wave velocity can be corrected as previously discussed. For an in situ sediment bulk modulus, the in situ bulk modulus of pore water (assumed to be that of the bottom water) requires computation. However, porosity, the aggregate bulk modulus of minerals, and the frame bulk modulus can be assumed to be the same as in the laboratory. A numerical example of such computations is in Appendix B of Hamilton (1971b).

All of the laboratory measurements (density, porosity, and compressional-wave velocity), and components of the system bulk modulus (moduli of pore water, minerals, and frame) used in computations of the elastic constants, have margins of error; some are known, and some are unknown. Consequently no attempt was made to statistically estimate variances, or errors, in the final computations of elastic constants. The numbers of decimal places shown in the examples and tables are for purposes of comparison between the various computations and sediment types, and should not be taken as the author's estimates of accuracy. The values of the elastic constants computed and listed should be considered as approximations, and as predictions for comparison with future measurements.

ACKNOWLEDGMENTS

This work was supported by the Office of Naval Research (Code 483), and the Naval Ship Systems Command (Code PMS 302).

APPENDIX: EQUATIONS FOR REGRESSION LINES AND CURVES

(ILLUSTRATED DATA)

Regression lines and curves were computed for those illustrated sets of (x,y) data that constitute the best indices (x) to obtain desired properties (y). Separate equations are listed, where appropriate, for each of the three general environments, as follows: continental terrace (shelf and slope), (T); abyssal hill (pelagic), (H); abyssal plain (turbidite), (P). The equations are keyed by figure numbers to the related scatter diagrams in the main text. The Standard Errors of Estimate, σ, opposite each equation, are applicable only near the mean of the (x,y) values, and accuracy of the (y) values, given (x), falls off away from this region (Griffiths, 1967, p. 448). Grain sizes are shown in the logarithmic phi-scale (ϕ = $-\log_2$ of grain diameter in millimeters).

It is important that the regression equations be used only between the indicated limiting values of the index property (x values), as noted below. These equations are strictly empirical and apply only to the (x,y) data points involved. There was no attempt, for example, to force the curves expressed by the equations to pass through velocity values of minerals at zero porosity, or the velocity value of sea water at 100% porosity.

The limiting values of (x), in the equations are:

(1) Mean grain diameter, M_z, ϕ
(T) 1 to 9 ϕ
(H) and (P) 7 to 10 ϕ

(2) Porosity, n, %
(T) 35 to 85%
(H) and (P) 70 to 90%

(3) Density, ρ, g/cm^3 $(kg/m^3 \cdot 10^{-3})$
(T) 1.25 to 2.10 g/cm^3
(H) 1.15 to 1.50 g/cm^3
(P) 1.15 to 1.70 g/cm^3

(4) Clay size grains, C, %
(H) and (p) 20 to 85%

(5) Density x (Velocity)^2, $\rho V_p{}^2$, dynes/cm^2 x 10^{10}
(H) 2.7 to 3.4 dynes/cm^2 x 10^{10}
(P) 2.7 to 3.8 dynes/cm^2 x 10^{10}

Porosity, n (%) vs. Mean Grain Diameter, M_z (ϕ) Figure 2

(T) $n = 31.05 + 5.52\ (M_z)$	$\sigma = 7.0$
(H) $n = 65.79 + 1.73\ (M_z)$	$\sigma = 4.8$
(P) $n = 42.47 + 4.43\ (M_z)$	$\sigma = 5.8$

Density, p (g/cm^3) vs. Mean Grain Diameter, M_z (ϕ) Figure 1

(T) $p = 2.191 - 0.096\ (M_z)$	$\sigma = 0.12$
(H) $p = 1.577 - 0.027\ (M_z)$	$\sigma = 0.09$
(P) $p = 1.933 - 0.069\ (M_z)$	$\sigma = 0.10$

Sound Velocity, V_p (m/sec) vs. Mean Grain Diameter, M_z (ϕ) Figures 7 & 8

(T) $V_p = 1927.2 - 75.82\ (M_z) + 3.21\ (M_z)^2$	$\sigma = 36.3$
(H) $V_p = 1581.6 - 8.3\ (M_z)$	$\sigma = 10.9$
(P) $V_p = 1628.0 - 12.7\ (M_z)$	$\sigma = 19.1$

Sound Velocity, V_p (m/sec) vs. Porosity, n (%) Figures 3 & 4

(T) $V_p = 2455.9 - 21.716\ (n) + 0.126\ (n)^2$	$\sigma = 34.9$
(H) $V_p = 1483.1 + 0.32\ (n)$	$\sigma = 12.8$
(P) $V_p = 1669.1 - 1.85\ (n)$	$\sigma = 19.2$

Sound Velocity, V_p (m/sec) vs. Density, ρ (g/cm^3) Figures 5 & 6

(T) $V_p = 2234.4 - 1129.3\ (\rho) + 448.1\ (\rho)^2$	$\sigma = 35.6$
(H) $V_p = 1542.7 - 25.3\ (\rho)$	$\sigma = 12.7$
(P) $V_p = 1387.0 + 99.7\ (\rho)$	$\sigma = 20.5$

Sound Velocity, V_p (m/sec) vs. Clay Size, C (%) Figure 9

(H) $V_p = 1552.6 - 0.67\ (C)$	$\sigma = 9.0$
(P) $V_p = 1569.3 - 0.88\ (C)$	$\sigma = 18.9$

Density, ρ (g/cm^3) vs. Porosity, n (%)

(T) $n = 155.4 - 56.5\ (\rho)$	$\sigma = 2.8$
(H) $n = 154.8 - 55.0\ (\rho)$	$\sigma = 0.8$
(P) $n = 162.9 - 61.4\ (\rho)$	$\sigma = 1.3$

Bulk Modulus, κ (dynes/cm^2 x 10^{10}) vs. Porosity, n (%) Figure 17

(T) $\kappa = 215.09467 - 133.1006\ (\log_e n) + 28.2872\ (\log_e n)^2 - 2.0446\ (\log_e n)^3$ $\sigma = 0.01146$

(H) and (P) $\kappa = 128.9909 - 72.0478\ (\log_e n) + 13.8657\ (\log_e n)^2 - 0.9097\ (\log_e n)^3$ $\sigma = 0.0100$

Bulk Modulus, κ (dynes/cm^2 x 10^{10}) vs.

Density x (velocity)^2, ρV_p^2 (dynes/cm x 10^{10}) Figure 18

(H) $\kappa = 0.32039 + 0.862\ (\rho V_p^2)$ $\sigma = 0.049$
$\kappa = 1.68823 + 0.134\ (\rho V_p^2)$ $\sigma = 0.069$

REFERENCES

Akal, T., The relationship between the physical properties of underwater sediments that affect bottom reflection, Mar. Geol., 13, 251-266, 1972.

Anderson, D. G., Strength properties of some Pacific and Indian Oceans sediments, U. S. Naval Civil Engineering Laboratory Tech. Note N-1177, 1971.

Breslau, L. R., Classification of sea-floor sediments with a ship-borne acoustical system, in Proc. Symp. "Le Petrole et la Mer", Sect. I, No. 132, pp. 1-9, Monaco, 1965. (Also: Woods Hole Oceanographic Institute Contrib. No. 1678, 1965).

Breslau, L. R., The normally incident reflectivity of the sea floor at 12 Kc and its correlation with physical and geological properties of naturally-occurring sediments, Woods Hole Oceanographic Institute Ref. 67-16, 1967.

Buchan, S., D. M. McCann, and D. T. Smith, Relations between the acoustic and geotechnical properties of marine sediments, Quart. J. Eng. Geol., 5, 265-284, 1972.

Bucker, H. P., Normal-mode sound propagation in shallow water, J. Acoust. Soc. Am., 36, 251-258, 1964.

Bucker, H. P., J. A. Whitney, G. S. Yee, and R. R. Gardner, Reflection of low-frequency sonar signals from a smooth ocean bottom, J. Acoust. Soc. Am., 37, 1037-1051, 1965.

Cernock, P. J., Sound velocities in Gulf of Mexico sediments as related to physical properties and simulated overburden pressures, Department of Oceanography Tech. Rept., Ref. 70-5-T, Texas A&M University, 1970.

Cole, B. F., Marine sediment attenuation and ocean-bottom-reflected sound, J. Acoust. Soc. Am., 38, 291-297, 1965.

Ewing, J. I., and J. E. Nafe, The unconsolidated sediments in the Sea, in The Earth Beneath the Sea, edited by M. N. Hill, 3, 73-84, Interscience Publ., N. Y., 1963.

Ewing, J. I., R. E. Houtz, and W. J. Ludwig, Sediment distribution in the Coral Sea, J. Geophys. Res., 75, 1963-1972, 1970.

Ewing, W. M., W. S. Jardetzky, and F. Press, Elastic Waves in Layered Media, McGraw-Hill, N. Y., 1957.

Ewing, M., R. Houtz, and J. Ewing, South Pacific sediment distribution, J. Geophys. Res., 74, 2477-2493, 1969.

Ferry, J. D., Viscoelastic Properties of Polymers, John Wiley and Sons, N. Y., 1961.

Fry, J. C., and R. W. Raitt, Sound velocities at the surface of deep sea sediments, J. Geophys. Res., 66, 589-597, 1961.

Gassmann, F., Uber die Elastizitat Poroser Medien Vierteljahrsschrift Naturforschenden Gesellschaft in Zurich, 96, 1-23, 1951.

Griffiths, J. C., Scientific Method in Analysis of Sediments, McGraw-Hill, N. Y., 1967.

Hamilton, E. L., Low sound velocities in high-porosity sediments, J. Acoust. Soc. Am., 28, 16-19, 1956.

Hamilton, E. L., Thickness and consolidation of deep-sea sediments, Bull. Geol. Soc. Am., 70, 1399-1424, 1959.

Hamilton, E. L., Consolidation characteristics and related properties of sediments from Experimental Mohole (Guadalupe Site), J. Geophys. Res., 69, 4257-4269, 1964.

Hamilton, E. L., Sound speed and related physical properties of sediments from Experimental Mohole (Guadalupe Site), Geophysics, 30, 257-261, 1965.

Hamilton, E. L., Shear wave velocities in marine sediments (abstract), Trans. Am. Geophys. Union, 51, 333, 1970a.

Hamilton, E. L., Sound velocity and related properties of marine sediments, North Pacific, J. Geophys. Res., 75, 4423-4446, 1970b.

Hamilton, E. L., Sound channels in surficial marine sediments, J. Acoust. Soc. Am., 48, 1296-1298, 1970c.

Hamilton, E. L., Reflection coefficients and bottom losses at normal incidence computed from Pacific sediment properties, Geophysics, 35, 995-1004, 1970d.

Hamilton, E. L., Elastic properties of marine sediments, J. Geophys. Res., 76, 579-604, 1971a.

Hamilton, E. L., Prediction of in situ acoustic and elastic properties of marine sediments, Geophysics, 36, 266-284, 1971b.

Hamilton, E. L., Compressional-wave attenuation in marine sediments, Geophysics, 37, 620-646, 1972.

Hamilton, E. L., Geoacoustic models of the sea floor, in Proc. Symp. Physics of Sound in Mar. Sediments, edited by L. D. Hampton, in press.

Hamilton, E. L., G. Shumway, H. W. Menard, and C. J. Shipek, Acoustic and other physical properties of shallow-water sediments off San Diego, J. Acoust. Soc. Am., 28, 1-15, 1956.

Hamilton, E. L., H. P. Bucker, D. L. Keir, and J. A. Whitney, Velocities of compressional and shear waves in marine sediments determined in situ from a research submersible, J. Geophys. Res., 75, 4039-4049, 1970.

Hamilton, E. L., D. G. Moore, E. C. Buffington, J. R. Curray, and P. H. Sherrer, Sediment velocities from sonobuoys: Bay of Bengal, Bering Sea, Japan Sea, and Northeast Pacific, J. Geophys. Res., in press.

Hastrup, O. F., Digital analysis of acoustic reflectivity in the Tyrrhenian Abyssal Plain, J. Acoust. Soc. Am., 70, 181-190, 1970.

Horn, D. R., B. M. Horn, and M. N. Delach, Correlation between acoustical and other physical properties of Mediterranean deep-sea cores, Lamont-Doherty Geological Observatory Tech. Rept. No. 2, 1967.

Horn, D. R., B. M. Horn, and M. N. Delach, Correlation between acoustical and other physical properties of deep-sea cores, J. Geophys. Res., 73, 1939-1957, 1968.

Horn, D. R., M. Ewing, B. M. Horn, and M. N. Delach, A prediction of sonic properties of deep-sea cores, Hatteras Abyssal Plain and environs, Lamont-Doherty Geological Observatory Tech. Rept. No. 1, 1969.

Houtz, R. E., and J. I. Ewing, Detailed sedimentary velocities from seismic refraction profiles in the western north Atlantic, J. Geophys. Res., 68, 5233-5258, 1963.

Houtz, R. E., J. I. Ewing, and X. LePichon, Velocity of deep-sea sediments from sonobuoy data, J. Geophys. Res., 73, 2615-2641, 1968.

Houtz, R., J. Ewing, and P. Buhl, Seismic data from sonobuoy stations in the northern and equatorial Pacific, J. Geophys. Res., 75, 5093-5111, 1970.

Keller, G. H., and R. H. Bennett, Variations in the mass physical properties of selected submarine sediments, Mar. Geol., 9, 215-223, 1970.

Kermabon, A., C. G. P. Blavier, and B. Tonarelli, Acoustic and other physical properties of deep-sea sediments in the Tyrrhenian Abyssal Plain, Mar. Geol., 7, 129-145, 1969.

Kinsler, L. E., and A. R. Frey, Fundamentals of Acoustics, John Wiley and Sons, N. Y., 1962.

Laughton, A. S., Sound propagation in compacted ocean sediments, Geophysics, 22, 233-260, 1957.

Laughton, A. S., W. A. Berggren, et al., Initial Reports of the Deep Sea Drilling Project, 12, U. S. Govt. Printing Office, Washington, D. C., 1972.

LePichon, X., J. Ewing, and R. E. Houtz, Deep-sea sediment velocity determination made while reflection profiling, J. Geophys. Res., 73, 2597-2614, 1968.

Ludwig, W. J., R. E. Houtz, and M. Ewing, Sediment distribution in the Bering Sea: Bowers Ridge, Shirshov Ridge, and enclosed basins, J. Geophys. Res., 76, 6367-6375, 1971.

McCann, D. M., Measurement of the acoustic properties of marine sediments, Acustica, 26, 55-66, 1972.

McCann, C., and D. M. McCann, The attenuation of compressional waves in marine sediments, Geophysics, 34, 882-892, 1969.

Morris, H. E., Bottom-reflection-loss model with a velocity gradient, J. Acoust. Soc. Am., 48, 1198-1202, 1970.

Nafe, J. E., and C. L. Drake, Variation with depth in shallow and deep water marine sediments of porosity, density and the velocities of compressional and shear waves, Geophysics, 22, 523-552, 1957.

Nafe, J. E., and C. L. Drake, Physical properties of marine sediments, in The Sea: Ideas and Observations on Progress in the Study of the Seas, edited by M. N. Hill, 3, 794-815, Interscience Publ., N. Y., 1963.

NAVOCEANO, Handbook of Oceanographic Tables, U. S. Naval Oceanographic Office SP-68, 1966.

Officer, C. B., Introduction to the Theory of Sound Transmission, McGraw-Hill, N. Y., 1958.

Schreiber, B. C., Sound velocity in deep-sea sediments, J. Geophys. Res., 73, 1259-1268, 1968.

Shepard, F. P., Nomenclature based on sand-silt-clay ratios, J. Sediment. Petrol., 24, 151-158, 1954.

Shumway, G., Sound speed and absorption studies of marine sediments by a resonance method, Part I; Part II, Geophysics, 25, 451-467, 659-682, 1960.

Sutton, G. H., H. Berckheimer, and J. E. Nafe, Physical analysis of deep-sea sediments, Geophysics, 22, 779-812, 1957.

Wu, T. H., Soil Mechanics, Allyn and Bacon, Boston, 1966.

MARINE GEOMECHANICS: Overview and Projections

ARMAND J. SILVA

Worcester Polytechnic Institute

ABSTRACT

Until recently, most geotechnical studies of deep-sea sediments have been necessarily concerned with the upper few meters, and comprehensive studies aimed at determining basic stress-strain parameters in relation to sediment processes are scarce. There are important differences between some deep-sea sediments and most fine-grained terrestrial soils including compositional differences and differences due to rate of deposition, pressure, and temperature. Some diagenetic processes are reviewed along with a brief discussion of the factors which affect microstructure. Detailed studies of sediment microstructure coupled with geochemical and engineering studies may yield valuable insight into the basic mechanisms involved in sediment diagenesis as well as stress-strain behavior.

Simple strength measurement systems such as vane shear and cone penetrometer can only be viewed as indicators of gross strength properties. The Mohr-Coulomb theory used in conjunction with triaxial methods of testing gives satisfactory results for engineering applications. In order to study basic stress-strain behavior, it is recommended that a more sophisticated yield theory, such as is embodied in the critical state method, and appropriate compatible testing procedures be used to develop predictive mathematical models of sediment behavior.

Compressibility of fine-grained sediments is reviewed with particular attention directed to secondary compression effects. Delayed compression under constant effective stress will contribute to a build-up of residual strength which results in pseudo-overconsolidation effect, and the secondary compression may continue for hundreds of years. The results of consolidation tests can aid in interpreting geological sea-floor processes involving deposition, erosion, slumping, etc. A procedure to determine a disturbance index is proposed to quantify sample disturbance effects.

The author's experiences with a large diameter long piston corer are reviewed. Data include properties of two long cores (up to 22 m) and some correlations with acoustic reflectors.

INTRODUCTION

This paper was prepared for presentation and discussion at a symposium/workshop on the geotechnical properties of ocean sediments. It was considered appropriate to include a fairly broad coverage of the state of the science as related to geotechnical properties and to highlight some of the more important aspects. Although a considerable literature review is included, no attempt was made to undertake an exhaustive review and much of the material is the author's perspective.

Marine sediment, as used in this paper, refers to the "unconsolidated" (i.e., not solidified) assemblage of minerals and organic materials, including interstitial pore fluid, which are found on the vast majority of sea floor. Until recently, marine geologists were principally interested in the morphology of ocean sediments as related to geological processes and therefore their studies were oriented toward description, identification, and dating of sediment strata. In addition to being essential for engineering applications, the study of geotechnical properties of sediments should be of interest to oceanographers interested in sea-floor processes. Due to the difficulty and expense of obtaining samples of deep-sea sediments, it would seem logical to coordinate the efforts of all concerned (physical, chemical, biological, and geological oceanographers as well as engineers) to obtain the maximum amount of information possible from recovered cores.

A knowledge of geotechnical properties is important to the understanding of sediment processes such as erosion, transport, deposition, slumping, consolidation, etc., whereas knowledge of sediment diagenesis and processes is important to understanding geotechnical behavior. For example, the stress-strain behavior (shear strength, compressibility) of fine-grained sediments is primarily controlled by physico-chemical inter-particle effects which in turn are largely related to the diagenesis of sediments, whereas many of the alterations that occur after the sediment is deposited (such as erosion, slumping, etc.) are in turn controlled by strength properties of the sediment. Detailed information on geotechnical properties is crucial in identifying correlations between acoustical and physical properties and may provide supplemental information on tectonic processes and dating of stratigraphy.

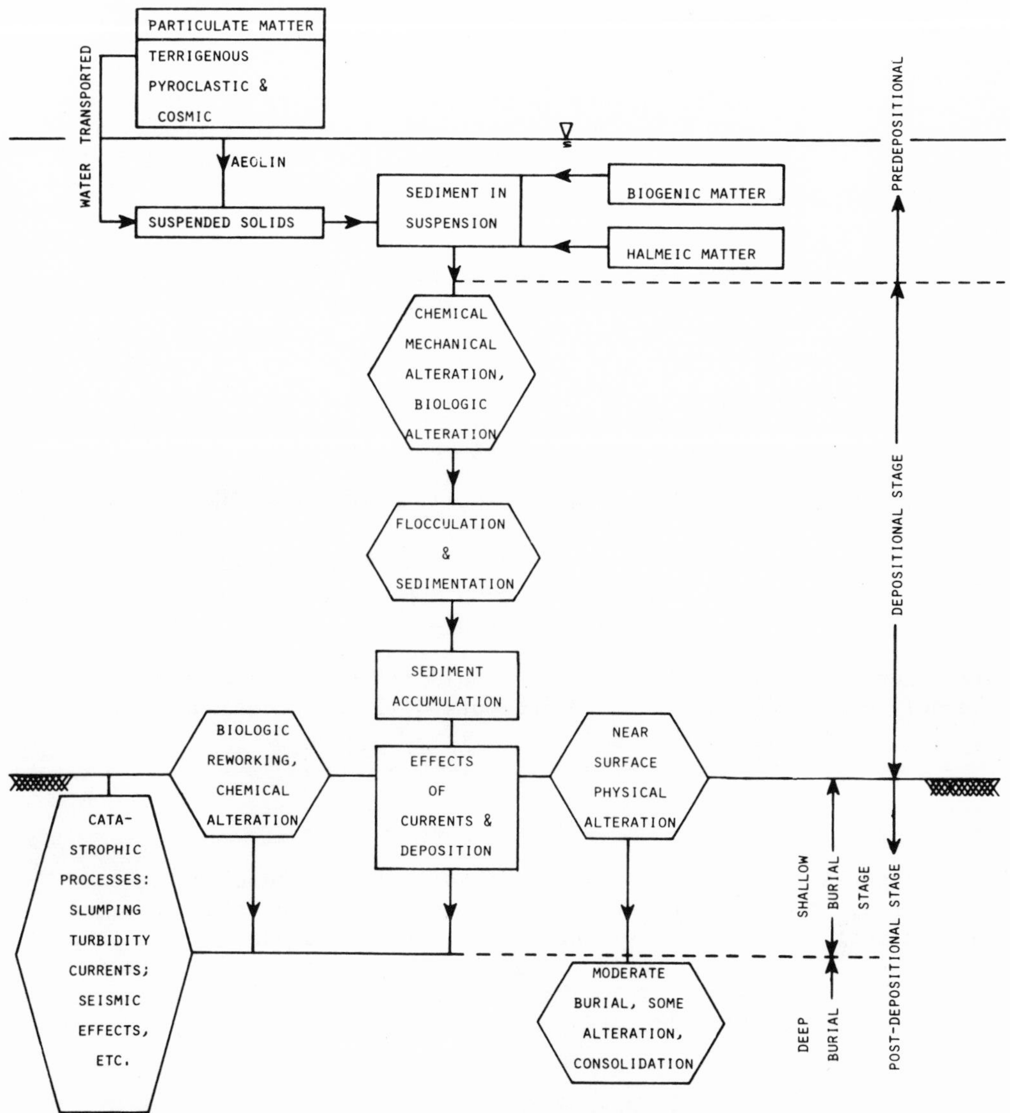

Figure 1. Flow chart of diagenetic processes of sediments

Some of the better known sediment diagenetic processes are indicated in Figure 1. The notion of mineral particles settling through a tranquil water column and being deposited on the bottom with little alteration is not valid except in rare situations. Rather, as indicated in Figure 1, it is probably more realistic to think in terms of a dynamic system involving formation, flocculation, alteration, reworking by animals, etc., as the sediments settle, and then a continuing alteration, movement, reworking, etc., once the material reaches the bottom. The effects on clay particles passing through the digestive tracts of animals in the water and in the near-bottom zone are virtually unknown but may be extremely important. Involvement of sediments in turbidity currents should have a significant effect on properties and the effect of turbidity currents on the sediment underlying affected areas (of both erosion and deposition) will be important.

Some of the possible environmental and compositional differences between terrestrial and deep-sea sediments should be mentioned:

a) Deep-sea sediments are saturated with saline pore fluid.
b) Compositions of deep-sea sediments can often be quite different than most terrestrial soils. For example, many marine sediments contain large percentages of skeletal material and other organic debris.
c) The rate of deposition is usually much slower in the deep sea.
d) The depositional environment (i.e., high pressure, low temperature) found in the deep sea may have important effects on sediment microstructure.

With these differences in mind, it should not be expected that all the soil mechanics theories and concepts will be successful in predicting marine sediment behavior.

SEDIMENT DIAGENESIS

Diagenesis encompasses all the changes that take place in a sediment at low temperatures and pressures without crustal movement being directly involved. Diagenetic processes include chemical and mineralogic changes due to weathering, transportation, and deposition in a sedimentary basin. The three stages of diagenesis are as follows (also see Fig. 1).

Pre-depositional Stage

Unweathered material (rock) is broken down to transportable material by in situ weathering either through chemical or mechanical process. Rivers, icebergs, and winds transport weathered and/or unweathered material into marine environment. There are other not so obvious sources of sediments such as volcanic, rafted, organic, etc. Some of the more important changes and processes are:

a) Aquatolysis, which is defined as the mechanical and chemical change of the material in the fresh-water transportation stage, forms the basic building units for many clays.

b) Halmyrolysis is the process of cation exchange in the marine water due to an abundance of free exchangeable cations.

Depositional Stage

After sediment particles have been formed, transported, and transformed, they are deposited to form a sedimentary soil (Fig. 1). The three main mechanisms of deposition in water are velocity reduction, solubility decrease, and electrolyte increase. When a stream reaches an ocean the velocity reduces and most of the material is deposited. In addition, the reduction in solubility coupled with an increase in electrolytes results in considerable flocculation. This aggregation of fine particles into large flocs increases the settling velocity.

Post-depositional Stage

Deep-sea sediments can be classified into five different groups, dependent primarily on color, texture, composition, and origin (see Table 1). After deposition there are catastrophic activities along the ocean floor, such as turbidity currents, contour currents, and slumping, which result in a reworking of the sediment (Fig. 1). Turbidity currents flow downhill and deposit hemipelagic sediments called "turbidites." Turbidites may range from gravel to clay sizes and a single bed may be greater than 100 cm thick. Contourites are more gentle deposits caused by bottom contour currents (along bathymetric contours) and rates of accumulation exceed 1 m/1000 years. The effect of reworking of the sediment by biological activity and the depth to which this reworking occurs has not been adequately investigated. All these causes and effects add to the complex picture of sediment deposition and transportation, and from an engineering point of view they may all have a significant effect on porosity and sediment microstructure.

Porosity of a saturated soil mass is directly related to water content and water content has a profound influence on the properties of clay sediments. Meade (1966) discusses in detail the clay-water interaction and concludes that the amount of water held in clayey sediments under effective overburden pressures between 0 and 4093 kPa (50 kg/cm^2) is influenced by particle size, type of clay minerals, exchangeable cations and interstitial electrolyte concentrations. Under overburden pressures greater than 4093 kPa, the water content is influenced by particle size, clay minerals, and temperature, although the cations and electrolyte concentration have some effect on the microstructure.

The compaction (consolidation) of sediments associated with porosity reduction is directly related to burial depth of the sedi-

TABLE 1

TYPES OF DEEP-SEA SEDIMENTS

Type of Sediment	Sedimentation Rate for 1000 Years	% Fines Less Than 30 μ m	Color	Remarks
Red clay	1/10 to 1 mm	greater than 80%	Chocolate brown	Usually contains less thab 10% of $CaCo_3$
Globigerina ooze	1 to 3 cm	less than 50%	Milky white to brown	Finer fraction consists of planktonic foraminifera
Diatom ooze	1 to 2 cm	greater than 50%	Yellow to cream	Siliceous sediment with 40% of diatom shell
Radiolarian ooze	1 to 2 cm	approx. 30%	Pale green to greenish yellow	40% of radialarian shell found in productive areas
Terrigenous or hemipelagic	5 to 100 cm	less than 50%	Green, black, or slightly red	30% of sand and silt grains

(from Hollister and Heezen, 1971)

ment (Engelhardt, 1963). The field porosity may give an indication of the presence or absence of loads.

MICROSTRUCTURE

The stress-strain properties of the sediment are dependent on the porosity (void ratio) and permeability which in turn are dependent on size, shape and arrangement of particles in the micro and macro scales. Microstructure is the fundamental building unit and a network of this microstructure forms the macrostructure. Microstructure is dependent on the type of clay mineral, environmental conditions and the imposed loads.

In an aqueous solution clay particles settle by coagulation, flocculation or aggregation in positions conforming to minimum energy requirements. This arrangement is dependent on the amount of water adsorbed on the surface of the clay mineral, and the amount of water adsorbed is inversely proportional to the size of the clay mineral (Meade, 1966; Engelhardt, 1963).

The structural arrangement which occurs during sedimentation is influenced by chemical factors such as electrolyte ion concentration, type of associated cations and anions, acidity and alkalinity, and associated organic matter (Muller, 1967). In addition to chemical factors, the physical and mechanical factors such as rate of deposition, kinetics of water, particle size distribution, and concentration of sediment being deposited, influence the spatial arrangement. One of the more important processes which may play an important role in the marine environment, is cementation at particle contacts. Soil properties such as residual strength, and pseudo over-consolidation may be directly related to cementation (Parker, 1972).

Loads imposed on a sediment fabric of low permeability create fluid pressure gradients and a resultant loss of fluid from the pores. This removal of water implies that the solid particles must move closer together into a more efficiently packed arrangement. It is suggested that detailed studies of sediment microstructure using electron microscopy coupled with engineering studies will yield valuable insight into the basic mechanisms involved in sediment diagenesis and stress-strain behavior.

SAMPLING METHODS

There are many problems associated with sediment sampling and testing procedures in the deep ocean, and a discussion of sediment properties would not be complete without at least some mention of methods currently used or proposed. One of the more recent publications dealing with sediment sampling and testing is the collection of papers published by ASTM (1972). The "State-of-the-Art Review" by Noorany (1972) has an excellent review of most sampling and testing methods. The length of deep-sea sediment samples (taken with tethered gravity or piston corers) is rarely longer than 6 m whereas many engineering projects require exploration well below 9 m. Most of the more sophisticated new corers (such as vibratory corers) have limited water depth capabilities and there is still a need for developing methods of obtaining relatively undisturbed continuous samples to significant depths (over 50 m) into the sediment column. Some recent developments with an in-situ wire line remote vane shear device and with penetrometers show promise of providing valuable information, but there will still be a need for obtaining some samples.

Some of the problems associated with piston-coring are presented by McCoy (1972). One of the major, still unsolved problems for deep water work is that of cable rebound after release of the corer. McCoy estimates that cable contraction will be approximately 2.5 m in water depths of 5000 m. This rebound together with vertical oscillations due to ship movements will tend to produce piston displacements which may result in intermittent coring, flow-in, and sample disturbance. However, one of the factors which has not been considered is the damping effect of water within the corer during penetration. Since most piston-corers have a constriction at the top end, there will be an "orifice" effect which prevents an instantaneous complete release of line tension when the corer is tripped. This has the effect of reducing vertical oscillations during corer penetration. On the other hand, the available driving energy can be significantly reduced due to the orifice effect.

Additional research is needed, using analytical approaches, model testing, and prototypes, to more accurately determine the dynamics of various coring techniques. Methods should be developed for isolating corers from winch line effects and immobilizing the piston of piston-corers. This would then immediately provide the means of obtaining fairly long, relatively undisturbed samples of soft sediments using some of the proven and available techniques. Development of new methods should be continued, but the objectives (in terms of length, diameter, quality) should be reassessed. Two of the most important limiting factors of most "deep-sea" research vessels are (a) the low winch load capacity, and (b) the length of available rail space for rigging a long corer. Means should be explored to upgrade handling capabilities of research vessels so that their full coring potentials can be realized.

STRENGTH

In accordance with theories of soil mechanics, "strength" of a sediment is taken to mean the resistance to shear deformation.

Because of the complex nature of sediment systems, strength cannot be easily defined as a basic material property nor is it simple to determine the "quantity" experimentally. Thus it is found that shear strength parameters of fine-grained sediments are dependent on:

a) void ratio, water content, porosity
b) electro-chemical interparticle forces, i.e., cohesive bonds
c) constraints on mode of failure and stress system during experimental tests.

It is therefore seen that shear strength of a given sediment is not an invariant property. It is quite obvious that a sediment can exist in various states ranging from a viscous fluid, with essentially no strength at time of deposition, to a solidified rock with high strength under the action of pressure. However, even at a given void ratio, laboratory strength measurements will depend, among other things, on the nature of loading.

It should be noted that the principle of effective stress is fundamental to the discussion of shear strength and unless otherwise noted the sediment stresses are effective stresses. It is difficult to characterize stress-strain behavior of sediments in terms of the more familiar mathematical models (elastic, plastic, elasto-plastic) since particulate systems exhibit some unique responses not found in homogeneous materials such as steel, rubber, etc. A typical stress-deformation curve for a sediment might reveal a fairly straight initial portion with a maximum resistance considerably greater than the ultimate resistance (Fig. 2a). Therefore, the material might be described as approximating elasto-plastic behavior but with strain-softening effects. The stress-strain curve will of course change as the porosity of the sediment changes.

Some Methods of Measurement

The strength parameter which is measured depends to a large extent on the measuring device and technique. A complete review of testing equipment and procedures is not intended but the principal ones which have or may have application in marine studies are as follows:

a) in situ tests - vane; cone penetrometer; accelerometers mounted on corers or penetrometer devices
b) laboratory tests: (usually require obtaining "undisturbed" samples) - vane; cone penetrometer; unconfined compression; direct shear; ring shear; triaxial compression.

There are several other devices which have been built as an attempt to refine techniques of measuring soil strength parameters and currently a few very sophisticated and complex instruments are being developed (see Parry, 1972). In an attempt to simplify the testing procedures but still be able to determine basic strength parameters, a consolidated-vane-shear device is being developed at WPI Marine Science Facility. The intent is to be able to use a miniature vane apparatus to test a given sample under varying vertical stresses and to also measure pore pressures on the failure surface (Beloff, 1973). In this way it should be possible to define the Mohr-Coulomb effective strength parameters and perhaps reduce the necessity for some of the more sophisticated tests. It is possible that one or more of these instruments will emerge as being practical for testing of deep-sea sediments and investigators should be encouraged to explore all possibilities. However, at the present state of development it would appear that the triaxial compression test with provisions for measurement of pore pressures and means of testing under anisotropic stress conditions offers a reasonably accurate, practical means for detailed studies of the stress-strain properties of ocean sediments.

Mohr-Coulomb Failure Criterion

The Mohr-Coulomb failure criterion is probably the most widely used due to simplicity and the fact that it is reasonably reliable for most soil engineering applications. In summary, according to this theory a sediment will fail along a plane when a critical combination of normal and shear stress occurs along that plane such that:

$$\bar{\sigma}_1 = \bar{\sigma}_3 \tan^2 (45 + \bar{\Phi}/2) + 2\ \bar{c}\ \tan (45 + \bar{\Phi}/2) \tag{1}$$

where $\bar{c}$ and tan $\bar{\phi}$ define the strength envelope on a plot of shear stress vs. normal stress (see Fig. 2b). The well-known equation defining the strength envelope is thus:

$$\tau_f = \bar{c} + \bar{\sigma}_n \tan\bar{\Phi} \tag{2}$$

However, as indicated in Figure 2b, this strength envelope is more complex and can be approximated by two straight line segments. In order to define the complete strength envelope, several triaxial compression tests must be conducted on the same material at different confining (lateral) stresses. The effects of (a) cementation, (b) "residual strength," and (c) crushing of skeletal material on the strength envelope are not known. The details of the Mohr-Coulomb theory can be found in most texts on soil mechanics (Lambe and Whitman, 1969; Wu, 1966; Terzaghi, 1943). The Mohr-Coulomb theory is an empirical convenience which yields satisfactory results for most engineering applications and in order to understand the compo-

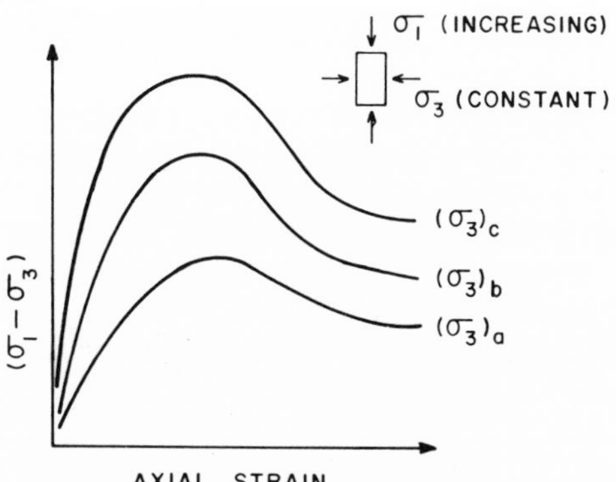

Figure 2a. Stress strain curves for a series of triaxial compression

nents of shear strength it will be necessary to study basic stress-strain behavior of sediments using a more comprehensive approach.

Cohesionless sediments are those in which the inter-particle forces (bonds) do not play a significant role in the stress-strain behavior. The predominant contributions come from (a) solid friction developed at inter-particle contacts, and (b) interlocking effects due to packing arrangement. One of the main problems is that it is virtually impossible to obtain undisturbed samples of cohesionless materials for laboratory testing and almost all test results are for artificially prepared samples. The effects of in situ

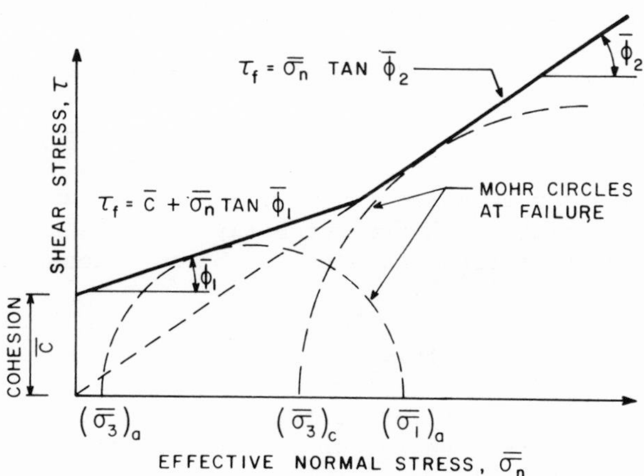

Figure 2b. Mohr-Coulomb analysis of results. The strength envelope is usually slightly curved but is shown here idealized as two straight line segments.

preferred particle orientations may be important and are difficult to reproduce in reworked test specimens. These materials are also difficult to test in situ.

Cohesive sediments are those in which inter-particle forces (bonds) play a significant and usually dominant role in stress-strain behavior. Since the electro-chemical inter-particle effects depend on several factors, including size and shape of particles, microstructural arrangement, nature of pore fluid, and cation exchangeability, the behavior of cohesive sediments is much more complex than cohesionless materials. Furthermore, the disruption of cohesive bonds due to disturbance or remolding is largely irreversible and it is essentially impossible to reproduce in situ microstructure by remolding sediments for preparation of test specimens. Fortunately, with careful sampling procedures it is possible to obtain reasonably "undisturbed" samples of many fine-grained sediments and it is feasible in many cases to conduct in situ measurements.

A cohesive sediment can be characterized as a particulate material possessing both cohesive and frictional components of strength. Furthermore, the cohesion is probably a combination of two effects (Bishop, 1972): one is present only in undisturbed sediments and is largely destroyed by moderate shear strain and the other is a function of strain, becoming small at large strains. The friction component also requires considerable strains for full mobilization (Nacci and Huston, 1970). The effect of the presence of large amounts of biogenic materials on stress-strain behavior has not been investigated nor speculated to any appreciable degree although it is generally felt that skeletal remains would tend to collapse during high inter-particle stresses. It is likely that organic debris such as large fragments of diatomaceous material will play a significant role in shear strength behavior and some of the concepts which have evolved for pure clays (clay minerals), as a result of detailed information from soil mechanics studies, will probably have to be significantly modified to account for the compositional differences in ocean sediments. However, for deep-sea sediments below the carbonate compensation depth with small amounts of biogenic material, the theories which are currently in use should be applicable to the marine environment with slight modification.

Yield Theories

The Hvorslev (1960) failure criterion is basically an extension of the Mohr-Coulomb theory, which incorporates a factor to take into account the void ratio at failure. One form of the Hvorslev equation is as follows:

$$\tau_f = \mu_0 \, \bar{\sigma}_f + \nu \exp\,(-Be_f) \qquad (3)$$

where μ_0, ν, and B are parameters for the particular soil and e_f is the void ratio on the failure plane.

In a series of papers (Roscoe et al., 1958; Poorooshasb and Roscoe, 1961; Roscoe and Poorooshasb, 1963; Roscoe et al., 1963a) a more comprehensive yield theory referred to as the critical state method, is advanced to predict stress-strain behavior of clays. The critical state methods are fully discussed and developed in a textbook by Schofield and Wroth (1968).

At the critical state there is a unique relationship between effective stress state and void ratio. The relationships determined for remolded saturated London clay described a three-dimensional space diagram similar to one reproduced in Figure 3 (Roscoe et al., 1963b). The p-e plane corresponds to one-dimensional consolidation and the N-N line is therefore the virgin consolidation curve. The line X-X is the critical void ratio (CVR) line which projects as a straight line (with slope M) through the origin on the p-q plane. When a sample is at a state corresponding to a point on the critical state line, X-X, it will continue to distort (yield) in shear. The state boundary surface NN-XX is described by the stress paths obtained from drained and undrained triaxial tests on saturated samples. It is thought (Roscoe et al., 1963) that similar surfaces will be obtained for overconsolidated clays. Unfortunately, almost all the detailed investigations of the type represented by Figure 3 have been conducted on remolded samples and the effects of different microstructures exhibited by normally and overconsolidated undisturbed clays have not been adequately investigated. In addition,

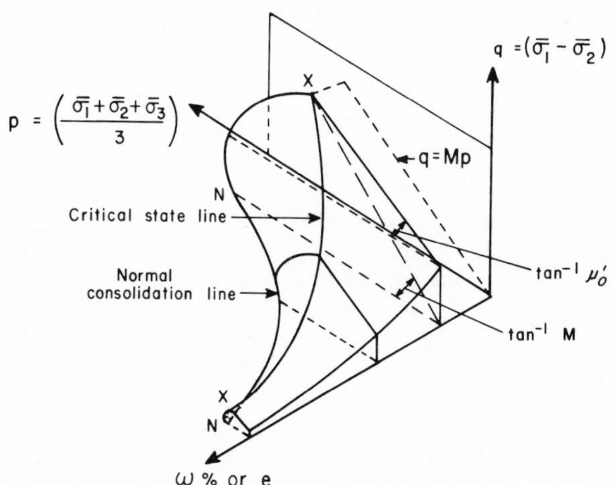

Figure 3. Isometric view of yield surface for weald clay (Ref. Roscoe, et al., 1963)

the detailed effects of organic materials on stress-strain behavior are largely unknown. There are various means of defining the shape of the state boundary surface (Fig. 3) and the reader is referred to the references for detailed discussions.

The predictive equations developed through a suitable stress-strain yield theory can be combined with methods of continuum mechanics to calculate deformations of sediment masses. It is suggested that a comprehensive experimental and analytical study of stress-strain behavior of ocean sediments using some of the more sophisticated apparatuses and techniques can be extremely beneficial in terms of determining basic shear strength parameters. When compared to some of the detailed studies made on "terrestrial" soil, it is likely that deep-sea sediments will show some significant differences in stress-strain response.

Complicating factors which have not been adequately considered in the yield theories discussed above are the effects of secondary compression and aging on stress-strain behavior. Studies by several investigators (Bjerrum and Wu, 1960; Leonards and Ramiah, 1960; Bjerrum and Lo, 1963) indicate that cohesive bonds increase with time. The strength increase is due principally to secondary compression effects, but Bjerrum and Lo (1963) suggest that cohesive bonds may increase with time even if secondary consolidation does not occur. The "aging" effect reported by Bjerrum (1967) may be associated with cementation effects previously mentioned. It is probable that secondary compression effects will be accounted for in the critical state methods but the "aging" effect under constant void ratio will be difficult to incorporate into the mathematical models.

Review of Some Data for Ocean Sediments

Most studies of ocean sediments (especially deep-sea sediments) have been concerned with vertical and areal variability of gross indicators of properties and practically all the definitive data is for the upper few meters of sediment. There has not been a great deal of sophistication applied to studying basic stress-strain properties of sediments and most of the investigators have attempted to arrive at empirical relationships and correlations among various parameters and measurements. A cursory review of data on physical properties of ocean sediments indicates that the sediment masses are far from homogeneous from the point of view of physical properties, but rather there is considerable vertical and horizontal variability. Even when sediments appear to be compositionally homogeneous, there can be important variations in properties, such as void ratio, compressibility, and shear strength.

Excerpts from a few of the more pertinent studies on shear

strength follow and additional review can be found in a state-of-the-art paper by Noorany and Gizienski (1970).

A great deal of data on shear strength has been published by A. F. Richards. For example, Richards (1962) made a detailed analysis of 35 sediment cores of lengths from 3 to 5 m, composed of silty clays and clayey silts. The cohesion ranged from 0.41 kPa to 23.8 kPa (4.2 to 234 g/cm^2), and there was considerable variation in shear strength with depth. As has been confirmed by most studies, the natural water content of surficial sediments is usually considerably greater than the liquid limit and therefore the void ratios are quite high. Other studies by Richards (1961, 1964) show similar results and because of vertical and areal variability of properties, even within a small geographic area, caution is urged in extrapolating property data.

Results of shear strength studies conducted on sediments from the Mohole (Guadalupe Site) were reported by Moore (1964) and reviewed by Noorany and Gizienski (1970). Moore found that the vane shear results were much lower (1/10 to 1/100) than triaxial results and attributed the difference to sample disturbance. However, as noted by Noorany and Gizienski (1970), the triaxial tests were conducted on samples isotropically consolidated to the vertical overburden pressure which would tend to produce lower void ratio and higher shear strength than the in situ condition.

Bryant et al. (1967) report vane shear test results for a number of cores from the Western Gulf of Mexico (maximum core length of 8.3 m). The results indicate extreme variation in shear strength (as much as a factor of 7) over small vertical distances (approximately 20 cm). They speculate that cementation effects in high carbonate sediments are very important in controlling geotechnical properties.

Results of samples from the Pacific in deep water (4298-5505 m), obtained with a spade corer, are reported by Noorany (1972). Triaxial compression tests (CIU) indicated a cohesion intercept of 1.96 kPa (20 g/cm^2) and an angle of internal friction of 0.559 rad to 0.576 rad (32° to 33°). Stress-strain data did not indicate appreciable reduction in strength at large strains (20%).

Abrupt changes in the geotechnical properties of a short core of deep-sea sediment was found to be due to relatively small changes in mineralogy and carbonate content (Parker, 1972). However, the reason for the changes remains undetermined.

Shear strength profiles have been used to determine the state of consolidation and geology of the continental shelf off Louisiana (Fisk and McClelland, 1959). The ratio of cohesion to overburden pressure and plasticity index for normally consolidated deposits was found to agree well with previously established relationships proposed by Bjerrum (1954).

Many of the concepts which were previously reviewed have been recognized by other investigators. A notable example is the paper by Nacci and Huston (1970) dealing with sediment structure and the relationship to sediment properties. In addition, the advantages of the consolidated anisotropic undrained (CAU) test over the isotropic (CIU) test were clearly demonstrated and problems with other test methods such as unconfined compression and vane shear are clearly stated. The unconfined compression strength tends to be too low compared with in situ conditions whereas the measured vane shear strength can be quite variable depending on rate of shear, disturbance, etc. In a recent paper Sangrey (1972) uses the critical state approach in analyzing strength properties of sediments and develops the following relationship for predicting the strength, S_u, profile (with depth):

$$\frac{S_\mu}{\bar{\sigma}} = \left[\frac{2.25 - 1.25 \sin \bar{\phi}}{\frac{3}{\sin \bar{\phi}} - 1} \right]$$

Based on some case studies, Sangrey concludes that the critical state approach underestimates strength for normally consolidated sediments but overestimates the strength of underconsolidated sediments.

Discussion of some of the writer's results of shear strength of sediments recovered with a large diameter long piston corer will be postponed to a later section.

Other Aspects

Some additional factors which need consideration are as follows:

1) Effects of different loading situations on the stress-strain behavior of sediments: some important situations which may produce dynamic effects are (a) earthquake loading, (b) slumping, (c) repetitive loading produced by large surface and subsurface waves;
2) Effects of high pore pressures on in situ behavior;
3) Effects of sediment reworking by animals;
4) Effects of physico-chemical alterations which may occur within the sediment column.

COMPRESSIBILITY

Compressibility is used here to refer to volumetric reduction of sediments due to compression and rearrangement of the constituents. If a sediment is completely saturated with water, elastic compression of the constituents (minerals, organic matter and pore fluid) is small compared to volume reduction due to expulsion of pore fluid and readjustment of the sediment fabric. The expulsion of pore liquid from a saturated sediment is commonly referred to as consolidation. Terzaghi (1943) presented a mathematical formulation of the consolidation phenomenon based on a number of simplifying assumptions. In this model, the rate of compression is controlled entirely by hydrodynamic effects and compression ceases when the excess pore pressure is dissipated.

It has long been recognized that many natural sediment deposits do not behave in the manner predicted by Terzaghi's consolidation theory and that compression continues under constant effective stress. Several investigators (Merchant, 1939; Taylor, 1942; Gibson and Lo, 1961; Ishii, 1951; Taylor and Merchant, 1940) have offered modifications to the Terzaghi consolidation theory in attempts to incorporate so-called secondary compression effects. The compression of clays which exhibit creep under constant effective stress was also described by Bjerrum (1967) in terms of compression (settlement) which consists of two parts:

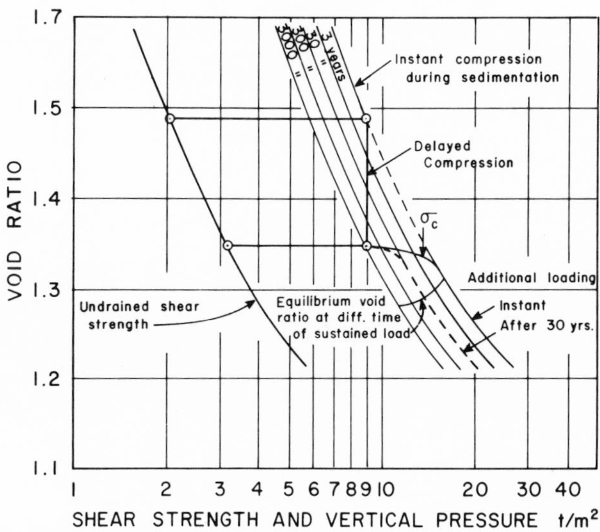

Figure 4. Compressibility and shear strength of a lay exhibiting delayed consolidation (Ref. Bjerrum, 1967)

a) an "instant compression" which occurs simultaneously with an increase in effective stress and causes a reduction in void ratio until an equilibrium value is reached at which the structure of the sediment effectively supports the pressure;
b) a "delayed compression" representing the reduction in volume at unchanged effective stress.

Bjerrum demonstrated how delayed compression can contribute to a significant reserve resistance against further compression if the build-up of pressure is interrupted (Fig. 4). This figure illustrates the principle for a gradually sedimented deposit and the parallel lines on the e-log $\bar{\sigma}$ diagram represent unique relationships between void ratio, pressure, and time. Units are the original author's. If sedimentation stops for a period of time, compression continues along a vertical line such that the void ratio continues to decrease. If another load is then applied the amount of "instant" compression is very small until a critical stress, $\bar{\sigma}_c$, is exceeded. This behavior helps explain the apparent "overconsolidation" which has been observed in some ocean sediment samples. It should be noted that even if the critical stress is not exceeded

TABLE 2

CLASSIFICATION OF SOILS BASED ON SECONDARY COMPRESSION

Coefficient of Secondary Compression, as a Percentage (1)	Secondary Compression (2)
0.2	very low
0.4	low
0.8	medium
1.6	high
3.2	very high
6.4	extremely high

(from Mesri, 1973)

during the second loading phase, the sediment will continue to compress along another vertical "delayed" compression line. The phenomenon described above has been observed in the laboratory and e-log $\bar{\sigma}$ curves similar to Figure 4 have been developed using laboratory consolidation techniques (see Bjerrum, 1967).

Recently, two investigators (Christie, 1972; Mesri, 1973) have published papers which go a long way toward sorting out the confusion regarding the rate of secondary compression. The paper by Christie (1972) explains the equivalence between Merchant's theory (Merchant, 1939; Taylor and Merchant, 1940; Taylor, 1942) and the theory proposed by Gibson and Lo (1961). The important modification to the Terzaghi theory of consolidation is the introduction of a "viscosity" term in which secondary compression is assumed to occur at a rate which is proportional to the amount of remaining or residual secondary compression. The mechanism of secondary compression is further explained by Mesri (1973) in terms of microstructural changes. Two useful secondary compression parameters obtained from laboratory compression test data are:

$$C_\alpha = \frac{\Delta e}{\Delta \log t}$$

$$\varepsilon_{\alpha p} = \frac{C\alpha}{1 + e_p}$$

where C_α = secondary compression index slope or the slope of the secondary compression line

$\varepsilon_{\alpha p}$ = secondary compression index

Soils are classified by Mesri in terms of amount of secondary compression (Table 2). Data presented by Mesri indicated that for normally consolidated clays, C_a continually decreases with consolidation pressure.

An interesting study of secondary compression effects and shear strength characteristics of a marine mud (San Francisco Bay Mud) has been recently published by Shen et al. (1973). This study confirmed other investigations indicating that an increase in pore-water pressure takes place during delayed compression. The results of isotropic triaxial compression tests showed excellent agreement with the critical-state theory (cam-clay model proposed by Roscoe et al., 1963a, and Schofield and Wroth, 1968). It is suggested that the use of critical-state methods be explored in more detail as a possible unified theory of explaining the stress-strain behavior of soft marine sediments.

Laboratory consolidation test results conducted on samples of ocean sediments can be useful from several points of view.

a) The e-log $\bar{\sigma}$ curve and strain-time relationships are essential in predicting the magnitude and rate of settlement of structures placed on or in the sediment column. It is important to determine consolidation properties to appreciable depths below the base of the foundation since even small stress increments may cause appreciable compression in ocean sediments exhibiting large void ratios and high compressibility characteristics. If it is desirable to have data to a point where the applied stress has been attenuated to a value of 10% or less, it will therefore be necessary to obtain information on geotechnical properties to depths of approximately four times the plan dimension.

b) The stress-strain data (e-log $\bar{\sigma}$ curve) can be used as a qualitative measure of the degree of sample disturbance due to sampling and handling procedures. The less disturbed samples will show a sharper curvature in the vicinity of the preconsolidation stress than samples which have considerable structural disturbance, and compression data from a highly remolded sample of clay will show little resemblance to an undisturbed sample of the same clay. Schmertmann (1955) has suggested a method of quantifying the degree of disturbance which entails conducting a second consolidation test on a completely remolded sample.

   A simpler method might be based on quantifying the shape of the e-log $\bar{\sigma}$ curve as indicated in Figure 5 to define a disturbance index (I_D) such that:

$$I_D = \frac{\Delta e}{\Delta e_o}$$

A classification based on disturbance index might be as follows:

Disturbance Index, I_D	Degree of Disturbance
.15	Very little disturbance ("undisturbed")
.15-.30	Small amount of disturbance
.30-.50	Moderate disturbance
.50-.70	Much disturbance
.70	Extreme disturbance (remolded)

   The author has not made an extensive study of degree of disturbance based on the proposed disturbance index, and the classification suggested above may have to be revised as more comparisons are made.

c) Results of consolidation tests conducted on "undisturbed" samples can be used to shed light on geological sea-floor processes. Therefore, if a sample indicates that the sediment is normally consolidated (preconsolidation effective stress equal to existing overburden effective stress) and delayed compression is small, this can be taken as fairly conclusive evidence that the area has not been eroded due to current activity or mass movements due to slump. On the other hand, sediment masses which have recently received large increases in overburden stresses should exhibit underconsolidation characteristics and a condition of overconsolidation would result if the area has undergone stress release (perhaps due to removal of glacial overburden, erosion, or mass movements). Other geologic processes might also be studied through proper exploitation of stress-strain signatures which are preserved in "undisturbed" samples. For example, if a sediment deposit has been laterally compressed due to tectonic movements, it should be possible to determine the magnitude of lateral stresses by conducting consolidation tests on horizontally oriented samples.

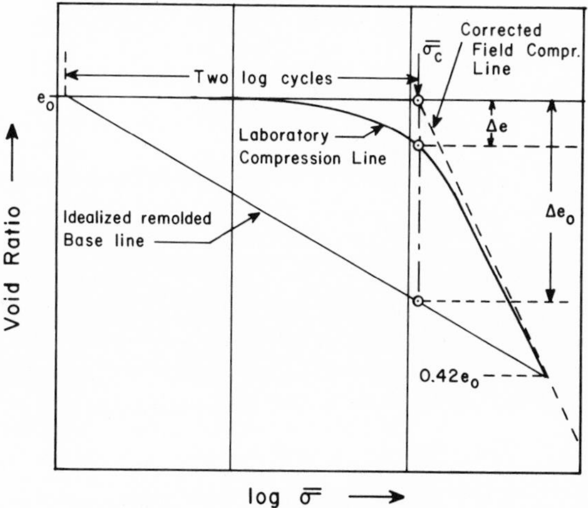

Figure 5. Proposed simplifying method of determining degree of disturbance

EXPERIENCE WITH A GIANT PISTON-CORER

Coring Program

In 1967 the Woods Hole Oceanographic Institution initiated a long coring program and in 1969 the Department of Civil Engineering of Worcester Polytechnic Institute became involved in a joint WHOI-WPI effort to further develop a large diameter (11.5 cm inside diameter) piston corer with an aim toward recovering up to 40 m of soft deep-sea sediment. Information on the corer is contained in a paper by Hollister et al. (1973), and data on geotechnical properties of recovered sediments are reported by Silva and Hollister (1973).

The first successful giant coring operation was conducted aboard the R/V KNORR over Stellwagen Basin, Gulf of Maine, in 81 m of water. The recovered sediment core was 21.74 m long and a recovery ratio was 0.78 (Core No. KN-10-1). A second core (Core No. KN-27-1) taken in the same general area was 19.65 m long with a recovery ratio of 0.87. The corer has been used successfully in deep water (5500 m depth) with a maximum recovered core length of 30.5 m.

Geotechnical Properties of Recovered Sediments

The results from core KN-10-1 and KN-27-1 are summarized in Figures 6 and 7. The SI units for shear strength are shown on Figure 7. Space does not permit detailed discussion of these results but some of the more important aspects will be described. Core KN-10-1 shows considerable vertical variability in physical property results and a distinct anomolous zone of high water content and concurrent high shear strength at about 1300 to 1450 cm. The usual quantitative measures of composition and physical properties did not fully explain the reason for this anomolous zone and therefore the scanning electron microscope (S.E.M.) was utilized to study the microstructure.

Two scanning electron micrographs are included in this paper (Fig. 8) to illustrate the microstructural differences which were detected from several zones studied with the S.E.M. The micrograph at 1860 cm (Fig. 8b) is typical of four samples taken above and below the anomalous zone. The four specimens did not contain any organic material and exhibited a fairly even distribution of voids and mineral fabric in a compact, sometimes oriented arrangement. The sample at 1380 cm (Fig. 8a) from within the anomalous zone shows an unusual abundance of diatoms together with an open flocculated structure, typical of other samples from the same zone. The clay-floc structure and orientation are considered to be related to the high shear strength of this particular zone and the

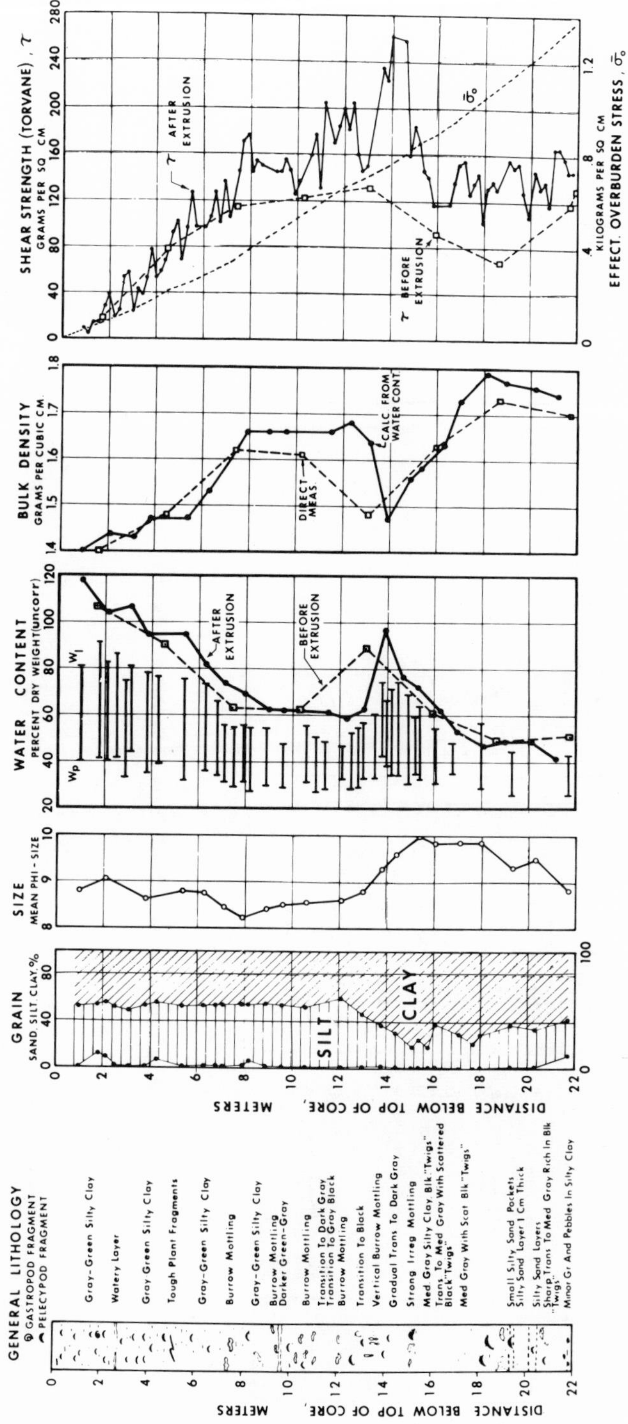

Figure 6. Geotechnical properties of Core KN-10-1

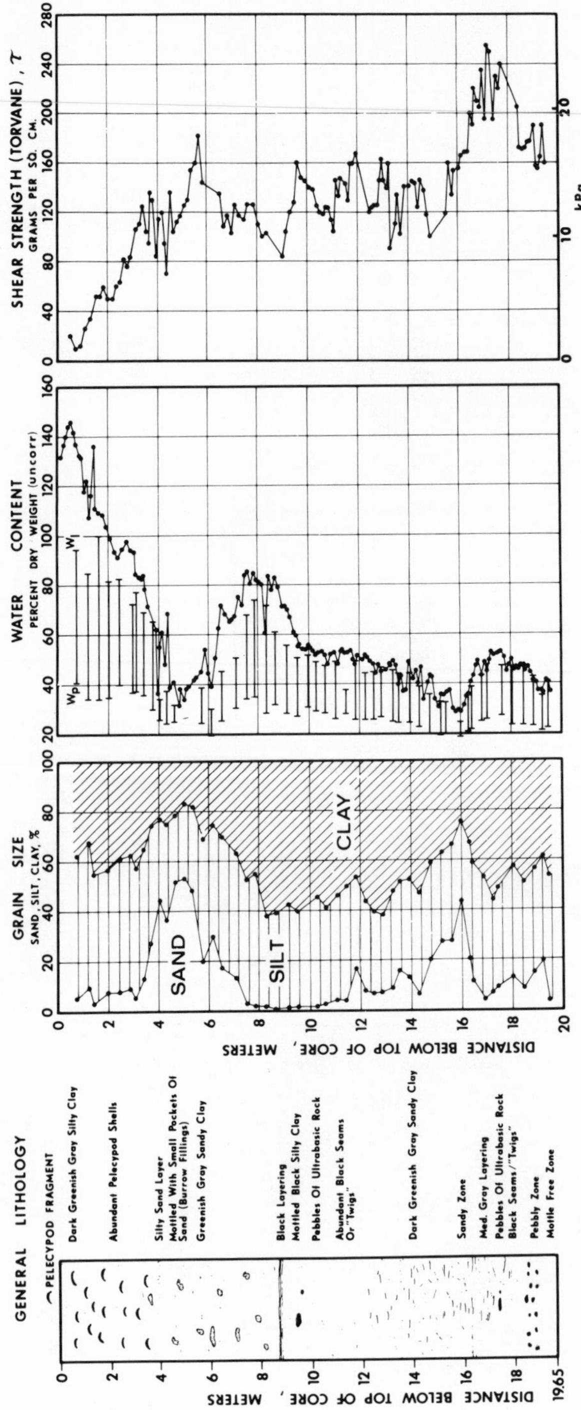

Figure 7. Geotechnical properties of Core KN-27-1

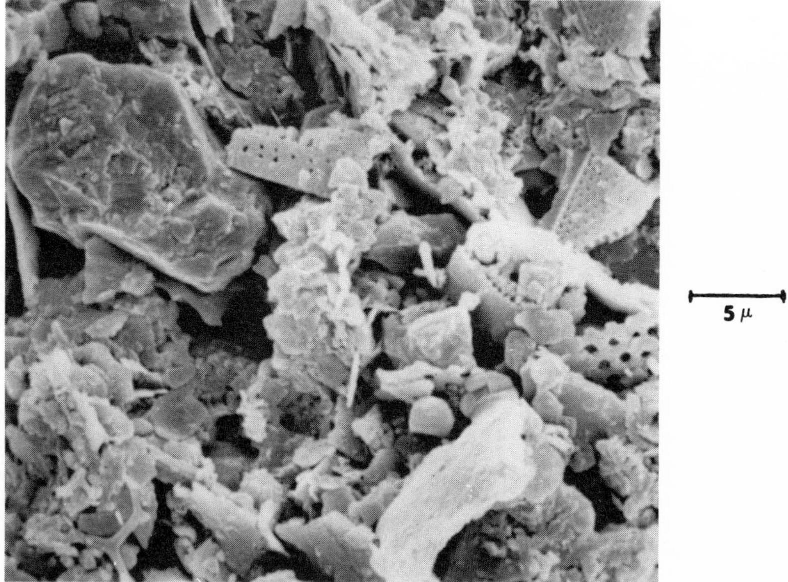

Figure 8a. S.E.M. at 1380 cms in Core KN-10-1

Figure 8b. S.E.M. at 1860 cms in Core KN-10-1

abundance of diatom fragments help explain the high water content.

The lithology of core KN-27-1 (see Fig. 7) is grossly similar to that of KN-10-1, but the two cores differ significantly in texture. The water content is somewhat more complex than that of KN-10-1, and shear strength variations were more pronounced than those in core KN-10-1. An anomalous zone of high water content-high shear strength is present at 16.4 to 18.8 m.

The scanning electron microscope was used to study the microstructure of core KN-27-1 also and the results indicated a more flocculant structure within the anomalous zone (16.4 to 18.8 m) than in other sections of the core.

Consolidation tests were conducted on several samples from both cores and the pre-compression stresses for all samples showed good agreement with overburden stresses. It is therefore concluded that the sediment in Stellwagen Basin is normally consolidated and was not subjected to compaction due to glacial overburden.

Geotechnical Properties and Acoustic Reflectors

At the coring site of KN-10-1 at least four distinct reflectors can be recognized. The first and third reflectors beneath the sea floor are distinct at the coring site (Fig. 9a), but discontinuous across the basin. These occur at 16 and 25 m sub-bottom on the 3.5 kHz records. Only the second and deepest reflectors, at 22 and 29 m, respectively, can be easily traced throughout the basin.

At the coring site of KN-27-1 there is one highly reverberant reflector about 4 m thick beginning at 16 m on the 3.5 kHz profile (Fig. 9b) which correlates with the zone of anomalously high water content and high shear strength at 16.5 to 19.0 m in the core. This reflector, at core site KN-27-1, can be traced back through the record to nearby crossings with KN-10 tracks to match up with the dark reflector at 22 m at core site KN-10-1. It is therefore believed that coring was continuous at KN-27-1 with penetration through the highly reverberant reflector whereas core KN-10-1 just reached the same reflector. One of the purposes of this continuing program is to obtain cores at the locations of high resolution reflection profiles in order to determine more certainly the correlation between physical or textural properties of sediment and the nature of the recorded reflectors.

The giant piston-corer provides the means of obtaining good quality cores to significant depths in soft sediments. Detailed studies of the engineering properties of recovered sediments coupled with mineralogical and microstructure studies can provide important data for engineering applications, provide insight into geological

Figure 9a. 3.5 kHz profile at site of Core KN-10-1

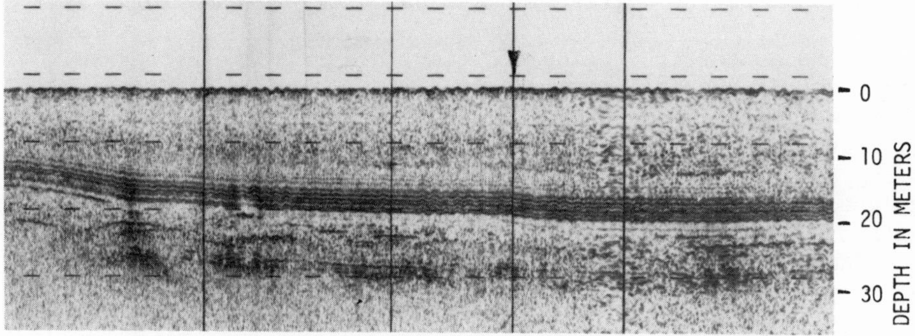

Figure 9b. 3.5 kHz profile at site of Core KN-27-1

processes, and help in correlating acoustic reflectors with changes in sediment properties. Analyses of several long cores show that there can be significant and important vertical variability in geotechnical properties such as water content and shear strength.

SUMMARY AND CONCLUSIONS

Sediment diagenetic processes in the ocean can be rather complex and resulting physical properties are influenced by a variety of factors including depositional environment; physical alterations due to consolidation, slumping, erosion, deposition; chemical alterations such as ion exchange, cementation; and alterations caused as a result of reworking by animals. Studies of microstructure can provide insight into sediment stress-strain behavior and lead to an understanding of changes in sediment properties and associated geological processes.

The Mohr-Coulomb method of shear strength analysis used in conjunction with triaxial methods of testing give satisfactory results for engineering applications. However, in order to understand basic stress-strain behavior it is recommended that more comprehensive yield theories be used. In particular, it is suggested that the critical state method may offer a rational approach in developing predictive mathematical models.

Many deep-sea sediments with high void ratios may exhibit large secondary compression effects. Therefore, mathematical models which incorporate a viscosity parameter will be more reliable than the classical (Terzaghi) theory for predicting rate of compression and pore pressures. Delayed compression can contribute to a significant reserve resistance which often results in a pseudo-overconsolidation effect on the e-log $\bar{\sigma}$ curve. Laboratory compression tests on "undisturbed" samples can provide insight into sediment processes. A disturbance index is proposed to quantify the shape of the e-log $\bar{\sigma}$ curve and indicate the degree of disturbance due to sampling.

There is a need to develop efficient and reliable methods of obtaining continuous, relatively undisturbed, samples of deep-sea sediments to appreciable depths into the sediment column (at least 50 m). Large diameter versions of the piston corer show promise of providing an interim solution to this problem but other modes should be explored and developed. Most available literature on shear strength of deep-sea sediments is for the upper few meters of sediments. The data indicates considerable vertical and horizontal variability in shear strength and other physical properties.

The author's experiences with a large diameter long piston corer (up to 40 m) indicate that good quality sediment cores can be obtained in deep water (up to 5600 m). The recovered cores (maximum length of 30.5 m) showed considerable vertical variation in sediment properties and anomalous zones of high water content coincident with high shear strength. These anomalous zones are related to microstructural changes observed with the scanning electron microscope and correlate with acoustical reflectors on the 3.5 kHz record.

ACKNOWLEDGMENTS

The writer extends special thanks to C. K. Satyapriya for his extensive aid in reviewing the literature and preparing the manuscript. Most of the writer's experimental research which is included in this paper was supported by grants from the Office of Naval Research (Research Contract N00014-72-A-0372) and the National Science Foundation.

REFERENCES

ASTM, Symp. Underwater Soil Sampling, Testing and Construction Control, Am. Soc. Test. Mat. Spec. Tech. Publ. 501, Phila., 1972.

Beloff, W. R., Consolidated Vane Shear Apparatus and Test, Masters thesis, Worcester Polytechnic Inst., Worcester, 1973.

Bishop, A. W., Shear strength parameters for undisturbed and remolded soil specimens, in Stress Strain Behavior of Soil, edited by R. H. G. Parry, pp. 3-58, G. T. Foulis and Co., Ltd., Oxfordshire, 1972.

Bjerrum, L., Geotechnical properties of Norwegian marine clays, Géotechnique, 4, 49-69, 1954.

Bjerrum, L., Engineering geology of Norwegian normally-consolidated marine clays as related to settlements of building, 7th Rankine Lecture, Géotechnique, 17, 81-118, 1967.

Bjerrum, L., and K. Y. Lo, Effect of aging on the shear strength properties of a normally consolidated clay, Géotechnique, 13, 147-157, 1963.

Bjerrum, L., and T. H. Wu, Fundamental shear strength properties of the Lilla Edet clay, Géotechnique, 10, 101-109, 1960.

Bryant, W. R., P. Cernock, and J. Morelock, Shear strength and consolidation characteristics of marine sediments from the western Gulf of Mexico, in Marine Geotechnique, edited by A. F. Richards, pp. 41-64, Univ. of Ill. Press, Urbana, 1967.

Christie, I. F., A re-appraisal of Merchant's contribution to the theory of consolidation, Géotechnique, 22, 309-320, 1972.

Engelhardt, V. W., and K. H. Gaida, Concentration changes of pore solutions during the compaction of clay sediments, J. Sediment. Petrol., 33 (4), 919-930, 1963.

Fisk, H. N., and B. McClelland, Geology of continental shelf off Louisiana: its influence on offshore foundation design, Bull. Geol. Soc. Am., 70, 1369-1394, 1959.

Gibson, R. E., and R. Y. Lo, A theory of consolidation for soils exhibiting secondary compression, Acta Polytech. Scandinavia, 296, reprinted by Norwegian Geotech. Inst., 1961.

Hollister, C. D., and C. B. Heezen, The Face of the Deep, Oxford Univ. Press, N. Y., 1971.

Hollister, C. D., A. J. Silva, and A. Driscoll, A giant piston corer, J. Ocean Eng., 2, 159-168, 1973.

Hvorslev, M. J., Subsurface exploration and sampling of soils for civil engineering purposes, Waterways Experiment Station, U. S. Army Corps of Engineers, Vicksburg, 1960.

Ishii, Y., General discussion, in Symp. Consolidation Testing of Soils, Am. Soc. Test. Mat. Spec. Tech. Bull. 126, pp. 103-109, 1951.

Lambe, T. M., and R. V. Whitman, Soil Mechanics, John Wiley & Sons, Inc., N. Y., 1969.

Leonards, G. K., and B. K. Ramiah, Time effects in the consolidation of clays, Am. Soc. Civil Engrs. Spec. Tech. Bull. 254, 116-130, 1960.

McCoy, F. W., Jr., An analysis of piston coring through corehead camera photography, in Symp. Underwater Soil Sampling, Testing, and Construction Control, Am. Soc. Test. Mat. Spec. Tech. Publ. 501, pp. 90-105, Phila., 1972.

Meade, R. H., Removal of water and rearrangement of particles during compaction of clayey sediments (review), U. S. Geol. Survey Prof. paper 450-E, E111-E114, 1966.

Merchant, W., Some Theoretical Considerations on the One Dimensional Consolidation of Clay, Masters thesis, Mass. Inst. of Tech., Cambridge, 1939.

Mesri, G., Coefficient of secondary compression, J. Soil Mech. and Fdn. Div., Proc. A.S.C.E., 99 (SM1), paper 9515, 123-137, 1973.

Moore, D. G., Shear strength and related properties of sediments from experimental Mohole (Guadalupe site), J. Geophys. Res., 69 (20), 4271-4291, 1964.

Muller, G., Diagenesis in argillaceous sediments, in Diagenesis in Sediments, edited by G. Larsen and G. V. Chilingar, pp. 127-177, Elsevier Publ. Co., 1967.

Nacci, V. A., and M. T. Huston, Structure of deep-sea clays, in Civil Eng. in the Oceans II, pp. 599-620, ASCE, N. Y., 1970.

Noorany, I., Underwater soil sampling and testing, a state-of-the-art review, in Symp. Underwater Soil Sampling, Testing, and Construction Control, Am. Soc. Test. Mat. Spec. Tech. Publ. 501, pp. 3-41, Phila., 1972.

Noorany, I., and S. F. Gizienski, Engineering properties of submarine soils: state-of-the-art review, J. Soil Mech. and Fdn. Div., Proc. ASCE, 96 (SM5), 1735-1741, 1970.

Parker, A., Minerology and geotechnical properties of a deep-sea carbonate sediment, Géotechnique, 22, 155-159, 1972.

Parry, R. H. G., Stress strain behavior of soils, edited from the Proc. Roscoe Memorial Symp., Cambridge Univ., G. T. Foulis & Co., Ltd., Oxfordshire, 1972.

Poorooshasb, H. B., and K. H. Roscoe, The correlation of the results of shear tests with varying degress of dilatation, Proc. 5th Intern. Conf. on Soil Mech. and Fdn. Eng., pp. 297-304, 1961.

Richards, A. F., Investigations of deep-sea sediment cores, I Shear strength, bearing capacity and consolidation, U. S. Navy Hydrographic Office Tech. Rept. 63, 1961.

Richards, A. F., Investigations of deep-sea sediment cores, II. Mass physical properties, U. S. Navy Hydrographic Office Tech. Rept. 106, 1962.

Richards, A. F., Local sediment shear strength and water content variability on the continental slope off New England, in Mar. Geol., Shepard Commemorative Vol., edited by R. L. Miller, pp. 474-487, The McMillan Co., N. Y., 1964.

Roscoe, K. H., and H. B. Poorooshasb, A theoretical and experimental study of strains in triaxial compression tests on normally consolidated clays, Géotechnique, 13, 12-38, 1963.

Roscoe, K. H., A. N. Schofield, and A. Thurairajah, A critical appreciation of test data for selecting a yield criterion for soils, in Symp. Laboratory Shear Testing of Soils, Ottawa, Canada, 1963a.

Roscoe, K. H., A. N. Schofield, and A. Thurairajah, Yielding of clays in states wetter than critical, Géotechnique, 13, 211-240, 1963b.

Roscoe, K. H., A. N. Schofield, and C. P. Wroth, On the yielding of soils, Géotechnique, 9, 22-53, 1958.

Sangrey, D. A., Obtaining strength profiles with depth for marine soil deposits using disturbed samples, in Symp. Underwater Soil Sampling, Testing, and Construction Control, Am. Soc. Test. Mat. Spec. Tech. Publ. 501, pp. 106-121, Phila., 1972.

Schmertmann, H. J., The undisturbed consolidation behavior of clay, J. Soil Mech. Fdn. Div., Trans. ASCE, 120, 1201-1233, 1955.

Schofield, A., and P. Wroth, Critical State Soil Mechanics, McGraw-Hill Co., N. Y., 1968.

Shen, C. K., K. Arulanandan, and W. S. Smith, Secondary consolidation and strength of a clay, J. Soil Mech. Fdn. Div., Proc. ASCE, 99 (SM1), 95-110, 1973.

Silva, A. J., and C. D. Hollister, Geotechnical properties of ocean sediments recovered with giant piston-corer: I - Gulf of Maine, J. Geophys. Res., 78 (18), 3597-3616, 1973.

Taylor, D. W., Research on consolidation of clays, Publ. Serial 82, Dept. of Civil and Sanitary Eng., Mass. Inst. of Tech., Cambridge, 1942.

Taylor, D. W. and W. Merchant, A theory of clay consolidation accounting for secondary compression, J. Maths and Physics, 19 (3), 167-185, 1940.

Terzaghi, K., Theoretical Soil Mechanics, Wiley, N. Y., 1943.

Wu, T. H., Soil Mechanics, Allyn and Bacon, Inc., Boston, 1966.

MARINE GEOTECHNICAL PROPERTIES: INTERRELATIONSHIPS AND RELATIONSHIPS TO DEPTH OF BURIAL

GEORGE H. KELLER

Atlantic Oceanographic & Meteorological Laboratories

ABSTRACT

Marine Geotechnique is a relatively new field which attempts to define and understand the mass physical and chemical properties of sea-floor deposits and the reponse of these materials to applied static and dynamic forces. Although only a relatively small portion of the ocean floor has yet been studied, it appears that it is feasible at this stage of our knowledge to define certain interrelationships among a number of the physical, electrical, and acoustical properties of these deposits. Most prominent of these correlations are unit weight (density) with water content, porosity with resistivity, mean grain size with porosity and density, sound velocity with unit weight and porosity, and reflectivity with porosity and unit weight. In some cases where a relatively large number of analyses have been made it has been possible to formulate predicting equations for such properties as unit weight, water content, void ratio, and shear strength. At best, most of the correlations provide only an approximation and are restricted to the sediment type or local area from which the basis for the correlation was developed.

INTRODUCTION

For a little more than 15 years there has been a concerted effort by a few to investigate the mass physical properties of deep-sea sediments. Much of the initial impetus for these studies was generated by the U. S. Navy's interest in defining the foundational and acoustic characteristics of ocean-floor deposits. The highlights of these early studies have been presented by Richards (1967) and Keller (1968a, 1968b).

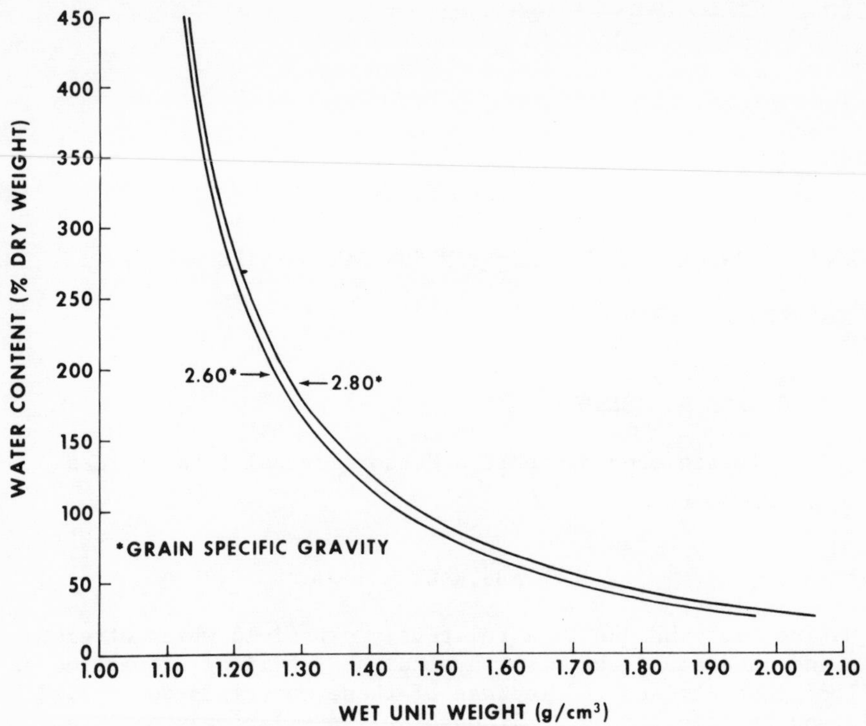

Figure 1. Fixed relationship of wet unit weight to water content for two grain specific gravities

This field of study, recently referred to as Marine Geotechnique, has advanced rapidly during these past fifteen years. It has advanced not only in regard to gaining a better understanding of deep-sea sediments, but by sizably increasing the number of researchers working in the field. Although there is as yet relatively little known about the mass physical properties of deep-sea sediments, a number of interrelationships (Hamilton, 1956; Moore and Shumway, 1959; Richards, 1961, 1962) and regional generalizations made on the distribution of these properties over the North Atlantic and North Pacific basins (Keller and Bennett, 1968) and the Mediterranean (Keller and Lambert, 1971) have been defined. With the advent of the MOHOLE Project and later the JOIDES Deep-Sea Drilling Project, a limited amount of mass property data have become available from depths as great as 1015 m below the sea floor (Hamilton, 1964; Moore, 1964; Creager et al., 1973).

Most of the basic relationships between the index properties of terrestrial deposits were established long ago by those working in soil mechanics. More recent studies of deep-sea sediments have shown that it may be feasible to develop additional interrelationships for submarine deposits, particularly in relating the physical, acoustical, and electrical properties to each other. Owing to the lack of data from many different depositional areas and from a number of sediment types, it is only possible to generalize on these relationships even though the correlation between certain parameters appears to be well established for the data at hand. This discussion is an attempt to summarize the most prominent of these interrelationships as proposed by various researchers as well as to mention a few cases where no relationships are found between certain major properties. For the sake of accuracy, where figures have been taken from other sources, they are presented with their original units.

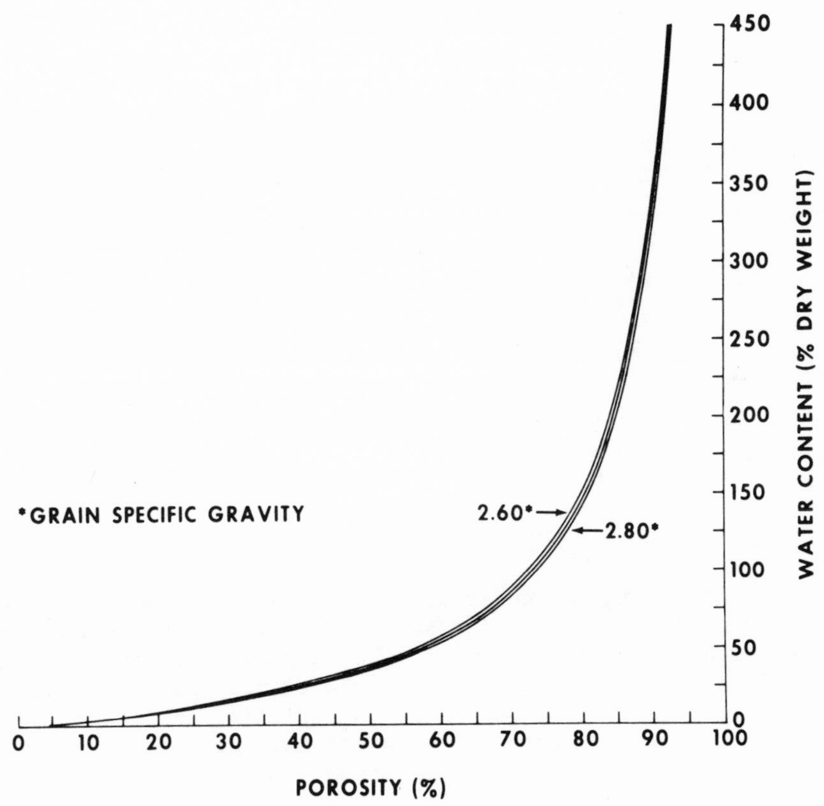

Figure 2. Fixed relationship of water content to porosity for two grain specific gravities

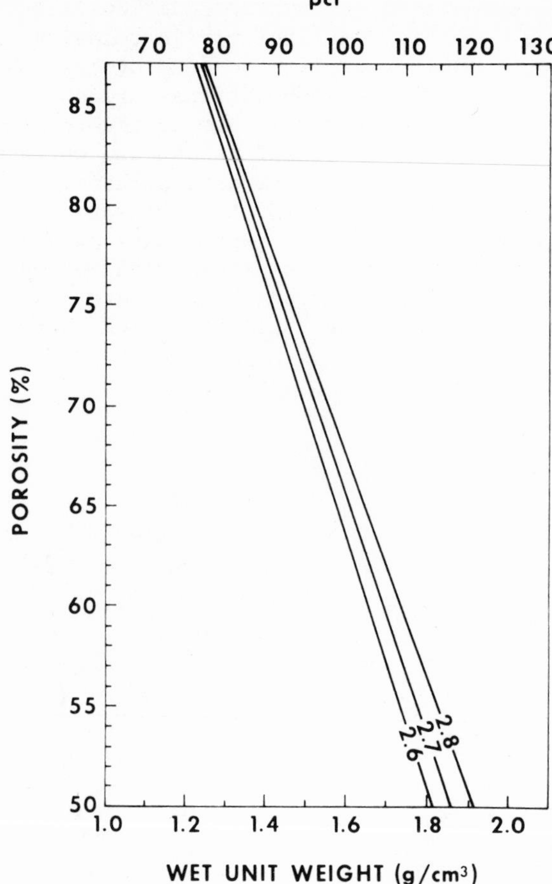

Figure 3. Fixed relationship of porosity to wet unit weight for three grain specific gravities

DISCUSSION

Before discussing a number of the correlations that have evolved from recent studies of submarine sediments, it must be pointed out that even with the establishment of what appear to be sound interrelationships, they can, at best, only be extrapolated to other areas in a very general sense. A graph depicting an interrelationship for two parameters from one area may have limited or no application in another depositional environment.

Certain relationships are well established, dating back to the early workers in soil mechanics. For each of the three cases shown, Figs. 1, 2, & 3) curves based on grain specific gravity are established for the three interrelationships. Figures 1 and 2 are based upon the author's work and Figure 3 is from Richards (1962). These correlations can be used with any sediment from any environment if the grain specific gravity is known. They provide a rapid method of determining porosity or wet unit weight if water content and grain specific gravity are known (Bennett et al., 1971; Lambert and Bennett, 1972) (Figs. 4 and 5). As shown in each case (Figs. 1, 2, & 3), considerable change in grain specific gravity is needed before the relationships between the other parameters are appreciably influenced. Despite this relatively small influence of grain specific gravity, care must be used (if specific gravity is not determined) before extrapolating from a series of curves established for one area or deposit to a distinctly different sediment.

Atterberg limits have long served as a simple means of classifying sediments as to their plasticity, compressibility, and activity (ratio of plasticity index to per cent clay) (Skempton, 1953). Although these limits are commonly determined in the routine analysis of submarine sediments, they are usually only used for defining plasticity characteristics.

In his study of the Mississippi Delta front, McClelland (1967) found the void ratio and liquidity index of the delta sediments to be functions of the liquid limit and the overburden pressure. Based on these findings, he developed a generalized family of pressure-void ratio curves which allowed the approximation of void ratio knowing the liquid limit, liquidity index, and overburden pressure. Liquidity index was thus shown to serve as an indicator of the state of consolidation. This same study also led him to devising a second family of curves whereby cohesion could be approximated based on liquid limit, liquidity index, and water content. Although these families of curves will undoubtedly not be applicable elsewhere, the concept is of interest for possible use with other sea-floor deposits.

Water Content

As noted above, water content is readily correlated with both unit weight and porosity. Based on 1480 analyses performed on 80 sediment cores collected from the major provinces of the Gulf of Mexico, Bryant and Trabant (1972) have established a curve for the water content-unit weight relationship (Fig. 6) which closely resembles Figure 1. Scatter in their data can be attributed to two things: variation of grain specific gravities, or error in determining either water content or wet unit weight.

Based on this same study of Gulf sediments a least squares curve was established showing the decrease of water content with depth. This inverse relationship of water content to depth is a

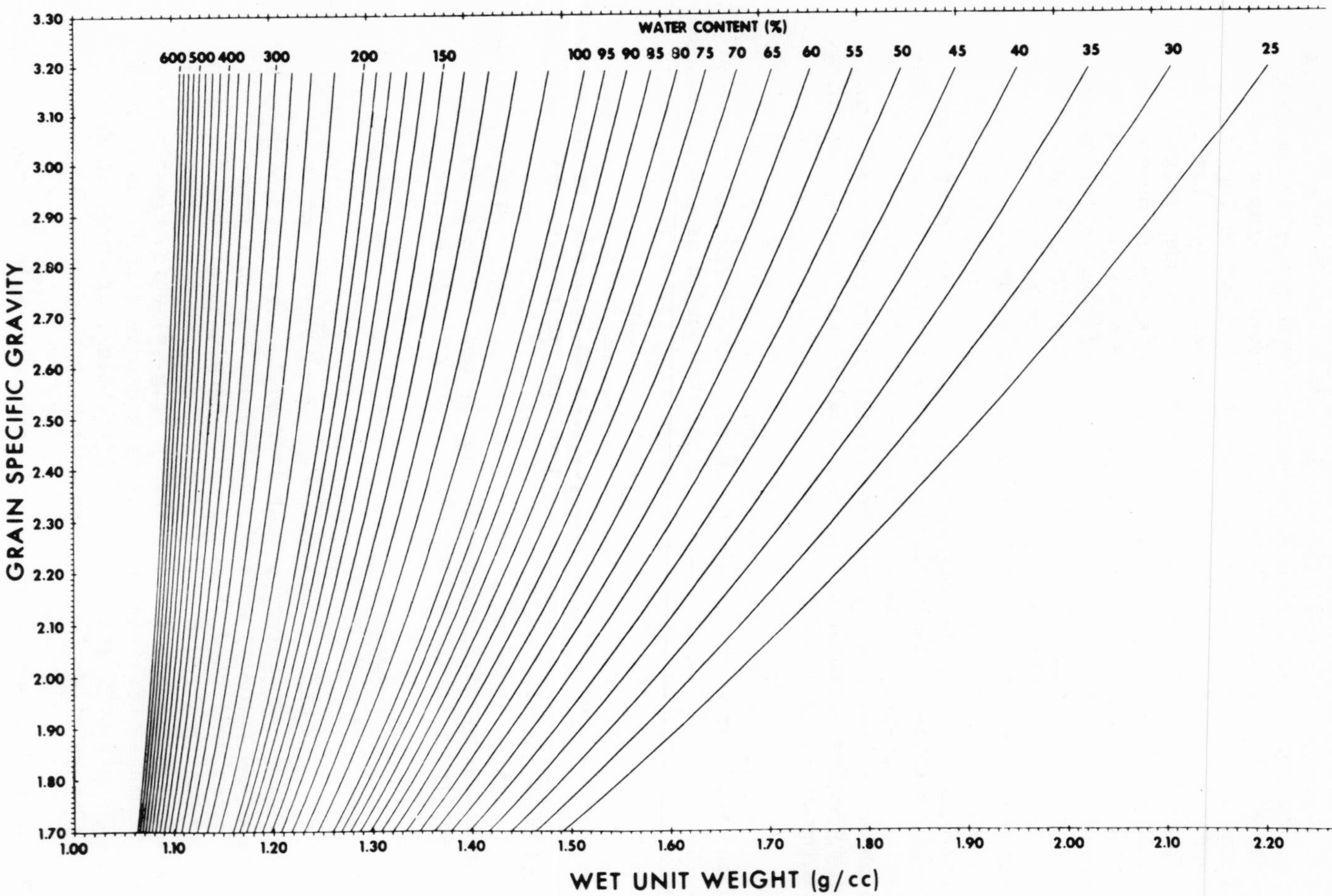

Figure 4. Nomogram for determining wet unit weight from grain specific gravity and water content (Bennet and Lambert, 1971)

common generalization and one that would be anticipated owing to increasing overburden pressure with depth.

From their study of shear strength and water content, Bryant and Trabant (1972) reported an inverse relationship between these two parameters. A similar relationship has been reported for remolded sediment by Rutledge (1947) and Bjerrum (1951, 1954). Yet, for natural occurring deposits, this statement can only be made in a very general sense.

Water content and sediment grain size are commonly inversely proportional, the coarser the sediment the lower the water content. An exception to this general rule is found in sediments rich in Foraminifera. Although a large number of forams will constitute a sandy texture, the framework of their test with its large central cavity is such that relatively large amounts of water are trapped within the foram thus resulting in a high water content being associated with this type of coarse-grained material.

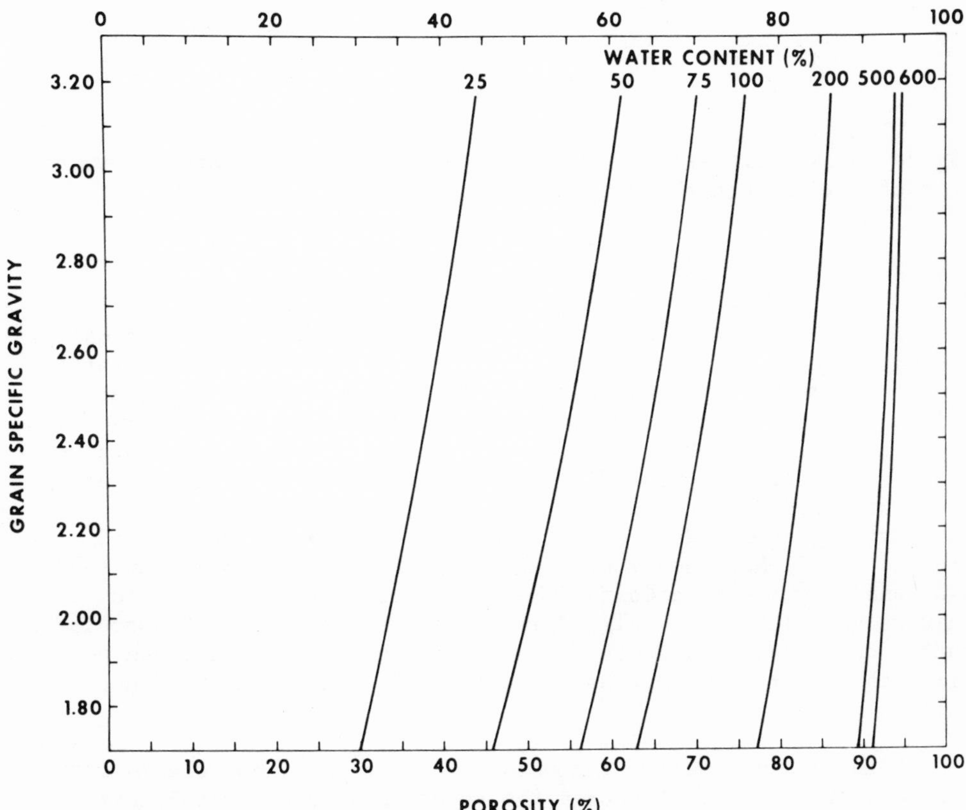

Figure 5. Nomogram for determining porosity from grain specific gravity and water content

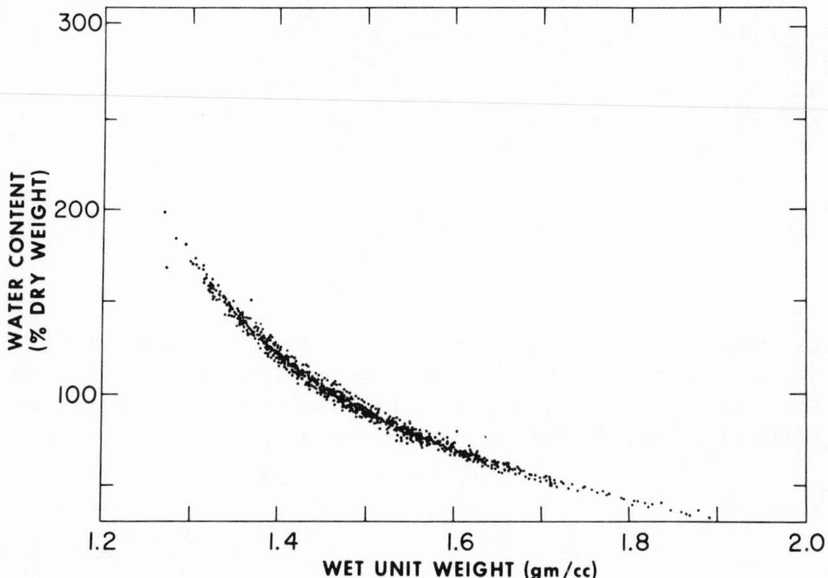

Figure 6. Relationship of wet unit weight to water content for the Gulf of Mexico (after Bryant and Trabant, 1972)

Wet Unit Weight and Porosity

The well defined correlation between wet unit weight, porosity, and water content for a given specific gravity was discussed earlier. Although not shown, the same inverse relationship of wet unit weight to porosity exists for wet unit weight and void ratio. It is commonly accepted that wet unit weight increases as depth below the sea floor increases. This is obviously not a linear function but one dependent on such factors and changes in grain size, cementation, overburden, depositional history, etc. Bryant and Trabant (1972) have developed a least squares curve, based on their 1480 analyses from the Gulf of Mexico, showing the increase in wet unit weight with depth (Fig. 7). It is clear from this figure that such a curve can serve only as a very rough rule of thumb and confirms the generalization that density increases as depth of burial increases.

Both wet unit weight and porosity are closely correlated with sediment grain size. Although the relationships are not linear throughout their respective limits, there is a definite inverse correlation between mean grain size and porosity (Hamilton, 1972) and a directly proportional relationship with wet unit weight.

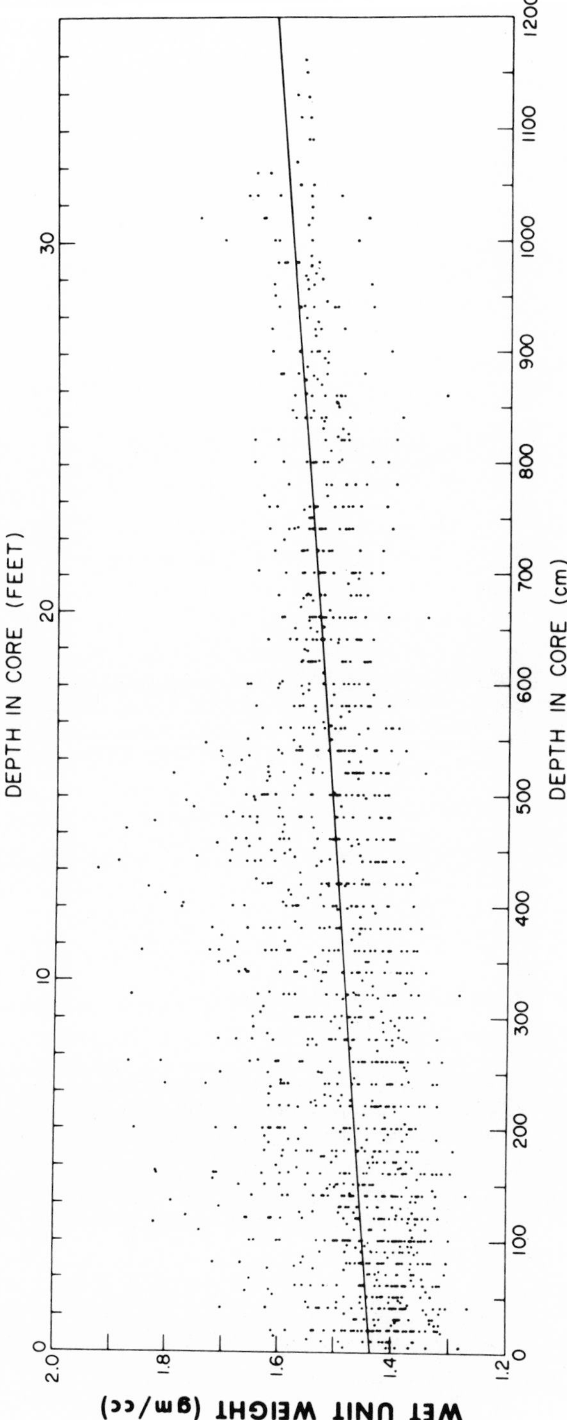

Figure 7. Variation of wet unit weight with depth in the Gulf of Mexico (after Bryant and Trabant, 1972)

Shear Strength

Shear strength, as wet unit weight, commonly increases with depth of burial due in part to overburden pressure. In the Gulf of Mexico, Bryant and Trabant (1972) have demonstrated this increase of shear strength down to a depth of 12 m.

Studies by both Richards (1962) and Inderbitzen (1969) indicate that there is some degree of correlation between shear strength and porosity. Their studies show that if data are taken from a very local area there is often an inverse relationship between these two parameters. It is clearly seen from Figure 8 that there is considerable scatter in the data if even two cores are compared and therefore whatever relationship does exist is rather weak and must be considered with caution. Others have not found even this clear a correlation between the two properties (Moore and Shumway, 1959; Moore, 1964).

There appears to be some question as to the relationship between rate of sedimentation and shear strength. Moore (1961, 1964) in comparing sediment from sites in the Gulf of Mexico and the Pacific has reported that the rate of shear strength increase is inversely proportional to the rate of sedimentation (Fig. 9). Inderbitzen (1969) in his study of the southern California continental borderland found no such correlation. Realizing the limitations of comparing sediments from different areas and environments, the validity of this relationship can only be ascertained from the study of similar sediments affected by different rates of deposition.

Although some work has been carried out to determine the effect of organic matter on shear strength, no definitive results are available. More recent studies by Hatcher (1974) offer a slight indication that some correlation exists between carbohydrate content and shear strength (Fig. 10). Hatcher's analysis of ten samples is far short of confirming such an interrelationship but it does offer an interesting problem for further study.

Moore and Shumway (1959), in their study of sediments off Pigeon Point, California, reported a weak positive correlation between sorting and shear strength. Well sorted sediments generally have higher strength than those poorly sorted. In addition to the relationships noted above there are indications that shear strength may well be correlated to some extent with a number of other properties, e.g., percentage sand-silt-clay and plasticity index. In most cases considerably more study is needed before these relationships can be taken seriously. At most, studies noted above serve only to point out crude and in some cases questionable correlations.

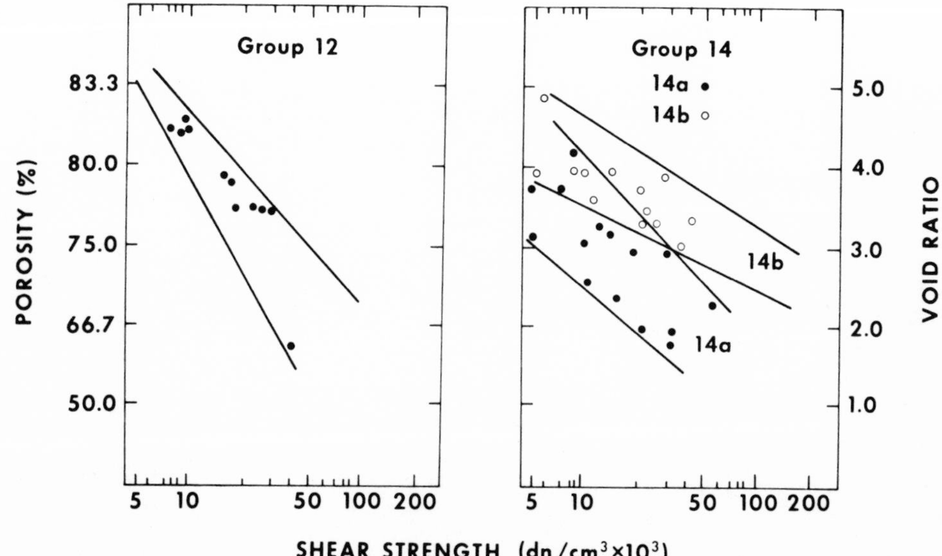

Figure 8. Relationship between shear strength and porosity or void ratio from sediments off southern California (after Inderbitzen, 1969)

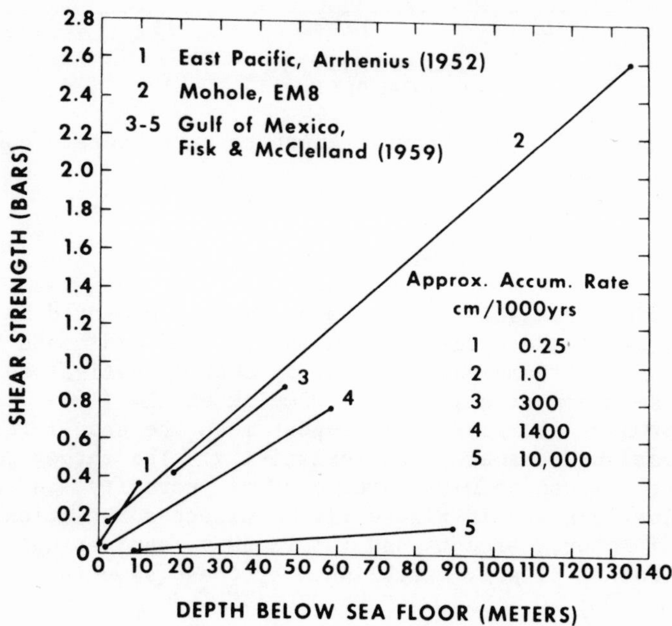

Figure 9. Variation of shear strength with depth for areas of varying rates of sedimentation (after Moore, 1964)

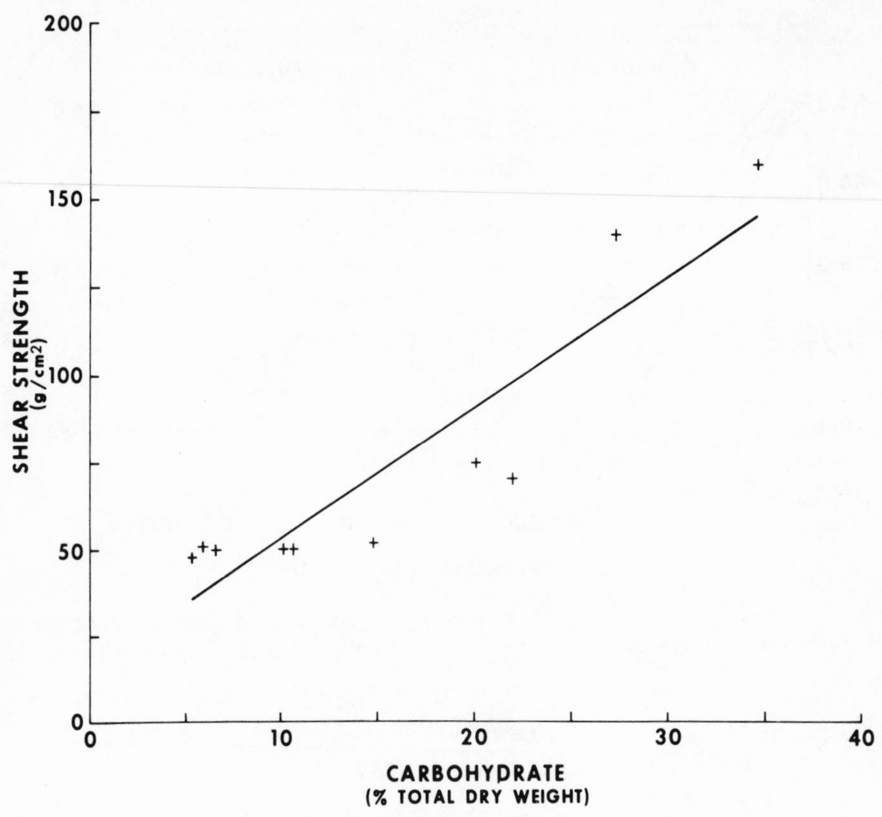

Figure 10. Relationship of shear strength to carbohydrate content (after Hatcher, 1973)

Resistivity

For some time, resistivity has been commonly measured in bore holes. It is readily determined in unconsolidated sediments both in situ and from core samples. Studies by such investigators as Boyce (1967); Kermabon et al. (1969), Chmelik et al. (1969), Sweet (1972), and Bouma et al. (1972) all report much the same interrelationships of various parameters to resistivity. The rather definitive inverse relationships with water content, porosity, and clay content are clearly shown in Figure 11. A direct correlation with grain median diameter, percent sand (up to 50%), and wet unit weight also has been rather clearly established by these same investigators.

Acoustics

In the past few years the acoustical properties of submarine sediments have received considerable attention. As studies have

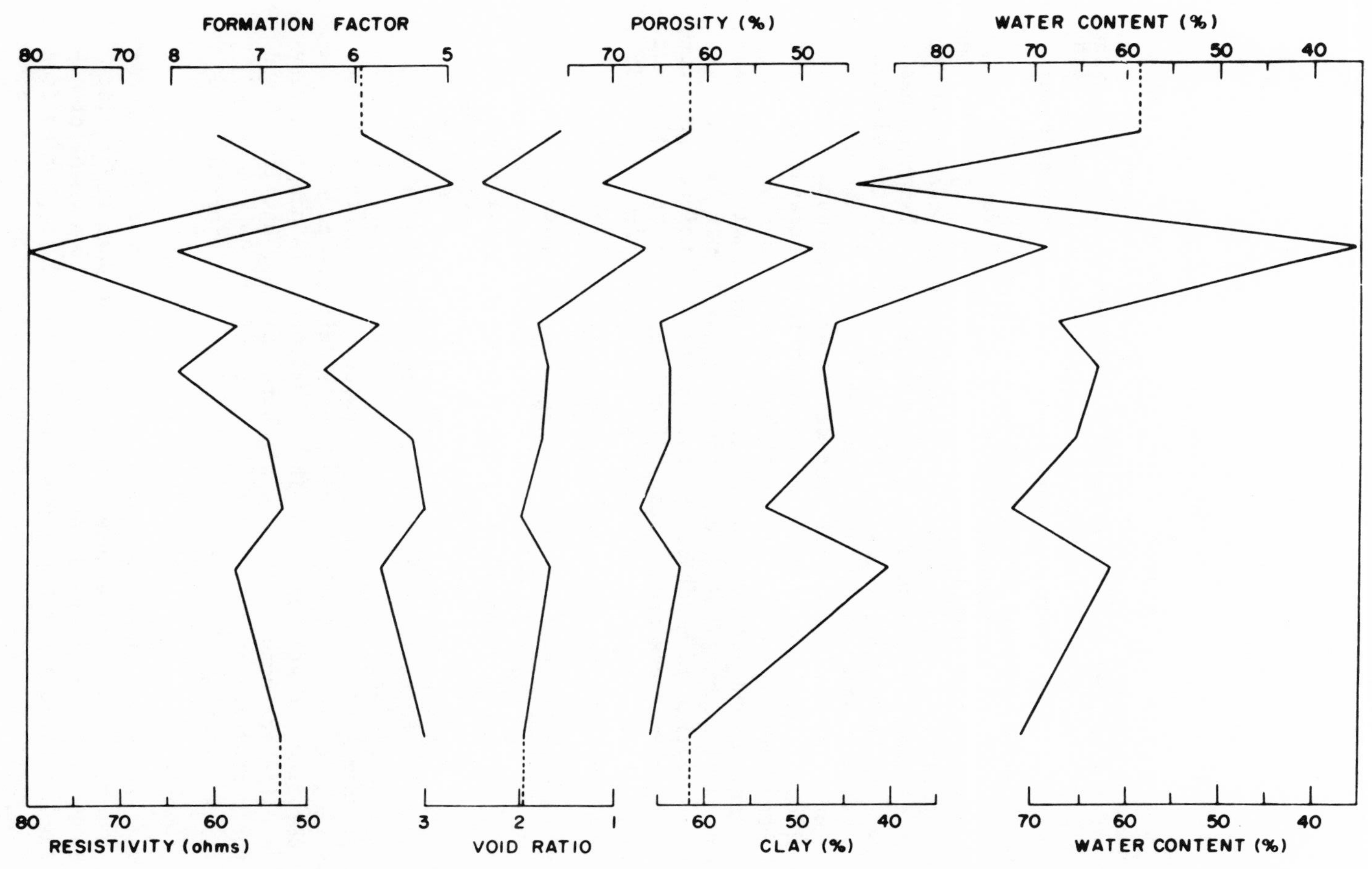

Figure 11. Relationship of resistivity to void ratio, porosity, water content, clay content, and formation factor for the Gulf of Mexico

progressed, interest has spread to the interrelationship between the acoustic and physical characteristics of submarine deposits. Rather well defined relationships have been established to the degree where it has been shown that mapping of sediment texture, wet unit weight, and porosity by acoustic profiling may be feasible over large areas.

Sound velocity. The correlation between sound velocity and sediment porosity has been known for many years; more recently a number of researchers have redefined this relationship for submarine sediments (Hamilton et al., 1956; Nafe & Drake, 1957; Shumway, 1960; Buchan et al., 1972). Although the velocity-porosity relationship is well defined, there is some degree of scatter among the data (Cernock, 1970). A correlation between velocity and wet unit weight is less well defined (Hamilton et al., 1956; Horn et al., 1968), but it does appear that velocity increases as unit weight increases. A somewhat similar relationship with grain size (mean diameter) has also been reported (Horn et al., 1968). A rather poor correlation between velocity and water content has been found by the same investigators from their study of deep-sea cores. The data scatter is considerable and even a crude correlation is questionable. Smith (1971) has shown from his analysis of both shelf and abyssal plain deposits that sound velocity is inversely proportional to such parameters as plasticity index, clay content, and liquid limit. Although the scatter is moderate, Smith reports a relatively good correlation between these properties. Of curious interest is the fact that very little if any correlation has been found between velocity and calcium carbonate, grain specific gravity, or shear strength.

Attenuation coefficient. The energy lost in passing an acoustic wave through sediment has provided a useful index which can be related to a number of physical properties. This index is the attenuation coefficient. Buchan et al. (1972) have clearly shown that as the grain size increases the attenuation increases (Fig. 12). Their study of textural composition also revealed that attenuation increases as the percent sand increases and as the percent clay decreases. Hamilton's (1972) in situ studies of sediments at water depths ranging from 4 to 1100 m revealed a definite correlation between porosity and attenuation. The relationship is complex, varying for different porosities. As a generalization, Hamilton found that for porosities of 52% or more the attenuation decreases as porosity increases. The reverse relationship was reported for porosities ranging from 36 to 50% (Fig. 13).

Reflectivity. The measurement of sound reflected off the sea floor has received considerable attention in the past eight years both in regard to military detection systems and as a means of developing a technique for mapping sediment physical properties from a moving platform. Early studies by Loring (1962) attempted to show that a qualitative acoustic classification of bottom types could be made based on the intensity of the acoustic bottom reflections recorded on a Precision Depth Recorder. More recently, Faas (1968) showed the distinct inverse relationship between porosity and the coefficient of reflectivity. Later Smith (1971) showed not only the same results but the equally distinct correlation with wet unit weight (Fig. 14). The degree of data scatter is so minor that reflectivity serves as a very good index property for determining either porosity or wet unit weight. Although not nearly as definitive as the above relationships, Smith (1971) also has shown there to be an inverse relation between reflectivity and plasticity index.

Predictor Equations

For those who must have an equation to define a correlation, there are even a few researchers who have gone out on this limb and offered such equations (e.g., Faas, 1968; Hamilton, 1970a, 1970b; Cernock, 1970; and Nacci et al., 1971). After studying eight cores

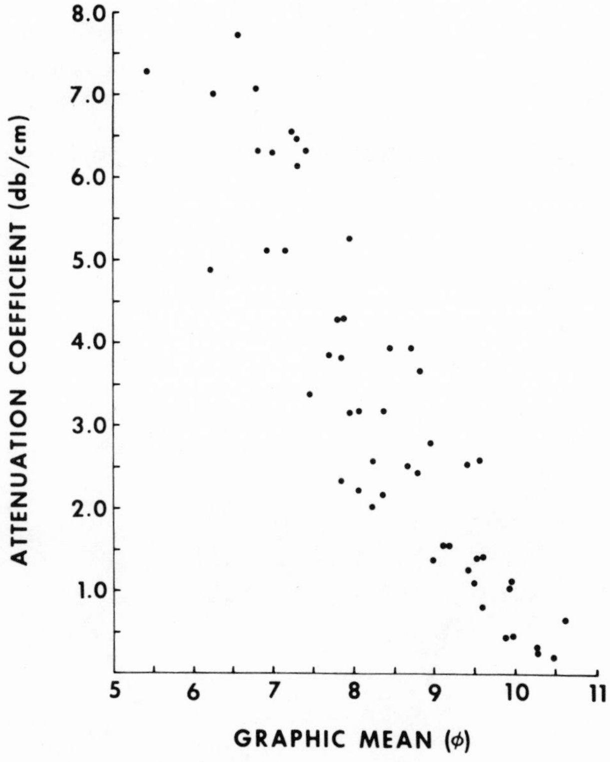

Figure 12. Relation of sediment mean grain size to the attenuation coefficient (after Buchan et al., 1972)

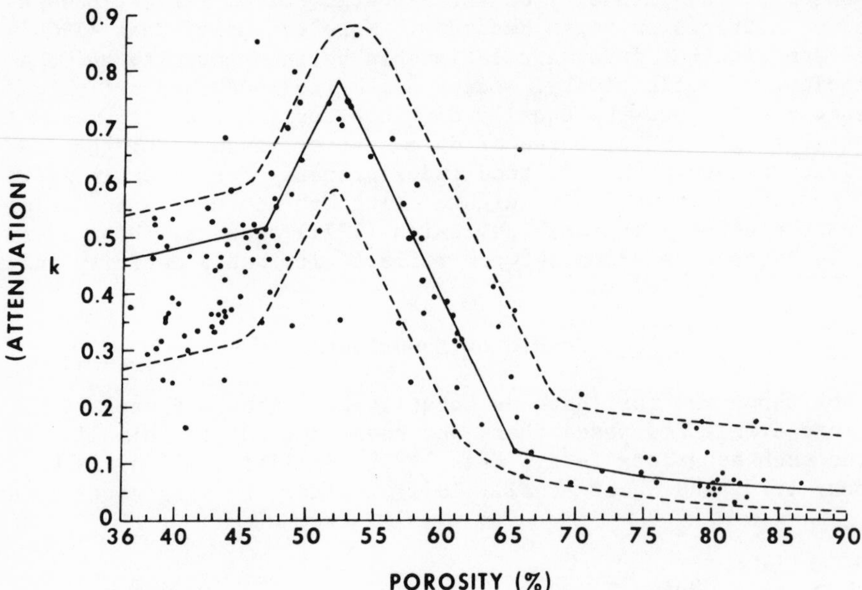

Figure 13. Relationship between sound attenuation and porosity (after Hamilton, 1972)

Figure 14. Relationship between reflectivity, porosity and wet unit weight (after Smith, 1971)

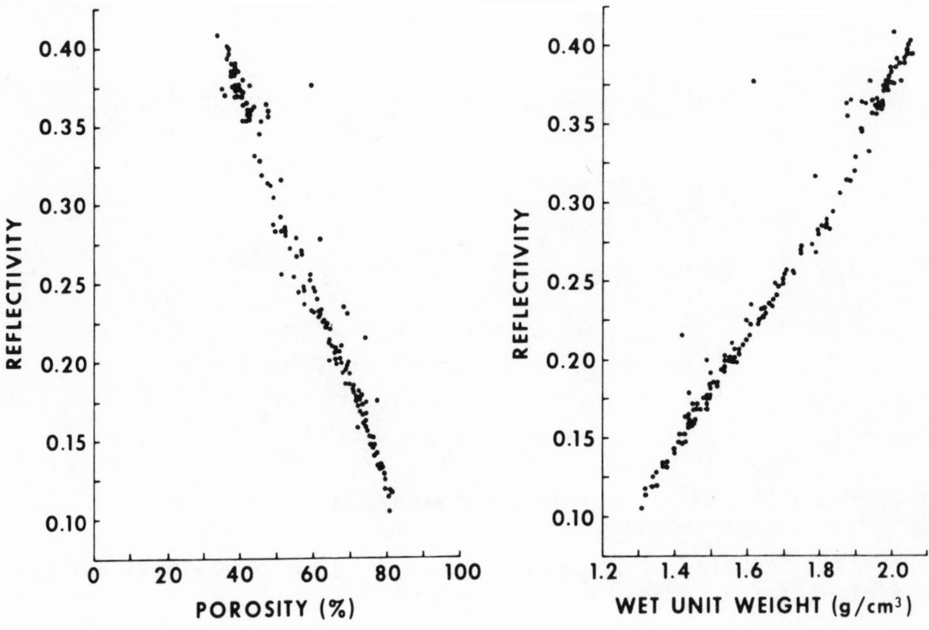

TABLE 1

PREDICTING EQUATIONS FOR SHEAR STRENGTH, WATER CONTENT AND BULK DENSITY WITH DEPTH FOR SEDIMENTS FROM THE GULF OF MEXICO[1]

	S	W	BD
For all data n = 1480	S = 82+(.2)D	W = 113.8-(.045)D	BD = 1.44+(.00015)D
Area I n = 264	S = 133+(.14)D	W = 97.0-(.037)D	BD = 1.52+(.00011)D
Area II n = 516	S = 65+(.2)D	W = 120.3-(.055)D	BD = 1.41+(.00018)D
Area III n = 343	S = 41+(.26)D	W = 121.4-(.050)D	BD = 1.40+(.00018)D
Area IV n = 87	S = 99+(.5)D	W = 128.1-(0.29)D	BD = 1.38+(.00008)D
Area V n = 112	S = 156+(.18)D	W = 104.4-(.009)D	BD = 1.45+(.00002)D
Area VI n = 151	S = 148+(.21)D	W = 102.7-(.026)D	BD = 1.48+(.0001)D

S = Shear strength in PSF

D = Depth below mudline in cm

W = Water content (% dry weight)

BD = Bulk density (gm/cc)

n = Number of analyses

[1]From Bryant and Trabant, 1972

from St. Andrew Bay, Florida, Holmes and Goodell (1964) attempted to determine the interrelationship of a number of the sediment physical properties. Based on their statistical analyses of these data, they derived the following equation for determining shear strength:

$$[\text{Shear strength } (g/cm^2) = -1.089 \text{ water content } (\%) + 0.254 \text{ depth in core (cm)} + 0.021 \text{ Kaolinite/Illite}]$$

Based on the study of 80 sediment cores from various sectors of the Gulf of Mexico, Bryant and Trabant (1972) formulated predicting equations for shear strength, water content, and wet unit weight versus depth below the bottom for each of six provinces of the Gulf (Table 1).

On yet a broader scale, A. F. Richards (personal communication, 1972) has statistically analyzed a large collection of physical properties data (18,000 analyses) from both the Atlantic and Pacific basins. This has lead to a series of equations for the determination of wet unit weight (γ sat), void ratio (e), and water content (W) (Table 2).

SUMMARY

This discussion has attempted to present the state of our knowledge concerning the interrelationships of various mass physical, electrical, and acoustical properties of submarine sediments. Obviously studies to date have not been able to examine all the various environments or sediment types found on the sea floor. Caution, therefore, is warranted in the acceptance of a number of the correlations presented here. The general pattern or relationship found in the various correlations will probably continue to be valid as more data become available, but shifting of curves can be expected.

Care must be exercised in attempting to apply a number of the curves presented here to areas other than those from which they

TABLE 2

PREDICTING EQUATIONS FOR WATER CONTENT, WET UNIT WEIGHT AND VOID RATIO FOR NORTH ATLANTIC AND PACIFIC SEDIMENTS[2]

$$\gamma_{sat} = 2.4329 - 2.0661 \times 10^{-2}W + 1.6645 \times 10^{-4}W^2 - 6.7536 \times 10^{-7}W^3 + 1.0663 \times 10^{-9}W^4$$

$$e = 0.01045 + 0.02725W$$

$$W = 4.0354 + 35.1216e$$

$$W = 2966.94 - 4481.679\gamma_{sat} + 2317.880\gamma^2_{sat} - 405.513\gamma^3_{sat}$$

[2]From A. F. Richards, personal communication

TABLE 3

VARIATION OF SELECTED SEDIMENT MASS PROPERTIES FROM THE NORTH ATLANTIC AND NORTH PACIFIC

	Wet Unit Weight kg/m^3 (g/cm^3)	Water Content % dry wt	Shear Strength kPa (g/cm^2)	Porosity %	Sensitivity	Grain Specific Gravity
ATLANTIC						
Maximum	2,650 (2.65)	217	90.71 (925)	85	88	2.86
Minimum	1,200 (1.20)	15	.098 (1)	32	1	2.45
Average	1,520 (1.52)	86	5.10 (52)	66	4	2.73
PACIFIC						
Maximum	2,030 (2.03)	423	54.52 (556)	91	57	3.83
Minimum	1,130 (1.13)	24	.098 (1)	39	1	2.32
Average	1,450 (1.45)	175	4.81 (49)	77	4	2.71

TABLE 4
VARIATION AMONG THREE SEDIMENT CORES LOCATED 20cm APART
(modified from Bennett et al., 1970)

	Sampling Interval (cm)			Water Content (% dry wt)			Unit Wet Weight (g/cm^3)=$kg/m^3 \cdot 10^{-3}$		
Horizon	Core 1	Core 2	Core 3	Core 1	Core 2	Core 3	Core 1	Core 2	Core 3
A	0–8	0–7	0–9	75	66	78	1.56	1.57	1.55
B	8–15	10–17	10–17	67	60	65	1.61	1.64	1.62
C	17–25	17–25	17–26	59	57	63	1.65	1.67	1.62

	Grain Specific Gravity			Shear Strength (kPa)			Porosity (%)		
Horizon	Core 1	Core 2	Core 3	Core 1	Core 2	Core 3	Core 1	Core 2	Core 3
A	2.78	2.77	2.78	2.06	2.26	1.37	70	69	70
B	2.79	2.78	2.78	5.10	3.92	3.43	67	65	66
C	2.78	2.80	2.79	7.06	14.61	5.30	64	64	66

	Sensitivity			Sand (%)			Silt (%)		
Horizon	Core 1	Core 2	Core 3	Core 1	Core 2	Core 3	Core 1	Core 2	Core 3
A	3	3	3	42	36	41	35	40	37
B	3	5	3	29	22	36	42	46	40
C	6	41	7	29	40	25	43	39	45

	Clay (%)		
Horizon	Core 1	Core 2	Core 3
A	23	24	22
B	29	32	24
C	28	21	30

were derived. A number of the correlations discussed will have their greatest accuracy when developed for a local area rather than on a regional basis. This is clearly shown in Table 1 where a series of predictor equations for the entire Gulf of Mexico can be compared to equations derived for separate provinces of the Gulf.

Depending upon the desired accuracy, one may find many of the correlations presented here suitable to either predict certain properties or to interpolate between available data. Needless to say there are many variations in the physical as well as chemical mass properties of sediments, both laterally and with depth below the sea floor, which indirectly influence the accuracy of any correlation of only two parameters. As shown in Table 3, the variability of certain physical properties within the upper 2.5 m of the sea floor can be sizeable. On a micro-scale it can also be seen that a number of these properties vary considerably in samples taken only 20 cm apart (Table 4). These points on variability are made to impress on the reader that the development and use of various interrelationships must not be taken for granted.

REFERENCES

Bennett, R. H., G. H. Keller, and R. F. Busby, Mass property variability in three closely spaced deep-sea sediment cores, J. Sediment. Petrol., 40, 1038-1043, 1970.

Bennett, R. H., D. N. Lambert, and P. J. Grim, Tables for determining unit weight of deep-sea sediments from water content and average grain density measurements, National Oceanic and Atmospheric Administration Tech. Memo, ERL AOML-13, 1971.

Bjerrum, L., Fundamental considerations on the shear strength of soil, Géotechnique, 2, 209-218, 1951.

Bjerrum, L., Theoretical and experimental investigations on the shear strength of soils, Norwegian Geotechnical Institute Publ. 5, 1954.

Bouma, A. H., W. E. Sweet, Jr., W. A. Dunlap, and W. R. Bryant, Comparison of geological and engineering parameters of marine sediments, 4th Ann. Offshore Tech. Conf. Preprints, 1, 21-34, 1972.

Boyce, R. E., Electrical resistivity of modern marine sediments from the Bering Sea, U. S. Naval Undersea Warfare Center Tech. Publ. 6, 1967.

Bryant, W. R., and P. K. Trabant, Statistical relationships between geotechnical properties of Gulf of Mexico sediments, 4th Ann. Offshore Tech. Conf. Preprints, II, 363-368, 1972.

Buchan, S., D. M. McCann, and D. T. Smith, Relations between the acoustic and geotechnical properties of marine sediments, Quart. J. Eng. Geol., 5, 265-284, 1972.

Cernock, P. J., Sound velocities in Gulf of Mexico sediments as related to physical properties and simulated overburden pressures, Texas A&M University Tech. Rept. 70-5-T, 1970.

Chmelik, F. B., A. H. Bouma, and R. Rezak, Comparison of electrical logs and physical parameters of marine sediment cores, Trans. Gulf Coast Assoc. Geol. Soc., 19, 63-70, 1969.

Creager, S. S., D. W. Scholl, et al., Initial Reports of the Deep Sea Drilling Project, 19, 913, U. S. Govt. Printing Office, Washington, D. C., 1973.

Faas, R. W., An empirical relationship between reflection coefficients and ocean bottom sediments, in Symp. Ocean Sci. and Eng. of the Atlantic Shelf, pp. 149-167, Phila., 1968.

Hamilton, E. L., Low sound velocities in high porosity sediments, J. Acoust. Soc. Am., 28, 16-19, 1956.

Hamilton, E. L., Consolidation characteristics and related properties of sediments for experimental Mohole (Guadalupe Site), J. Geophys. Res., 69, 4257-4269, 1964.

Hamilton, E. L., Sound velocity and related properties of marine sediments, North Pacific, J. Geophys. Res., 75, 4423-4446, 1970a.

Hamilton, E. L., Reflection coefficients and bottom losses at normal incidence computed from Pacific sediment properties, Geophysics, 35, 995-1004, 1970b.

Hamilton, E. L., Compressional wave attenuation in marine sediments, Geophysics, 37, 620-646, 1972.

Hamilton, E. L., G. Shumway, W. H. Menard, and C. J. Shipek, Acoustic and other physical properties of shallow-water sediments off San Diego, J. Acoust. Soc. Am., 28, 1-15, 1956.

Hatcher, P., Diagenesis of Organic Matter in the Sediments of Mangrove Lake, Bermuda, Masters thesis, Univ. of Miami (in preparation), 1974.

Holmes, C. W., and H. G. Goodell, The prediction of strength in the sediments of St. Andrew Bay, Florida, J. Sediment. Petrol., 34, 134-143, 1964.

Horn, D. R., B. M. Horn, and M. N. Delach, Correlation between acoustical and other physical properties of deep-sea cores, J. Geophys. Res., 73, 1939-1957, 1968.

Inderbitzen, A. L., Relationship between sedimentation rate and shear strength in recent marine sediments off Southern California, Ph.D. thesis, Stanford Univ., 1969.

Keller, G. H., Shear strength and other physical properties of sediments from some ocean basins, in Civil Engineering in the Oceans, pp. 391-417, Am. Soc. Civil Engrs., N. Y., 1968a.

Keller, G. H. and R. H. Bennett, Mass physical properties of submarine sediments in the Atlantic and Pacific basins, in Proc. 23rd Intern. Geol. Cong., 8, 33-50, Prague, 1968b.

Keller, G. H. and D. N. Lambert, Geotechnical properties of submarine sediments, Mediterranean Sea (abstract) in Proc. VIII Intern. Sediment. Cong., Heidelberg, 1971.

Kermabon, A., C. Gehin, and P. Blavier, A deep-sea electrical resistivity probe for measuring porosity and density of unconsolidated sediments, Geophysics, 34, 554-571, 1969.

Lambert, D. N., and R. H. Bennett, Tables for determining porosity of deep-sea sediments from water content and average grain density measurements, National Oceanic and Atmospheric Administration Tech. Memo. ERL AOML-17, 1972.

Loring, D. H., Bottom analysis of sediments of the Gulf of St. Lawrence and the Atlantic Continental Shelf from Newfoundland to Georges Bank in respect to acoustical reflection, Fisheries Res. Board of Canada Manuscript Rept. Series, 127, 1962.

McClelland, B., Progress of consolidation in delta front and prodelta clays of the Mississippi River, in Marine Geotechnique, edited by A. F. Richards, pp. 22-40, Univ. of Ill. Press, Urbana, 1967.

Moore, D. G., Submarine slumps, J. Sediment. Petrol., 31, 343-357, 1961.

Moore, D. G., Shear strength and related properties of sediments from experimental Mohole (Guadalupe site), J. Geophys. Res., 69, 4271-4291, 1964.

Moore, D. G., and G. Shumway, Sediment thickness and physical properties: Pigeon Point Shelf, California, J. Geophys. Res., 64, 367-374, 1959.

Nacci, V., L. Lewis, J. Gallagher, and M. Huston, Correlation of geotechnical and acoustical properties of ocean sediments, in Proc. Intern. Symp. on the Eng. Properties of Sea Floor Soils and their Geophys. Ident., pp. 268-278, Univ. of Washington, Seattle, 1971.

Nafe, J. E., and C. L. Drake, Variation with depth in shallow and deep water marine sediments of porosity, density and the velocities of compressional and shear waves, Geophysics, 22, 523-552, 1957.

Richards, A. F., Investigations of deep-sea sediment cores, I. Shear strength, bearing capacity, and consolidation, U. S. Navy Hydrographic Office Tech. Rept. 63, 1961.

Richards, A. F., Investigation of deep-sea sediment cores, II. Mass physical properties, U. S. Navy Hydrographic Office Tech. Rept. 106, 1962.

Richards, A. F., Preface, in Marine Geotechnique, edited by A. F. Richards, Univ. of Ill. Press, Urbana, 1967.

Rutledge, P. C., Review of the cooperative triaxial research program of the War Department, Corps of Engineers, covering the period February 1940 to May 1944, in Soil mechanics fact finding survey, progress report: triaxial shear research and pressure distribution studies on soils, pp. 1-178, U. S. Army Corps of Engineers, Waterways Experiment Station, Vicksburg, 1947.

Shumway, G., Sound-speed and absorption studies of marine sediments by a resonance method - Part II, Geophysics, 25, 659-682, 1960.

Skempton, A. W., Soil mechanics in relation to geology, Proc. Yorkshire Geol. Soc., 29, 33-62, 1953.

Smith, D. T., Acoustic and electrical techniques for sea-floor sediment identification, in Proc. Intern. Symp. on the Eng. Properties of Sea Floor Soils and their Geophys. Ident., pp. 235-267, Seattle, 1971.

Sweet, W. E., Jr., Electrical resistivity logging in unconsolidated sediments, Texas A&M University Sea Grant Rept. 72-205, 1972.

MECHANICS PROBLEMS AND MATERIAL PROPERTIES

LOUIS J. THOMPSON

Texas A&M University

ABSTRACT

Rational design methods can be developed for submarine engineering problems, only if nonconservative mechanics problems can be solved. This requires the development of material constitutive equations for submarine soils that can be combined with the classical field equations to yield a unique solution for a given set of boundary and initial conditions.

The field equations require that the constitutive equations be written for stress, heat flux, internal energy and entropy as functions of displacement and temperature. The axioms of constitutive theory restrict the functions to the gradients of displacement and temperature in both space and time. The constants in the constitutive equations are the material properties and they do not depend on material state, history of the material state, test equipment or test procedure. Shear strength, tensile strength, density, moisture content, color, etc. are not material properties.

The properties are invariant for a given material and not functions of stress, time, temperature, etc. If the kind of material can be changed, as by changing the moisture content or type of mineral, then the material properties must change. How a material responds depends on the amount and rate of energy addition, the initial state of the material, the boundary conditions and the material properties.

To evaluate material properties, constitutive equations must be assumed and a boundary value problem solved for a test's imposed stress and heat flux boundary and initial conditions. The predicted boundary displacement and temperature, in terms of stress and heat flux, can be compared to measured results to evaluate constants. However, in order to evaluate material properties, the test boundary value problem must be solved because no amount of boundary constraints will insure that the specimen will deform uniformly.

When simple constitutive equations are assumed, additional restrictions must be imposed such as weightlessness, steady state, incompressibility, etc. to obtain a unique non-trivial solution for a boundary value problem, and it is these solutions that are required to evaluate material properties.

Limiting equilibrium, ultimate strength theory, plasticity and fracture mechanics are all efforts to get around the problems of large and/or discontinuous motion and large strains where the simple constitutive equations do not adequately describe the material. These theories lead to indeterminate problems. Failure must be described by motion and temperature. Failure theories of Mohr, Coulomb, Rankine and Fellenius cannot lead to the solution of a mechanics problem and shear strength parameters determined by arbitrary tests are not material properties, but the theories have been invaluable in the practice of engineering as an art.

There are many reasons to solve mechanics problems - especially those where energy is not conserved, but the major reason is that the solutions become an analysis method in engineering. The performance of a design can be tested by an analysis method; otherwise, model testing or full scale testing must be used. With analysis methods, design optimization procedures are developed.

Some of the mechanics problems that need to be solved and the submarine engineering problems where the solutions would be applicable as analysis methods are as follows:

A. Interaction between a body and a static half space
   1. Penetrating body - applicable to pile driving, anchor penetration, coring devices, etc.
   2. Horizontal motion of body on or in surface - applicable to bottom trafficability, plowing, ditching, etc.
   3. Motion of embedded body - applicable to anchor resistance and loads on buried pipelines
   4. Motion of flexible body partially embedded - applicable to offshore structures

B. Interaction between a body and a moving fluid
   Lift and drag forces on a body resting on fixed boundary - applicable to erosion and scour of bottom, especially around pipelines

C. Motion of an irregular surface or slope due to change in

material properties or change in boundary conditions - applicable to submarine slope stability problems

D. Convective motion of a fluid about a linear heat source - applicable to dissipation of heat from a buried submarine pipeline.

The solution to a mechanics problem consists of the displacements and the change in temperature for each point in a material. These four variables are functions of both space and time. They are called the independent state variables or the material response variables.

When the response of a material can be predicted, then a thermo-mechanical problem has been solved.

The response of a material depends on four conditions:

1. The initial state of the material
2. The constraints or boundary conditions for the material
3. The rate of energy addition into the material
4. The amount of energy added relative to the material critical energy (melting, vaporization, etc.).

The solution of a mechanics problem must satisfy the field or conservation equations, which are (Truesdell, 1969, pp. 25-40):

1. The one equation for conservation of mass
2. The three equations for conservation of linear momentum
3. The three equations for conservation of moment of momentum
4. The one equation for conservation of energy
5. The one inequality for dissipation of energy.

The field equations relate functions of the response variables and the dependent state variables. The dependent state variables are from the field equations:

1. The stresses (tensor)
2. The heat flux (vector)
3. The density (scalar)
4. The internal energy (scalar)
5. The entropy (scalar)
6. The dissipation (scalar).

The field equations are equally applicable to all materials. Being insufficient in number to yield a solution, additional equations are needed to characterize the material absorbing or transmitting energy.

Certain conditions or axioms can be used to restrict constitutive equations for materials (Eringen, 1962, pp. 145-162):

A. Axiom of Causality - The displacement and temperature change of the material points are considered as self-evident observable effects as energy is absorbed by or transmitted through a body. These are the independent variables of the constitutive equations. The remaining state variables that enter the field equations are the causes of the dependent constitutive variables.
The results are functions of time and space. The causes are functions of the results. The causes must be uniquely related to results.
B. Axiom of Determinism - The value of the dependent constitutive variables or the dependent state variables, at all points in a material at any time is determined by the history of the motion and temperature of all the points of the body.
C. Axiom of Equipresence - All constitutive functionals should be expressed in terms of the same list of independent variables until the contrary is deduced.
This precautionary measure prevents prejudice or favor for certain classes of variables.
D. Axiom of Objectivity - Constitutive equations must be invariant in form with respect to rigid motions of the spatial frame of reference in either time or space.
Physically, material properties cannot depend on the motion of the observer or the coordinate transformation. Essentially, this means that the state variables cannot be explicit functions of time or position but must be functions of the gradients of the displacement and temperature. Time, place, motion, velocity, acceleration are not admissible as variables in a constitutive equation.
E. Axiom of Admissibility - Constitutive equations must be consistent with the field equations and the dissipation inequality. This axiom prevents formulation of equations that violate the basic laws and can be used to eliminate dependence on some variables.

Other axioms are possible to restrict the material to be studied. These are:

F. Material Invariance
G. Neighborhood
H. Memory type
I. Etc.

The principles allow elimination of certain variables and require continuity for space and time derivatives. Types of material symmetry are invoked under the axiom of material invariance.

Constitutive equations and the constants or material properties are not laws of nature. They are the results of a logical effort to describe material behavior in a mechanics problem. A set of constitutive equations for a material consists of (Eringen, 1962, p. 143):

1. The six independent components of the stress tensor
2. The three independent components of the heat flux vector
3. The internal energy of the material
4. The entropy of the material

all of which are functions of the gradients and rates of change of the displacements and temperature. They cannot be functions of stress rate, position, time, velocity or acceleration if they are to be objective (Eringen, 1962, p. 111).

Since constitutive equations characterize materials, the constants that occur in these equations are called the material properties. Material properties are invariant for a given material; they are independent of:

1. Material state variables such as stress, strain, rate of strain, temperature, etc.
2. Paths the state variables are constrained to follow
3. Test equipment or how the boundaries of the material are constrained
4. Test procedure
5. Size of object tested
6. Shape of object tested.

For multiphase materials or mixtures such as soil, there are also phase variables. These relate volumes or weights of the constituent phases. As these phase variables are changed, the material changes and the material properties must change. For multiphase materials, the material properties are functions of the phase variables. All equations and concepts that apply to a material apply to a mixture and also to each phase. Examples of phase variables are: moisture content, void ratio, percent saturation, dry unit weight, etc.

The constitutive equations should contain as many constants as are needed to characterize a material for a given range of state variables. The number of constants may be as few as one or two for simple materials with a limited range of applicability. If necessary, the equations may be left open-ended and the number of constants determined by a material test data fit.

The number of test measurements must at least be equal to the number of properties to be evaluated in the constitutive equation. If extra measurements are made it may be found the calculated "material properties" have changed. How little the "constants" change determines the usefulness or limitation of the assumed constitutive equation.

Properties can only be evaluated for the range over which the state variables are allowed to vary in a test. The larger the range of behavior the constitutive equation fits, the better the constitutive equation. Determining how well the constitutive equation fits is a statistical problem.

As the range of the constitutive equation is extended, and it is found that the equations no longer describe the material behavior, new constitutive equations must be assumed, more complicated equations with a larger number of material properties.

Ordinarily in a material test, stresses, heat flux, entropy, density, internal energy displacements and temperature are assumed to be uniform within the specimen. Most often gross effects are measured on the boundaries and the boundary conditions are assumed to vary, at most, linearly.

Critical review will show that only displacements and temperature can be measured directly at discrete points on the boundaries of a specimen and nothing can be measured directly inside the specimen. If another material with known material properties is used to construct the boundaries then the normal stresses and the normal heat flux can be determined on the boundaries. Tangential components of stress and heat flux on boundaries are almost impossible to measure with present equipment. Usually, total shear force and the magnitude of the heat flux vector can be measured.

The discrete boundary data for the test must be generalized for the boundary by fitting curves through the data. Then these functions must be generalized for the whole specimen. With a little reflection, it will be seen that an infinite number of displacement fields will satisfy the measured boundary conditions. This is also true for temperature, stresses, and heat flux. At this point, the field equations can contribute in a global manner and with reference only to the boundary conditions. No amount of constraints on the boundary can insure that the specimen will deform uniformly. The response depends on the material as well as the boundary conditions. It might be added that if the measured boundary data show that energy is being extracted from a material in a cycle of application and removal of additional forces, the data is suspect (Drucker, 1966).

Some of the above problems are circumvented by combining assumed constitutive equations with the field equations and solving for the response caused by the boundary stresses and heat flux. The predicted response can then be compared to the measured boundary response to evaluate material constants. The predicted response must exist, it must be non-trivial and above all, it must be unique.

Ordinarily when simple constitutive equations are combined with the field equations a system of nonlinear partial differential equations results. Equations of this type may have no solution or an infinite number of solutions (Ames, 1965; Ladyzhenskaya, 1969). Therefore, it is important to choose constitutive equations that, when combined with the field equations, yield a system that has a unique solution; otherwise, it will not be possible to make predictions.

At the present time, only very simple problems can be solved if simple linear constitutive equations are used. Invariably the solution procedure involves some simplifying assumptions to reduce the four non-linear differential equations to linear ones. Three dimensional solutions are available for rigid bodies. Two dimensional solutions are available for inviscid liquids in steady state flow. Solutions are available for incompressible steady-state pipe flow of linear viscous fluid. Two dimensional or axisymmetric solutions are available for linear elastic statics problems if the material is weightless.

This list of solutions nearly exhausts the available solutions for simple materials and none of these idealized materials start to approximate soil behavior for large strains or for high rates of strain. Also, the restricted situations do not approximate the situations needed for solution of submarine engineering problems.

Examples of material properties that develop out of simple constitutive equations are:

1. The two Lame constants in the theory of linear elasticity which can be calculated from Young's Modulus and Poisson's ratio
2. The tensor of permeability in the theory of flow through porous media
3. The coefficient of consolidation in the consolidation theory
4. The compression index in settlement analysis
5. The coefficient of viscosity in the theory of viscous fluids
6. The thermal conductivity in heat transfer theory.

When the constitutive equations that utilize these properties are used in a mechanics problem, predictions can be made even though they may be only for static problems or two dimensional situations. The constitutive equation may not fit the material very well; general three dimensional dynamics problems may be impossible to solve and the material properties calculated from tests may vary over a large range, but at least there is some fundamental basis.

If simple constitutive equations must be used so that uniqueness of solution is not assured, a new principle is needed to select the appropriate solution out of a multiplicity of solutions. Historically such principles have been developed for conservative systems where they are not needed. The principle of least action is an example. Others are Hamilton's principle or Lagrange's principle (Sokolnikoff, 1964, p.231).

It may be that the rate of change of total energy for a material must be a minimum in any process, and on the other hand, it may be that nothing new can be wrung out of such a principle. If so, another device might be found to select the unique solution from infinitely many solutions.

The best procedure may be to look for more complicated constitutive equations that when combined with the field equations will produce a unique solution which models the material's behavior.

The development of limiting equilibrium theory, ultimate strength design, plasticity theory, and fracture mechanics are all efforts to get at the problem of large and/or discontinuous motions of a material. But, in anything but a one-dimensional problem, motion is ignored by establishing criteria that relate stresses at failure and the problems are all indeterminate (Lambe and Whitman, 1969, p. 357). Failure must be defined. This can be done with relative motion and temperature, because they are the only measurable quantities on a body boundary. Thus, we are forced back to the original mechanics problem, predicting motion and temperature change. By no other means can any theory be checked.

Along with the development of these arbitrary analysis methods has been the development of arbitrary tests. Usually these test data depend on the test apparatus and test procedure as well as being dependent on the material being tested. Usually these data do not relate state variables, but are determined by some sort of defined criteria. These are _material test values_. Examples of these are: tensile strength, cohesive strength, angle of internal friction, hardness, etc. The Atterberg limits are material test values, but they are involved with changes in phase variables and changes in state variables.

Similar schemes make the geometry of a material in a test the major influence. Examples of these _geometric test values_ are: drag coefficient, stability number, etc. If the test conditions are the major influence, test condition values result. An example of this type number is the Reynolds number.

When these arbitrary test values are correlated with experience they can become useful as long as the limitations are remembered.

Often these test values have been used to relate variables statistically when a relationship may not exist. Eventually if they are used long enough they begin to have the flavor of a law of nature. No test value will aid in the solution of a mechanics problem. No motion and temperature changes can be predicted by their use. However, test values have been invaluable in the practice of the art of engineering and advances in civil engineering usually develop through art rather than science. Test values have been most useful in showing that real material response is complicated and that constitutive equations of the simple linear type have no hope in representing material whose shear resistance is sensitive to hydrostatic or spherical stress.

There has been much time, effort and money spent to obtain data on submarine soils and their properties. Much of the effort has been a repeat of onshore experience. Many facts have been assembled, facts from tests that are usually arbitrary. Precious few problems have been solved. Advances in ocean-bottom engineering have progressed by scaling up successes and beefing up failures. This is a trial and error procedure, and it's expensive. We have been long on facts and short on theory.

Three years ago, Treasury Secretary, George Schultz, concluded a speech on economic policy by abruptly breaking into song (Anonymous, 1973). To the lively tune of "Silver Dollar," he belted forth in full voice:

A fact without a theory
Is like a ship without a sail,
Is like a boat without a rudder,
Is like a kite without a tail.
A fact without a figure
Is a tragic final act,
But one thing worse
In this Universe
Is a theory without a fact.

Have we in soil mechanics drifted along with the failure concepts of Mohr, Coulomb, Rankine, Prandtl and Fellenius without the rudder of rationality and theory? Is it possible that some of the effort being made to gather data on the wide variety of bottom soils should be redirected toward evaluation of the sophisticated but arbitrary testing methods, and even the arbitrary analysis methods that utilize these test data? It may be that none of these facts can be used to solve problems; indeed, it may be that the facts have no significance, and, in Dr. Schultz's words, we are participating in a tragic final act.

REFERENCES

Ames, W. F., Nonlinear Partial Differential Equations in Engineering, Academic Press Inc., N. Y., 1965.

Anonymous, Time, p. 80, Feb. 26, 1973.

Drucker, D. C., Concept of path independence and material stability for soils, in Rheology and Soil Mechanics, edited by J. Krautchenko and P. M. Sirieys, pp. 23-43, Springer-Verlag, N. Y., 1966.

Eringen, A. C., Mechanics of Continua, John Wiley & Sons, Inc., N. Y., 1962.

Ladyzhenskaya, O. A., The Mathematical Theory of Viscous Incompressible Flow, 2nd Edition, Gordon and Breach, London, 1969.

Lambe, T. W., and R. V. Whitman, Soil Mechanics, John Wiley & Sons, Inc., N. Y., 1969.

Sokolnikoff, I. S., Tensor Analysis, Theory and Applications to Geometry and Mechanics of Continua, 2nd Edition, John Wiley & Sons, Inc., N. Y., 1964.

Truesdell, C., Rational Thermodynamics, McGraw-Hill Book Company, Inc., N. Y., 1969.

Determination of Mechanical Properties in Marine Sediments

THE ROLE OF LABORATORY TESTING IN THE DETERMINATION OF DEEP-SEA SEDIMENT ENGINEERING PROPERTIES

HOMA J. LEE

U. S. Naval Civil Engineering Laboratory

ABSTRACT

Engineering properties are needed primarily for engineering activities such as the design of foundations or anchors. They may be evaluated either through direct in situ testing or laboratory testing of samples. Laboratory testing provides a degree of flexibility, control, and economy not usually achieved with in situ testing and is, therefore, given primary attention in this paper. The area of sediment property determination for direct embedment anchor design is considered specifically and two examples are given of possible uses of laboratory testing. The first deals with quantitatively correcting laboratory strength results for sample disturbance. The procedure is evaluated using in situ and laboratory vane shear and residual pore pressure measurements. The second example deals with using triaxial testing to investigate shear strength variation with sub-bottom depth and drainage. Results of tests performed on "red" clay are given. The two examples given provide information for predicting the holding capacity of an embedment anchor. The examples are followed by a discussion of suggested additional research.

INTRODUCTION

Engineering properties are basically those sediment characteristics needed to predict the engineering performance of foundations, anchors, or other facilities. These characteristics may occasionally be used to analyze certain geologic and underwater sound problems, but the principal purpose of engineer-properties is engineering. Therefore, advances in the area of property determination should be directed primarily toward items which are important in terms of engineering design.

Engineering problems which could require sediment properties are relatively numerous and include direct embedment anchor holding capacity, bearing capacity and settlement of footings, trafficability of bottom-crawling vehicles, dynamic object penetration, and breakout of partially embedded objects. It is beyond the scope of this paper to discuss all of the characteristics of the engineering properties which are required for each of these problems. Instead, the one problem of estimating direct embedment anchor holding capacity will be taken as an example so that some of the ramifications of engineering property determination can be discussed in detail. One currently used procedure for estimating anchor-holding capacity is given by Taylor and Lee (1972) and also by Valent (this volume). The procedure is not finalized and continuing research is currently underway at the Naval Civil Engineering Laboratory (NCEL). However, at present it appears fairly certain that holding capacity is not a unique characteristic of a site but is instead strongly dependent upon time, anchor embedment depth, and possibly nature of loading. Virtually all previous research indicates that the controlling engineering property is the soil shear strength. Therefore, an evaluation must be made of how this parameter varies with time, depth, and load.

The three ways in which shear strength may be obtained are in situ testing, laboratory testing of samples, and empirical correlation with index properties, geophysical characteristics, or sediment type. Since empirical correlations require direct measurements for their development, initial emphasis must be given to in situ and laboratory tests. Each of these two approaches has advantages and disadvantages. Laboratory tests can be performed with better control and can be continued for longer periods of time. However, all samples are disturbed to a greater or lesser extent and results obtained may not represent in situ conditions. In situ testing removes most of the problems of disturbance. However, most in situ tests are expensive and short-term. Also, their results may be difficult to analyze because of a lack of boundary condition control (Lambe and Whitman, 1969, p. 450).

Both laboratory and in situ tests are needed to develop a better understanding of the behavior of sea-floor sediments. The tests should be used complementarily, each serving as a check on the other. Much recent emphasis has been placed on in situ testing with laboratory testing being downgraded somewhat. This paper is specifically directed toward laboratory testing and is intended to indicate some of the favorable characteristics of this approach.

APPROACH

This paper presents results of two investigations which were selected as examples of ways in which laboratory testing can be used productively to obtain estimates of engineering properties for use in predicting anchor holding capacity. While this is a specific situation, the procedures and framework can serve as a model for other investigations.

The first example deals with sample disturbance, a problem which must be approached if laboratory results are to be used effectively. A procedure for correcting laboratory results for disturbance is developed and quantitatively evaluated for one geographic area. The second example deals with the variation of shear strength with sub-bottom depth and time (drainage). These variations were analyzed for one deep-ocean sediment (pelagic or "red" clay) using results of triaxial tests. The results are expressed in terms of a predicted undrained shear strength profile and a set of drained strength parameters. The sort of information obtained from these two examples could be used in evaluating the holding capacity of embedment anchor installations.

Following a discussion of these two examples, a listing is made of unsolved problems relevant to the overall problem of sea-floor sediment engineering property determination.

SAMPLE DISTURBANCE

Most deep-ocean sediment samples have been obtained either with typical oceanographic corers or through a drilling process. A comparison of the geometrics of these corers with the criteria for good samplers, given by Hvorslev (1949), indicates that the samples obtained are of relatively poor quality. Drilling aggravates an already bad situation. In addition to mechanical disturbance introduced by the corer itself, there are other factors which can alter the quality of sea-floor sediment samples. These are water and gas expansion produced when the high hydrostatic pressures of the sea floor are removed, organic material growth accelerated by pressure and temperature changes, and the rough treatment which samples may receive on shipboard and during transit. A good discussion of the various disturbance mechanisms is given by Richards and Parker (1968). It is possible to reduce mechanical sampler disturbance by using special types of samplers, for example, a spade-type box corer (Rosfelder and Marshall, 1967) or a coring device mounted on a bottom-sitting platform (Demars and Taylor, 1971; Hironaka and Green, 1971). Disturbance can also be reduced by sample refrigeration and careful handling. Under no conditions, however, can a sample be considered to be completely undisturbed; the properties are always altered from their in situ state prior to testing.

One approach to the evaluation of shear strength changes produced by sample disturbance can be based on the concepts of Hvorslev (1960). In this extensive work, it is shown that the undrained shear strength of a saturated cohesive soil is primarily dependent upon two items: (1) the normal effective stress acting on the failure plane at failure, and (2) the water content. In turn, the normal effective stress acting on the failure plane at failure is usually directly related to the effective stress state before shear. Therefore, a corollary to the Hvorslev concept would be that the strength is most dependent upon the initial effective stress state and the water content.

This concept can be used to analyze disturbance as follows. During sampling most fine grained soils do not expel or imbibe a significant amount of water by virtue of their low permeability and the relatively short time involved in sampling. Therefore, the sample water content may be assumed equal to the in situ water content. Therefore, by the Hvorslev hypothesis, any changes in strength can be accounted for largely on the basis of changes in effective stress state. It should be noted, however, that water content changes often occur during storage and that care must be taken to assure that samples are well sealed.

Assuming good sealing practices, it is necessary only to analyze changes in effective stress state to determine approximately the changes in strength produced by disturbance. This analysis begins by referring to the effective stress principle:

$$\bar{\sigma} = \sigma - u_w \tag{1}$$

where $\bar{\sigma}$ = effective normal stress
σ = total normal stress
u_w = pore water pressure.

An unconfined sample in a laboratory has virtually no total stresses acting on it. Therefore, the effective stress is equal to the pore water pressure with a reversed sign; negative (relative to atmospheric) pore pressures produce positive (compressive) effective stresses.

The effective stresses of the sample are by necessity isotropic, or equal in all direction. In situ, the effective stresses are anisotropic, or direction-dependent. Therefore, it is difficult to make a direct comparison between the laboratory and in situ effective stresses. The usual procedure followed is to invoke a concept which has been referred to as "perfect sampling" in the literature (Ladd and Lambe, 1963). This hypothetical situation would result if the in situ anisotropic total stresses could be removed without introducing any additional disturbance. The term "perfect sampling" is unfortunate since the concept does imply some disturbance. In the remainder of this report, the phrase "in situ shear stress removal" will be used to describe the same concept as that usually denoted by "perfect sampling." In situ shear stress removal has been simulated in the laboratory; and as a result, it is possible to predict empirically the isotropic effective stress state which would result from this removal. In this report the residual pore pressure resulting from in situ shear stress removal (normal effective stress with opposite sign) is termed the "reference pore pressure." It may be predicted by the following equation (Ladd and Lambe, 1963):

$$u_{ps} = -\bar{\sigma}_{vo} \{K_o + A_u (1 - K_o)\} \tag{2}$$

where u_{ps} = reference residual pore pressure

$\bar{\sigma}_{vo}$ = in situ vertical effective stress

K_o = coefficient of lateral earth pressure at rest (ratio of lateral to vertical in situ effective stress)

A_u = pore pressure parameter relevant to in situ shear stress removal.

The in situ vertical effective stress at any point beneath the sea floor can be calculated by integrating the buoyant unit weights of the sediment above that point with respect to sub-bottom depth. The parameter, A_u, can be related to sediment type and varies between -0.1 and +0.3 (Ladd and Lambe, 1963). A typical value for cohesive sediment is about +0.1. The parameter, K_o, also varies with sediment type with a typical value of 0.5 for cohesive sediment. It is possible to estimate K_o more accurately from an anisotropically consolidated triaxial test or empirically (Brooker and Ireland, 1965).

Once the reference residual pore pressure, u_{ps}, is known, it is necessary to measure the residual pore pressure, u_r, retained by a real sample. Techniques for measuring this quantity are given by Lambe (1961), Gibbs and Coffey (1969), and Lee (1973a). The residual pore pressure ratio, u_r/u_{ps}, is a measure of the disturbance to which a sample has been subjected (Ladd and Lambe, 1963).

While the residual pore pressure ratio is a disturbance measure, the desired measure is the undrained shear strength ratio, S_F/S_L, where S_F is the in situ strength and S_L is a measured laboratory strength. A direct correlation between the ratios should exist; however, the numerical characteristics of this correlation need to be evaluated empirically for the various types of sediment found in the oceans.

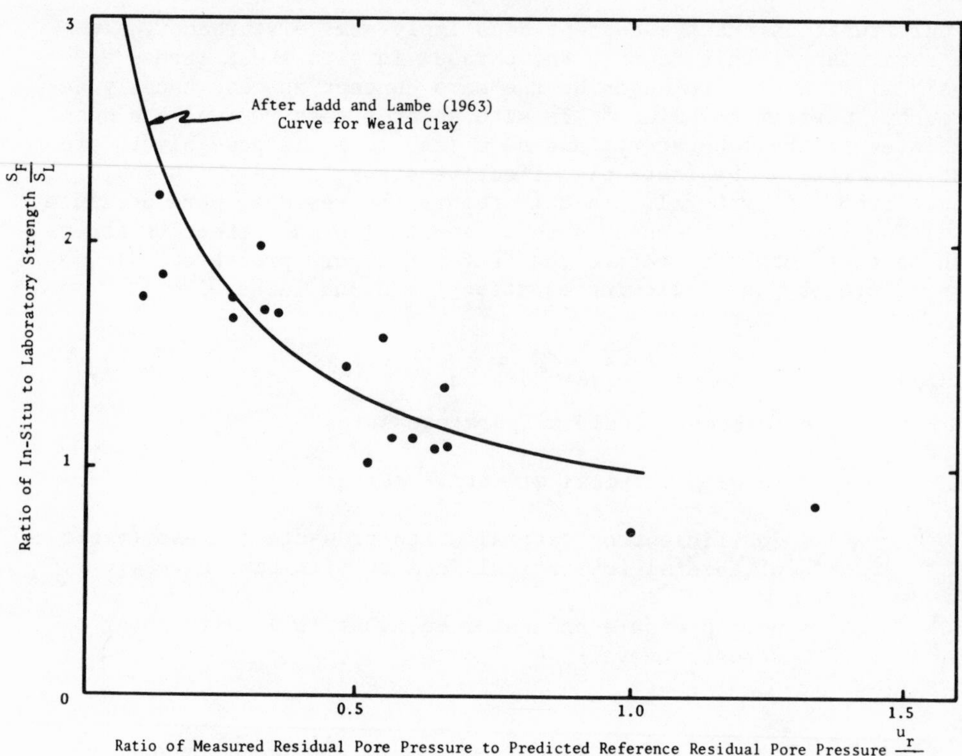

Figure 1. Correlation of data for predicting strength changes resulting from sample disturbance Santa Barbara Channel sediments

One direct evaluation of the correlation between vane strength and pore pressure ratios has been made to date (1973). The NCEL Deep Ocean Test-in-Place and Observation System (DOTIPOS), described in more detail by Demars and Taylor (1971), was used at three sites in the Santa Barbara Channel of California. This bottom-sitting platform can perform an in situ vane shear test and obtain a fixed piston core within about a meter (3 ft) of each other, each to a sub-bottom depth of 3.29 m (10 ft). Vane shear tests were run, and cores were obtained at each of the three sites. Laboratory vane shear and residual pore pressure tests were performed on the samples, and the reference residual pore pressures were estimated. The strength and pore pressure ratios (S_F/S_L and u_r/u_{ps}) were formed and plotted against each other (Fig. 1). The data correlated approximately and could be represented by a curve for Weald clay derived in a different manner by Ladd and Lambe (1963). It is necessary to obtain similar curves for deep-ocean oozes and clays by direct measurement. An investigation of this nature is planned.

Additional information is also needed on how in situ vane

shear strengths relate to actual foundation and anchor response. This information can be gained only through the monitoring of the behavior of installed structures or small scale models. Research of this nature is underway at NCEL.

One important conclusion which can be drawn from Figure 1 is that the laboratory vane strength can differ from the field vane strength by as much as a factor of 2 even if the sampler is designed to reduce disturbance. Without some correction for disturbance, the results obtained are either misleading or harmful, depending on the engineering application. In terms of direct embedment anchors, an undrained shear strength which is too low could lead to an unconservative estimate of penetration depth. This would be somewhat balanced by an overconservative estimate of holding capacity, and the final results might actually agree. However, it is dangerous to rely on compensating errors.

The literature contains other examples of disturbance correction procedures. In the area of consolidation, the Schmertmann (1955) procedure is often used to estimate the in situ compression characteristics. This procedure was applied by this author in analyzing cores obtained by the Deep-Sea Drilling Project Leg XIX (in Creager and Scholl et al., 1973). In triaxial testing, disturbance effects may be partially eliminated by consolidating specimens to pressures well above the in situ effective stresses prior to shear. The resulting strengths are obviously higher than the in situ strengths. However, the strength parameters, $\bar{c}$ and $\bar{\Phi}$, and the pore pressure parameter, $\bar{A}$, more closely approximate the field parameters than would parameters obtained from an unconsolidated, disturbed sample. This procedure is discussed in detail by Ladd and Lambe (1963) and serves as a partial basis for the next section of this paper.

STRENGTH VARIATIONS WITH SUB-BOTTOM DEPTH

Undrained shear strength almost always increases with sub-bottom depth as a result of increasing effective overburden stresses and decreasing water contents. Since anchor holding capacity varies directly with strength, it is usually desirable to obtain penetrations into this higher strength material. Unfortunately, it is often difficult to determine how rapidly the strength increases with depth. Because sample disturbance often increases with core length it is difficult to obtain a good, deep sample. The procedure for disturbance correction suggested by the preceeding section offers one means for obtaining an approximate estimate of the in situ undrained shear strength variation with sub-bottom depth. Another possibility is to obtain a high quality shallow core and subject it to a consolidated-undrained triaxial test with pore pressure measurements. The stress and water content characteristics of deeper sediments are simulated and since all results can be expressed in terms of effective stresses, an estimate of both drained and undrained (long-term and short-term) behavior can be obtained (Lambe and Whitman, 1969, p. 427). Disturbance effects are relatively well eliminated through the consolidation process (Ladd and Lambe, 1963).

Unfortunately, two major assumptions must be made in order to use triaxial test data to obtain strength distributions. First, it must be assumed that the basic sediment type remains relatively constant with sub-bottom depth. This may be true in many deep-ocean basins and, in any case, can be checked by obtaining a long, disturbed core. Second, it must be assumed that consolidation time effects are of secondary importance. Consolidation in the field may continue for millions of years whereas laboratory consolidation is accomplished in a few days or less. Some differences in behavior would be expected but it is not clear whether the in-situ material would be denser or looser than the simulated material. This is a very controversial topic and almost any point of view can be substantiated by at least some data. For the purposes of this paper, the tentative assumption will be made that consolidation time effects are indeed of secondary importance and that triaxial test data can be used to predict strength profiles. This assumption will be re-examined at a later date through an experimental program which will involve obtaining corrected strengths of long piston cores and consolidated triaxial strengths from short box cores.

Recently two spade-type box corer samples containing deep-ocean pelagic ("red") clay were obtained from the sea floor between Hawaii and California. Preliminary tests were performed which yielded vane shear strengths of 6.9 kPa (1 psi) and water contents of 150%. This strength is considerably higher than that previously reported in the literature for pelagic clay (Keller, 1967), probably because of the reduction in disturbance achieved by using a box corer. The sensitivity of the soil is high, averaging about 6.0. There was little variation in properties either over the .49 m (1.5 ft) length of the cores or between the cores (Lee, 1973b).

Five consolidated undrained triaxial shear tests with pore pressure measurements were performed using procedures described by Bishop and Henkel (1962) with modifications listed in Lee (1973a). The results are given as a series of stress paths in Figure 2. A stress path (Lambe and Whitman, 1969, p. 112) is a concise way of expressing effective stress information obtained from triaxial, loading tests. The ordinate of these paths is the quantity, $(\bar{\sigma}_1 - \bar{\sigma}_3)/2$, and the abscissa is $(\bar{\sigma}_1 + \bar{\sigma}_3)/2$. The parameter, $\bar{\sigma}_1$, is the major (usually vertical) principal effective stress and $\bar{\sigma}_3$ is the minor (usually horizontal) principal effective stress. The definition of the coordinates is such that each point on the stress path is the top of a Mohr circle representing a complete, axisymmetric state of stress. Each path represents the infinite number of stress states developed during the loading and failure of a triaxial test specimen.

The stress paths appear to converge relatively well upon the failure envelope shown in Figure 2, which requires a break to adequately define the stress paths. The paths to the left of the break display characteristics of overconsolidation (slight leftward path curvature and failure envelope with cohesion intercept) while those to the right indicate normal consolidation (strong leftward curvature and no cohesion intercept). One-dimensional consolidation tests performed on this material also indicated apparent overconsolidation at low stress levels with an apparent maximum past pressure of 28 kPa (4 psi). This sort of behavior has been noted by numerous sea-floor sediment researchers in the past and is apparently not a result of true overconsolidation but rather some currently unidentified form of interparticle bonding. The cause is not especially important from a practical point of view as long as it is noted that the sediment near the sea floor is much stronger than it would be if it were a typical normally consolidated material.

The strength parameters defining the two segments of the failure envelope are given in Figure 2. These could be used directly to estimate the long-term (drained) behavior of a direct embedment an-

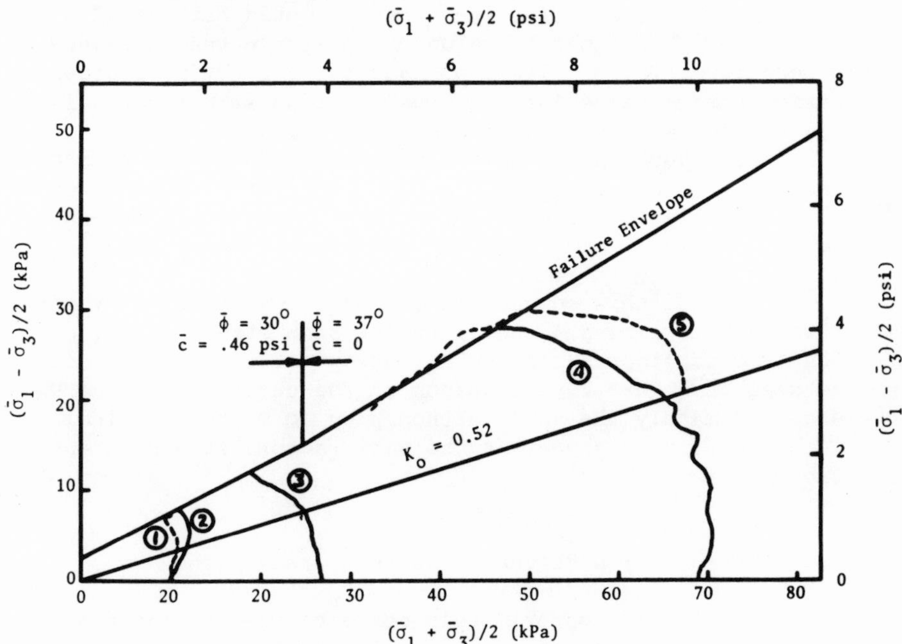

Figure 2. Stress paths resulting from triaxial testing of "red" clay

chor using methods given by Taylor and Lee (1972).

Obtaining an undrained (short-term) shear strength profile requires some additional analysis. Specifically, two items need to be obtained: (1) a means of evaluating the influence of anisotropic initial in situ stresses, and (2) a procedure for interpolating between the measured stress paths.

Influence of Anisotropic Initial Stresses

In situ, the horizontal effective stress seldom equals the vertical effective stress. Therefore, it would be desirable to consolidate samples anisotropically in the laboratory prior to shear and thereby more nearly correctly model the field situation. Unfortunately, anisotropic consolidation is difficult and time consuming. Instead, samples are usually consolidated isotropically, i. e., with all effective stresses equal. In this experimental study, four specimens were consolidated isotropically, and one was consolidated anisotropically under no lateral strain (K_o) conditions. The anisotropic test (Number 5 of Fig. 2) yielded a ratio of horizontal to vertical effective stress (K_o) of 0.52, a typical value. It was assumed that this value of K_o was applicable to all stress levels for this sediment, that is, that if the other samples had been consolidated with no lateral strain, the same ratio of horizontal to vertical effective stresses would have resulted. A line indicating this constant value of K_o is plotted on Figure 2. A common technique (Taylor, 1948, p. 383) for converting isotropic consolidation stress paths into K_o consolidation stress paths is to remove the portion of the stress path below the K_o line. The portion of the path above the line is then assumed to be the same as the stress path which would have been obtained for a specimen consolidated originally to the point at which the isotropic stress path crosses the K_o line.

The validity of this conversion technique was checked by performing two tests (Numbers 4 and 5) in the same stress range, one isotropically consolidated and one anisotropically consolidated. As may be seen in Figure 2, the shapes of the curves above the K_o line are approximately the same, although by no means identical. However, the comparison appears to be satisfactory as a good approximation.

Interpolation Between Measured Stress Paths

With the series of approximately correct stress paths available, it is next necessary to obtain an interpolation relationship for the variation of strength between the measured values. This may be done by using an expanded version of an equation provided by

Ladd (1965):

$$\frac{S_u}{\bar{\sigma}_{vo}} = \frac{[K_o + \bar{A}_f(1 - K_o)] \sin \bar{\phi} + (\bar{c}/\bar{\sigma}_{vo}) \cos \bar{\phi}}{1 + (2A_f - 1) \sin \bar{\phi}} \qquad (3)$$

where S_u = undrained shear strength

$\bar{\sigma}_{vo}$ = vertical effective stress during K_o consolidation (equal to field overburden pressure)

$\bar{A}_f$ = a pore pressure parameter

$\bar{c}$, $\bar{\phi}$, and K_o are as defined previously.

To use this equation it is necessary to determine a distribution for the pore pressure parameter, $\bar{A}_f$, as defined by Skempton (1954). This was done by obtaining $\bar{A}_f$ from the five triaxial test stress path segments above the K_o line and plotting it versus the assumed vertical consolidation stress, $\bar{\sigma}_{vo}$ (Fig. 3). A broken straight line interpolation between the data points is drawn with the break again representing a transition between apparent overconsolidation and normal consolidation.

The values of $\bar{c}$, $\bar{\phi}$, K_o, and $\bar{A}_f$ were substituted into Equation 3 to yield the strength, S_u, as a function of overburden pressure, $\bar{\sigma}_{vo}$. A one-dimensional consolidation curve was integrated to yield $\bar{\sigma}_{vo}$ as a function of sub-bottom depth. These two functions were then combined to yield the undrained shear strength, S_u, as a function of sub-bottom depth. This relation is shown in Figure 4.

The profile of Figure 4 represents the extent to which triaxial test data can be used to define behavior below the level of sampling. It has several characteristics which strengthen its credibility: the strength near the surface approaches the measured vane strength of 6.9 kPa (1 psi) and at depth the ratio of strength to overburden pressure, $S_u/\bar{\sigma}_{vo}$, is equal to about 0.3, a reasonable value for a clay. To progress beyond this level, soil from greater sub-bottom depths would need to be tested.

DISCUSSION AND RECOMMENDATIONS

In the preceeding sections, two example investigations were summarized to illustrate the role that laboratory testing can play in developing parameters for use in engineering analysis. Many assumptions were made in the development of those examples and additional work is clearly needed to evaluate these assumptions and to

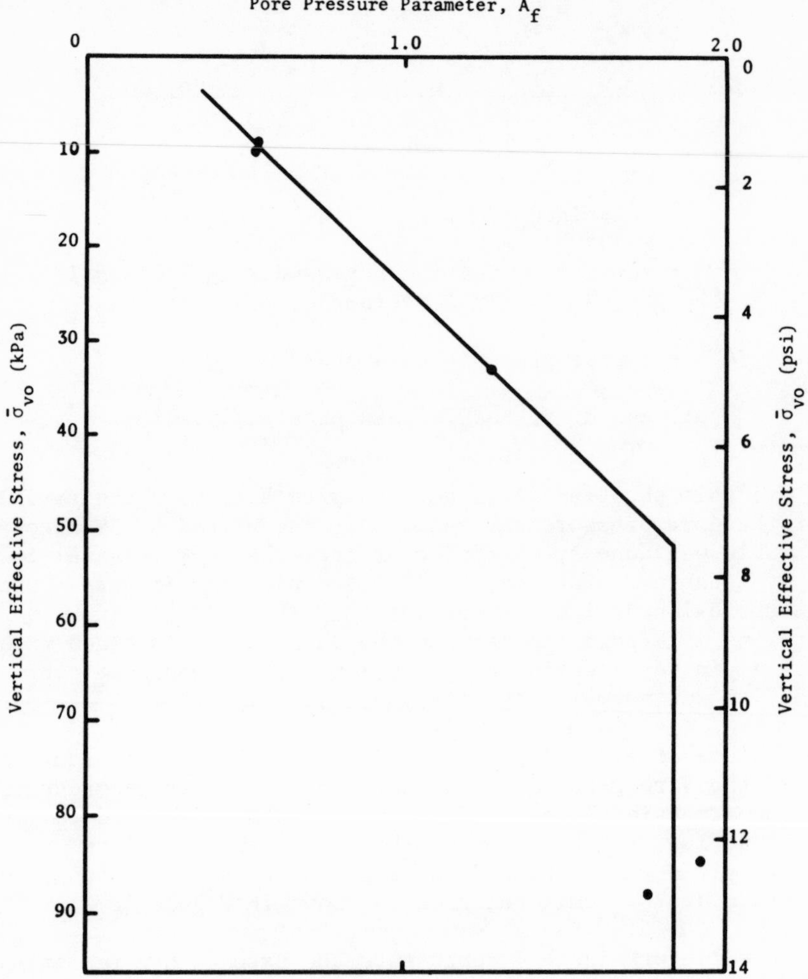

Figure 3. Variation of pore pressure parameter, $\bar{A}_f$, with vertical effective stress - triaxial testing of "red" clay

apply the procedures to different situations. The intention in presenting this work at this time was to indicate a direction toward which laboratory testing can be aimed to obtain information for use to engineers. The need to identify specific engineering property requirements for specific efforts is emphasized as is the need to evaluate the differences between laboratory and in situ response. Much earlier work, for example, vane tests on short, disturbed cores, density measurements, and some index property testing, does not adequately satisfy these needs. This is not to imply that simple tests do not have a proper place in sea-floor engineering. Indeed, most future sea-floor foundation and anchorage systems will probably be designed on the basis of simple tests and experience, as are most foundations on land at present. Currently, however, sufficient experience with deep-ocean sediments as engineering materials does not exist. In order to advance the state-of-the-art to the level at which experience can play a greater role, it is necessary to rationally evaluate the correct properties under the correct conditions. Index property and other simple tests also should be conducted so that ultimately empirical correlations and experience can suffice for most routine operations.

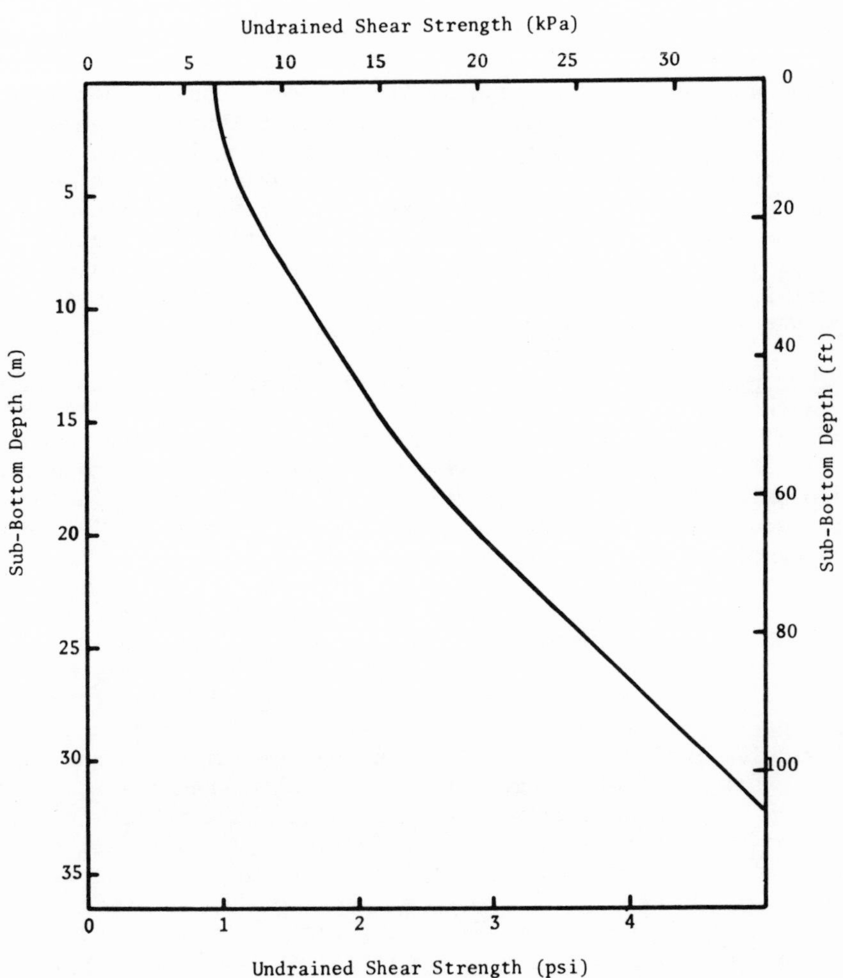

Figure 4. Estimated variation of undrained shear strength of a typical "red" clay deposit -- obtained from triaxial tests of short core samples

A number of appropriate research areas can be suggested on the basis of this framework, some resulting from the two examples and some relating to other fields. As an example, means for evaluating sample disturbance need to be expanded. One possible course is to develop curves such as Figure 1 for a variety of sea-floor sediments and then employ residual pore water pressure principles in correcting for disturbance. A coordinated effort involving both in-situ and laboratory testing is needed to achieve this goal. Other disturbance correction procedures are probably also possible. Properties other than undrained shear strength should be considered as well.

From the second example given in this paper, the major problem which arose was the simulation of geologic compaction through laboratory consolidation. The correctness of this simulation can be evaluated by selecting a relatively homogeneous deposit, accurately measuring the in-situ strength and density distributions, and comparing these with the results of triaxial and consolidation tests performed on shallow samples. If a poor comparison is obtained, a different procedure for evaluating property variations with subbottom depth will need to be developed. This procedure might include applying disturbance corrections to long piston cores, in-situ testing through the string of a drilling vessel, or developing an improved long corer for deep-ocean use.

Other problems which arose in the two examples were associated with the determination of pore pressure parameters (A_u and A_f) and the lateral pressure coefficient K_o. These parameters are needed to reduce test data into effective stress terms, which in turn form a bridge between short-term (undrained) and long-term (drained) behavior. Additional knowledge concerning these parameters is needed and can probably be gained through triaxial testing.

The two examples were directed toward predicting the holding capacity of direct embedment anchors. However, since the important properties obtained were drained and undrained shear strength, the work is applicable to other areas as well, e. g., bearing capacity of footings and piles and slope stability. Shearing characteristics not considered include strength loss during remolding, thixotropic strength regain following remolding, and dynamic shearing resistance during high speed penetration. Each of these characteristics has application to certain engineering activities. However, all of them represent complex problems which are not easily evaluated by any one form of test. For example, a laboratory sensitivity can be obtained relatively easily but it is not clear how this sensitivity relates to field problems. If the original strength of a typical pelagic clay is about 6.9 kPa (1 psi) and the laboratory sensitivity is 6.0, as noted previously, is the appropriate strength for use in dealing with bottom-crawling vehicles 1.15 kPa (1/6 psi)? There is little reason to believe that it would be; the vehicle almost certainly remolds in a manner different from a laboratory technician. Unfortunately, it is impossible to state with any degree of accuracy what the appropriate value should be. This and the other topics listed above require further research; uncertainty factors ranging into the hundreds of percent currently exist.

Consolidation has not been considered previously in this paper with the exception of its influence on shear strength. Since it relates to relatively small displacements as opposed to rupturing, sediment compression is probably not as significant as shear strength. Therefore, the principal research associated with consolidation should be directed toward its interrelationship with shear strength: for example, evaluation of K_o, determination of volume change-shear strength change relations, and comparison of consolidation during geologic time with laboratory consolidation.

Laboratory testing should play a major role in evaluating the various properties and behavioral characteristics discussed above. It offers a degree of flexibility, control, and economy which cannot be achieved through in situ testing alone. Clearly, in situ testing, possibly full-scale field testing, and monitoring of existing structures is also needed so that results can be adequately checked.

When a supply of good engineering property data is available, an attempt should be made to correlate the data with index properties, sediment type, and perhaps environmental characteristics. It should be possible ultimately to design routine facilities on the basis of these correlations.

As a final requirement, there is a need to develop improved techniques for analyzing certain soil mechanics problems which are significant in sea-floor engineering. Included are long- and short-term anchor holding capacity, trafficability, and penetration. Since sea-floor sediments often possess properties which change rapidly with sub-bottom depth, some existing procedures may need to be modified. For example, in evaluating the bearing capacity of a sea-floor footing or mat, it is often unclear as to what shear strength should be inserted into one of the standard equations.

In general, the major sea-floor sediment engineering requirements are in the area of strength determination, including both sub-bottom depth and time effects, and in the area of prediction technique development. These requirements should be met through the development of a cohesive research framework which has as its major goal the acquisition of information which is directly applicable to practical engineering problems.

REFERENCES

Bishop, A. W., and D. J. Henkel, The Measurement of Soil Properties in the Triaxial Test, Edward Arnold Ltd., London, 2nd edition, 1962.

Brooker, Elmer W., and H. O. Ireland, Earth pressures at rest related to stress history, Canadian Geotech. J., 11 (1), 1965.

Creager, J. S., D. W. Scholl et al., Initial Reports of the Deep Sea Drilling Project, 19, U. S. Govt. Printing Office, Washington, D. C., 1973.

Demars, K. R., and R. Taylor, Naval sea floor sampling and in-place equipment: a performance evaluation, U. S. Naval Civil Engineering Laboratory Tech. Rept. R-730, 1971.

Gibbs, H. J., and C. T. Coffey, Application of pore pressure measurements to shear strength of cohesive soils, U. S. Bureau of Reclamation Rept. No. EM-761, 1969.

Hironaka, M. C., and W. C. Green, A remote controlled sea-floor incremental corer, 3rd Ann. Offshore Tech. Conf. Preprints, Paper No. 1325, 1971.

Hvorslev, M. J., Subsurface exploration and sampling of soils for civil engineering purposes, U. S. Army Corps of Engineers, Vicksburg, 1949.

Hvorslev, M. J., Physical components of the shear strength of saturated clays, in Am. Soc. Civil Engrs. Res. Conf. Shear Strength of Cohesive Soils, Boulder, Colorado, 1960.

Keller, G. H., Shear strength and other physical properties from some ocean basins, in Civil Eng. in the Oceans, pp. 391-418, Am. Soc. Civil Engrs., N. Y., 1968.

Ladd, C. C., Stress-strain behavior of anisotropically consolidated clays during undrained shear, in Proc. 6th Intern. Conf. Soil Mech. and Fdn. Eng., 1, pp. 282-286, 1965.

Ladd, C. C., and T. W. Lambe, The strength of "undisturbed" clay determined from undrained tests, in Am. Soc. Test. Mat. Std. Tech. Publ. No. 361, pp. 342-371, 1963.

Lambe, T. W., Residual pore pressures in compacted clay, in Proc. 5th Intern. Conf. Soil Mech. and Fdn. Eng., 1, pp. 207-211, Paris, 1961.

Lambe, T. W., and R. V. Whitman, Soil Mechanics, John Wiley and Sons, N. Y., 1969.

Lee, H. J., In situ strength of sea-floor soil determined from tests on partially disturbed cores, U. S. Naval Civil Engineering Laboratory Tech. Note N-1295, 1973a.

Lee, H. J., Engineering properties of a deep sea brown clay, U. S. Naval Civil Engineering Laboratory Tech. Note N-1296, 1973b.

Rosfelder, A. M., and N. F. Marshall, Obtaining large, undisturbed and orientated samples in deep water, in Marine Geotechnique, edited by A. F. Richards, pp. 243-263, Univ. of Ill. Press, Urbana, 1967.

Richards, A., and H. W. Parker, Surface coring for shear strength measurements, in Civil Engineering in the Oceans, pp. 445-489, Am. Soc. Civil Engrs., N. Y., 1968.

Schmertmann, J. M., The undisturbed consolidation of clay, Trans. Am. Soc. Civil Engrs., 120, 1201-1227, 1955.

Skempton, A. W., The pore-pressure coefficients A and B, Géotechnique, 4 (4), 143-147, 1954.

Taylor, D. W., Fundamentals of Soil Mechanics, John Wiley & Sons, N. Y., 1948.

Taylor, R. J., and H. J. Lee, Direct embedment anchor holding capacity, U. S. Naval Civil Engineering Laboratory Tech. Note N-1245, 1972.

STRENGTH AND STRESS-STRAIN CHARACTERISTICS OF CEMENTED DEEP-SEA SEDIMENTS

VITO A. NACCI, WILLIAM E. KELLY, MIAN C. WANG, AND KENNETH R. DEMARS

University of Rhode Island

ABSTRACT

It has become increasingly apparent that the engineering properties of ocean sediments differ significantly from land sediments. The suspicion is that interparticle bonding may be the major cause of the difference.

Although the nature and stability of these bonds are unknown, various tests are made to infer the brittle nature of the bonding and indicate stress-strain-pore pressure parameters.

Consolidation and shear tests, as well as standard classification tests, have been carried out on boomerang and piston deep-sea cores. Consistent results have been obtained such that the low pore pressure parameter, A_f, and the high ratio of undrained shear strength to overburden pressure, c/p, are credible.

Recommendations as to sampling and laboratory testing for soft ocean sediments are made.

INTRODUCTION

For some time it has been apparent to the authors that the laboratory strength behavior of many deep-sea sediments differs significantly from that of apparently similar land sediments. While there are obvious differences in the sampling, transporting and storage techniques of ocean and land sediment cores, not all the behavioral discrepancies can rightly be attributed to disturbance factors. Some progress has been made in studying the mechanisms controlling the shear strength of deep-ocean sediments and future plans include consideration of the effects of chemistry, temperature and pressure upon engineering behavior.

The U. S. Navy recently took about 18.3 m (60 ft) of piston and boomerang cores in over 3658 m (12,000 ft) of water in the North Atlantic. The cores were transported in maximum lengths of 1.52 m (5 ft) using standard procedure to maintain the in situ temperature at 4°C and minimize vibrational disturbance. Conventional laboratory soil analyses including vane shear, fall cone penetrometer, grain size, index properties, triaxial compression and consolidation were performed at the University of Rhode Island's Geotechnical Laboratory. Unfortunately, the test program was planned at U. R. I. with little information as to its intended use as a foundation material - information essential for establishing a realistic test program.

The results of published laboratory consolidation tests (Noorany, 1971; Bryant et al., 1967; Richards and Hamilton, 1967) on sea-floor soil samples suggest that the consolidation behavior of deep-sea soils follows a pattern which may be attributed in part to cementation. The general consolidation behavior of sea-floor soils will be discussed using actual laboratory data to demonstrate this behavior. Degree of disturbance of samples obtained from gravity, piston and other oceanographic core samples appears to be predictable from consolidation tests.

Unfortunately, adequate strength data are not available for soft marine soils and are much needed for an estimate of foundation bearing capacity, anchor and pile resistance, and foundation scour. Most strength testing has been confined to vane shear and cone penetrometer techniques which provide only an estimate of the undrained shear strength. Few triaxial compression tests have been performed on marine soils to obtain the effective strength parameters, $\bar{c}$ and $\bar{\phi}$, and the stress-strain behavior. In addition, triaxial test data may be used to infer soil structure, cementation and stress history. In this paper, the general strength behavior of a cemented deep-ocean sediment will be discussed using actual laboratory data.

SOIL STRUCTURE

In order to understand the strength and stress-strain behavior of a deep-sea sediment, it is necessary to consider the sedimentation processes and soil structure. Submarine sediments are comprised mainly of pelagic materials and biogenic remains, with nearshore deposits containing increasing amounts of terrigenous material. Pelagic material originates in the water column and settles slowly to the sea floor. Although the resulting sedimentation rates are miniscule, they have produced an average sediment thickness of about 1 km in the Atlantic (Ewing et al., 1973).

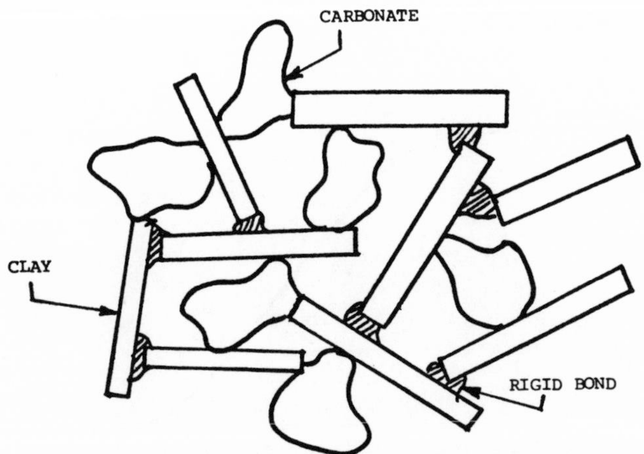

Figure 1. Cemented flocculant structure

The combination of (1) an extremely slow rate of deposition, given as 1-1.5 cm per 1000 years for the Atlantic (Ewing et al., 1973), (2) a very small particle size with a medium diameter less than 1 micron, (3) the presence in solution of silica, potassium, calcium, carbonate, iron and manganese, and (4) a saline environment, encourages the formation of a cemented flocculant soil structure. There is some evidence from published vane shear and cone penetrometer values (Richards, 1962; Keller, 1968) that cementation is present in some deep-sea sediments. Figure 1 shows a possible structure of deep-sea sediments containing clay particles and clay-

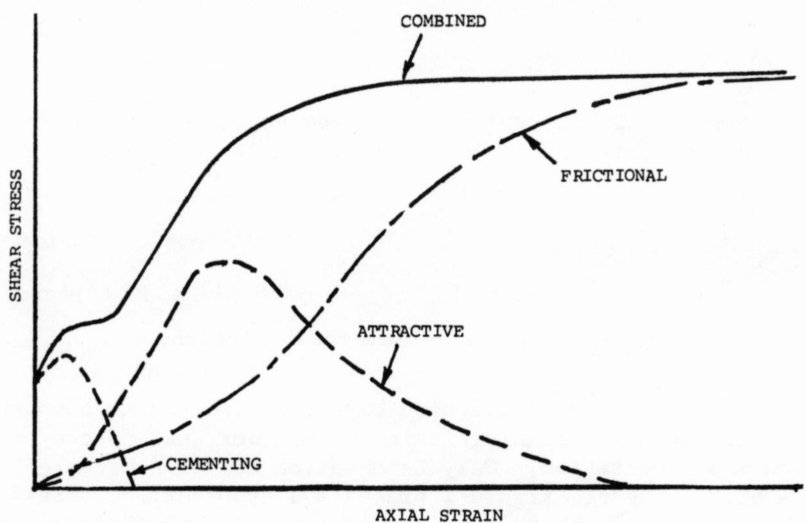

Figure 2. Components of shear resistance

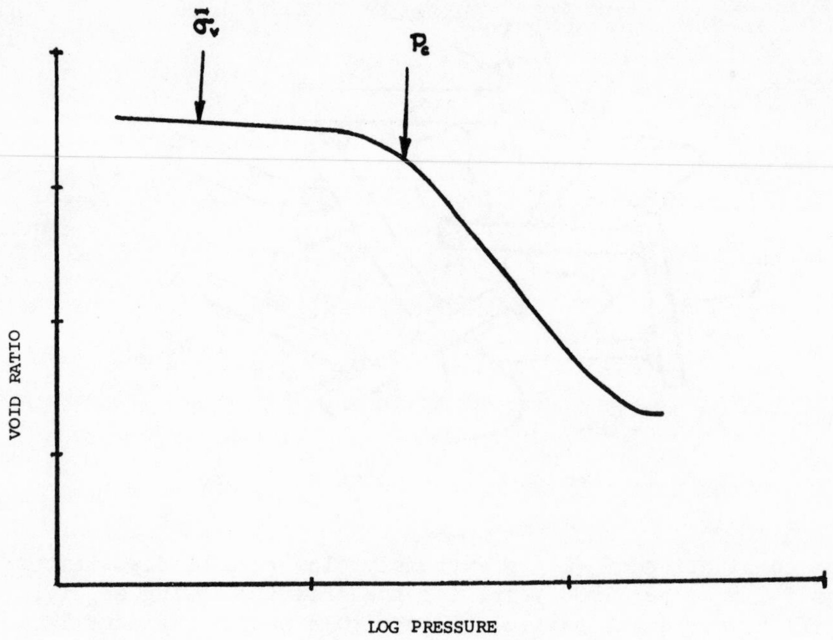

Figure 3. Apparent overconsolidation

size carbonates. If cementing is present, it may be inferred that cementing materials precipitate around points of contact resulting in rigid bonding.

SHEAR STRENGTH

The shear strength of this soil may be expressed by the Mohr-Coulomb relationship:

$$s = \bar{c} + \bar{\sigma} \tan \bar{\phi}$$

where $\bar{c}$ = cohesion

$\bar{\sigma}$ = effective normal stress on the plane of failure

$\bar{\phi}$ = effective angle of internal friction.

Following Lambe (1960) cohesion may be considered as composed of several components including ionic and van der Waal forces of attraction and cementation. Only cementation may be considered to promote a brittle bond. Figure 2 presents a hypothetical stress-strain relationship demonstrating the contribution of three major components, i. e., cementation, attraction, and intergranular friction, to the shear resistance. The stress-strain relationship shows that the strength at low strain is chiefly dependent upon cohesion, namely attraction and cementation, at the contact points. At greater strains the shear resistance is increasingly frictional in nature. Cementation, when present, may also be manifested as an increased $\bar{\phi}$.

Most significant is that the shear resistance contributed by cohesion (cementation), $\bar{c}$, and intergranular friction, $\bar{\phi}$, mobilizes and decays at different rates and that the maximum shearing resistances contributed by $\bar{c}$ and $\bar{\phi}$ occur at different values of strain.

INDICATORS OF RIGID BONDING

Laboratory tests on cemented clays show behavior very different from non-cemented clays of similar mineralogical composition, plasticities and porositities. A method proposed by Kenney (1967) to evaluate cementation is to dissolve the cementing agent and compare the behavior of the sample before and after treatment. However, when the suspected cementing agent is calcium carbonate, iron or silica and it occurs in large quantities, its removal would completely destroy the sediment fabric.

Alternative methods may be used to infer cementation without modifying the sediment structure. These include the following tests.

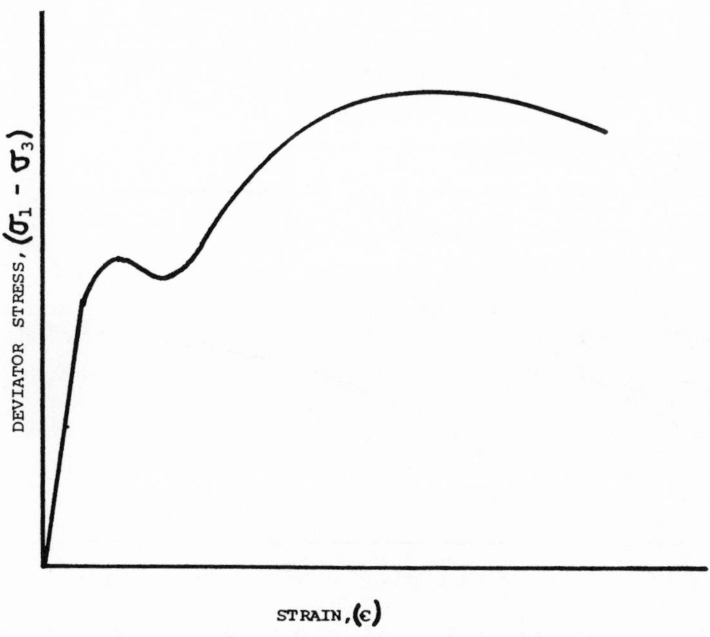

Figure 4. Influence of cementation on stress-strain relationship

Consolidation Test

For normally consolidated soils, i.e., at applied stresses equal to or greater than the effective overburden pressures, $\bar{\sigma}_v$, the increase in effective stress results in an increased resistance to deformation. The familiar curve of void ratio versus log-effective-stress virgin compression results. However, nearly all consolidation test results for deep-sea sediments indicate an apparent preconsolidation pressure, p_c, which exceeds the effective overburden pressure, $\bar{\sigma}_v$, as illustrated in Figure 3. Possible causes for the resulting preconsolidation pressure include preconsolidation and cementation. Since preconsolidation in the deep-sea sediments is unlikely although not entirely impossible, it is concluded that the presence of an apparent preconsolidation pressure greater than the effective overburden pressure is the result of cementation.

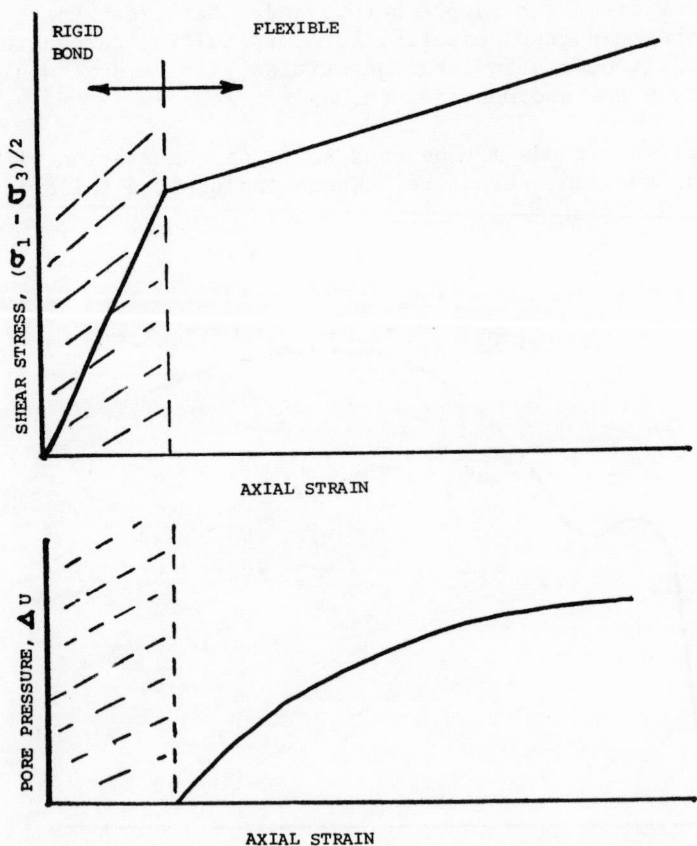

Figure 5. General stress-strain and pore pressure strain response of cemented soil. Note behavior of rigid and flexible portions of soil.

Triaxial Compression Tests

Stress-strain Characteristics. As already mentioned, shear resistance at very low strains is dominated by cementation (Fig. 2). When cementation is present, due to its brittle nature, the strain will steadily and sharply increase with increasing stress until the bonding is ruptured. Consequently, a peak stress (Fig. 4) at very small strain appears. Sangrey et al., (1969) studied undisturbed Canadian quick clay and substantiated this result. Therefore, this stress-strain characteristic at very small strain range can be used to indicate the presence of cementation.

Pore Pressure. The strength of a cemented clay may be attributed to a structural rigidity that reaches a maximum value at low strain and then rapidly decays. The frictional component of strength increases at a slow rate reaching its maximum value only at large strains. One should expect dramatically different stress-strain and pore pressure-strain relationships for each component of strength.

Figure 5 shows the general expected behavior of shear stress with strain and pore pressure with strain, indicating a very rapid strength increase while rigid bonds are intact. Before the rupture of rigid bonds, strain is primarily due to deformation of individual particles, therefore very little excess pore pressure will be induced. As strain increases and the cementing bonds break and slip, the cementation component decays rapidly while the dilational and frictional strength continues to increase (Fig. 2). Most noteworthy is the rapid increase in induced pore pressure at strains beyond the peak stress due to cohesion. During this stage there is an increasing tendency for the soil volume to decrease or, in an undrained test, for the pore pressure to increase. The maximum structural resistance is mobilized at small strain, less than 2%, whereas the interparticle sliding resistance may increase for strains of 5 to 15%.

Other Parameters. Table 1 summarizes other soil parameters that may be used to indicate the presence of cementing bonds. The liquidity index, $LI = (W - PL)/(LL - PL)$, in which W is the natural water content, LL the liquid limit and PL the plastic limit, is widely used to establish states of comparable consistencies. For an uncemented normally consolidated single grain structure, a maximum liquidity index of 1 may be expected. With cementing and a salt water flocculated soil structure, higher liquidity indices are likely suggesting that a cemented structure is capable of sustaining a high porosity.

Vane shear and fall cone penetrometer values are often used to indicate the undrained shear strength of marine clays. Since the shear strength characteristics of cemented soils depend upon structural factors, both the arrangement of particles and the nature of

TABLE 1

INDICATORS - CEMENTING BOND

TEST	RESULTS	
Atterberg Limits	Liquidity Index	> 1.5
Vane Shear, Cone Penetrometer, etc.	Shear Strength, C_u	> 13.79 kPa (2 psi)
	Sensitivity	> 4
Consolidated Undrained Triaxial Test	Ration Undrained Shear Strength to Effective Overburden Pressure, c/p	> 0.5
	Pore Pressure Parameter, A_f	< 0.5

the interparticle forces will have a great effect upon the undrained shear strength. Therefore, near-surface undrained strength values exceeding 13.79 kPa (2 psi) suggest cementation.

When consolidated undrained triaxial compression tests with pore pressure measurements are performed, two important parameters are usually determined: The ratio of undrained shear strength to the consolidation pressure, c/p, and the pore pressure parameter at failure, A_f. Undisturbed normally consolidated clays may have values of $c/p \simeq 0.25$ and $A_f \simeq 1.0$. When these parameters are: $c/p \geq 0.5$ and $A_f \leq 0.5$, one may conclude that the sample is either overconsolidated or the sample is greatly disturbed. However, by its very nature, an undisturbed cemented flocculant soil would also exhibit anomalous values for c/p and A_f.

To summarize, most of the factors used to indicate cementing of deep-sea soils may also infer overconsolidation or disturbance. However, with a knowledge of the geologic history it is believed

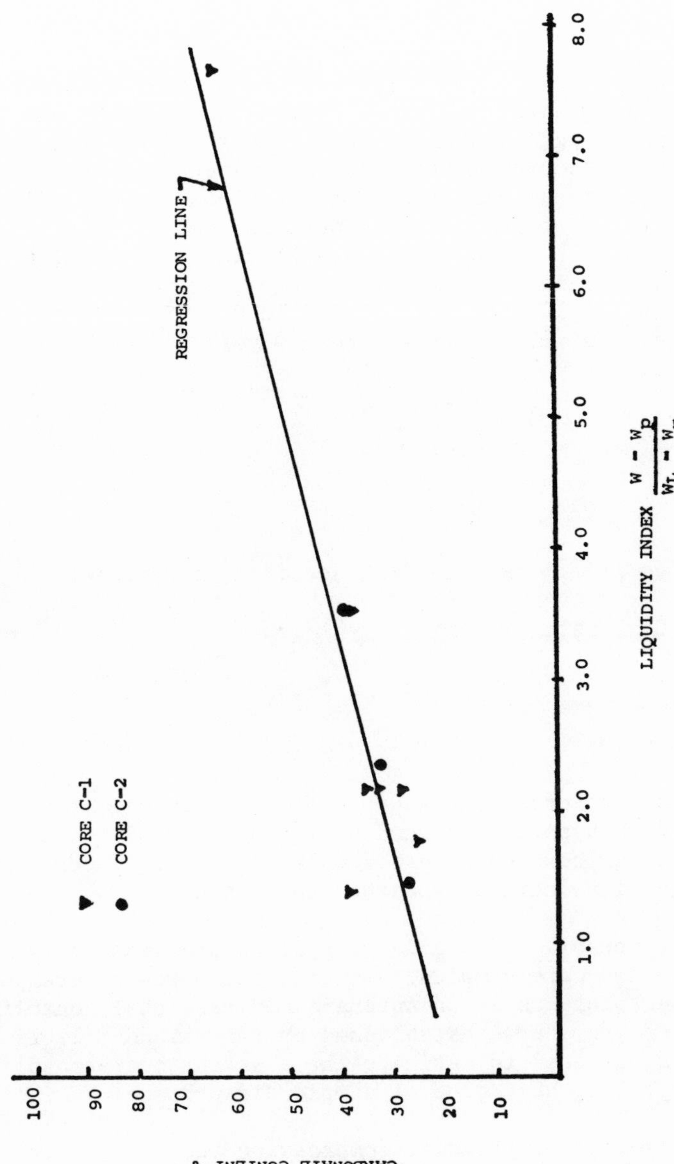

Figure 6. Relation between liquidity index and carbonate content

that cementation may frequently explain the engineering behavior of marine soils.

TEST MATERIAL AND TEST PROCEDURES

The U. S. Navy's 18.3 m (60 ft) of deep-sea cores were subjected to routine laboratory analyses. The tests included the determination of unit weight, water content, index properties, grain size distribution, carbonate content and undrained shear strengths and sensitivities from the vane shear and fall cone tests. Geotechnical data for two of the cores to be discussed in this paper are reported in Table 2. These cores are considered typical of a deep-sea sediment. It is interesting to note that much of the clay and silt size material is carbonate as can be seen in this breakdown from a typical wet sieve grain size analysis:

Sand Size	=	18.9%
Silt and Clay Size	=	81.1%
Total	=	100.0%

Treatment of these fractions with 0.3N acetic acid revealed:

Sand Size Carbonates	=	17.7%
Silt and Clay Size Carbonates	=	18.5%
Total Carbonates	=	36.2%

This breakdown affords some support for the cemented flocculated clay structure hypothesized and shown in Figure 1. Another interesting trend is shown in Figure 6 where liquidity index was found to increase with increasing carbonate content.

In addition, consolidation and triaxial compression tests were performed on selected core samples. Consolidation tests were performed without backpressure in an Anteus one-dimensional consolidometer according to procedures established by Parker and Miller (1970). The data was used to estimate the apparent over-consolidation pressure, p_c, and the degree of sample disturbance.

Approximately fifteen triaxial compression tests were performed according to procedures established by Andersen and Simons (1960). Most tests were performed at confining pressures slightly greater than the effective overburden pressure, $\bar{\sigma}$, as recommended by Kenney (1967) but less than the apparent Casagrande preconsolidation pressure to minimize structural disturbance as suggested by Sangry et al. (1959). Samples were back-pressured to 196 kPa

TABLE 2

GEOTECHNICAL DATA ON MARINE SEDIMENT

Labrador Basin - Approximately 54° N Lat., 44.6 W Long. (Cores C-1, C-2); Depth 3700 m, 6.1 m Penetration	
Clay 2 Micron	30-60%
Silt	30-40%
Sand	20-30%
Natural Water Content	70-200%
Activity (PI/% Clay)	.6-1.5
Liquidity Index	1.5-8.0
Clay Minerals	Illite, Chlorite, Kaolinite, (Montmorillonite)
Carbonate Content	25-65%
Vane Shear	3.83-11.01 kPa (80-230 psf)
Cone Penetrometer	2.87-13.41 kPa (60-280 psf)
Sensitivity	3-6
C/P	.5-.7
A_f	0.3-0.6
$\bar{c}$	2.07-6.89 kPa (0.3-1.0 psi)
$\bar{\phi}$	.54-.64 rad (31°-37°)

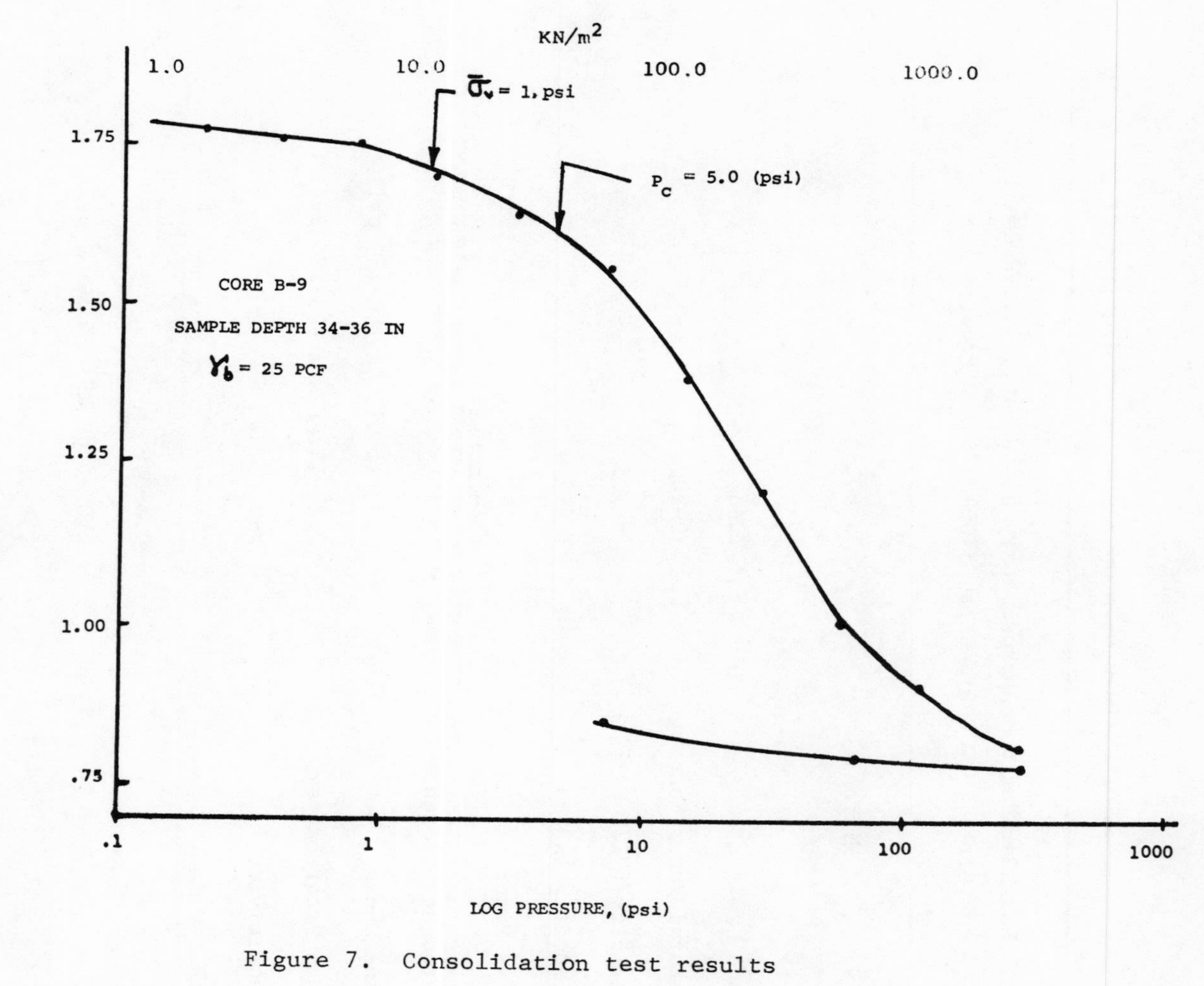

Figure 7. Consolidation test results

($2\ kg/cm^2$) to insure saturation. Pore pressures were measured with a null indicator-type device.

RESULTS

Consolidation Tests

Figure 7 presents the compressibility curve that is typical of six other consolidation tests. The apparent preconsolidation pressure as determined by Casagrande's technique, is p_c = 34.47 kPa (5.0 psi) and the vertical effective overburden pressure is $\bar{\sigma}_v$ = 6.89 kPa (1.0 psi). This apparent increase in overburden pressure of 27.58 kPa (4.0 psi) is attributed to cementation. The possibility that approximately 7.32 (24 ft) of overburden had been removed prior to sampling is not considered a likely alternative.

Sampling disturbance has received little attention since the work of Hvorslev (1949) in the forties. Most researchers have attempted to develop new sampling tools using Hvorslev's design criteria rather than consider the degree of disturbance to samples taken with the conventional geologic samplers. We believe that consolidation test results may be used to measure degree of disturbance. Undisturbed and completely remolded void-ratio versus log-

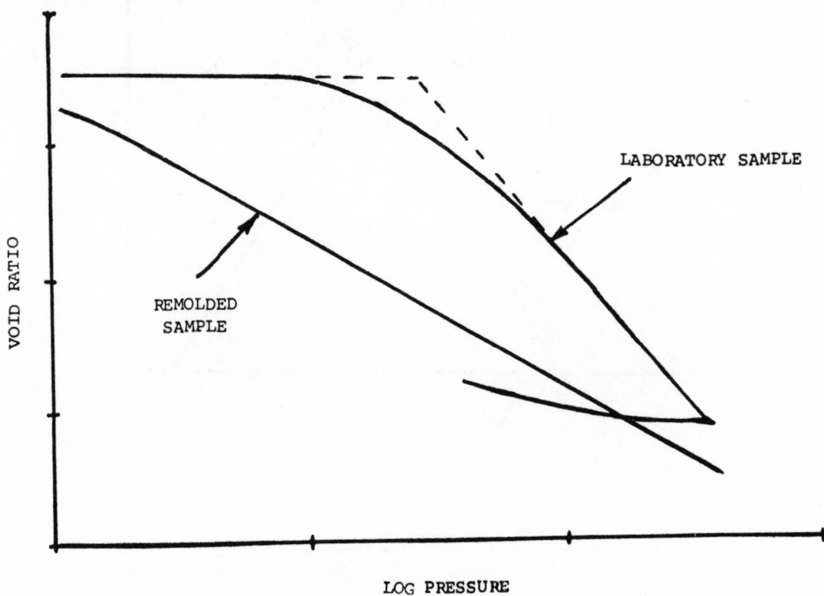

Figure 8. Consolidation -- sample disturbance

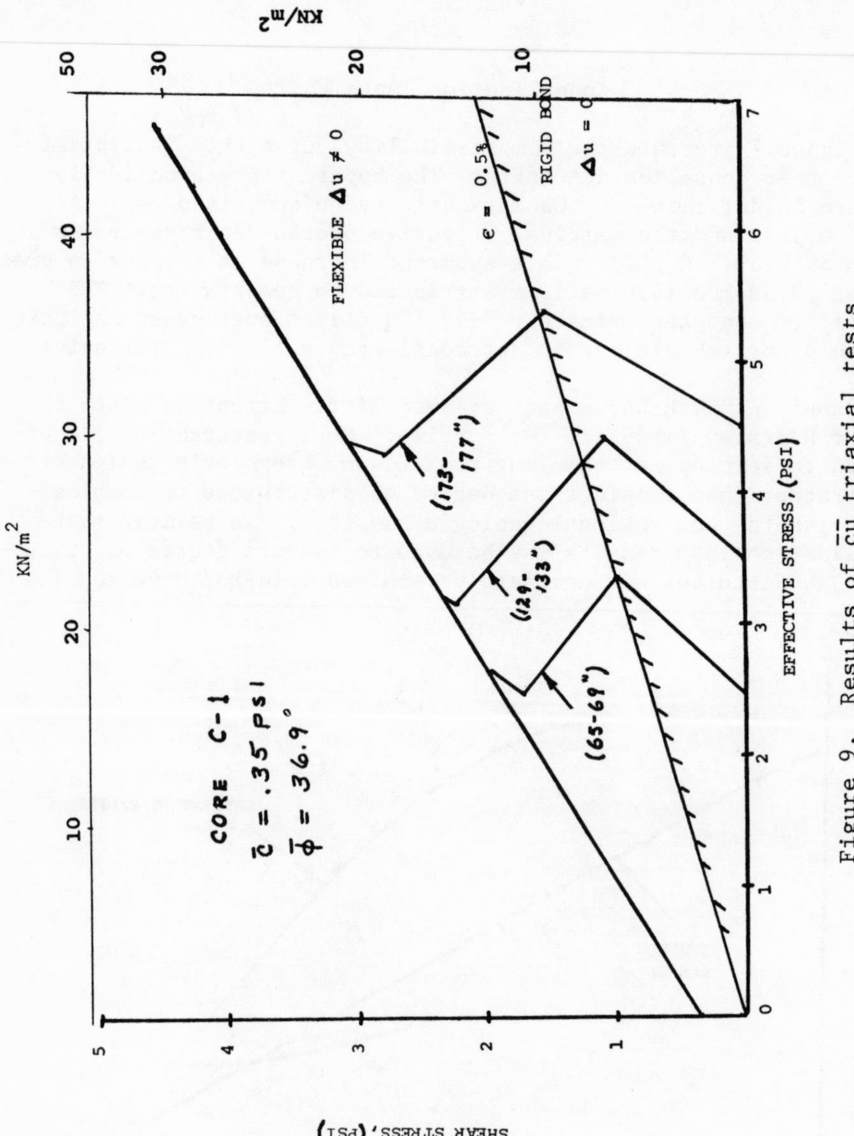

Figure 9. Results of $\overline{cu}$ triaxial tests

pressure curves are shown in Figure 8 for laboratory consolidation tests. Ideal test results are shown by the dotted lines. With this in mind, it is only necessary to locate the laboratory consolidation curve with respect to the ideal and remolded consolidation curves as a measure of degree of disturbance. Empirical techniques are available to establish the remolded curve for terrestrial soils and only need be established for marine clays.

Consolidated Undrained Compression Tests

For the determination of drained and undrained strength parameters a number of triaxial tests were performed. Samples were consolidated approximately to a pressure between the effective and the apparent preconsolidation pressure.

The results are plotted in the conventional Mohr-Coulomb plot, shear stress versus effective normal stress on the failure plane. The range of effective strength parameters for all samples were: cohesion, $\bar{c}$, 2.07 to 6.89 kPa (0.3 to 1.0 psi) and angle of internal friction, $\bar{\Phi}$, .54 to .64 rad (31° to 37°). Figure 9 shows the results of consolidated undrained tests on core C-1 with samples obtained from depths of 1.65 m (65 in), 3.28 m (129 in), and 4.39 m (173 in). This plot requires the assumption of vertical homogeneity within the core.

A study of the effective stress paths shows graphically the response of the soil structure to strain. Up to 0.5% strain, all of the applied stresses were resisted by the intact structure since the change in pore pressure is zero and the total and effective stress paths coincide. With further strain, particle cementation begins to break down and some of the applied stress is carried by the pore water. During this stage the effective stress path departs from the total stress path. The overall A-factor is 0.5. Some insight is gained by noting that the A-factor is zero during strain less than 0.5% and is 1.2 at a strain of approximately 5.0%.

The high effective angle of internal friction, $\bar{\phi}$ = .64 rad (37°), is mobilized at large strains and indicates the frictional capacity of a carbonate soil matrix. According to Lambe (1960) high friction angles may be due to agglomeration of soil particles as a result of cementation.

The behavior of one typical triaxial test at low strain is shown again in Figure 10. At low strain the applied stress $(\sigma_1 - \sigma_3)$ is carried by the rigid interparticle bond and the induced pore pressure is negligible. At strains greater than 0.5%, rigid bonds are destroyed and the shearing resistance decreases as the soil structure deforms with increasing pore water pressure.

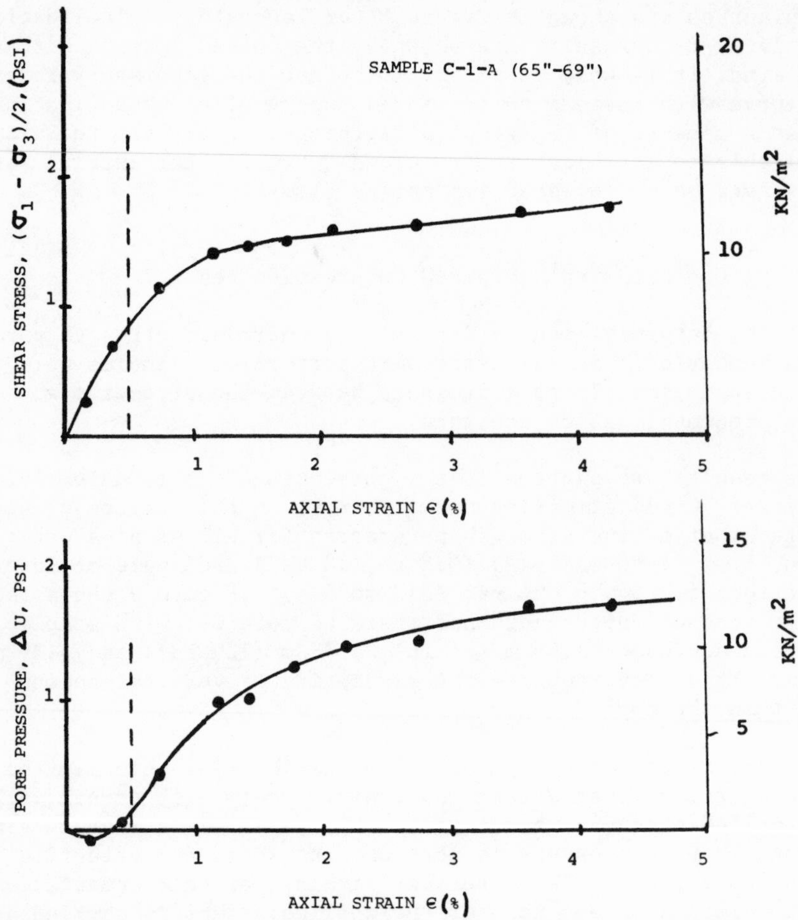

Figure 10. Actual stress-strain and pore pressure response of carbonate clay

Vane Shear

Vane shear and fall cone penetrometer tests were performed on all samples. Figure 11 shows the vane shear value versus overburden pressure. By regression analysis one gets approximately 4.79 kPa (100 psf) near the soil surface with an increase in vane shear strength of 22% with depth pressure. For near surface cemented soils the undrained vane shear strengths are primarily due to cementation. To test this theory, the undrained shear strengths at 0.5% strain of the samples shown in Figure 9 have been plotted with excellent correlation. This data is admittedly limited but it does encourage our efforts to separate the rigid and flexible strength contributions.

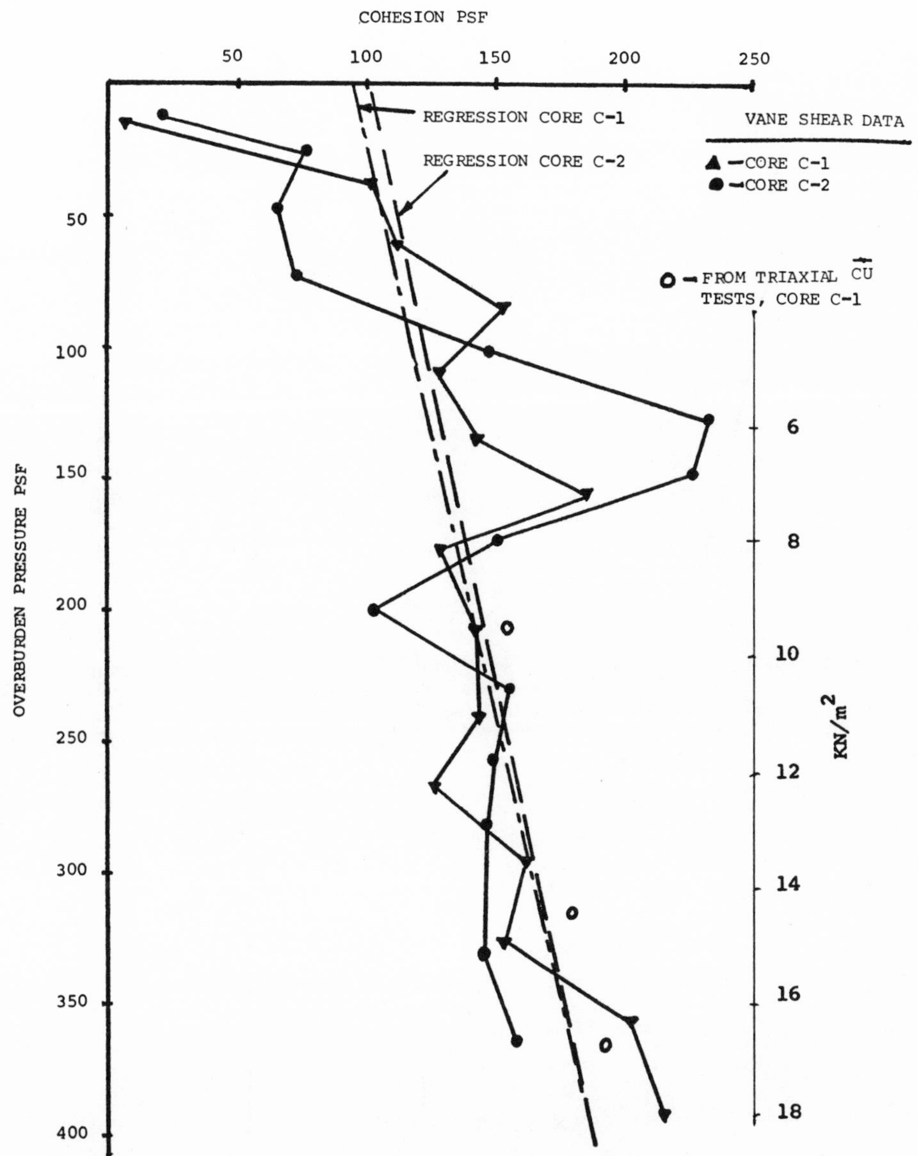

Figure 11. Vane shear strength, triaxial cohesion versus depth relationship

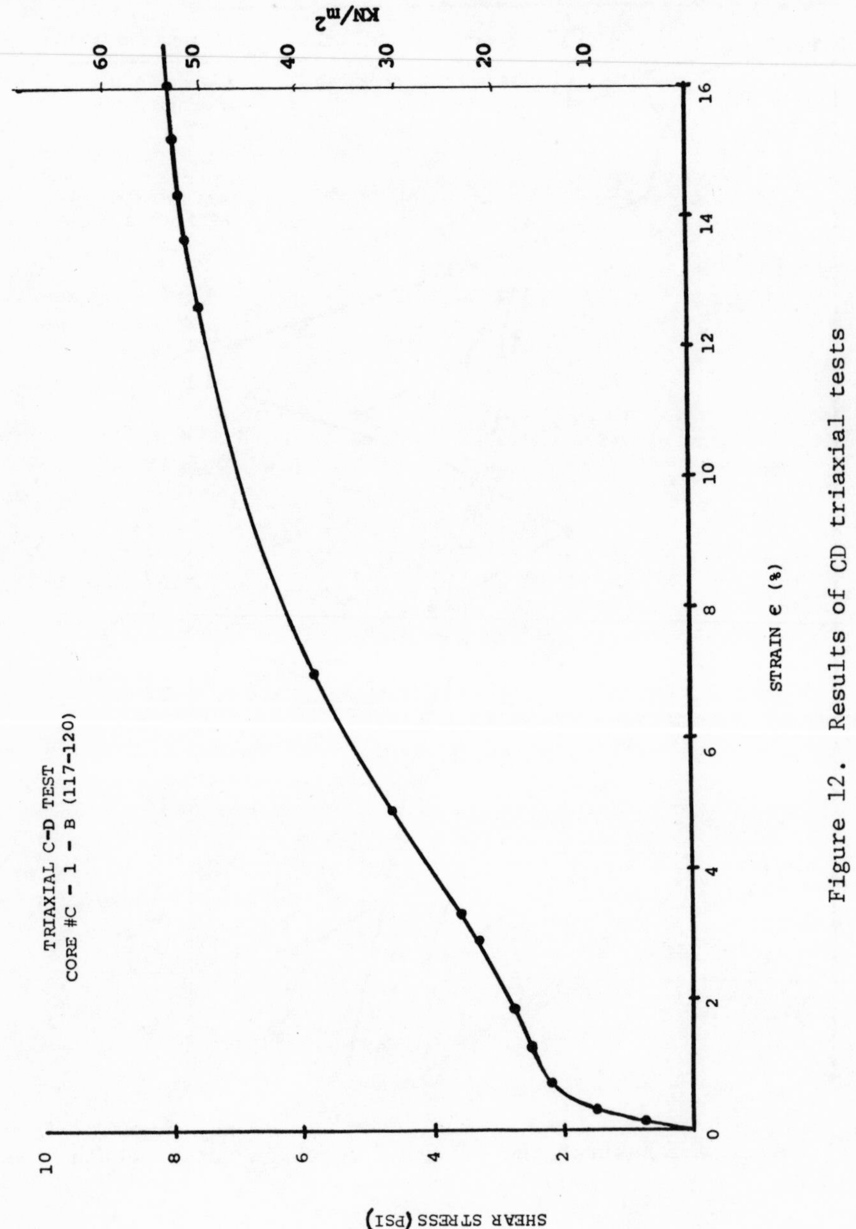

Figure 12. Results of CD triaxial tests

Consolidated Drained Triaxial Test

Very few drained shear strength tests were performed; however, they did confirm the strength parameters, $\bar{c}$ and $\bar{\phi}$, obtained from undrained tests. Figure 12 is included here to show the definite division in stress-strain behavior at 0.5% strain. This tends to confirm the rigid and flexible response theory.

SUMMARY AND RECOMMENDATIONS

A review of submarine soil mechanics shows a paucity of data on strength properties and some lack of agreement in the published data. This paper attempts to show that some deep-sea soils have an inherent, although limited, strength capable of sustaining structural integrity so long as attempts are made to minimize disturbance. In view of the costs of obtaining and transporting deep-sea cores, maximum information can be obtained from carefully conducted back-pressured consolidated undrained compressions tests with pore pressure measurements. Further, as noted in Figure 8, if the sample is consolidated to its overburden pressure and not the apparent overburden pressure, additional disturbance will be minimized. Our philosophy is that conventional piston and gravity cores are capable of providing the necessary engineering properties by adhering to these principles, and using the approach that it is the behavior of the sample during testing that is important.

One of the purposes of the seminar was to provide guidance to the Office of Naval Research as to the future needs and priorities for research on engineering properties of ocean sediments. Table 3 summarizes, in decreasing order of importance, some recommendations for a national ocean soils program. Greatest emphasis is placed on the need to improve the sensitivity of testing equipment, particularly in the very low confining pressure range. Research on the effects of temperature and the nature of interparticle bonds also is needed. Considering the large expenditure in money and time in order to procure samples, it seems reasonable that high priority be given to developing a standardized method to evaluate core disturbance. Finally, new methods and instrumentation that may be more adaptable to the study of the behavioral aspects of sediments under load should be investigated.

TABLE 3

RECOMMENDATIONS FOR DEVELOPMENT - LABORATORY TESTING

1. Triaxial Compression

   a. Load Cell and Pore Pressure Transducers (Increment .69 $\pm$ .14 kPa or .1 $\pm$ .02 psi) back pressure

   b. Reduction of Vibration Associated with Testing Equipment

   c. Consolidate to Pressure Less Than the Apparent Overburden Pressure

   c. Preferably Use Anisotropic Consolidation Pressures

2. Temperature Corrections

   a. Transportation and Storage

   b. Testing - Research Needed on Influence of Temperature on Engineering Properties

3. Physico - Chemistry of Ocean Sediment

   a. Theoretical and Experimental Research Needed

   b. Osmotic Effects During Testing

4. Degree of Disturbance

   a. X-Ray (Radiography)

   b. Consolidation Tests

   c. Stress Strain Corrections

   d. Electron Microscopy - Soil Fabric

5. Development of Other Instrumentation

   a. Conductivity

   b. Stress-Strain "Noise"

   c. Shear Wave Propagation

ACKNOWLEDGMENTS

The authors are grateful for the core samples and support provided by Naval Facility Engineering Command, Chesapeake Division and the Naval Underwater Systems Center.

The laboratory analyses of the cores were coordinated by M. Huston, F. Stevenson, C. Katesetos, R. Michniewica, P. L. Tan, M. Bennett and K. Soccia assisted with the laboratory testing.

REFERENCES

Andersen, A., and N. E. Simons, Norwegian triaxial equipment and technique, in Proc. Res. Conf. Shear Strength of Cohesive Soils, Am. Soc. Civil Engrs., N. Y., 1960.

Bryant, W. R., P. Cermock, and J. Morelock, Shear strength and consolidation characteristics of marine sediments from the western Gulf of Mexico, in Marine Geotechnique, edited by A. F. Richards, pp. 41-62, Univ. of Ill. Press, Urbana, 1967.

Ewing, M., G. Carpenter, C. Windish, and J. Ewing, Sediment distribution in the oceans: the Atlantic, Geol. Soc. Am. Bull., 84, 77-87, 1973.

Hvorslev, M. J., Subsurface exploration and sampling of soils for civil engineering purposes, U. S. Army Corps of Engineers, Vicksburg, 1949.

Keller, G. H., Shear strength and other physical properties of sediments from some ocean basins, in Civil Eng. in the Oceans, pp. 391-418, Am. Soc. Civil Engrs., N. Y., 1968.

Kenney, T. C., Shear strength of soft clay, in Proc. Geotech. Conf. pp. 49-55, Norwegian Geotech. Inst., Oslo, 1967.

Lambe, T. W., A mechanistic picture of shear strength in clay, in Proc. Res. Conf. Shear Strength of Cohesive Soils, pp. 555-580, Am. Soc. Civil Engrs., N. Y., 1960.

Noorany, I., Engineering Properties of submarine calcareous soils from the Pacific, in Proc. Intern. Symp. Properties of Sea Floor Soils and their Geophys. Ident., pp. 130-139, Univ. of Washington., Seattle, 1971.

Parker, H. W., and D. G. Miller, Jr., Operation manual for model A

Anteus back pressure consolidometers, Lehigh Univ., Marine Geotechnical Laboratory Int. Rept., 1970.

Richards, A. F., Investigations of deep-sea sediment cores, U. S. Navy Hydrographic Office Tech. Rept. 106, 1962.

Richards, A. F., and E. L. Hamilton, Investigations of deep-sea sediment cores, III. Consolidation, in Marine Geotechnique, edited by A. F. Richards, pp. 93-117, Univ. of Ill. Press, Urbana, 1967.

Sangrey, D. A., and C. L. Towsend, Characteristics of three sensitive Canadian clays, Dept. of Civil Eng. Res. Rept. 63, Queen's Univ., Ontario, 1969.

AN ANALYSIS OF THE VANE SHEAR TEST AT VARYING RATES OF SHEAR

NEIL T. MONNEY

U. S. Naval Academy

ABSTRACT

Vane shear tests were performed at varying rates of shear on three types of cohesive sediment: a clayey silt, a calcareous ooze, and a red clay. The shear strength was found to vary significantly for the range of shear rates typically used by researchers and practicing engineers. A standard rate of 0.0262 rad/s (90 deg/min) is recommended.

INTRODUCTION

The great majority of laboratory and in situ measurements to determine sediment strength utilize the vane shear test. This test can be performed quickly and with a minimum of disturbance, and the equipment is relatively simple and inexpensive. Vane shear tests have several serious deficiencies, however (Monney, 1971):

(1) The tests are not applicable to granular sediment;
(2) The vane failure surface is predetermined and vertical;
(3) The vane failure surface that is assumed (a cylinder) for sediment strength calculations is not accurate;
(4) Confining stresses are not known in laboratory tests;
(5) The size of the vane is not standardized;
(6) Conditions of drainage are not known;
(7) The rate of shear is not standardized.

The last three deficiencies are related and their significance can be evaluated by laboratory experiments. Only rate of shear is considered in detail in this study.

One would expect shear strength of a sediment to vary to some extent as rate of shear is varied. A clayey sediment behaves as a viscoplastic material (Monney, 1967) and should exhibit an increase in strength as rate of shear is increased. Changes in porewater pressures at the failure surface will also influence shear strength. If the porewater pressure decreases (as with a dense sediment where the grains must move apart to fail in shear), the effective stress between the grains increases and the shear strength increases. Conversely, if porewater pressure increases (as with a loose-structured sediment where the grains are squeezed together in a shear failure), the shear strength decreases. The rates of shear that are normally used in testing sediment vary from 0.0002909 rad/s (1 deg/min) to 0.0262 rad/s (90 deg/min), with 0.00175 rad/s (6 deg/min) being the most commonly used rate. The question is whether shearing resistance will vary significantly within these rates.

THE VANE SHEAR TEST

The vane shear test device was developed and tested in Sweden (Cadling and Odenstad, 1950). The measured torque on the vane shaft is taken as a direct function of the shear strength, τ_f, utilizing the following assumptions:

(1) The surface of failure is in the form of a right circular cylinder, with dimensions equal to those of the vane blade.
(2) The stress distribution at maximum torque is everywhere equal and uniform about the surface of the cylinder.

With these assumptions, the equation follows (see Fig. 1):

$$T = \pi \left(H\frac{D^2}{2} + \frac{D^3}{6}\right) \tau_f$$

where
T = maximum torque
H = height of vane blade
D = diameter of vane blade
τ_f = shear strength at maximum torque.

With the Wykeham-Farrance Laboratory Vane Tester, the device most commonly used in laboratory tests, the torque is applied through torsion springs. Therefore, the stress cannot be applied at a predetermined rate as the angular strain in the vane blades will affect the applied stress. This has been a common criticism of the vane shear test (Wilson, 1963).

For this experiment, a Wykeham-Farrance Laboratory Vane Tester was modified to be driven by a variable speed D. C. motor with di-

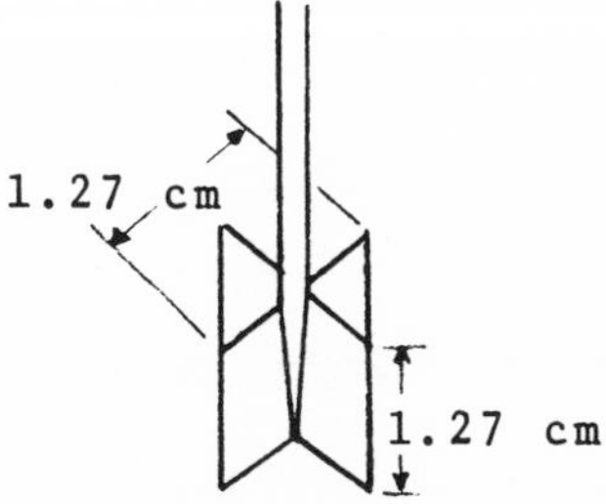

Figure 1. Vane blade

rect chain drive to the vane shaft. With this device, the rate of applied strain was rigidly controlled at predetermined levels. A bonded strain gauge was coupled to the shaft of the vane for direct electrical printout of the torque on the shaft.

EXPERIMENTAL PROCEDURE

Vane shear tests were performed on sediment from relatively undisturbed core samples, and on sediment that had been totally remolded and then reconsolidated to form an artificially uniform sample.

The former tests were performed on a 10.61 cm (4 in) diameter core sample taken in 197 m (600 ft) of water in the Santa Barbara Channel off the coast of California. It was composed of 3% sand,

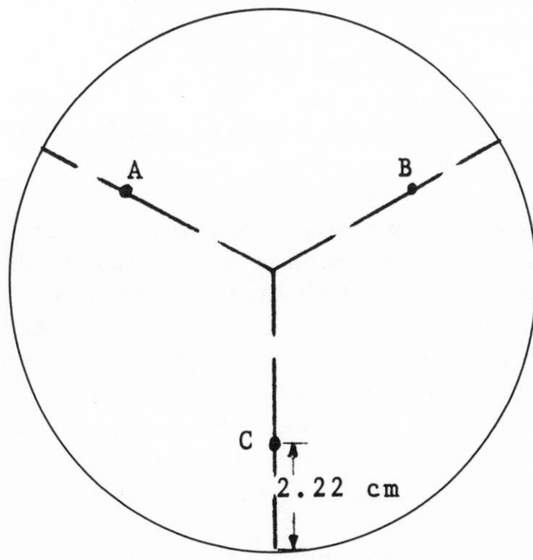

Figure 2. Position of vane shear tests in each core section

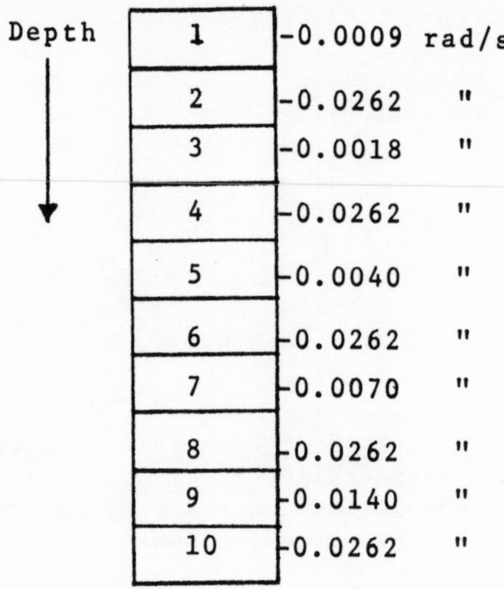

Figure 3. Rate of shear in each core section

25% clay, and 72% silt. The sediment was tested in ten sections of equal thickness from a point 1.64 m (5 ft) in sediment depth to a depth of 2.13 m (6.5 ft). Three vane shear measurements were made at the same rate in each section at the points shown in Figure 2. The orientation was maintained for the tests on each section such that all tests were performed vertically in the core at points directly below position "A," "B," and "C". Tests were run at rates varying from 0.000873 rad/s (3 deg/min) to 0.0262 rad/s (90 deg/min), with a rate of 0.0262 rad/s at alternating sections (see Fig. 3). At the conclusion of the vane shear tests on each undisturbed section, the sediment was totally remolded and the vane shear strength measured at the same rate. The water content was then measured, and on sections #1, #4, #7, and #10 the liquid and plastic limits were determined.

The vane shear tests performed on the artificially uniform samples were done in a similar manner using standard large consolidation cups with a cross sectional area of 100 cm^2. Tests were run on clayey silt from the same location as the sediment used in the undisturbed tests. Tests were also run on a calcareous ooze obtained from a depth of about 3280 m (10,000 ft) off the coast of Puerto Rico, and a red clay obtained from an area between California and Hawaii at a depth of nearly 4920 m (15,000 ft). Only six tests were run on the red clay because of time limitations.

The sediment was prepared for testing by thorough mechanical remolding, adding enough sea water to make a thick slurry. The fixed ring consolidometers were then filled and loaded in a consolidation apparatus. The final consolidation pressure was varied to give a range in permeability. The samples were then tested for permeability using a technique that permitted measurements to be made without removing the sample from the consolidation cup. With this test completed, a very thin wall stainless steel divider was placed in the specimen container to make three pie-shaped equal areas in each container for testing. The divider allowed three runs in each sample with its characteristic permeability. A reference rate of 0.0262 rad/s (90 deg/min) was used on one side of the divider for each sample, and on the other two sides other rates were used. The vane blade was inserted to a 2.54 cm (1 in) depth for every sample to maintain uniformity.

RESULTS

The data obtained from the tests on the undisturbed sediment are presented in Tables 1, 2, and 3. These data are presented in graphical form in Figures 4, 5, 6, and 7. The data obtained from the tests on the consolidated uniform sediment are presented in Tables 4, 5, and 6. These data are presented in graphical form in Figures 8, 9, 10, 11, and 12.

The data in Table 1 represent 40 separate shear strength measurements on the core sample of clayey silt. Table 4 represents 59 shear strength tests on the consolidated uniform clayey silt. The shear strengths of 39 sediment samples (13 consolidation cups divided into three sections) were measured to obtain the data in Table 5 for the calcareous ooze. Because the remolded samples were consolidated before testing, it took a great deal of time to obtain the data for Tables 4, 5, and 6. As noted previously, only six shear strength measurements were made on the red clay (Table 6) because of time limitations. Work is continuing on a research project at the U. S. Naval Academy to obtain more data on the shear strength behavior of red clay as rate of shear is varied. The values of shear strength and permeability were averaged at varying rates of shear in Tables 4 and 5 because of the quantity of data involved. The shear strength data have been normalized to the average value at the 0.0262 rad/s (90 deg/min) rate of shear in Table 4 and in Figures 6, 8, and 10. The technique used for normalization of the data can be best explained by an example. Section 7 of the sediment core was sheared at a rate of 0.007 rad/sec (24 deg/min), which produced an average undisturbed shear strength of 21030 pascals (3.05 psi). The average undisturbed shear strength throughout the core at the shear rate of 0.026 rad/s (90 deg/min) is 17306 pascals (2.51 psi). From Figures 4 and 5, it can be seen that this value is 8% lower than the shear strength at 0.026 rad/sec (90 deg/min) for section 7 when this latter value is interpolated between sections 6 and 8. The 8% must be subtracted from the value of 21030 pascals to obtain the 19582 pascals (2.84 psi) value plotted in Fig-

TABLE 1

MAXIMUM VANE SHEAR STRENGTHS OF THE SECTIONS

FROM THE UNDISTURBED CORE

Section No.	Shear Rate Rad/s (deg/min)	Maximum Shear Strength Pascals (psi)			
		A	B	C	Remolded
1	0.000873 (3)	9929 (1.44)	14962 (2.17)	12066 (1.75)	5288 (0.767)
2	0.0262 (90)	13445 (1.95)	17996 (2.61)	15927 (2.31)	7998 (1.16)
3	0.00175 (6)	15927 (2.31)	16893 (2.45)	15927 (2.31)	4916 (0.713)
4	0.0262 (90)	19444 (2.82)	15100 (2.19)	17100 (2.48)	6192 (0.898)
5	0.00349 (12)	22616 (3.28)	24891 (3.61)	20340 (2.95)	6633 (0.962)
6	0.0262 (90)	17996 (2.61)	13445 (1.95)	21650 (3.14)	8826 (1.28)
7	0.00698 (24)	25512 (3.70)	21581 (3.13)	15927 (2.31)	11170 (1.62)
8	0.0262 (90)	23374 (3.39)	16479 (2.39)	19375 (2.81)	8895 (1.29)
9	0.01396 (48)	18961 (2.75)	13445 (1.95)	11446 (1.66)	7102 (1.03)
10	0.0262 (90)	11653 (1.69)	23443 (3.40)	13307 (1.93)	9791 (1.42)

ure 6. This technique accounts as well as possible for the natural strength variation throughout the length of the core which masks the actual variation in shear strength due to changing rate of shear. The same procedure was used in normalizing the data for the consolidated uniform sediments (except the red clay), with the average value of shear strength at the 0.026 rad/s (90 deg/min) rate used as the base value.

Figure 4 (opposite). Vane shear tests on a core sample of clayey silt sediment

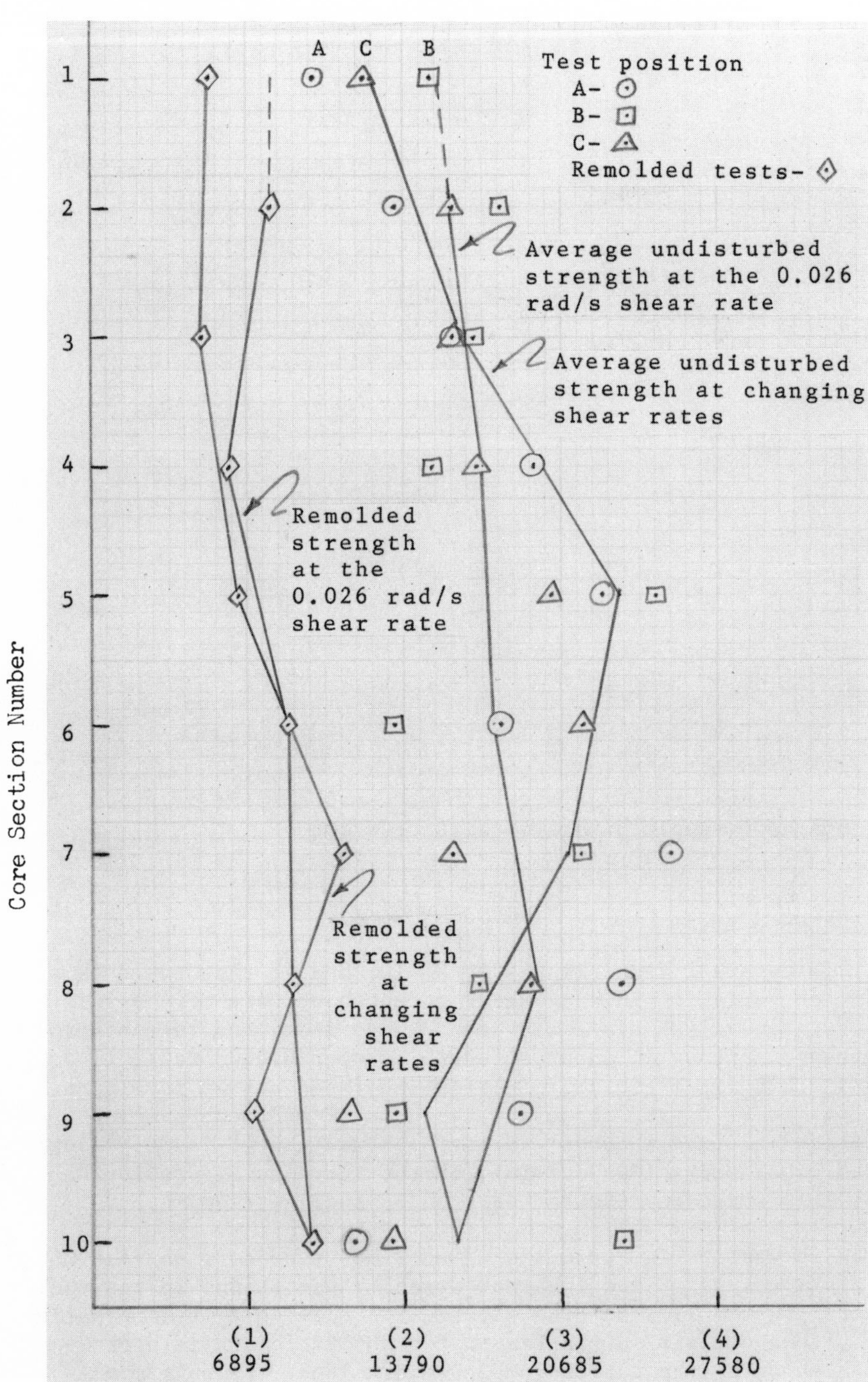

Shear Strength in Pascals (psi)

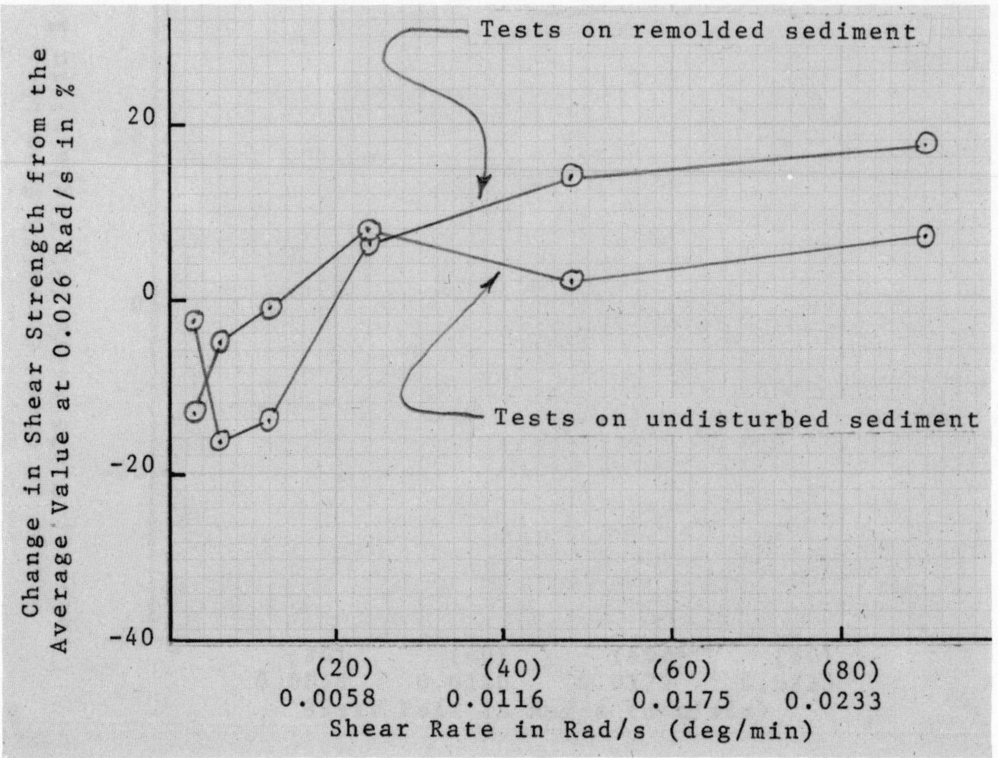

Figure 5. Changes in shear strength at varying rates of shear for tests on a core sample of clayey silt sediment

DISCUSSION

It is apparent from Figures 6, 8, 10, and 12 that there is a significant variation in vane shear strength as rate of shear is varied between values typically used by engineers and researchers. The average increase in shear strength between 0.0003 rad/s (1 deg/min) and 0.026 rad/s (90 deg/min) for the three curves in Figure 8 is 28.7%. These curves represent tests on 59 sediment samples. The average increase in shear strength between 0.002 rad/s (6 deg/min) and 0.026 rad/s (90 deg/min) for the curve in Figure 10 is 30.5%. This curve represents tests on 39 sediment samples.

The curves on Figures 6 and 12 must be regarded with some skepticism because of the inherent strength variations in the sediment core and the limited data for the red clay. It is interesting, however, to speculate on the reason for the variation in shear strength shown by the curves in Figure 6. There are only small changes in the water content and Atterberg Limits throughout the core, so one would expect the reversal in slope of the curves to be a function of changing porewater pressure rather than major changes in viscoplastic behavior. It appears that the rate of shear is slow enough to permit drainage up to a shear rate of 0.0035 rad/s (12 deg/min). At rates greater than this value, the porewater pressure may be increasing in the failure zone, reducing the effective intergranular stress and, hence, the shear strength. At some point between 0.0035 rad/s (12 deg/min) and 0.026 rad/s (the only data point is at 0.014 rad/s) it would appear that the increased rate of shear has resulted in an undrained failure condition, and the shear strength again begins to increase due to viscoplastic behavior. (Research is continuing on this project to attempt to measure porewater pressure at the failure surface simultaneously with shear strength measurements. If this is successful, it will be possible to separate shear strength variations due to viscoplastic behavior from those due to porewater pressure changes.) The inflection in the curve for the remolded sediment (Fig. 6) occurs at a higher shear rate and less dramatically than for the undisturbed sediment. This would be expected because the original grain structure has been destroyed, and there would be less volume change at the failure surface during shear. The general characteristics of

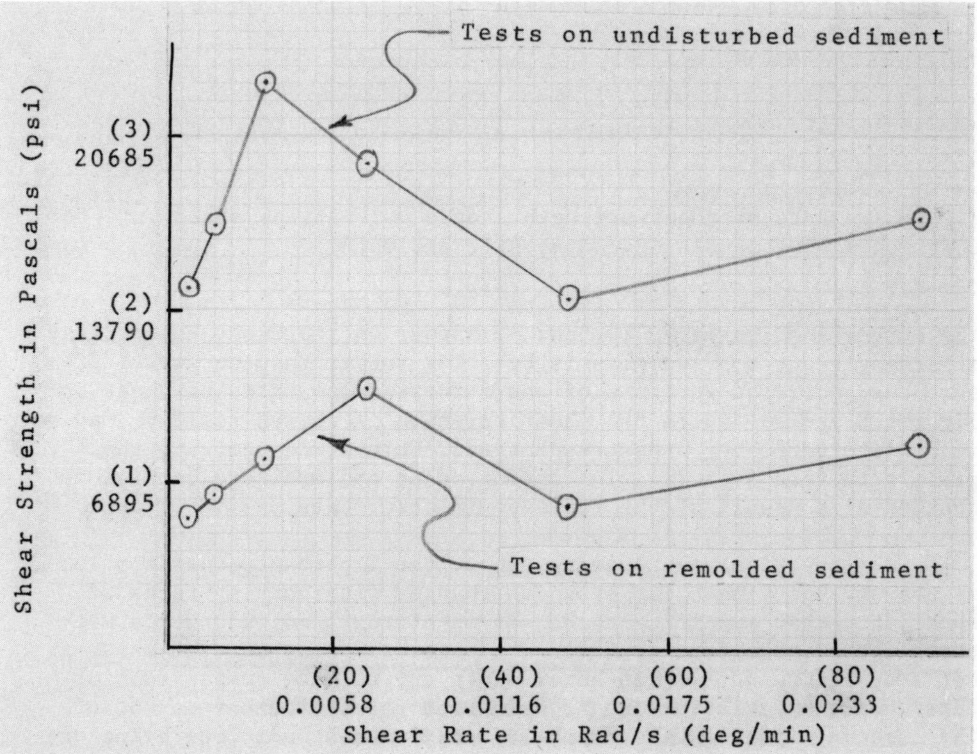

Figure 6. Normalized shear strength at varying rates of shear for tests on a core sample of clayey silt sediment

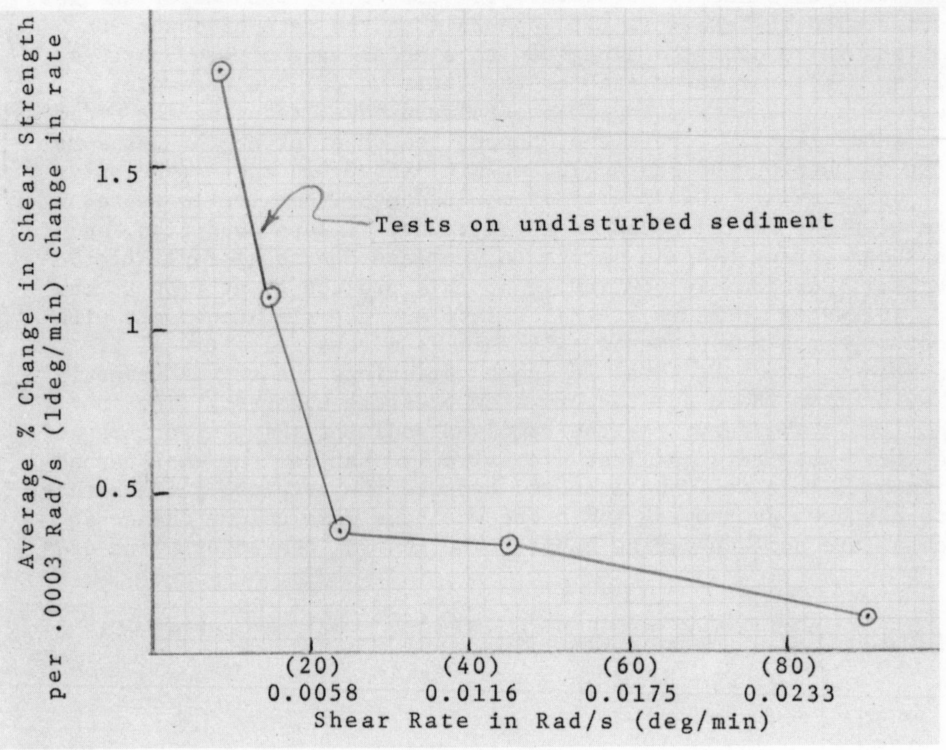

Figure 7. Rate of shear strength change vs. varying shear rate for tests on a core sample of clayey silt sediment

the curves in Figures 8, 10, and 12 appear to represent an increase in strength due to viscoplastic behavior rather than porewater pressure changes. The portions of the curves at low rates of shear in Figures 8 and 10, where the shear strength increases rapidly, may be related to porewater pressure changes. This would occur if the grain structure were dilating during shear and drainage conditions changed as a result of the increase in shear rate.

In a practical engineering sense, the important consideration is not why the vane shear strength changes with varying rates of shear, but that significant changes in strength do occur. In many cases where shear strength data are presented, the rate of shear is not even mentioned. It is hopeless to try to make comparisons of data when the variation due to different rates of shear may be of the same magnitude as the change caused by the conditions being examined. All vane shear strength measurements should be made at the same rate of shear to build a consistent base of data and engineering experience that can be relied upon.

Although the majority of vane shear tests are now made at 0.002 rad/s (6 deg/min), Figures 7, 9, and 11 indicate that the standard shear rate should be 0.026 rad/s (90 deg/min). There are three reasons for this choice:

(a) For sediment for which the vane shear test is applicable, a shear rate of 0.026 rad/s will probably always result in undrained shear failure. Rates on the order of 0.002 rad/sec (6 deg/min) may result in partial drainage in some cases.

(b) The change in strength with small changes in shear rate is very small near the rate of 0.026 rad/s. The changes in exhibited strength can be quite large for small changes in shear rate near the rate of 0.002 rad/s. For example, Figure 9 shows that a variation of plus or minus 0.0009 rad/sec (3 deg/min) from the shear rate

TABLE 2

ROTATION ANGLE AT WHICH MAXIMUM SHEAR STRENGTH OCCURRED

Section No.	Angle at Maximum Shear Strength Radians (degrees)			
	A	B	C	Remolded
1	0.4188 (24)	0.4275 (24.5)	0.5235 (30)	0.5061 (29)
2	0.3926 (22.5)	0.5235 (30)	0.5235 (30)	0.4101 (23.5)
3	0.349 (20)	0.4363 (25)	0.4537 (26)	0.2269 (13)
4	0.5235 (30)	0.5235 (30)	0.5759 (33)	0.5759 (33)
5	0.4886 (28)	0.5061 (29)	0.4188 (24)	0.4188 (24)
6	0.5235 (30)	0.5235 (30)	0.5235 (30)	0.6544 (37.5)
7	0.5584 (32)	0.4886 (28)	0.4537 (26)	0.5933 (34)
8	0.4014 (23)	0.4188 (24)	0.4014 (23)	0.4188 (24)
9	0.4363 (25)	0.4363 (25)	0.4188 (24)	0.6631 (38)
10	0.5235 (30)	0.4014 (23)	0.4363 (25)	0.4363 (25)

of 0.002 rad/s would cause a shear strength variation of approximately 5%. This same variation from the shear rate of 0.026 rad/s would cause a shear strength variation of less than 0.1%. This is important because, as noted previously, most vane shear devices are not accurately rate controlled.

(c) The tests at 0.026 rad/s can be performed quickly. This is particularly important when tests are being performed on board ships, and long cores are being taken.

CONCLUSIONS

Vane shear strength will vary significantly in cohesive sediment as rate of shear is varied within the range of values typically used by researchers and practicing engineers. The shear rate should be standardized to provide consistent data upon which engineering experience can be established. A shear rate of 0.0262 rad/s (90 deg/min) is recommended as the standard.

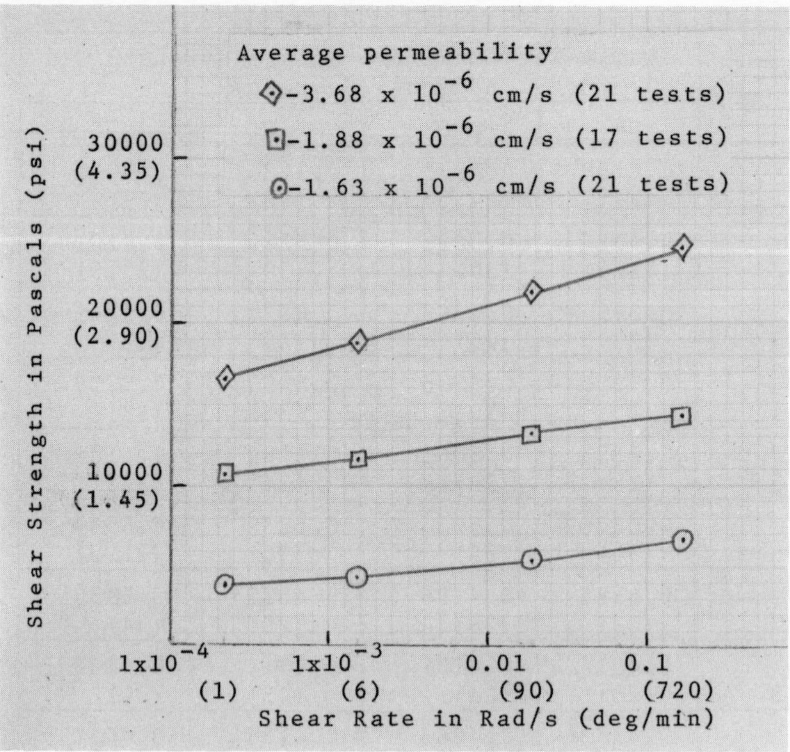

Figure 8. Normalized shear strength at varying rates of shear for tests on a remolded consolidated sample of clayey silt sediment

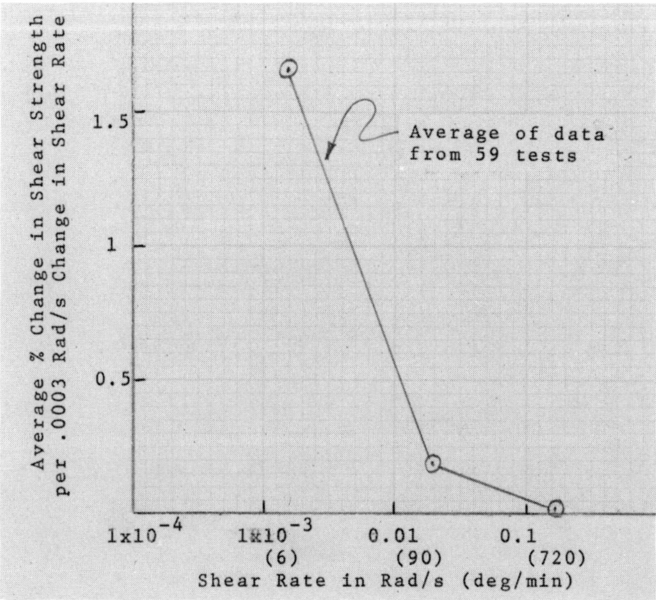

Figure 9. Rate of shear strength change vs. varying shear rate for tests on a remolded consolidated sample of clayey silt sediment

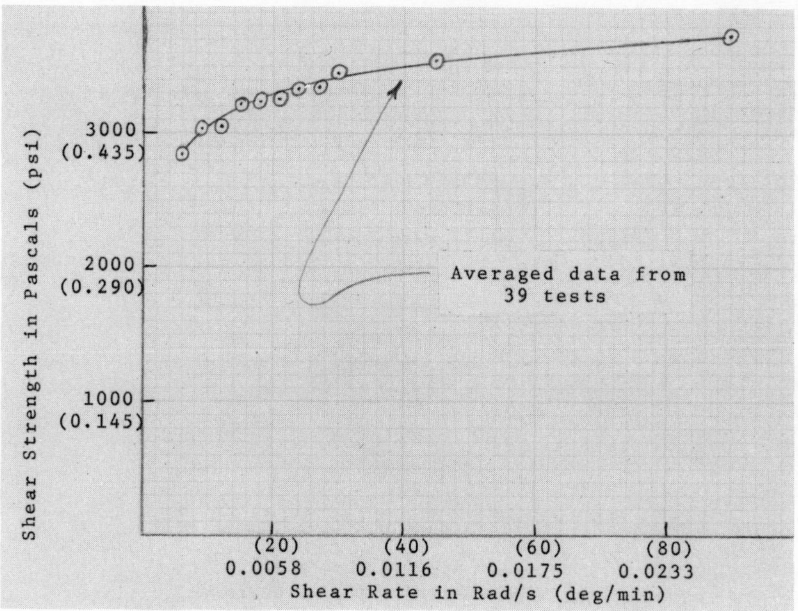

Figure 10. Normalized shear strength at varying rates of shear for tests on a remolded consolidated sample of calcareous ooze

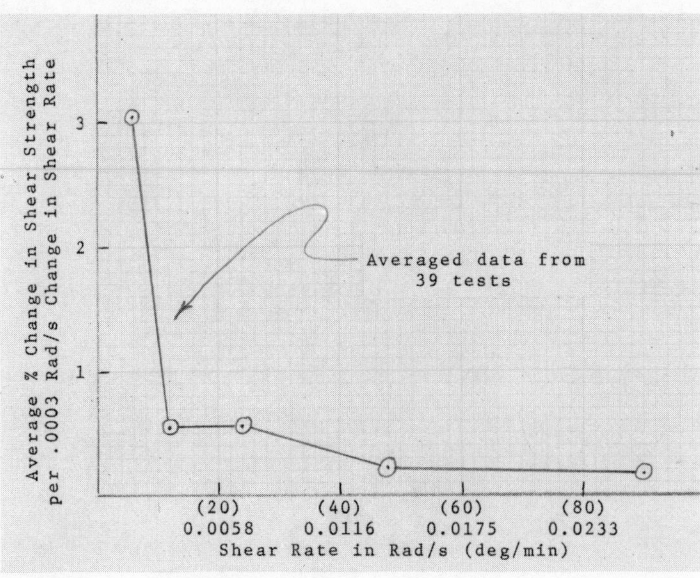

Figure 11. Rate of shear strength change vs. varying shear rate for tests on a remolded sample of calcareous ooze

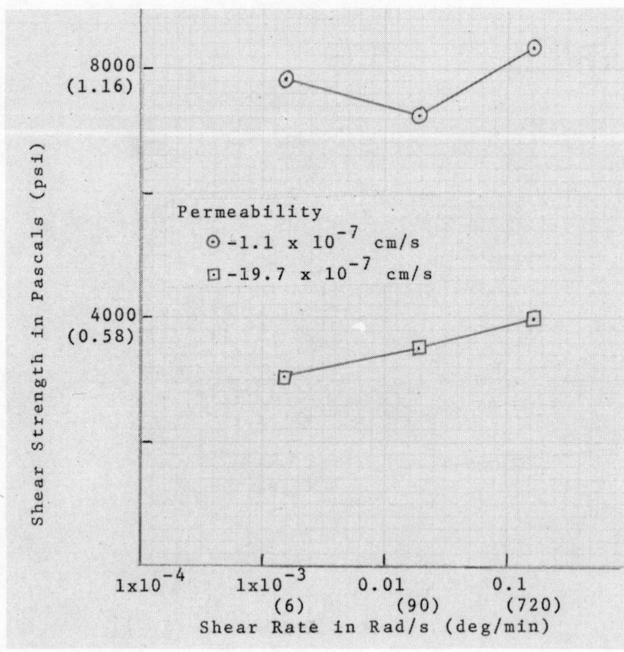

Figure 12. Shear strength at varying rates of shear for tests on a remolded consolidated sample of red clay

TABLE 3

WATER CONTENT AND ATTERBERG LIMITS (%)

Section No.	Water Content	Liquid Limit	Plastic Limit	Plasticity Index	Liquidity Index
1	72	79	53	26	73
2	74				
3	75				
4	67	79	46	33	64
5	67				
6	73				
7	73	83	41	42	76
8	73				
9	76				
10	72	80	46	34	76

TABLE 4

MAXIMUM VANE SHEAR STRENGTHS OF CONSOLIDATED UNIFORM CLAYEY SILT sediment samples, averaged and normalized to the value obtained at the 0.0262 rad/s (90 deg/min) rate of shear

Shear Rate Rad/s (deg/min)	Shear Strength pascals (psi)	Average Permeability cm/s	Number of tests
0.0003 (1)	4000 (0.58)	3.68×10^{-6}	4
0.0018 (6)	4275 (0.62)	3.68×10^{-6}	5
0.0262 (90)	5378 (0.78)	3.68×10^{-6}	7
0.2094 (720)	6826 (0.99)	3.68×10^{-6}	5
0.0003 (1)	10963 (1.59)	1.88×10^{-6}	4
0.0018 (6)	11790 (1.71)	1.88×10^{-6}	4
0.0262 (90)	13169 (1.91)	1.88×10^{-6}	6
0.2094 (720)	14273 (2.07)	1.88×10^{-6}	3
0.0003 (1)	16755 (2.43)	1.63×10^{-6}	5
0.0018 (6)	18823 (2.73)	1.63×10^{-6}	5
0.0262 (90)	21995 (3.19)	1.63×10^{-6}	7
0.2094 (720)	24684 (3.58)	1.63×10^{-6}	4

TABLE 5

MAXIMUM VANE SHEAR STRENGTHS OF CONSOLIDATED UNIFORM CALCAREOUS OOZE
average of three series of tests

Sample No.	Shear Rate Rad/s (deg/min)	Shear Strength pascals (psi)	Average Permeability cm/s
1,2,3	0.0018 (6)	2310 (0.335)	1.8×10^{-6}
1,2,3	0.0026 (9)	2468 (0.358)	1.8×10^{-6}
1,2,3	0.0262 (90)	3185 (0.462)	1.8×10^{-6}
4,5,6	0.0035 (12)	3420 (0.496)	1.3×10^{-6}
4,5,6	0.0044 (15)	3592 (0.521)	1.3×10^{-6}
4,5,6	0.0262 (90)	4116 (0.597)	1.3×10^{-6}
7,8,9	0.0052 (18)	2441 (0.354)	1.4×10^{-6}
7,8,9	0.0061 (21)	2462 (0.357)	1.4×10^{-6}
7,8,9	0.0262 (90)	2937 (0.426)	1.4×10^{-6}
10,11,12	0.0070 (24)	3496 (0.507)	1.7×10^{-6}
10,11,12	0.0079 (27)	3516 (0.510)	1.7×10^{-6}
10,11,12	0.0262 (90)	3909 (0.567)	1.7×10^{-6}
13	0.0087 (30)	4282 (0.621)	0.6×10^{-6}
13	0.0131 (45)	4378 (0.635)	0.6×10^{-6}
13	0.0262 (90)	4558 (0.661)	0.6×10^{-6}

TABLE 6

MAXIMUM VANE SHEAR STRENGTHS OF CONSOLIDATED UNIFORM RED CLAY

Sample No.	Shear Rate Rad/s (deg/min)	Shear Strength Pascals (psi)	Permeability cm/s
1	0.0018 (6)	7860 (1.14)	1.1×10^{-7}
1	0.0262 (90)	7240 (1.05)	1.1×10^{-7}
1	0.2094 (720)	8343 (1.21)	1.1×10^{-7}
2	0.0018 (6)	3034 (0.44)	19.7×10^{-7}
2	0.0262 (90)	3516 (0.51)	19.7×10^{-7}
2	0.2094 (720)	3999 (0.58)	19.7×10^{-7}

REFERENCES

Cadling, L., and S. Odenstad, The vane borer, in Proc. No. 2, Roy. Swedish Geotec. Inst., Stockholm, 1950.

Monney, N. T., Engineering Aspects of the Ocean Floor, Ph.D. thesis, Univ. of Washington, Seattle, 1967.

Monney, N. T., Measurements of the engineering properties of marine sediments, Mar. Technol. Soc. J., 5 (2), 21-30, 1971.

Wilson, N. E., Laboratory vane shear tests and the influence of porewater stresses, in Laboratory Shear Testing of Soils, pp. 371-385, ASTM Spec. Publ. 361, 1963.

USEFULNESS OF SPADE CORES FOR GEOTECHNICAL STUDIES AND SOME RESULTS FROM THE NORTHEAST PACIFIC

ROYAL HAGERTY

Deepsea Ventures, Inc.

ABSTRACT

The next best thing to in situ measurements of sediment engineering properties is probably data collected from cores taken with spade corers. Included in the discussion are the advantages and disadvantages of spade cores compared to conventional small-diameter cores.

The "undisturbed" nature of spade cores is also illustrated. Changes in temperature or storage in water for periods up to 7.5 months do not significantly alter the shear strength of the sediment.

Despite popular beliefs, pelagic sediments are not homogeneous over extended distances. Significant differences in shear strength occur even within the horizontal confines of a single spade core. These differences are ascribed to bioturbation. Drastically different strength characteristics also occur over distances of a mile or less due to different depositional histories. Characteristics of deep-ocean sediments may prove to be as variable as shelf sediments.

INTRODUCTION

For the past decade the major effort in equipment design for marine geotechnical studies has been directed towards development of in situ devices. The need for such devices for use in basic studies is beyond question, especially in lieu of laboratory or shipboard tests on cores obtained by conventional small-diameter cores. The high costs, complexity of operation, and ship time

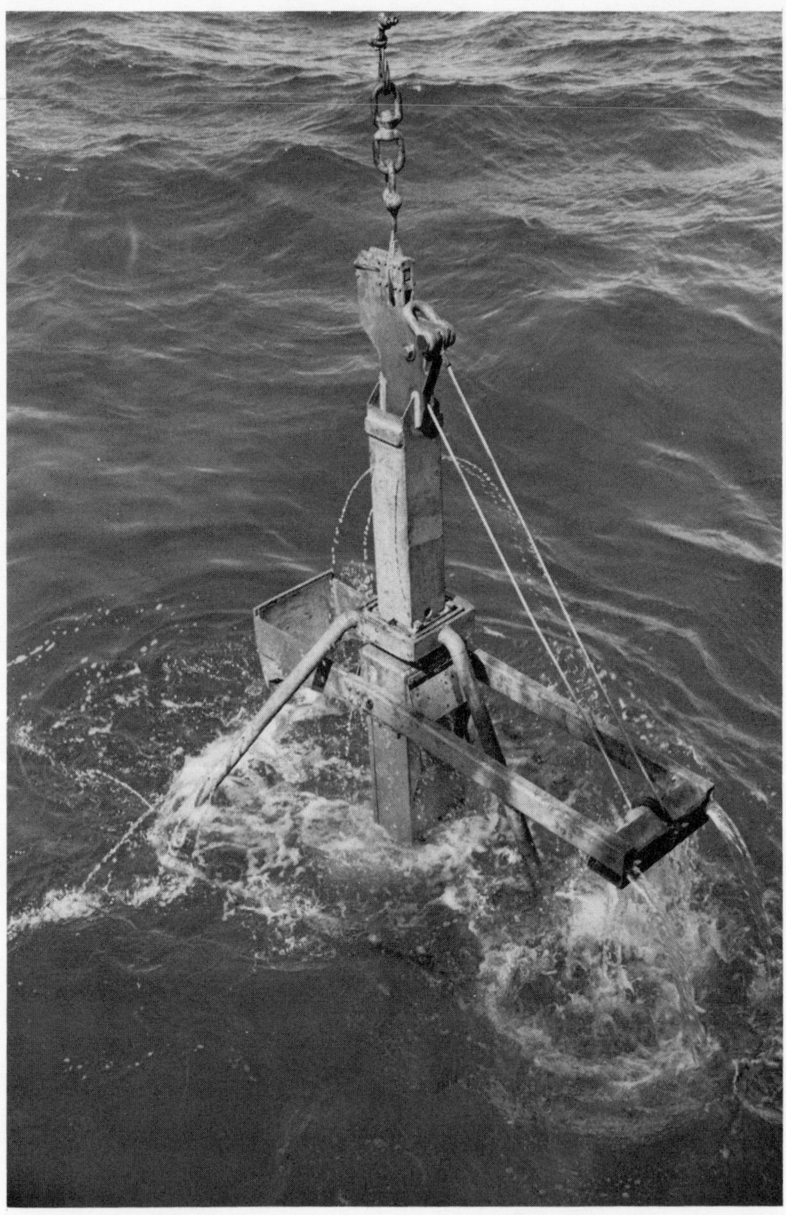

Figure 1. An open spade corer in its sampling configuration

requirements, however, limit the practical use of in situ devices at present. The day when cores will not be required for geotechnical studies is not within the forseeable future.

The major problem in using cores for geotechnical studies is the mechanical disturbance of the sediment during the coring operation. Numerous methods have been devised to reduce this disturbance. The easiest and perhaps most successful method has been to use thin-walled, large-diameter core barrels (Rosfelder and Marshall, 1967). The spade corer is one such device. Surprisingly, to this writer's knowledge, no published comparison has been made between in situ measurements and those from spade cores.

THE SPADE CORER

A shallow water spade corer was developed by Reineck (1958). The basic design was adapted to deep water by Bouma and Marshall (1964) and during the mid-sixties the Naval Electronics Laboratory made further improvements resulting in a corer similar to the one illustrated in Figures 1 and 2. Bouma (1969) presents a full operational description of the corer. The basic principle of operation involves driving a thin-walled, 20 x 30 cm (8 x 12 in) stainless steel box either 61 or 91 cm long (24 or 36 in) into the sediment by gravity. As pullout is initiated, a spade rotates under the box, cutting off the core and retaining it in the box during pullout and recovery of the corer. If desired, a locking compass and tilt indicator can be attached to the corer for orientation of the core.

Advantages

A spade core has numerous advantages over conventional small-diameter cores. Among these are:

1. There is limited sediment disturbance. The thin walls of the box and its large area limit the effect of wall friction. The slicing action of the space limits disturbance during pullout and the sediment-water interface is not disturbed by a piston, as occurs with piston cores. The core length is not noticeably compressed during coring.

2. The large volume of sediment and horizontal area have several benefits. (a) Averages of multiple measurements or subsamples can be obtained from a specific horizon or depth in the core. (b) Subsamples collected for different analyses can all be collected from the same horizon. (c) Large bulk samples for consolidation

tests, triaxial shear tests, sound velocity, etc. can be obtained.

3. Visual inspection of the sediment, without significantly disturbing it, is possible by removing the side of the box before subsamples are selected.

4. Sedimentary structures are well preserved for study.

5. The core provides an excellent sample for biological studies of the infauna (Smith and Howard, 1972).

6. Meaningful direct measurements of manganese or phosphate nodules concentrations can be obtained.

7. The core can be oriented both with respect to north and the vertical plane.

8. With a supply of boxes, rapid turn around times are possible and the core can be stored in the box.

9. Service and maintenance are minimal.

Disadvantages

No corer is perfect and the spade corer is no exception. Its major disadvantages are:

1. Core lengths are limited to 61 or 91 cm (24 or 36 in) depending on the corer used.

2. The corer is awkward to handle on deck.

3. Premature closing of the spade is a severe problem in rough weather. A 10-degree ship roll can reduce the cable tension enough to release the spade.

"UNDISTURBED" NATURE OF THE CORES

Aside from the theoretical aspects which suggest that spade cores should be free from major mechanical disturbance, visual inspection of the cores also demonstrates this fact. Figure 3 shows the top of a core from the northeast Pacific. The surface sediment is a red clay with a shear strength of 0.7 to 1.4 kPa (0.1 to 0.2 psi) and a water content of 250 to 300 percent. Despite the very soupy nature of this sediment, the manganese nodules are in place and two burrows are visible. Two unidentified benthic animals are also visible. A worm is partly extending from its burrow. This

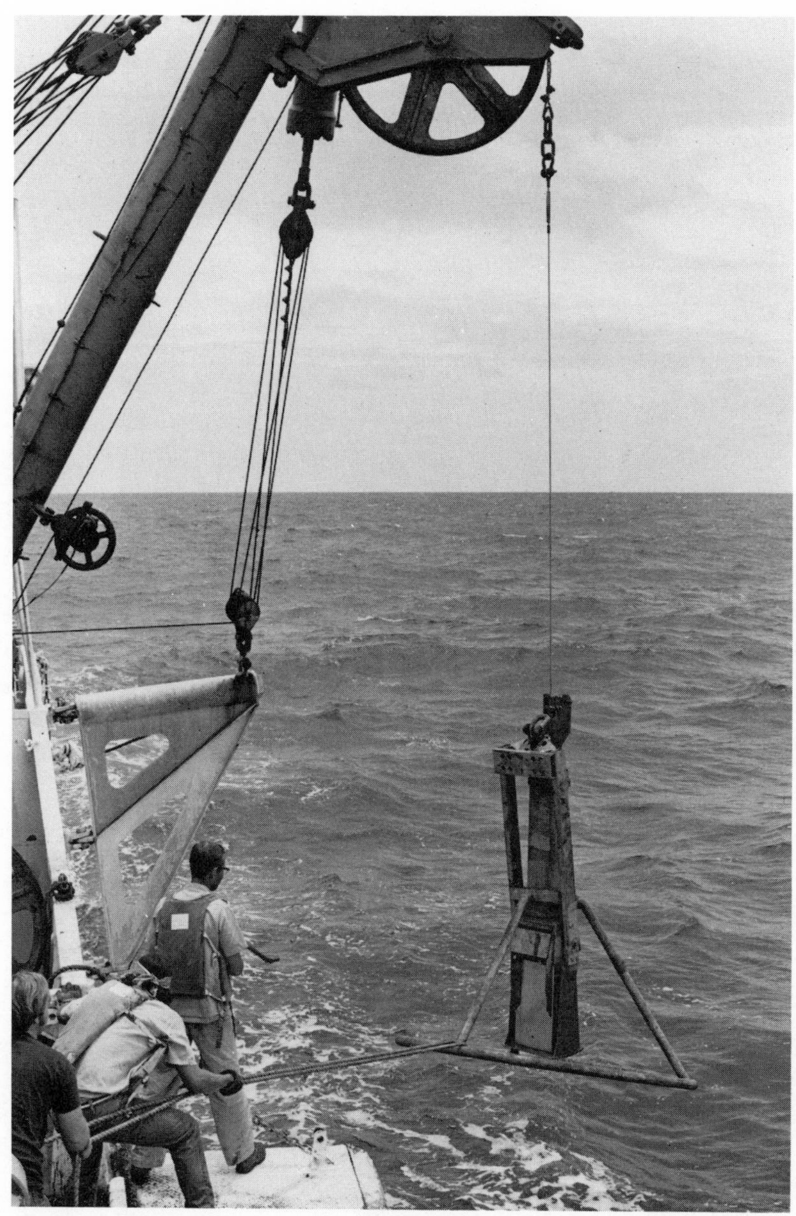

Figure 2. A closed spade corer after sampling

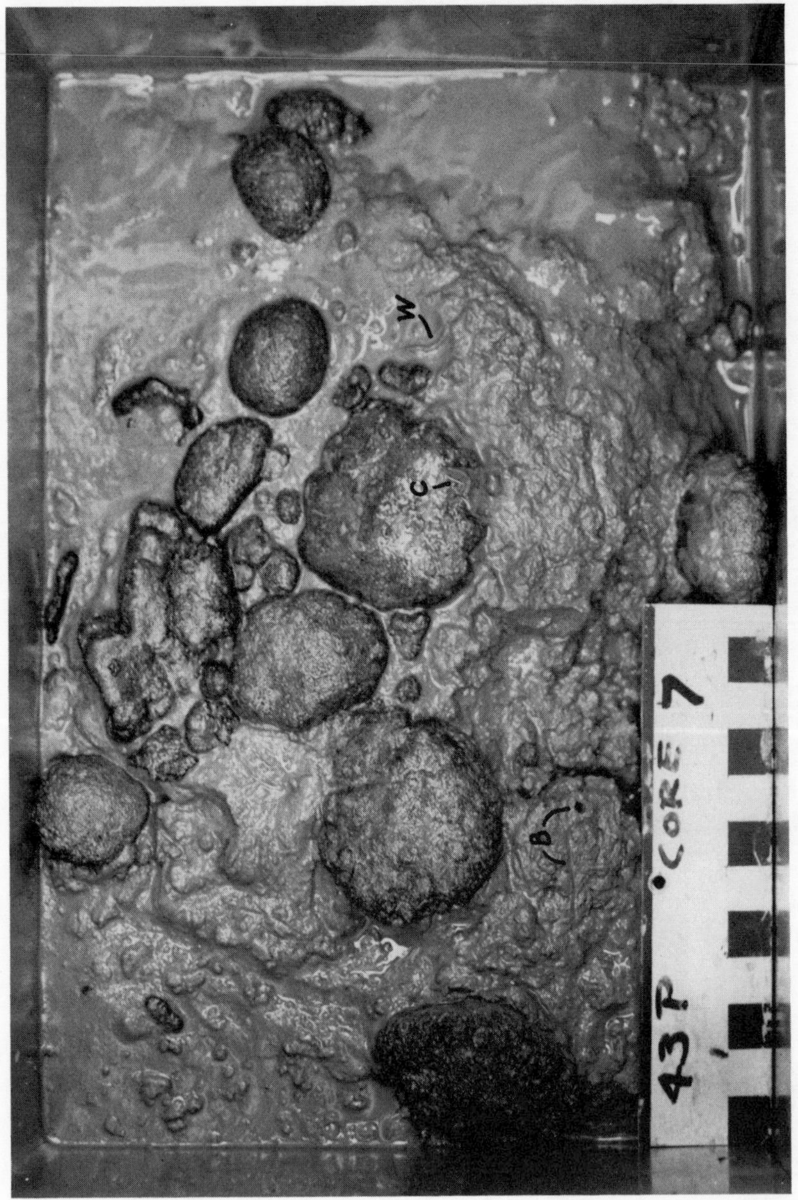

Figure 3. "Undisturbed" surface of a spade core from the northeast Pacific. Note the relatively undisturbed nature of the manganese nodules, the burrows (B), worm partly out of burrow (W), and a collapsed organism attached to nodule (C).

worm was so fragile that it immediately tore apart when an attempt was made to extract it from the burrow. Also present is a very soft and delicate organism attached to a nodule. When covered with water this organism was fully extended and in perfect condition. Upon draining the water from the top of the core, the organism collapsed into a mass of jelly. These factors indicate a minimum of disturbance at the sediment-water interface in this core which was taken at a water depth exceeding 4 kilometers.

The side of the same core is shown in Figure 4. The sediment in sharp contact with the soft surface clay is a very firm red clay with shear strengths of about 27 kPa (4 psi). The contact is an unconformity which accounts for its undulating form. The important feature to note is the lack of distortion of this contact at the edges of the core.

The side of another core is shown in Figure 5. Again, the lack of distortion at the edge of the core is apparent, especially at the contact between the upper slightly burrowed clay and the lower, highly burrowed clay. Shear strengths in the vicinity of this contact are about 7 kPa (1 psi).

A series of vane shear measurements were made on the core shown in Figure 5 in order to establish if any correlation existed between the shear strengths and the distances from the edge of the core. The lack of any consistent correlation is illustrated in Figure 6. Measurements were made at selected depths with the center of a 2 cm (0.8 in) diameter vane 3, 4.6 and 5.6 cm (1.2, 1.8 and 2.2 in) from each edge of the core. The four resulting values for each depth and distance from the edge were then averaged to arrive at the curves in Figure 6. (A hand-held vane shear similar to that described by Dill and Moore (1965), was used to obtain the shear strengths for these and other measurements discussed in this paper.) If significant mechanical disturbance had occurred due to the penetration of the core box into the sediment, a decrease in strength with decreasing distance from the edge of the core would be expected. Such a relationship is not apparent in this core.

TEMPERATURE INSENSITIVITY OF SHEAR STRENGTHS

Numerous authors have speculated on what effect temperature changes have on the shear strength of sediments. Their major concern is that when a core warms up from its in situ temperature, the expansion of fluids, and perhaps the solids, might disrupt the original texture and lower the shear strength.

A series of measurements were made on a typical red clay core to evaluate this possibility; however, no consistent change in shear strength with increasing temperature was found. The results

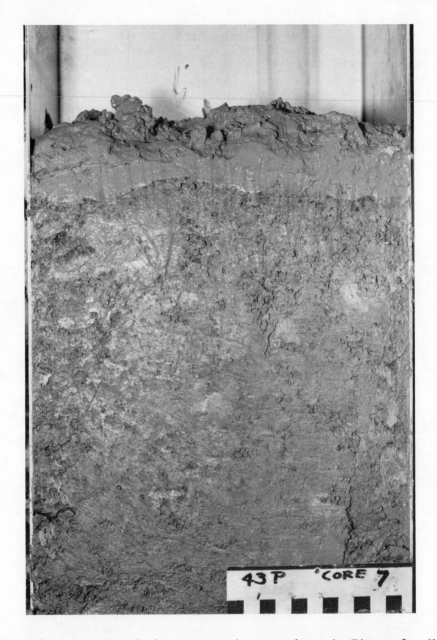

Figure 4. The side of the same spade core shown in Figure 3. Note the lack of distortion of the sharp unconformity at the edges of the box. The unconformity separates very soft surface sediment from the underlying firm clay. The upper surface has been disturbed from removal of the manganese nodules and taking vane shear measurements.

Figure 5. A highly burrowed red clay spade core obtained from the northeast Pacific. Note the lack of disturbance along the box edges.

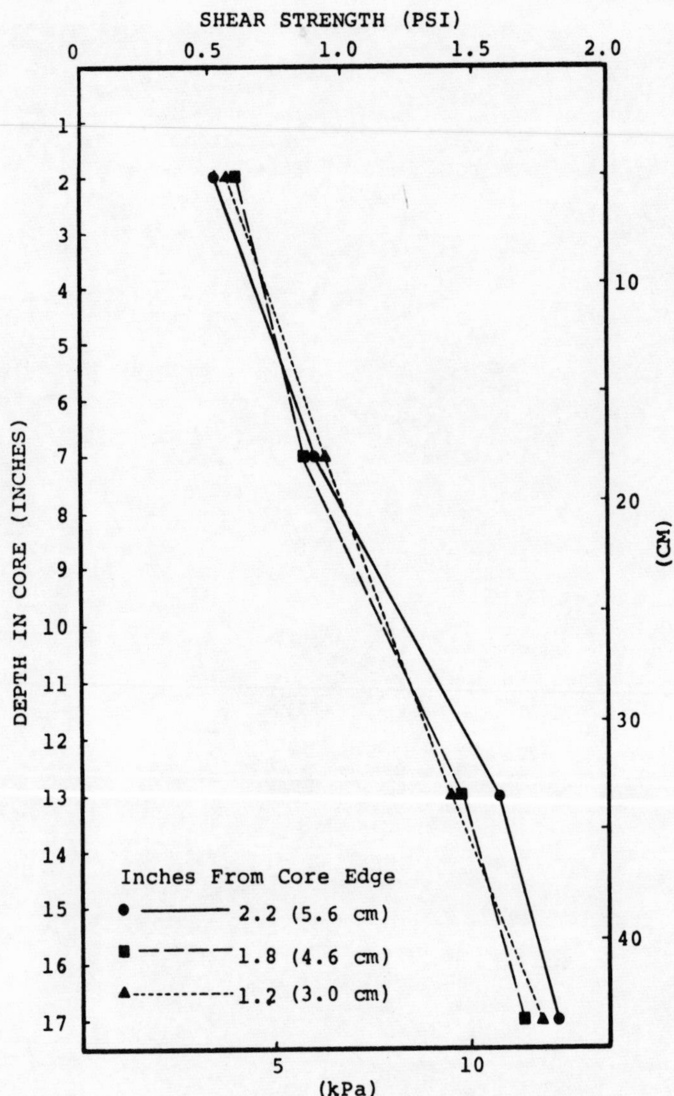

Figure 6. Shear strengths in a Pacific red clay spade core at various distances from the edge of the core. Distances are measured from the center of a 0.8 inch (2 cm) vane to the nearest core edge. Each strength value is the average of four measurements.

are given in Figure 7. The first three measurements were made shortly after the core arrived on deck and exactly 2 hours after it was cut. The sediment temperature at the time of these first measurements was 1.1°C or very close to the in situ temperature. Three additional series of three measurements each were made at two hour intervals until the sediment temperature reached 21.1°C. During this time the core was immersed in a container of seawater to prevent drying and gravity drainage.

While the results of one experiment do not prove that warming of a core does not change the shear strength of the sediment, they do point in that direction.

STORAGE AND TIME EFFECT ON SHEAR STRENGTH

Another concern of scientists in the study of geotechnical properties is what effect does time and the method of storage have on shear strengths. Data presented in the previous section did not indicate any consistent change during the initial six hours a core is warming to ambient temperatures. However, many laboratory studies of cores cannot be made for weeks or months after the core is cut.

Results of work on five almost identically appearing Pacific red clay spade cores suggest that if cores are stored immersed in seawater, even at normal air temperatures, their shear strengths do not change significantly for at least 7.5 months. However, cores sealed in plastic and stored at 2° to 3°C can have significant changes.

Shear strengths of one of the five cores were determined at sea immediately upon its recovery to serve as a reference. Two cores were placed in large plastic bags after covering the top of the cores with wet rags. The bags were sealed and the cores were stored in the ship's walk-in chill box. Periodically the bags were opened and small amounts of water were added to the top of the cores. Two more cores were submerged in seawater in a 55 gallon (208 l) drum and left on deck. All four of the cores were stored in their core boxes.

Approximately 2 months later, the shear strengths of the two cores from the chill box and one of the water-immersed cores were measured aboard ship after the ship returned to Virginia. The second water-immersed core was left submerged in the drum until 7.5 months after it was recovered. Its shear strengths were then measured.

Figure 8 shows the results of the shear strength measurements on the five cores. Each data point is the average of three measure-

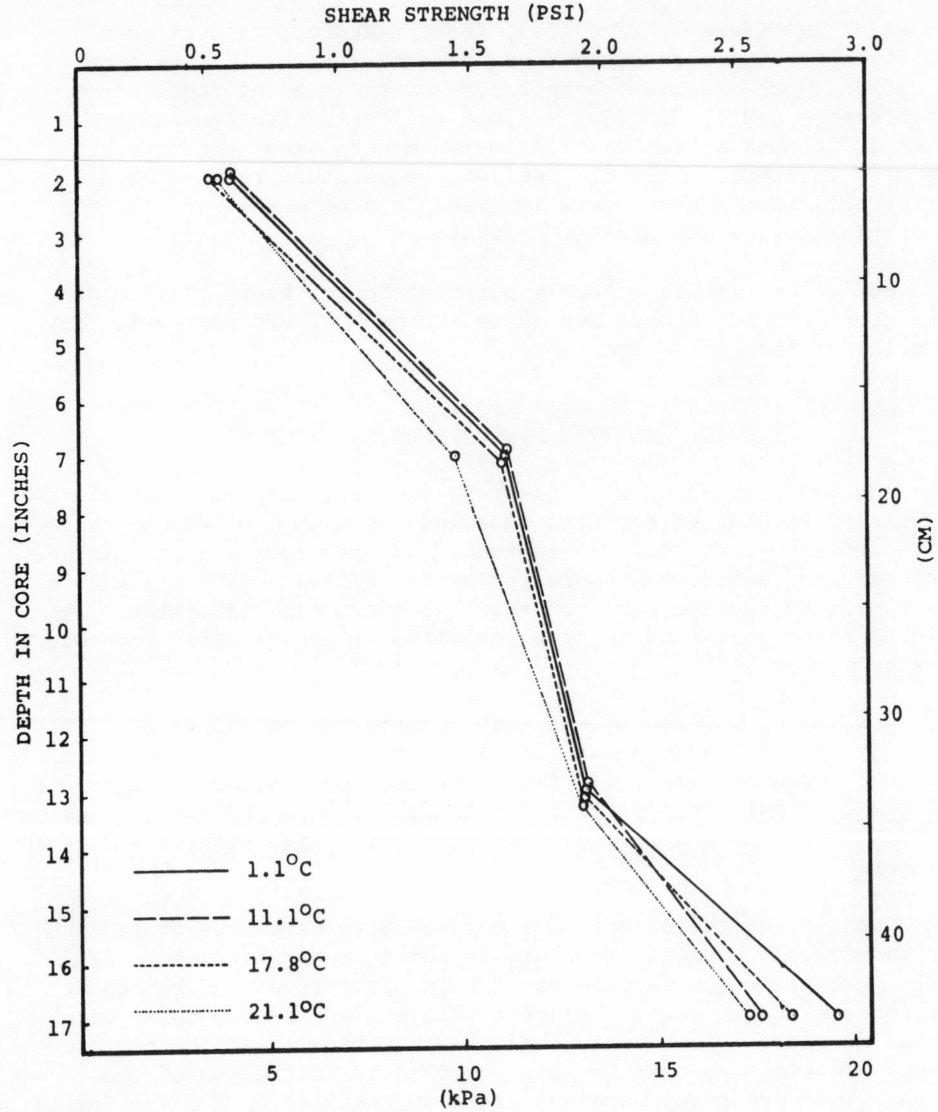

Figure 7. Shear strengths in a Pacific red clay spade core as the core warms from 1.1°C to 21.1°C. Each shear strength value is an average of three measurements.

ments. The agreements between the reference core and the two cores stored in water is excellent except at a core depth of 41 cm (16 in). Experience with other cores suggests that the high strength at 41 cm (16 in) in the reference core is anamolous. The two cores stored in the chill box had significantly higher strengths in the

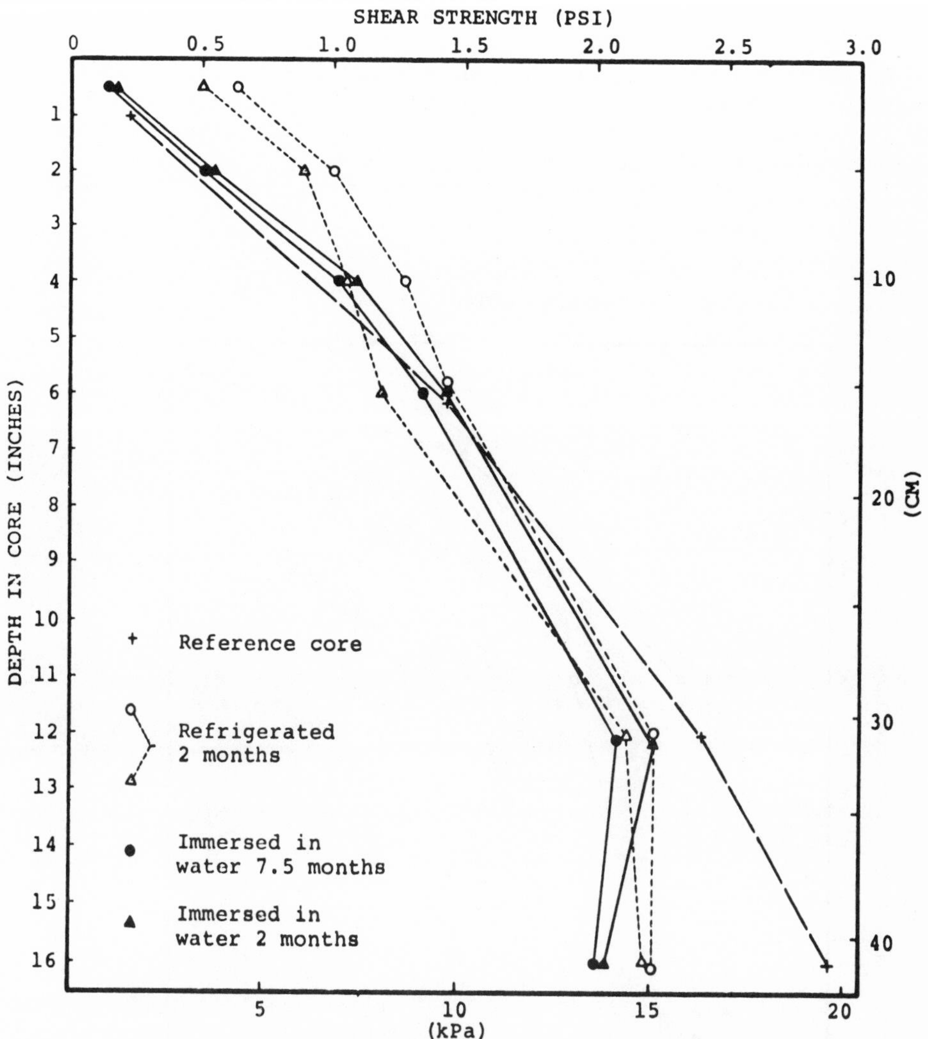

Figure 8. Comparison of shear strengths of Pacific red clay spade cores stored immersed in water or refrigerated. Reference core was analyzed immediately upon recovery. Each shear strength value is the average of three measurements.

upper portion of the core. This was a direct result of desiccation. Shrinkage cracks were prominent on the surface of these cores when opened and their water contents were considerably lower in the upper 10 cm (4 in) than in the cores immersed in water (Figure 9). Below 10 cm (4 in), the water contents of cores from the chill box were higher than those stored in water. Unfortunately no water content values were made for the reference core.

Evaporation accounts for some of the water loss but it is believed that gravity drainage is responsible for the majority of the loss in the upper 10 cm (4 in) of the cores stored in the chill box. It appears some of the original water from the upper 10 cm (4 in), as well as that added to the cores, may have migrated to the lower

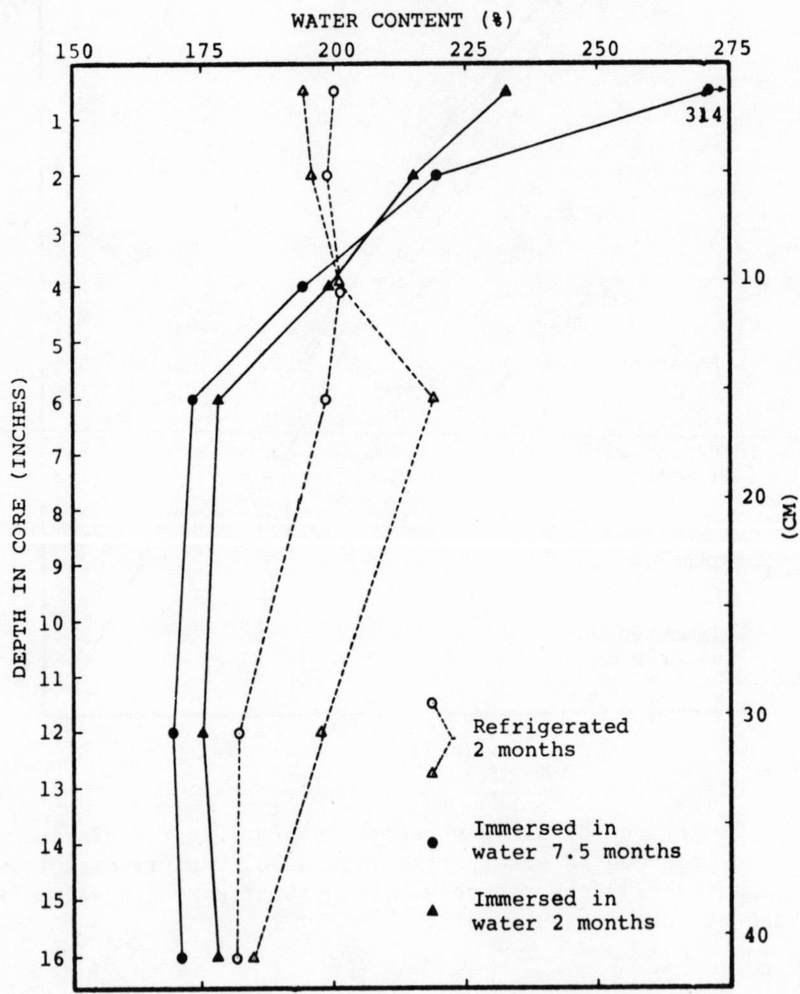

Figure 9. Comparison of water contents of the Pacific red clay spade cores stored immersed in water or refrigerated whose shear strengths are given in Figure 8. Water contents are on a percentage of dry sediment weight basis.

portions of the cores, thereby increasing their water contents. Exactly how this gain in water content can occur in a saturated core without lowering the shear strength is not understood.

Results from this experiment suggest that the best way to store spade cores, or any other sediment sample intended for geotechnical studies, is immersed in seawater. It is also of interest to note that in these particular cores any chemical or biological activity which may have occurred did not materially change the shear strengths of the sediment which was immersed in water. These cores were subjected at times to temperatures of at least 32°C during their storage. Since all sediments may not be as immune to chemical and biological activity, it probably would be wise to store the immersed samples near their in situ temperatures.

HORIZONTAL VARIATIONS

One of the important and unique features of spade cores is that multiple measurements can be made at the same horizon. Average values can then be obtained to smooth out certain random experimental errors as well as small-scale horizontal changes in the property being measured. One cause for rapid horizontal changes in sediment characteristics is biological activity. Bioturbation, as evident in Figure 5, creates marked changes in shear strength over distances similar to the dimension of the normal laboratory vane. Some burrowing activity has a cementing effect on the sediment while other activity reduces the strength of the sediment. Burrows 1 to 3 cm in diameter, filled with soupy sediment but surrounded by sediment with a shear strength on the order of 14 kPa (2 psi), have been observed in Pacific red clays. Hollow tubes have also been found 30 to 60 cm below the sediment-water interface.

This bioturbation often results in vane shear measurements which differ by factors of 1.5 to 2 in horizontal distances of a few centimeters. Unless several measurements are made at each depth in a core, erroneous conclusions may be drawn. Some of the differences between in situ and laboratory measurements probably arise because of these small-scale variations. Vanes used with in situ devices are usually much larger than those used in the laboratory. The in situ vanes are, therefore, less subject to the local effects of bioturbation or other local variations.

From a practical engineering standpoint, a more serious horizontal variation in shear strength exists in the deep ocean than the small-scale variations within a core. The popular concept of deep-ocean sediments is one of homogeneity over vast area; however, surface sediment within a few square kilometers can have drastically different strength characteristics due to differences in depositional history. Shear strengths of seven spade cores obtained

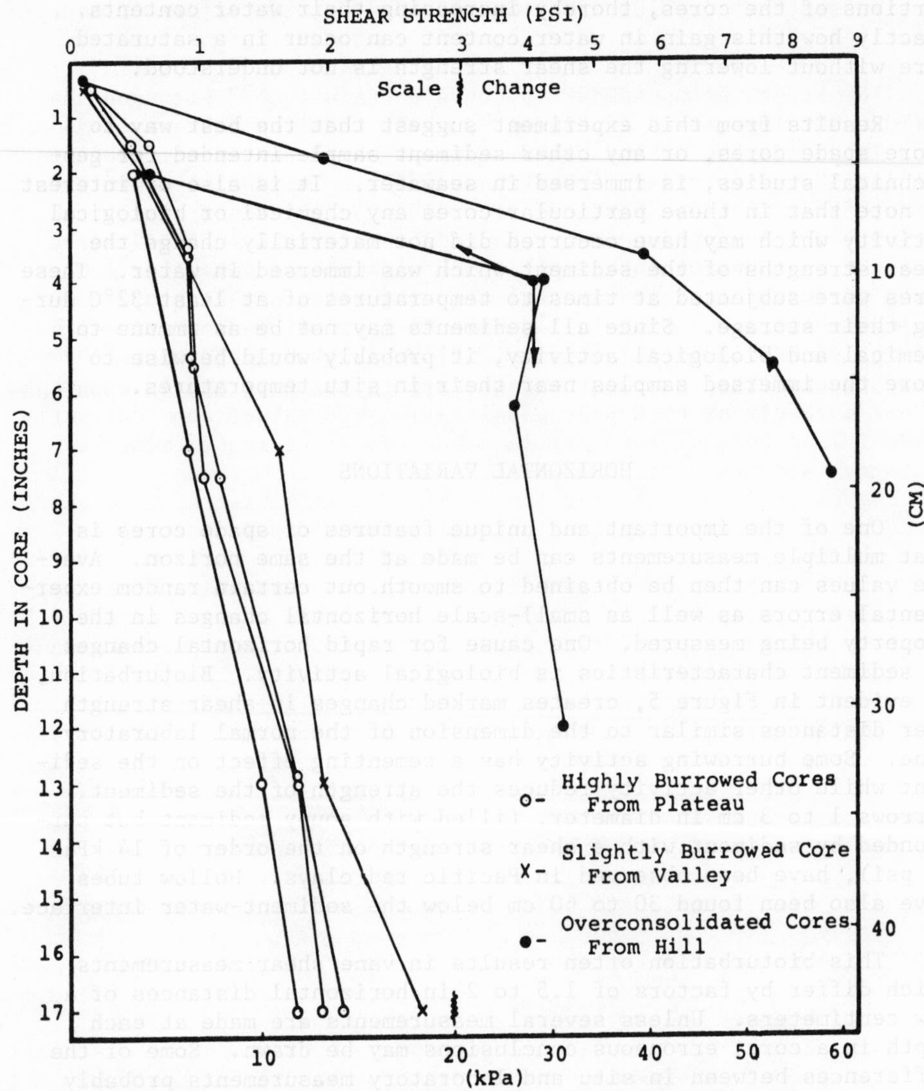

Figure 10. Variations in shear strengths of 7 Pacific red clay cores with various depositional histories. All cores are from within a 9-km radius area.

from within a radius of 9 km are presented in Figure 10. The area was one of abyssal hills in the northeast Pacific. Each core had a thin veneer of very soft sediment at the surface. Three of the cores from the flank and crest of a hill contained stiff, crumbly red clay immediately below this veneer. The core shown in Figure 4 is one of these. Shear strengths of this material ranged from

28 to 55 kPa (4 to 8 psi) within a few centimeters of the surface. This sediment apparently at one time had been buried much deeper and its overburden subsequently removed either by slumping or erosion.

Three more cores were located on a plateau-like feature and were highly burrowed as shown in Figure 5. Shear strength of these cores ranged from about 3.4 kPa (0.5 psi) 5 cm (2 in) below the surface, to about 14 kPa (2 psi) at 43 cm (17 in). A single core obtained from a valley floor was only slightly burrowed and had shear strengths that averaged about 3.4 kPa (0.5 psi) higher than the burrowed cores.

All seven cores would be considered pelagic red clays but their engineering properties are vastly different. Lateral variations in sediments properties may be as rapid and common in the deep ocean as they are in shallow shelf waters.

SUMMARY

Cores cut with spade corers are probably the next best thing to in situ measurements in the upper meter of sediment. They are far superior to small-diameter cores for all types of studies of the upper surface layers. Limited disturbance and the large volume of sample available for multiple measurements are the major attributes of spade cores.

If possible, storage of cores by immersion in seawater and refrigeration appears to be preferable to refrigeration alone.

The problems and complexity of the geotechnical study of deep-ocean sediments are just beginning to be understood. Industry, government, and the academic community must coordinate their efforts, standardize their techniques and communicate freely if meaningful progress is to be made.

REFERENCES

Bouma, A. H., Methods for the Study of Sedimentary Structures, Wiley, N. Y., 1969.

Bouma, A. H., and N. F. Marshall, A method for obtaining and analyzing undisturbed oceanic sediment samples, Mar. Geol., 2, 81-99, 1964.

Dill, R. F., and D. C. Moore, A diver held vane-shear apparatus, Mar. Geol., 3, 323-327, 1965.

Reineck, H. E., Kastengreifer und lotröhre "schnepfe", Senckenbergiana Lethaea, 39, 42-48 & 54-56, 1958.

Rosfelder, A. M., and N. F. Marshall, Obtaining large, undisturbed and oriented samples in deep water, in Marine Geotechnique, edited by A. F. Richards, pp. 243-263, Univ. of Ill. Press, Urbana, 1967.

Smith, K. L., and J. D. Howard, Comparison of a grab sampler and large volume corer, Limnol. and Ocean., 17, 142-145, 1972.

SETTLEMENT CHARACTERISTICS OF SEA-FLOOR SEDIMENTS SUBJECTED TO VERTICAL LOADS

HERBERT G. HERRMANN

U. S. Naval Civil Engineering Laboratory

ABSTRACT

Cores from five sites were analyzed in the laboratory to determine index and compression/consolidation properties. Based on these data, predictions were made for the long-term settlement behavior of small footing-type foundations for each of the sites. The foundations were placed at four of the sites and their behavior monitored. Interrelationships of sediment properties, including consolidation and secondary compression coefficients, are discussed along with the procedure used to predict settlements based on the sediment properties. Predicted and measured settlements are compared and discussed.

INTRODUCTION

Fine-grained sea-floor sediments subjected to a compressive vertical load, either from a negatively buoyant sea-floor installation or an increased overburden resulting from sedimentation or other causes, will compress over time resulting in settlement of the original sediment surface or supported structure. Analysis of such loadings and resulting settlements on land is usually accomplished through a one-dimensional analysis of three consecutive phases: (a) immediate settlement also called plastic deformation (non-time-dependent); (b) primary compression or consolidation (compression that is time-dependent because of hydrodynamic effects on the pore water being squeezed out of the pore spaces); and (c) secondary compression (time-dependent compressions resulting from relative particle compressions, bending or reorientations).

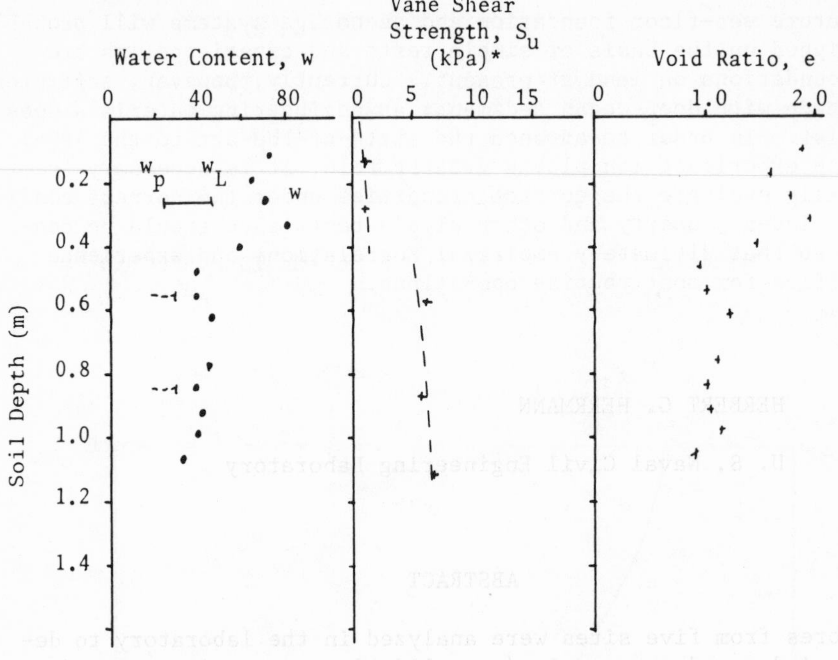

* 1 pound per square inch equals 6.9 kilopascals

Figure 1. Soil properties at the 4-foot Site

The purpose of this study is to investigate the time-dependent compressions of sea-floor sediments to determine whether they behave similarly to terrestrial soils and if one-dimensional analysis of such settlements is applicable. Laboratory analysis of core samples and in situ measurements of foundation settlement were utilized in this investigation of four sites. The non-time-dependent immediate settlement will not be considered here although it has been investigated in conjunction with the studies reported here (Herrmann et al., 1972).

TEST SITE PROPERTIES

Samples were taken at four offshore sites and one lagoon site, all within 190 km of Port Hueneme, California. The soil properties at each site are summarized in Figures 1 through 5. The sediments at each site would be classed as fine-grained sediments, that is, more than 50% by weight of the particles are smaller than 74 microns (7.4×10^{-5} m in diameter). They are largely terrestrial in origin and exhibit organic carbon contents ranging from

0.4 to 2.4% by weight. The carbonate carbon content ranged from 0.12 to 2.99% by weight. The exact locations, topographic region, and types of samplers used at each site are summarized in the following table.

Site Province	Latitude (N) Longitude (W)	Sampler
4-foot Site (Mugu Lagoon) Lagoon (depth of 1.21 m)	34°06.0' 119°06.0'	Modified Hydroplastic
120-foot Site (Pitas Point) Continental Shelf (depth of 36.58 m)	34°16.7' 119°24.3'	DOTIPOS Platform Corer (Demars and Taylor, 1971 and Hydroplastic Corer
600-foot Site (SEACON I) Continental Shelf (depth of 182.88 m)	34°14.6' 119°44.5'	DOTIPOS Platform Corer
1200-foot Site (Santa Barbara Channel) Basin (depth of 365.76 m)	34°09.5' 119°45.5'	DOTIPOS Platform Corer and Hydroplastic Corer
6000-foot Site (STU-1) Continental Slope (depth of 1828.8 m)	33°43.1' 120°43.3'	Ewing Corer

All samples were sealed in their core barrel or liner, transported vertically, stored in a humidity room, and are considered to be of good quality and relatively undisturbed.

LABORATORY INVESTIGATION

Twenty-four one-dimensional consolidation tests were performed using an Anteus and a modified Karol-Warner Consolidometer. The Anteus unit was utilized with a back pressure of 344 kPa (50 psi) to insure sample saturation. Load increments were started at 0.38 kPa (8 psf or 0.0039 kg/cm^2) and continued to 770 kPa (16,000 psf or 7.8 kg/cm^2). The Karol-Warner unit was modified to use a dead weight loading system at low loads (see Herrmann et al., 1972, for a detailed description of this equipment). Loading increments were started at 0.19 kPa (4 psf or 0.0020 kg/cm^2) and continued to 1500 kPa (32,000 psf or 15.6 kg/cm^2). Loading increments equal to 100% of the previous load were used for all tests. A minimum of 24 hours was allowed for each increment. Specialized sample preparation techniques, handling procedures and equipment were developed

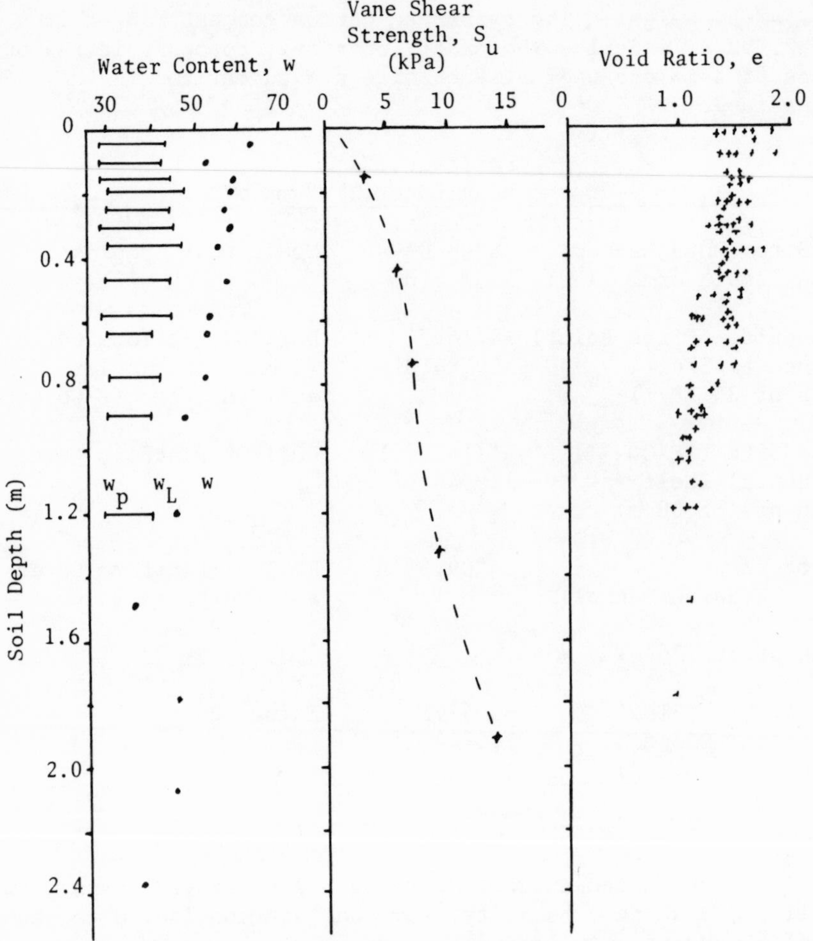

Figure 2. Soil properties at the 120-foot Site

for each unit. Experience with the two units showed that the back pressure capability was not required for meaningful results from these saturated samples and that the dead weight loading system was much more precise at the lower applied pressure levels.

Several tests were performed in the Anteus unit with bottom drainage of the sample prevented and measurements of excess pore pressure made at this bottom surface. Figures 6 and 7 are representative of the data obtained and illustrate the relationship of the times to 100% and 90% complete consolidation (as determined by the logarithm-of-time and square-root-of-time graphical methods) compared to the measured excess pore pressure. In examining Figure 7 it should be remembered that the time to 90% complete consolidation determined by the graphical technique is for a 90% average for the entire sample and thus, is not directly comparable to a 90% dissipation of excess pore pressure at the undrained side of a sample with single drainage. Figures 6 and 7 do illustrate the general applicability of the graphical techniques' results to these sea-floor sediments.

Unfortunately, the majority of the logarithm-of-time plots do not exhibit the double curvature shape (referred to as a type I curve) shown in these figures. For this reason the square-root-of-time method was used exclusively for data reduction in this study. Singh and Yang (1971) have pointed out that one means of obtaining a type I curve is the use of larger load increment ratios -- 200% for example rather than the 100% ratio used in this study.

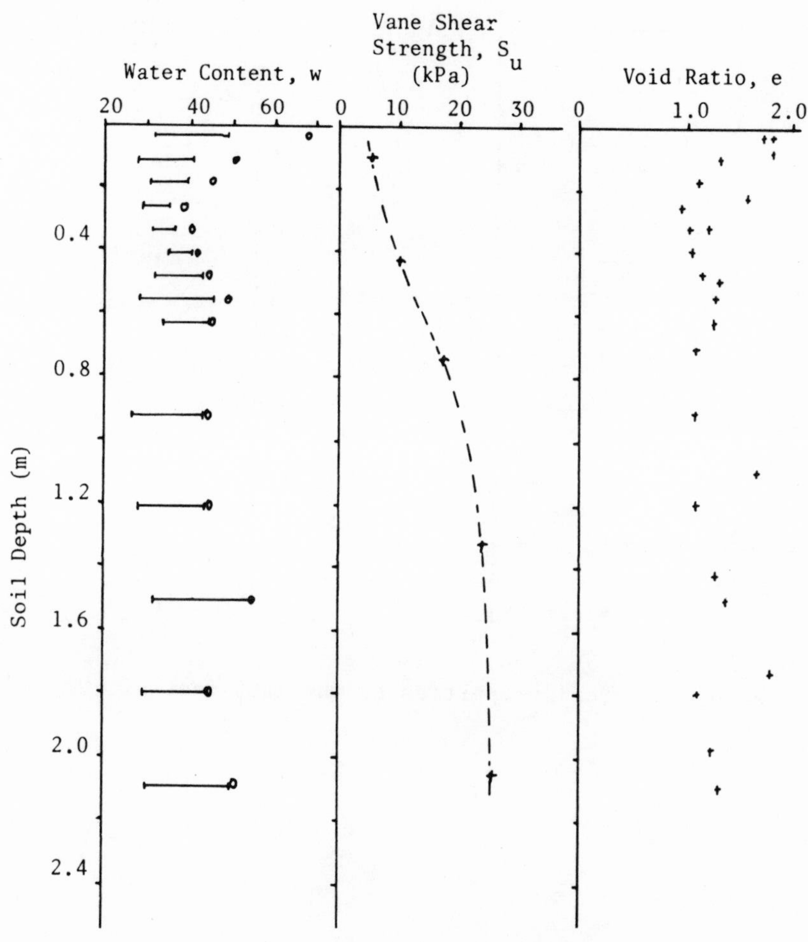

Figure 3. Soil properties at the 600-foot Site

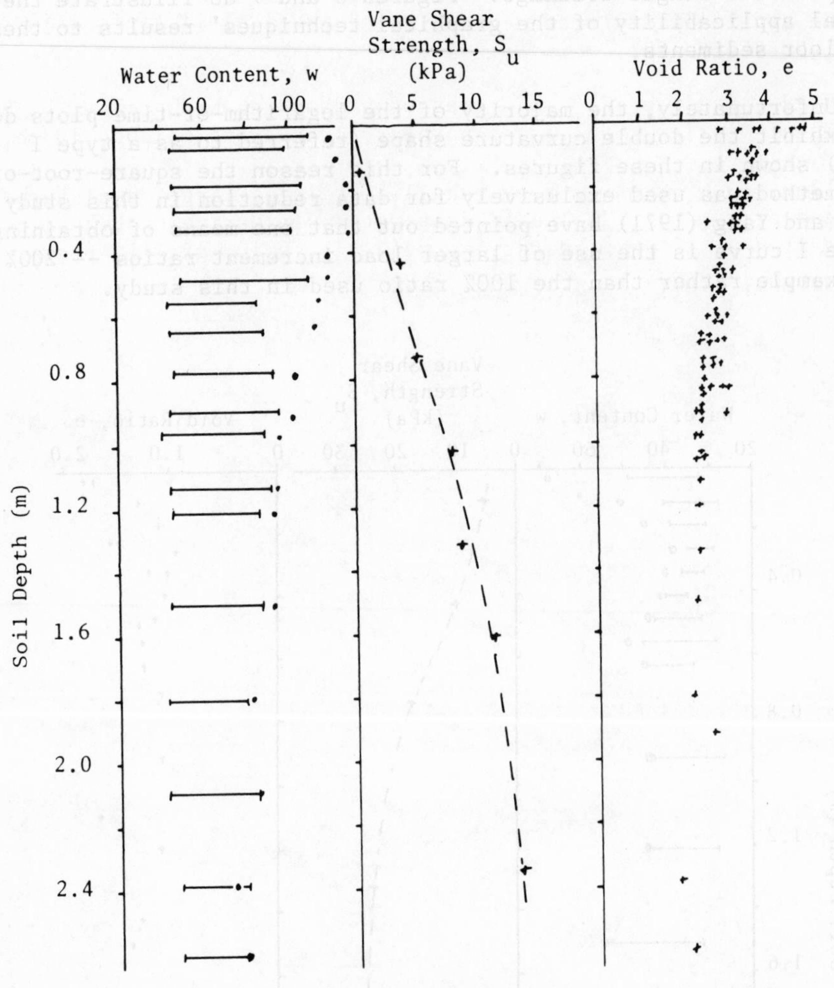

Figure 4. Soil properties at the 1200-foot Site

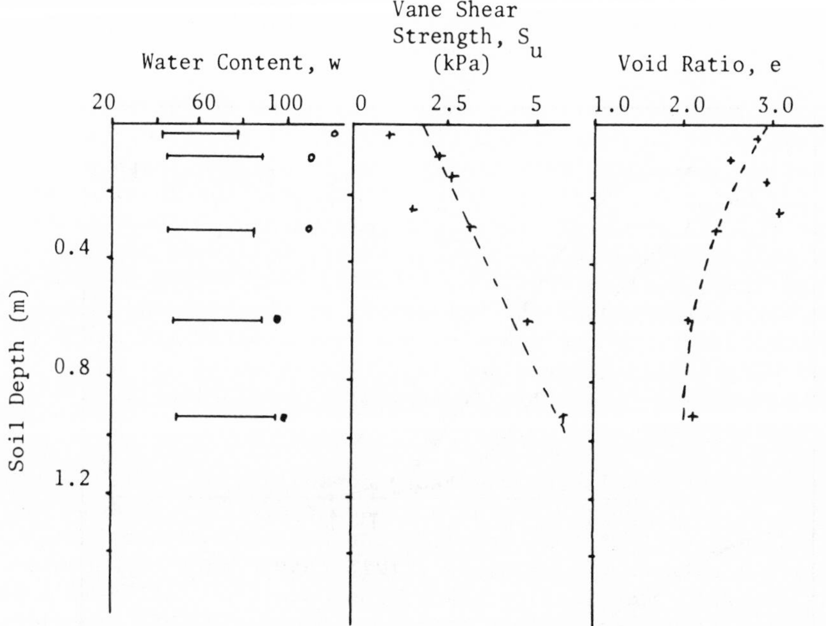

Figure 5. Soil properties at the 6000-foot Site

CONSOLIDATION TEST RESULTS

Figure 8 shows typical consolidometer data. This test was carried out on the modified Karol-Warner unit with a sample from a soil depth of 0.097 m (3.8 in) measured from the top of the soil retained in the core barrel to the center of the 0.0058 m (2.3 in) diameter by 0.020 m (0.78 in) thick sample used in the consolidometer, at the 600-foot Site. The Compression Index, C_c, apparent preconsolidation pressure, p_c', and effective preconsolidation pressure (based on the submerged unit weight and assumed thickness of overburden), $\bar{\sigma}_v$, are indicated on the figure. As has been pointed out by Noorany and Gizienski (1970), most sea-floor sediments exhibit an apparent overconsolidation. The five sites investigated in this study exhibited an average overconsolidation ratio, OCR, equal to 4.2. There is no reason to expect (with the possible exception of the lagoon, the 4-foot Site) that any of the sites have actually been over-consolidated by a larger overburden which has since been removed. This apparent overconsolidation is ascribed to thixotropic and physicochemical effects.

The values of the Compression Index, C_c, determined from the 24 tests ranged from .19 to .90. The relationships of the Compression Index to several index properties were investigated. Figures 9, 10, and 11 show the results of three of these comparisons. A

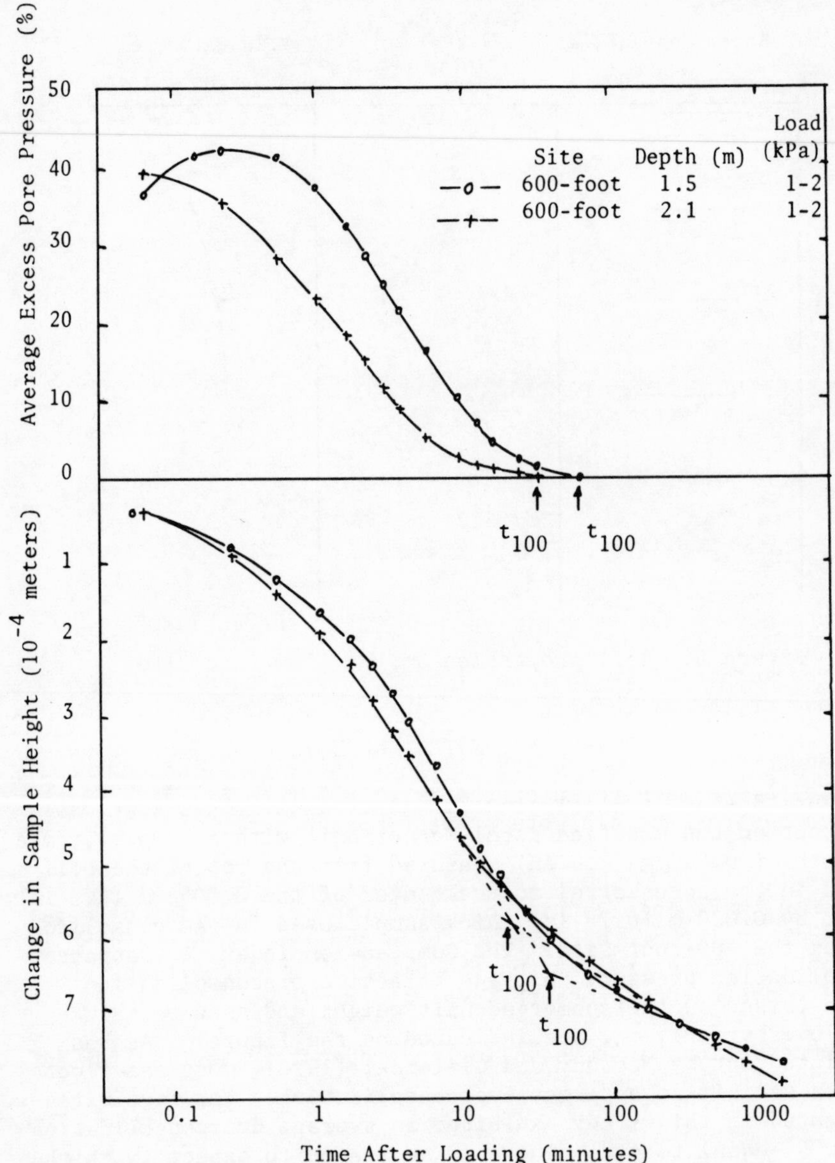

Figure 6. Excess pore pressure dissipation compared to the logarithm-of-time method

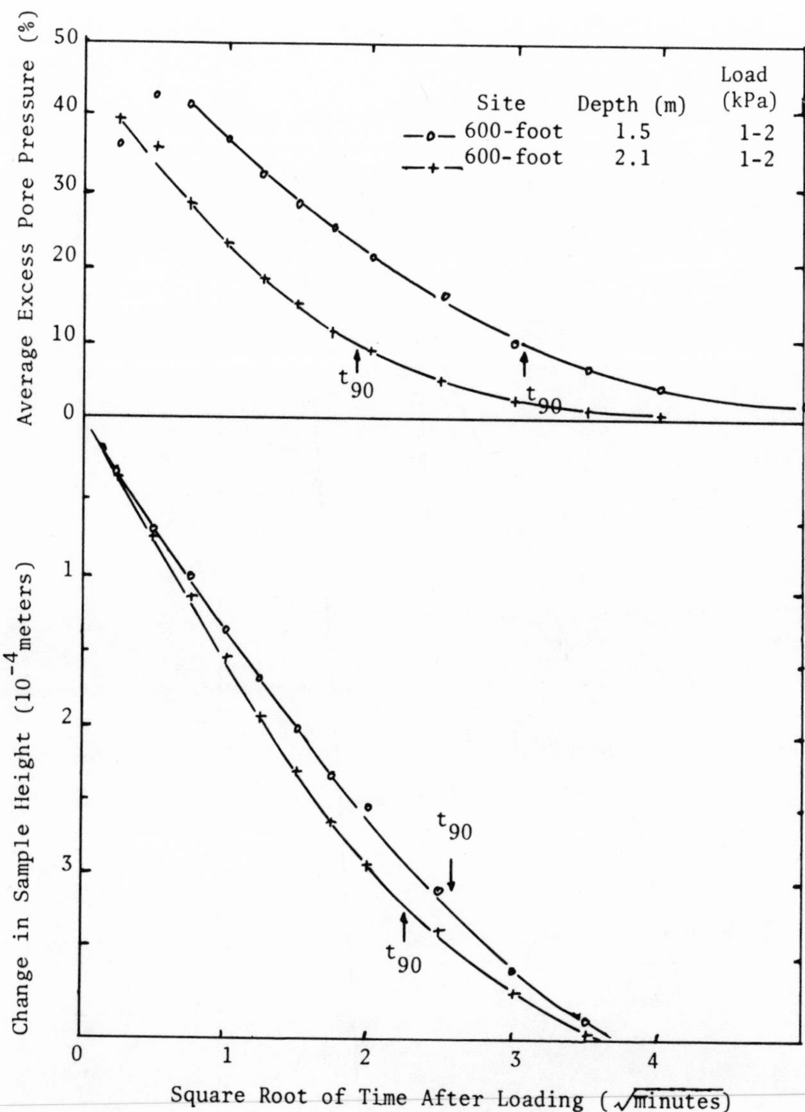

Figure 7. Excess pore pressure dissipation compared to the square-root-of-time method

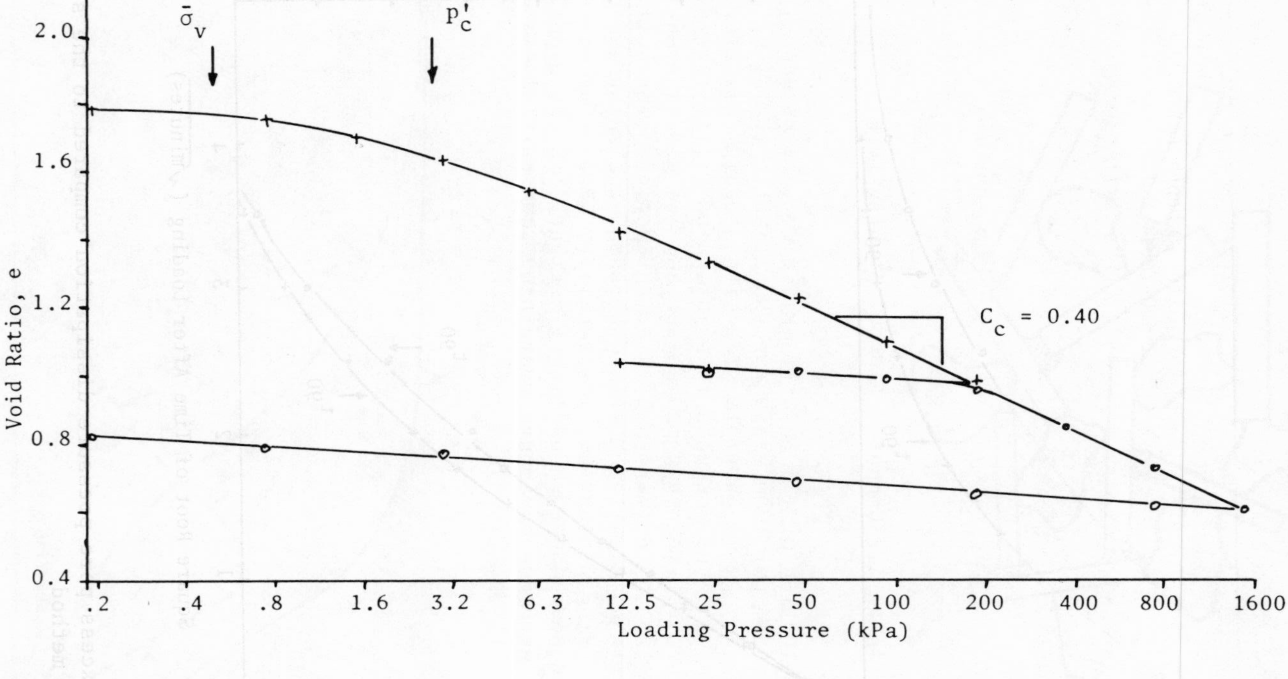

Figure 8. Typical consolidometer data plotted as void ratio versus logarithm of pressure

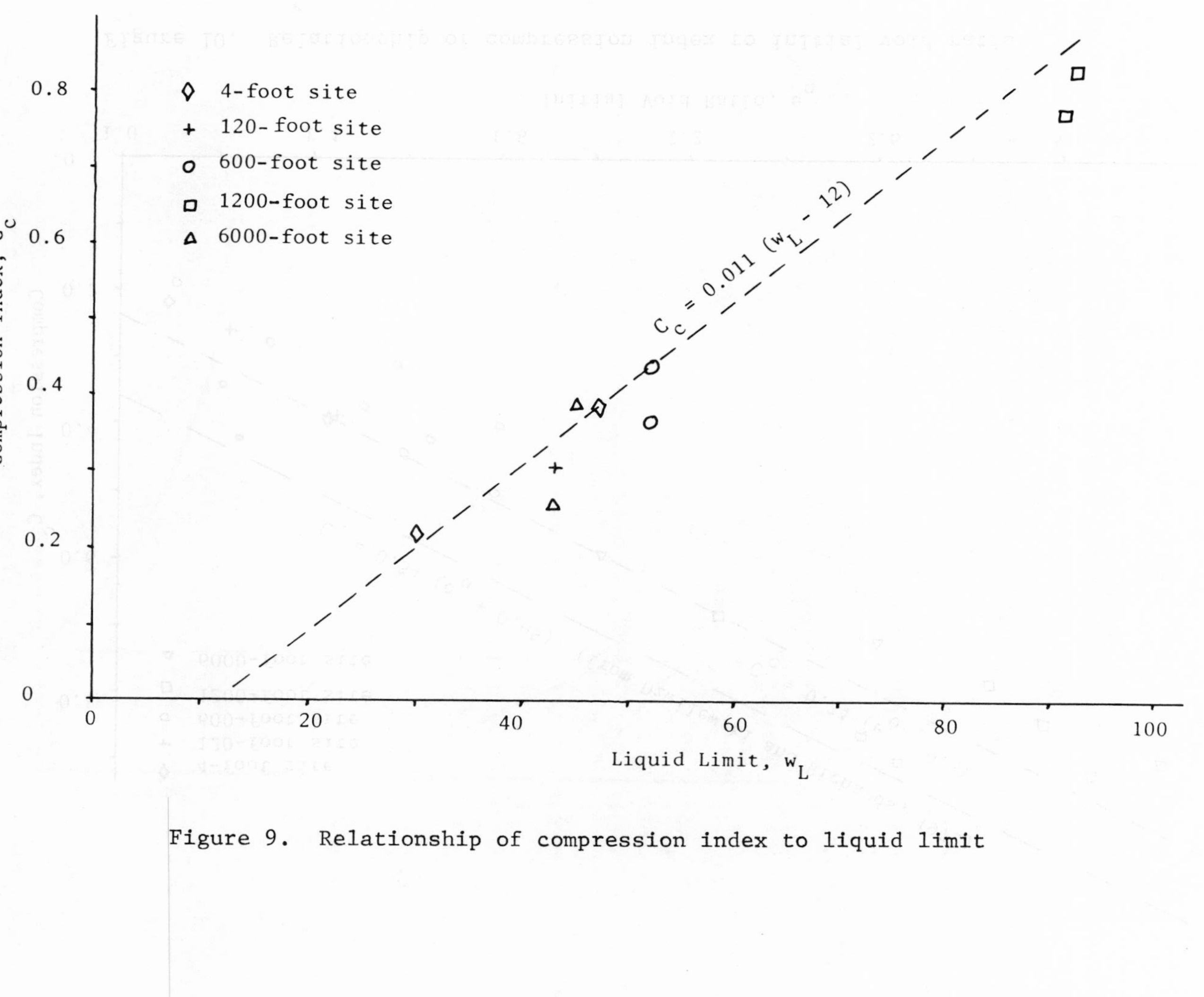

Figure 9. Relationship of compression index to liquid limit

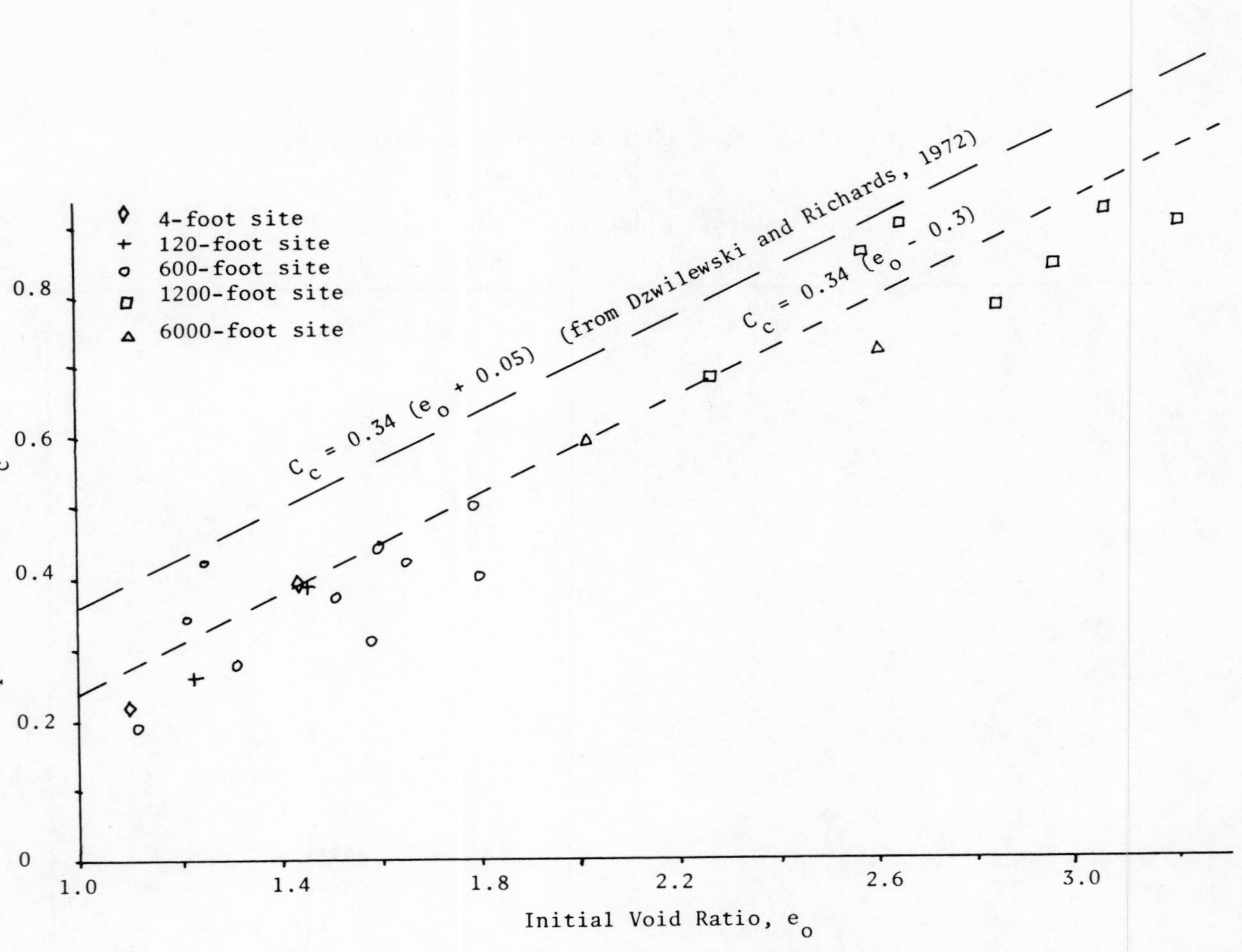

Figure 10. Relationship of compression index to initial void ratio

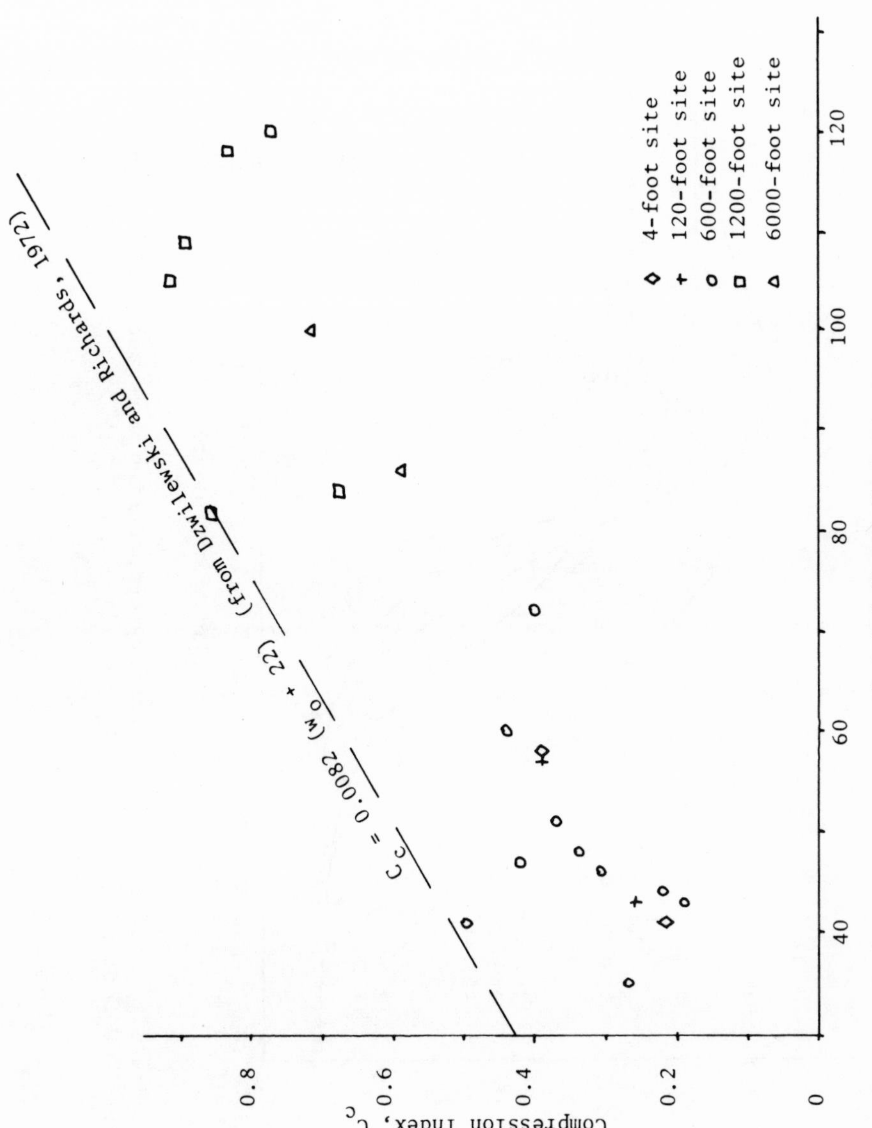

Figure 11. Relationship of compression index to initial water content

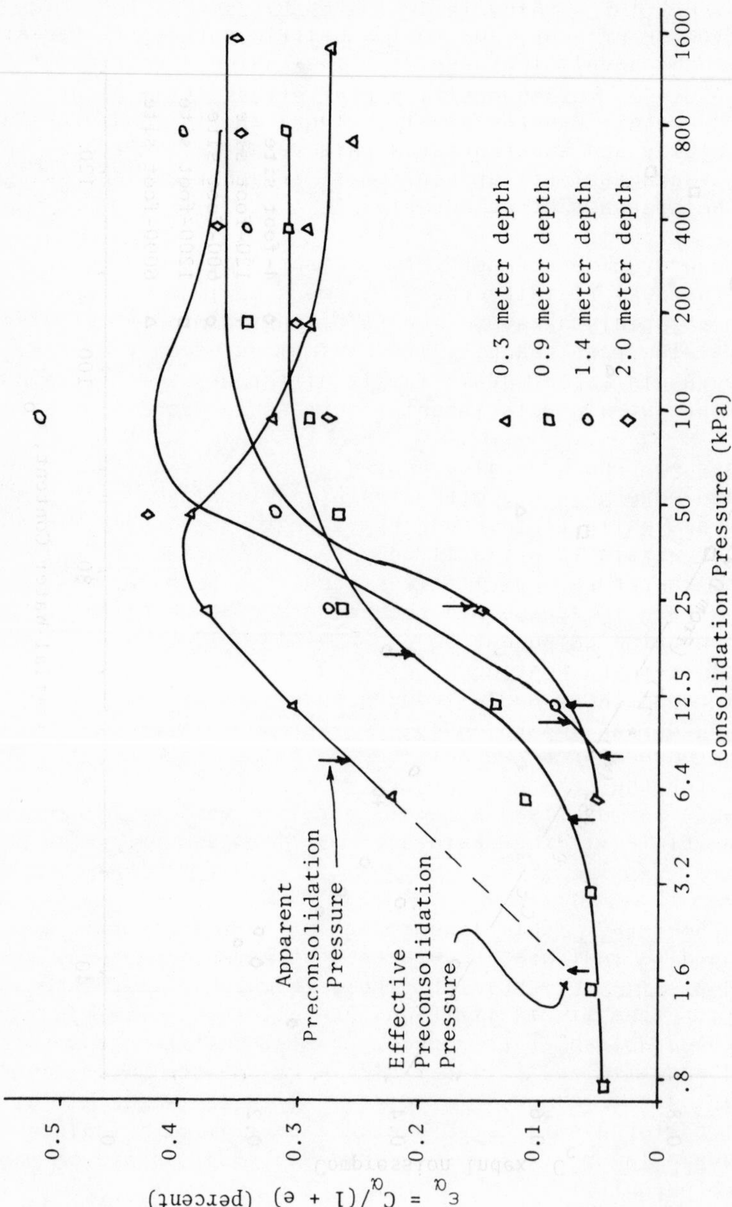

Figure 12. Coefficient of secondary compression versus consolidation pressure at the 1200-foot Site

general relationship between the Compression Index and Liquid Limit of a sample has been presumed by a number of investigators. Figure 9 shows the data from consolidation samples for which a Liquid Limit was determined on an immediately adjacent specimen and shows the numerical relationship developed by Rocker (Herrmann et al., 1972). This relationship:

$$C_c = 0.011\ (w_L - 12) \tag{1}$$

was developed from data which include those in this study. The relationship was tested against data published by Richards and Hamilton (1967) and Richards (1962) from the Atlantic and Mediterranean and found to be similarly valid.

The relationships of Compression Index to initial void ratio and water content are shown in Figures 10 and 11, respectively. The data in Figure 10 form a rather consistent band bounded on the upper side by the numerical relationship developed by Dzwilewski and Richards (1972):

$$C_c = 0.34\ (e_o + 0.05)$$

A better fit for these data would be:

$$C_c = 0.34\ (e_o - 0.3)$$

The relationship developed by Dzwilewski and Richards between original water content and Compression Index forms a similar upper bound for the sediments investigated here. As shown on Figure 11, the data are somewhat more scattered. Dzwilewski and Richards point out that their two relationships were developed for one Atlantic site and were more reliable for samples with larger void ratios and water contents. Figure 10 indicates that a general empirical relationship of Compression Index to initial void ratio may be possible. This would require the examination of a good deal more data and further validation studies.

Values for the Coefficient of Consolidation varied from 5.7 x 10^{-8} to 2.4 x 10^{-6} m^2/sec in the loading regions for virgin compression. The individual data points appeared to have a wide scatter; however, the average values proved to be relatively constant with increasing stress levels in both the short-term (as determined from the consolidometer loading increments on a given specimen) and long-term (as determined from comparison of results from consolidometer tests on specimens from different soil depths in the same core and loaded to stress levels of similar multiples of their effective overburden pressure, $\bar{\sigma}_v$). A relationship of the Coefficient of Consolidation to Liquid Limit is often assumed (NAVFAC, 1971). The data from this investigation were widely scattered and did not suggest any trend or qualitative relationship.

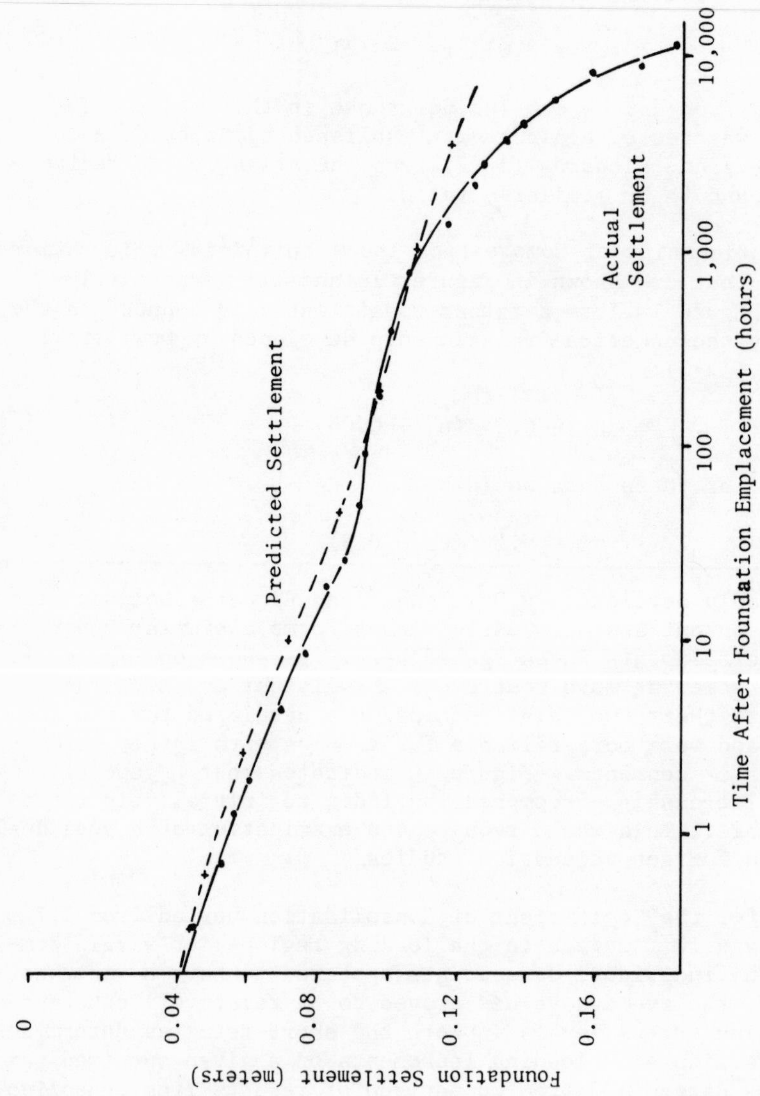

Figure 13. Predicted and actual footing settlement at the 4-foot Site

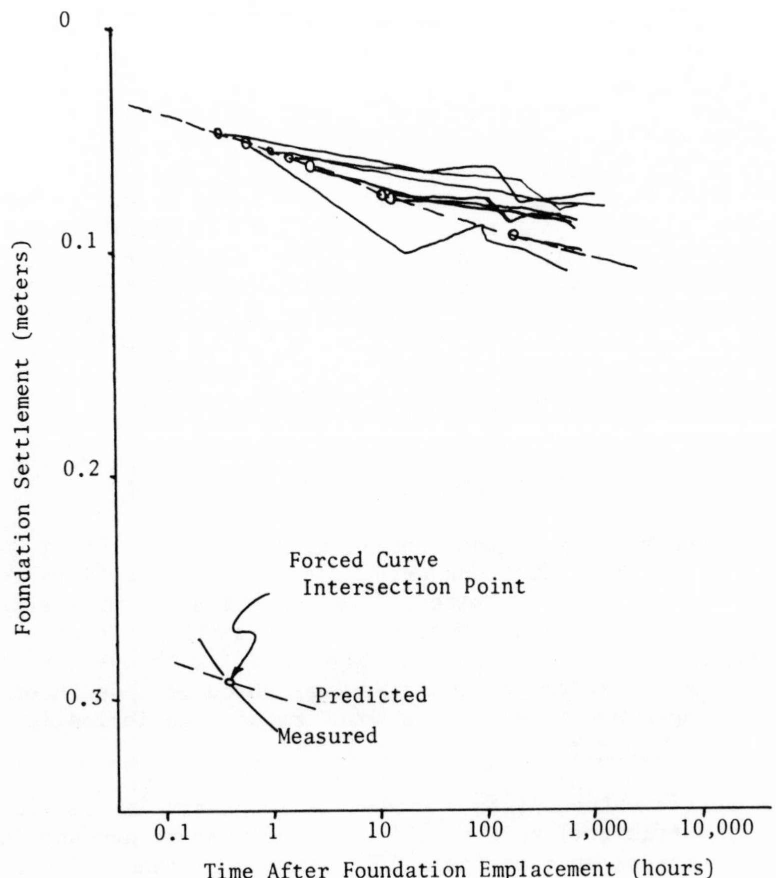

Figure 14. Predicted and measured settlements at the 120-foot Site

The Coefficient of Secondary Compression, C_α, was determined from the relatively linear final portion of the sediment compression versus logarithm-of-time data plot for each loading increment. Values for loading increments in virgin compression ranged from 0.0044 to 0.036. A second definition of Coefficient of Secondary Compression is:

$$\varepsilon_\alpha = \frac{C_\alpha}{1 + e}$$

Values for ε_α for virgin compression ranged from 0.0021 to 0.0113. This range of values would classify the sediments as having low, medium, or high secondary compression (Mesri, 1973). Results of this study showed that, as has been suggested elsewhere (NAVFAC,

1971), larger values of ε_α were characteristic of those sediments with larger Liquid Limits; however, the data from this study were too scattered to suggest an obvious quantitative relationship. Figure 12 shows data representative of the relationship of ε_α to applied pressure. These data are for consolidation tests at four sediment depths at the 1200-foot Site. Mesri (1973) suggests that the value of ε_α should be relatively constant as stresses increase above some level just beyond the maximum preconsolidation pressure, or the apparent preconsolidation pressure for the case of these sea-floor sediments. Examination of the data from all five sites shows that approximately one-half of the specimens tested showed this relatively constant value. Virtually all data show that a maximum value of ε_α is reached within two or three loading increments of the apparent preconsolidation pressure.

SETTLEMENT PREDICTIONS

Based on the results of the consolidation tests and other core data, settlement predictions were made for model footings to be placed at four of the five sites. The consolidation properties were sufficiently constant with depth at three of the sites to allow the assumption of no geological layering. At the 4-foot Site a distinct change in sediment properties was noted at a depth of 0.46 m. This depth was taken as the boundary of a geological layer.

In the settlement analysis a number of imaginary layers (10 to 12 is usually sufficient) were established in order to account for the variations in soil properties and loading conditions which are normally present in a soil profile. These factors which vary fairly uniformly with depth include the following: initial void ratio, initial vertical effective stress, stress increment due to applied surface loading, drainage path length, and time to completion of primary compression-consolidation. These layers were thin (0.076 m, 3 in) at the sediment surface where conditions vary more rapidly with depth, and thicker at greater soil depth. Sediment compressions were calculated to a depth where the stress increase due to the applied loading was less than 10% of the previously existing vertical effective stress. Insignificant sediment compressions can be expected below this depth (Hough, 1969).

A one-dimensional settlement analysis was performed. Settlement of a foundation of limited dimensions is the result of sediment compressions and deformations in more than one dimension; however, the laboratory testing procedures required for a proper three-dimensional analysis are extremely difficult to perform (Moore and Spencer, 1969) and were beyond the scope of this investigation. The one-dimensional analysis is fairly accurate for the case where much of the sediment compression occurs in sediments at a shallow soil depth, immediately beneath the footing. This is largely the case for the conditions investigated here where, because of the high stress increase ratios and compressibilities of the surficial sediments, the majority of the predicted settlement resulted from compressions in the upper foot or two of the soil profile. Stress increases with depth were determined using Boussinesq (Lambe and Whitman, 1969) for an average point located at a distance from the footing center equal to .707 times the radius of the circular footing. A triangular initial excess pore pressure distribution (Lambe and Whitman, 1969) was selected, as the best model of the Boussinesq stress distribution, to be used for determining the time rate of consolidation. The compression of each sediment layer due to primary compression, ΔH_p, was calculated using the following equation:

$$\Delta H_p = H.C_c \frac{\log\left(1 + \frac{\Delta\sigma}{\bar{\sigma}_v}\right)}{1 + e_o}$$

The quality of the samples appeared to be good; thus, no correction of the laboratory data to field conditions was deemed necessary. The effects of apparent preconsolidation were not taken into account in the settlement analysis.

In determining the rate of settlement, a drainage path equal to the sediment depth was used. This was modified at shallow sediment depths to take into account the average non-vertical distance to foundation weep holes which were spaced on either 0.3 m or 0.6 m centers, depending on the model footing used. Secondary compressions were assumed to begin once primary compression had reached completion, which for practical reasons was taken as Degree of Consolidation, U, = 95%. The amount of Secondary Compression, ΔH_s, at any given time, t, subsequent to the completion of primary compression, t_o, was determined from the following:

$$\Delta H_s = HC_\alpha \frac{\log(t/t_o)}{1 + e}$$

Where the values for layer thickness, H, and the void ratio, e, are determined at the completion of primary compression.

The predicted time-rates of settlement are simply the total of the primary and secondary compression for all layers summed for various times subsequent to foundation placement. These predicted time rates are shown in comparison to the actual measured settlements for the model footings in Figures 13, 14, 15, and 16.

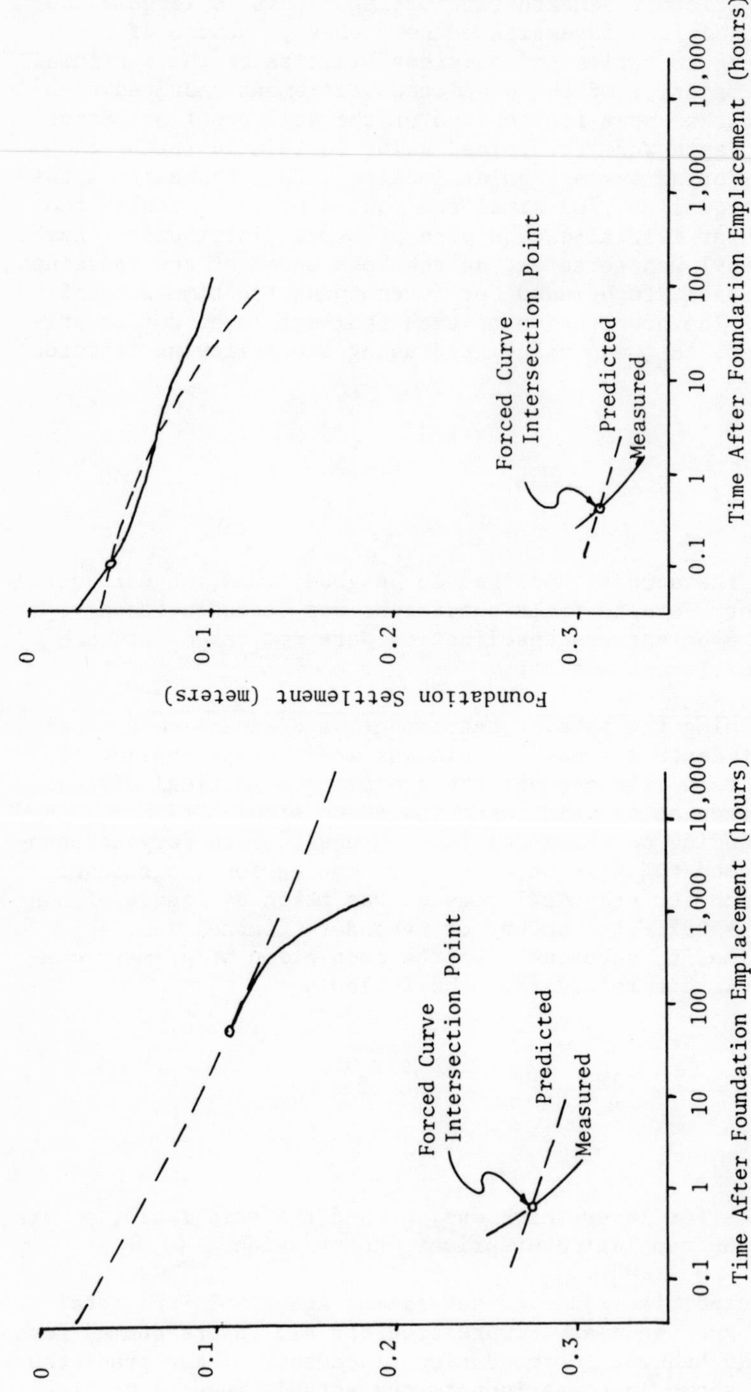

Figure 15. Predicted and measured settlement at the 600-foot Site

Figure 16. Predicted and measured footing settlement at the 1200-foot Site

MODEL FOOTING TEST RESULTS

Footings 1.22 m in diameter were deployed at each of the four shallower sites, and a 1.83 m diameter footing was also deployed at the 600-foot Site. These were monitored either manually or by data logging systems such as the Foundation Monitor System or the LOBSTER (Herrmann et al., 1972).

As can be seen on Figures 13 through 16, there is relatively good agreement between the predicted and actual time-rate of settlements. The 1.83 m diameter footing at the 600-foot Site is somewhat of an exception to this conclusion. The discrepancy is attributed to the areal variability of the site in combination with the fact that the footing had to be located approximately 400 m from the location of the core sample. The method of settlement analysis outlined here appears generally applicable to high void ratio, compressible sediments of the continental margins.

RECOMMENDATIONS

As a result of this and related sea-floor foundation investigations, it is concluded that several areas within sea-floor soil mechanics are in need of future investigative effort. These include: (a) the collection and analysis of case histories of long-term sea-floor foundation settlement; (b) the investigation of three-dimensional compressions occurring beneath a surface load of limited lateral dimension; (c) the effect of lateral loads on short- and long-term foundation displacements and stability; (d) the bearing capacity of footings on non-uniform sediment profiles; and (e) methods of reducing foundation settlements.

ACKNOWLEDGMENTS

The author wishes to acknowledge the help of Mr. Karl Rocker, Jr. in analyzing a major portion of the data presented here. The thought-provoking technical discussions with, and encouragement of, Messrs. P. Valent, H. Lee, H. Gill, and Dr. A. Inderbitzen are also gratefully acknowledged.

REFERENCES

Demars, K. R., and R. J. Taylor, Naval sea floor soil sampling and in-place test equipment: a performance evaluation, U. S. Naval Civil Engineering Laboratory Tech. Rept. R-730, 1971.

Dzwilewski, P. T., and A. G. Richards, Consolidation Properties of Wilkinson Basin Soils, preprint of a paper submitted to the J. Soil Mech. Fdn. Div., Am. Soc. Civil Engrs., 1972.

Herrmann, H. G., K. Rocker, Jr., and P. H., Babineau, LOBSTER and FMS: devices for monitoring long-term sea floor foundation behavior, U. S. Naval Civil Engineering Laboratory Tech. Rept. R-775, 1972.

Hough, B. K., Basic Soils Engineering, Second Edition, The Ronald Press, N. Y., 1969.

Lambe, T. W., and R. V. Whitman, Soil Mechanics, Wiley, N. Y., 1969.

Mesri, C., Coefficient of Secondary Compression, J. Soil Mech. Fdn. Div., Am. Soc. Civil Engrs., 99, (SM1), 123-137, 1973.

Moore, P. J., and G. K. Spencer, Settlement of building on deep compressible soil, J. Soil Mech. Fdn., Am. Soc. Civil Engrs., 95, (SM3), 769-790, 1969.

NAVFAC DM-7, Design Manual: Soil Mechanics, Foundations and Earth Structures, U. S. Naval Facilities Engineering Command, 1971.

Noorany, I., and S. F. Gizienski, Engineering properties of submarine soils: state-of-the-art review, J. Soil Mech. Fdn. Div., Am. Soc. Civil Engrs., 96, (SM5), 1735-1762, 1970.

Richards, A. F., Investigations of deep-sea sediment cores, II: mass physical properties, U. S. Navy Hydrographic Office Tech. Rept. TR-106, 1962.

Richards, A. F., and E. L. Hamilton, Investigation of deep-sea sediment cores, III: Consolidation, in Marine Geotechnique, edited by A. F. Richards, pp. 93-117, Univ. of Ill. Press, Urbana, 1967.

Singh, A., and Z. Yang, Secondary compression characteristics of a deep ocean sediment, in Proc. Intern. Symp. Eng. Properties of Sea Floor Soils and their Geophys. Ident., pp. 121-129, Seattle, 1971.

CONSOLIDATION OF MARINE CLAYS AND CARBONATES

WILLIAM R. BRYANT, ANDRE P. DEFLACHE, AND PETER K. TRABANT

Texas A&M University; Lamar Technical University;
Texas A&M University

ABSTRACT

Consolidation tests performed on a large number of marine sediments from the Gulf of Mexico indicate that high void ratio marine clay sediments exhibit a linear void ratio-pressure relation in contrast to the non-linear relation as ordinarily observed in clay soils. The use of this linear relation will provide a more accurate evaluation of preconsolidation pressures of marine sediments.

Testing of deeply buried marine sediments obtained from the Deep-Sea Drilling Project indicates that most Gulf of Mexico sediments tested were underconsolidated and those from the Pacific Ocean were both normally and underconsolidated.

Consolidation testing of carbonate sediments from the Gulf of Mexico and Bahama Bank indicate that fine-grained carbonate muds have consolidation characteristics similar to those of silty clays. The main difference between noncarbonate silty clays and fine-grained carbonate sediments is that under equal loads the carbonate sediments do not consolidate to as low a final void ratio as noncarbonates.

The permeability of the sediments undergoing consolidation testing were determined by the use of the time-compression curves. The relationship between void ratio and coefficient of permeability was found to be $k = 10^{-9}e^{5}$ cm/sec.

INTRODUCTION

Skempton (1970) defined consolidation as "the results of all processes causing the progressive transformation of an argillaceous

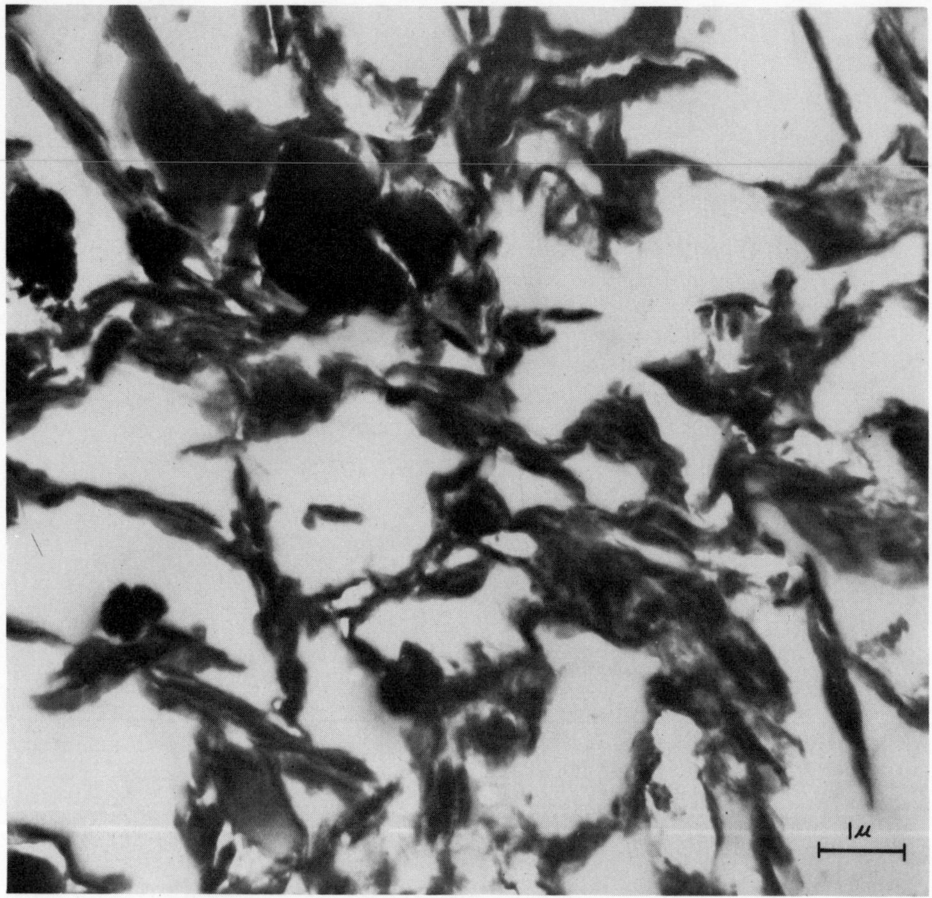

Figure 1. Electron micrograph showing ultrathin section of undisturbed sample

sediment from a soft mud to a clay and finally to a mudstone or shale." He suggested that the processes of consolidation include inter-particle bonding, dessication, cementation and the squeezing out of pore water under increasing weight of overburden. The latter process, essentially the reduction of void ratio with increasing imposed load, is the same as is accomplished in the laboratory by the use of an oedometer. The time factor in the laboratory study of the consolidation characteristics of a sediment sample necessarily precludes changes in inter-particle bonding and cementation that may take place with time in the natural environment.

Thus the reduction of the void ratio of a sediment under a rapidly applied load most certainly does not have the same effect as extremely slow continuous loading occuring over long periods of time.

Figure 1 is a micrograph of an ultrathin section of an unconsolidated sediment and reveals that the microstructure is a loose, open, random arrangement of particles. Figure 2 shows the microstructure of this material consolidated by an imposed load of 392.3 kPa (4.0 kg/cm^2). This resulted in a much denser packing of particles and a tendency for the particles to be oriented in packets but without as high a degree of preferred orientation as expected. Under a load of 6276.3 kPa (64 kg/cm^2), a high degree of preferred particle orientation occurs (Fig. 3). It is debatable whether such orientation is obtained in nature at an equivalent in situ pressure.

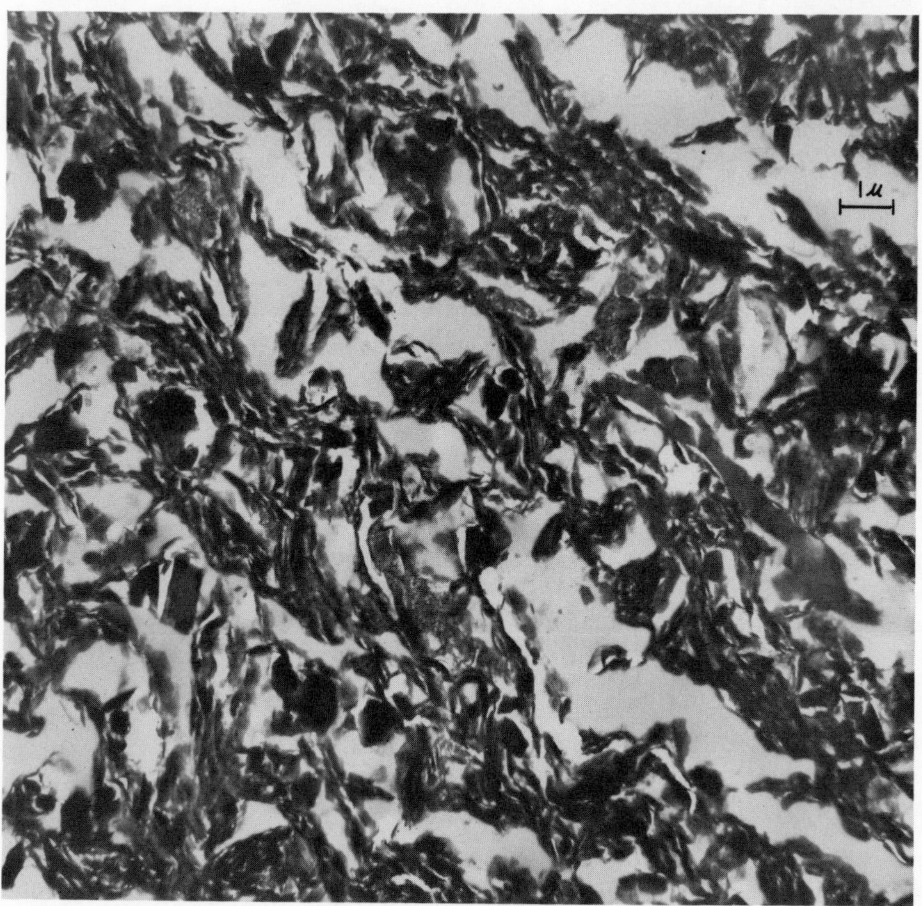

Figure 2. Electron micrograph showing ultrathin section of sediment sample subjected to a normal load of 4.0 kg/cm^2

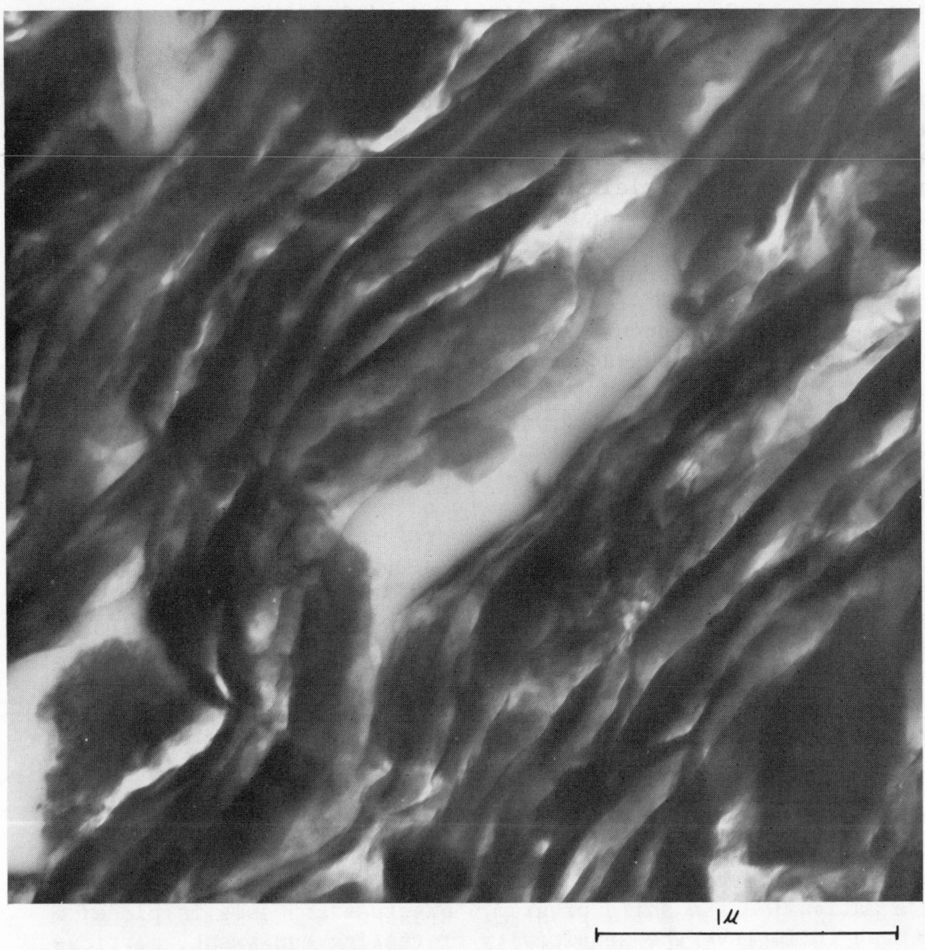

Figure 3. Electron micrograph showing ultrathin section of sediment sample subjected to a normal load of 64 kg/cm^2

It is safe to say that testing sediments by an oedometer furnishes information to the engineers for construction purposes but perhaps leaves a lot to be desired in obtaining a knowledge of the processes of consolidation in the natural environment.

Even with the severe limitations of oedometer testing and the impossibility of obtaining undisturbed samples from most of the areas covered by the world's oceans, an examination of the results of such testing may help us gain insight into the process of consolidation of marine clays and oozes.

AN ANALYSIS OF THE CONSOLIDATION OF CLAY SOILS

In a laboratory compressibility test developed by Terzaghi (1925), a clay specimen is encased in a ring and sandwiched between two porous stones. A vertical pressure applied to the specimen places the water in compression so that initially the pore water carries all the vertical pressure. As water leaves the voids, the pore pressure decreases and the intergranular pressure increases, causing a compression of the specimen. When the compression has virtually ceased, all the applied vertical pressure is carried by the soil particles and the pore pressure is nil. The primary or hydrodynamic stage of consolidation has ended.

The vertical pressure is then increased (usually doubled) and this is repeated for the range of pressures under investigation. Since the compression of the soil specimen is caused by the pressure actually carried by the soil particles only, this pressure is called the "effective pressure, $\bar{p}$" ($\bar{p} = p - u$).

Results of laboratory compressibility or consolidation tests performed on soil specimens can be presented in the form of conventional strain - effective pressure, true strain - effective pressure, or void ratio - effective pressure curves. The conventional strain, ε, is the ratio of the total compression to the original height of the specimen. The true or natural strain, ε', is the ratio of the compression increment to the actual height of the specimen. The void ratio, e, of a soil is the ratio of the volume of voids to the volume of solids. Since, in the consolidation test, the cross-sectional area of the sample remains constant, the void ratio, e, is also the ratio of the height of voids to the height of solids. The height of solids remains constant throughout a given consolidation test.

The sensitivity of a clay is expressed by the ratio between the unconfined compressive strength of an undisturbed sample and the strength of the same sample at the same water content but in a remolded state. For medium to low sensitive clays, the sensitivity is less than 4. Extrasensitive clays are known with sensitivities greater than 8.

For clays with medium to low sensitivity, the void ratio-pressure relation revealed by consolidation tests can be described as follows (Terzaghi and Peck, 1948).

As shown in Figure 4A, when the $\varepsilon - \bar{p}$ or $e - \bar{p}$ curve is plotted to arithmetic scales, the shape of the "compression curve" is concave upward, i.e., the slope of the curve decreases with increasing pressure and decreasing void ratio (curve a).

The slope of the $\varepsilon - \bar{p}$ curve defines the coefficient of volume

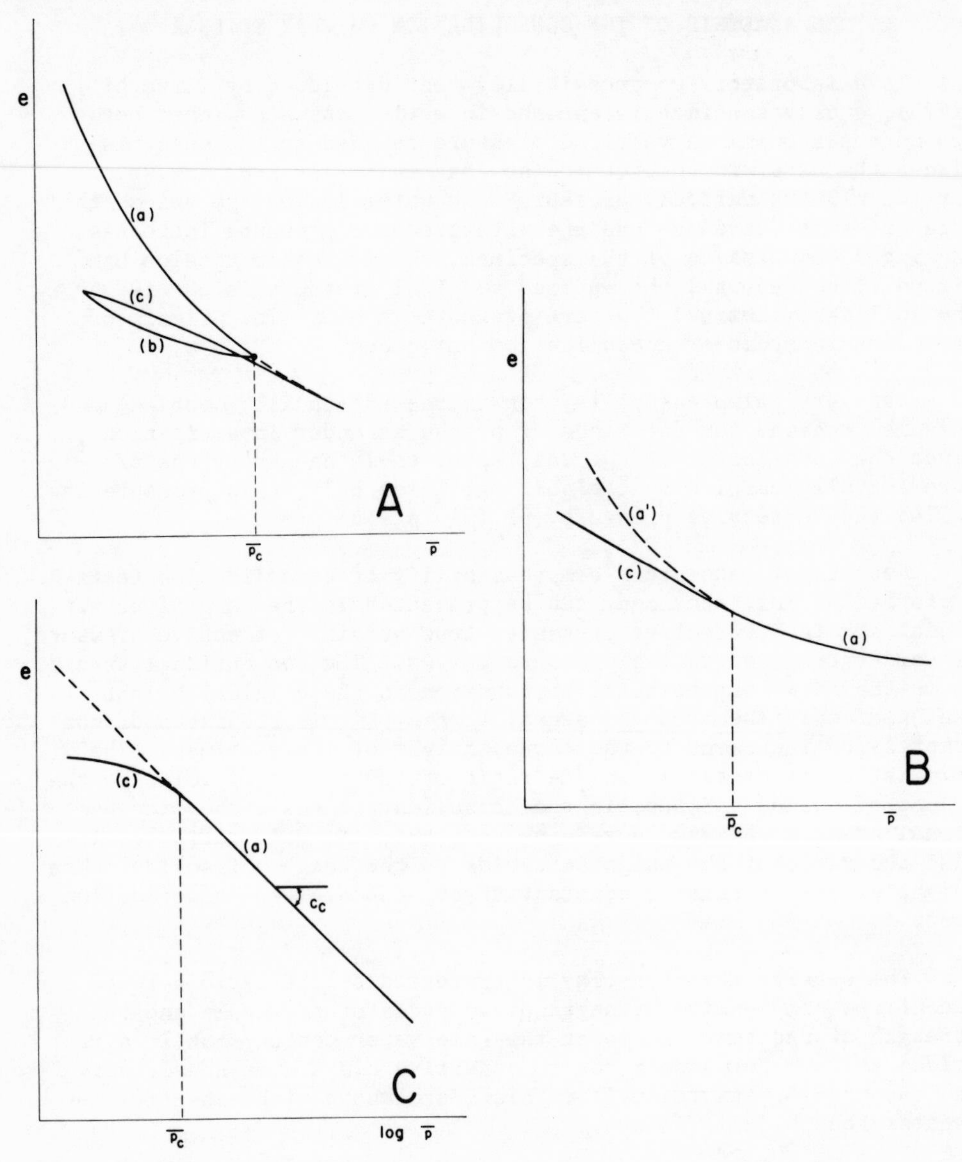

Figure 4a. e – $\overline{p}$ curves of ordinary clay soils

Figure 4b. Compression curves of ordinary clay soil samples in e – $\overline{p}$ diagram

Figure 4c. Compression curves of ordinary clay soil samples in e-log $\overline{p}$ diagram

decrease, m_v, which is the decrease in unit volume per unit increase of pressure. The slope of the e - $\bar{p}$ curve is the coefficient of compressibility, a_v, the decrease in void ratio per unit increase of pressure.

If the pressure is increased to a certain value $\bar{p}_c$ and then reduced, the soil sample will not rebound to its initial void ratio along the same curve. It will instead rebound along the "decompression curve" (curve b), which is also concave upward but has a slope smaller than that of the compression curve. If the pressure is increased again, the void ratio-pressure relation follows the "recompression curve" (curve c).

The first part of the recompression curve is often nearly straight, i.e., the coefficients of volume decrease, m_v, and of compressibility, a_v, are constant in recompression. If the pressure is increased beyond $\bar{p}_c$, the recompression curve (curve c), joins the initial compression curve a, called the virgin curve.

When a soil sample is removed from the ground, the pressure caused by the overburden is reduced and the sample rebounds. The first part of the laboratory compression curve corresponds, therefore, to a recompression relative to in situ conditions (Fig. 4B, curve c), the second part is the laboratory virgin compression curve (Fig. 4B, curve a). For any point on the recompression curve, $\bar{p}_c$ is the maximum previous consolidation pressure or the preconsolidation pressure.

As shown in Figure 4C when the pressure is plotted to a logarithmic scale, the part of the e-log $\bar{p}$ curve which represents recompression is concave downward (curve c). It is followed by the virgin curve which forms a nearly straight line, the slope of which is the compression index C_c (line a). Therefore, the virgin compression of clays with medium to low sensitivity occurs according to a linear relationship between the conventional strain, ε, and the logarithm of $\bar{p}$.

The virgin void ratio-pressure relation can be extrapolated to pressure values smaller than $\bar{p}_c$ by extending the laboratory virgin straight line to the left (Fig. 4C, curve a).

As shown in Figure 4B when these extrapolated values are transferred to the e - $\bar{p}$ diagram (curve a'), the recompression curve (curve c) is tangent to the virgin curve a and a'.

Several methods have been proposed to estimate the preconsolidation pressure $\bar{p}_c$ from e-log $\bar{p}$ diagrams. When the preconsolidation pressure and the calculated effective overburden pressure are the same, the clay sample is said to be normally consolidated. When the preconsolidation pressure is greater or smaller than the calculated effective overburden pressure the clay is said to be over- or under-consolidated, respectively.

CHARACTERISTICS OF ABYSSAL PLAIN AND CONTINENTAL SLOPE SEDIMENTS

More than 120 consolidation tests were performed on samples taken from 70 cores most of which were from the abyssal plain in the Gulf of Mexico.

An extensive examination of the sediment samples disclosed the following geological facts:

1. Most of the sediments are Pleistocene or older.
2. No erosion or removal of overburden has taken place.
3. The sediments are homogeneous silty clays with an average medium diameter of 9 ϕ units.
4. The predominate clay mineral is montmorillonite.
5. The organic content averages about 1% and the calcium carbonate content about 14%.
6. The environment where the sediments were deposited has undergone little change since deposition of the sediment samples.
7. The sediments have always been in a saturated state.
8. Very little biomechanical disturbance has taken place.

The age of the sediments and the stable conditions of the environment since deposition of the samples indicate that the sediments should be in a state of normal consolidation. In fact, testing indicates that sediments from the abyssal plain of the Gulf of Mexico could be used as a standard whose conditions could characterize a normally consolidated fine-grained marine sediment.

To illustrate the compressional behavior of such marine sediments, compression curves obtained on core 7-9-18E recovered from 3639 m (1990 fms) on the floor of the abyssal plain, are presented in Figure 5. The location and associated geotechnical properties of samples used in this paper are given in Table I. These compression curves were determined from consolidation tests run in an Anteus Back Pressure consolidometer. The main advantage of the Anteus consolidometer is that it allows complete saturation of the sample by the application of back pressure. Gas bubbles in a partially saturated sediment are highly compressible compared to the relatively incompressible pore water and particles. The presence of gas bubbles within the pore water also impedes the flow of water through voids reducing the effective permeability of the sediment. In order to redissolve the gases within the pore water a back pressure is applied and maintained on the sample throughout the test. The application of back pressure produces an in-place pressure on the pore water prior to loading. For samples from high hydrostatic pressure

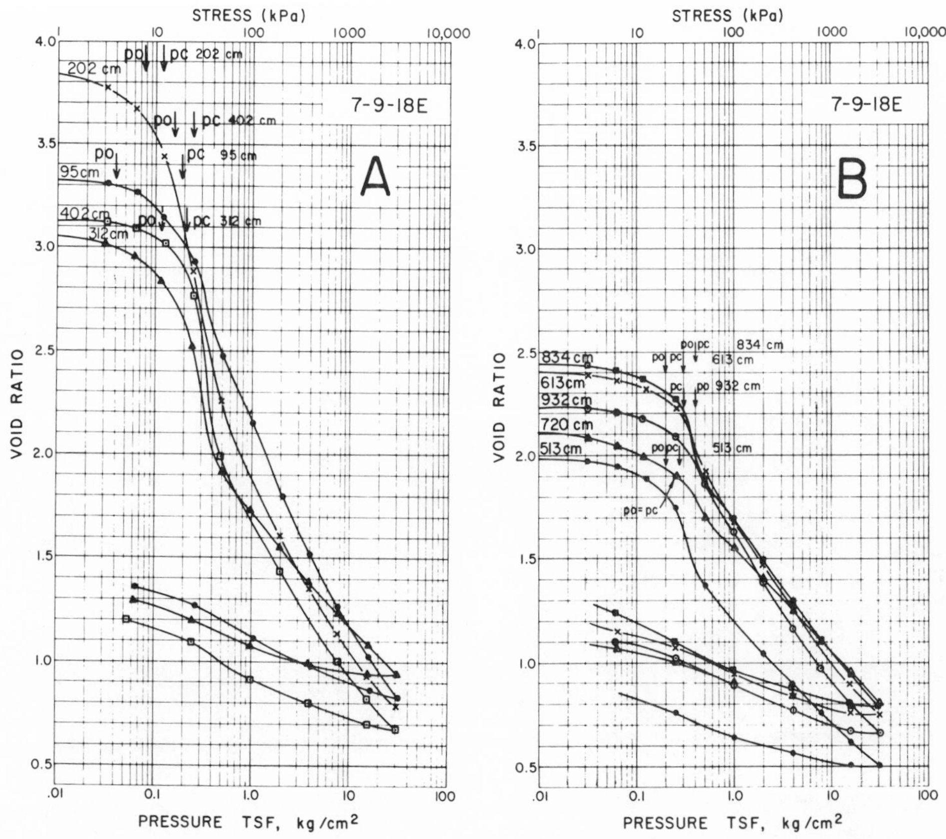

Figures 5a & b. e-log $\overline{p}$ curves of abyssal plain sediments from Core 7-8-18E

environments the in-place pressure cannot be duplicated but tests indicate that back pressure of 1034.2 kPa (150 psi) insures 99.99% saturation.

COMPRESSION CHARACTERISTICS OF SEDIMENTS FROM THE DEEP WATER PORTIONS OF THE GULF OF MEXICO

The results of oedometer tests of sediment samples from the abyssal plain and continental rises of the Gulf of Mexico are shown in Figures 5 and 6.

Figures 4A and 4B constitute tests of samples taken from a single core (No. 7-9-18) from which nine sediment samples ranging in depth below the sediment water interface from 95 cm to 932 cm,

TABLE 1

LOCATION AND GEOTECHNICAL PROPERTIES OF CONSOLIDATED SAMPLES

Sample No.	Depth In Core cm.	Location Long. W	Lat. N	Water Depth meters	Median Dia. phi units	Carbonate %	Water Content % dry wt.	Unit Weight gm/cc	Shear Strength p.s.f.	Shear Strength kPa
64-9-7	290	92°27.9'	23°46'	3465	10.0	50	118	1.41	-	-
65-4-12	70	92°42.5'	20°46.5'	2545	9.5	51	100.5	-	204	9.77
65-4-18	90	92°35'	20°33'	2371	9.6	27	115.9	1.42	101	4.79
65-4-18	180	92°35'	20°33'	2371	8.5	53	76.0	1.59	142	6.80
65-4-29	40	95°47.5'	20°51.5'	2262	9.3	57	99.8	1.48	198	9.48
66-3-6	267	87°38'	25°07'	3303	10.2	-	113.2	1.42	90	4.31
66-3-12	264	88°03'	26°17'	2739	10.1	-	106.2	1.43	51	2.44
66-3-13	230	88°05'	26°29'	2598	-	-	-	-	-	-
66-3-25	217	88°50'	27°55'	1625	10.3	-	135.9	1.36	70	3.35
66-13-30	276	88°59'	28°14'	1225	10.0	-	108.1	1.43	86	4.12
66-13-73	260	92°12'	27°41'	345	9.8	-	60.8	-	492	25.56
66-13-73	266	92°12'	27°41'	345	10.0	-	-	-	Very Stiff	
66-13-73	422	92°12'	27°41'	345	9.6	-	-	-	Very Stiff	
66-13-73	527	92°12'	27°41'	345	9.8	-	-	-	Very Stiff	
67-3-8	195	89°48'	25°46'	3195	8.3	6	144	1.35	60	2.87
67-3-8	506	89°48'	25°46'	3195	8.9	9	111	1.41	143	6.85
67-3-8	812	89°48'	25°46'	3195	8.7	9	75	1.57	173	8.28
67-3-8	916	89°48'	25°46'	3195	8.6	9	85	1.52	173	8.28
67-3-9	278	89°21'	27°03'	2441	7.2	15	109	1.42	90	4.31
67-3-9	572	89°21'	27°03'	2441	8.0	13	86	1.51	137	6.56
67-3-9	888	89°21'	27°03'	2441	10.0	8	79	1.54	223	10.68
67-9-12	692	94°10'	23°11.5'	3642	10.1	11	80	1.54	155	7.42
67-9-12	780	94°10'	23°11.5'	3642	10.1	11	80	1.53	177	8.47
67-9-12	884	94°10'	23°11.5'	3642	8.9	11	80	1.54	217	10.39
67-9-18	202	93°50'	24°40'	3630	10.3	7	-	-	57	2.73
67-9-18	402	93°50'	24°40'	3630	10.6	7	122	1.40	103	4.93
67-9-18	513	93°50'	24°40'	3630	10.3	11	98	1.48	185	8.86
67-9-18	613	93°50'	24°40'	3630	10.3	10	86	1.52	164	7.85
67-9-18	720	93°50'	24°40'	3630	9.9	45	77	1.56	267	12.78
67-9-18	932	93°50'	24°40'	3630	10.1	10	80	1.55	230	11.01
68-11-8	35	82°19'	23°40'	1536	8.5	83	84	1.52	276	13.2
68-11-8	230	82°19'	23°40'	1536	9.0	75	85	1.52	~300	14.36
68-11-8	535	82°19'	23°40'	1536	7.2	85	65	1.62	300	14.36
68-11-8	765	82°19'	23°40'	1536	7.8	82	71	1.59	375	17.96
70-1-1G	295	94°37'	26°09'	2743	10.3	15	99	1.46	146	6.99
70-1-1G	298	94°37'	26°09'	2743	10.3	15	100	1.46	148	6.98
Site 91	195 meters	92°20.7'	23°46.4'	3763	9.9	22	29	1.99	Stiff	
Site 91	306 meters	92°20.7'	23°46.4'	3763	9.2	14	30	2.01	Very Stiff	
Site 91	415 meters	92°20.7'	23°46.4'	3763	9.4	28	24	2.11	Very Stiff	

TABLE 1

(CONTINUED)

Sample No.	Depth In Core cm.	Location Long. W	Lat. N	Water Depth meters	Median Dia. phi units	Carbonate %	Water Content % dry wt.	Unit Weight gm/cc	Shear Strength p.s.f.	Shear Strength kPa
Site 91	541 meters	92°20.7'	23°46.4'	3763	9.9	42	28	2.00	Very Stiff	
Site 92	34 meters	91°49.3'	25°50.7'	2573	10.3	16	48	1.86	Stiff	
Site 92	130 meters	91°49.3'	25°50.7'	2573	9.7	16	29	2.01	Very Stiff	
Site 92	182 meters	91°49.3'	25°50.7'	2573	10.0	11	35	1.92	Very Stiff	
Test 1	54	28°56.5'	29°01.8'	92	9.8	2.0	130	1.38	48	2.29
Test 2	128	93°50.0'	24°45.8'	3639	10.1	8.0	148	1.37	63	3.02
Test 3	208	88°45'	27°37'	1824	10.5	5.7	137	1.37	72	3.44
Test 4	118	94°06.3'	26°55.8'	1935	10.2	21.1	131	1.37	93	4.45

were collected and tested. At all levels tested except one (932 cm) the sediment could be, for all practical purposes, normally consolidated, even though in all cases p_o was less than p_c. The same is true for the four sediment samples of core 7-3-8E shown in Figure 6A, the three samples of core 7-3-9E (Fig. 6B), and the three samples of core 7-9-12 (Fig. 7A). In each case only the deepest sample had a value of p_c larger than p_o.

Seven cores were taken on a traverse of the Mississippi Fan from the abyssal plain to the continental shelf edge immediately south of the Mississippi Delta. Samples taken from these cores at approximately 250 cm below the sediment water interface were tested and the resulting e-log $\bar{p}$ curves are shown in Figures 8A and 8B. Core 6-3-6E is in the deepest water (3600 m) and at the south end of the traverse, 444.5 km (240 nautical miles) distance from core 6-3-30E. The similarity of the e-log $\bar{p}$ curves for six of the cores in the deeper water shows the uniformity of sediment characteristics in the area.

Samples of continental slope sediments are shown in Figure 9A. Four sediment samples were taken and tested from core 6-13-73 which was taken in 700 m of water. Certain sediments of the northern continental slope in the Gulf of Mexico have much smaller void ratios than in most other parts of the Gulf. Two samples (260 cm and 266 cm) show almost identical e-log $\bar{p}$ curves attesting to the quality of the oedometer tests and the similarity of sediment type.

The same similarity of e-log $\bar{p}$ curves for samples taken at close intervals is also shown in Figures 10A and 10B. These four samples were taken from core 70-1-1G. The samples tested, taken at 285 cm and 289 cm, were similar while those taken at 295 cm and 298 cm were almost identical. These curves demonstrate that the determination of p_c by the standard methods is at best only a good estimate; they also display the reproducibility of the testing procedure.

A series of high void ratio sediments subjected to very high normal pressures is shown in Figure 9B. Four samples of void ratios above 3.5 were compressed under final pressures from 11768 kPa (120 kg/cm^2) up to 24517 kPa (250 kg/cm^2). It is interesting to note that the void ratio at 24517 kPa was 0.19 and the virgin curves do not appear to becoming asymptotic with the pressure axis.

CONSOLIDATION OF MEDIUM TO HIGH SENSITIVITY CLAYS

Laboratory consolidation tests conducted on Holocene and Pleistocene clays of the Gulf of Mexico indicate that medium to high sen-

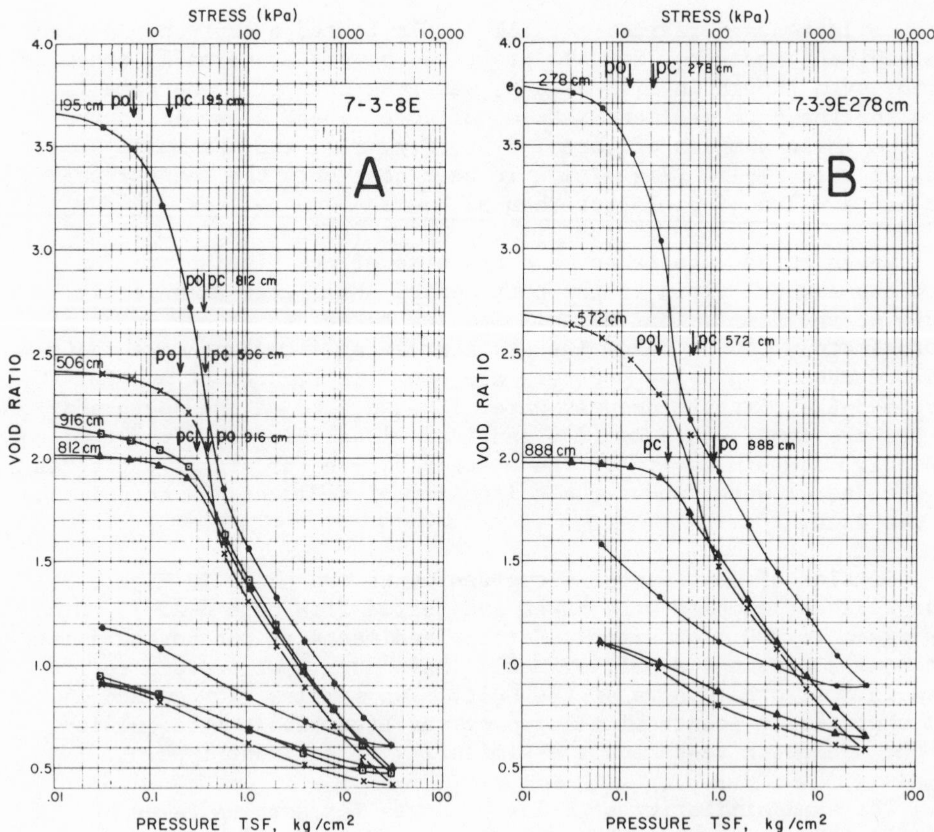

Figures 6a & b. e-log $\bar{p}$ curves of abyssal plain sediments from Cores 7-8-8E and 7-3-9E

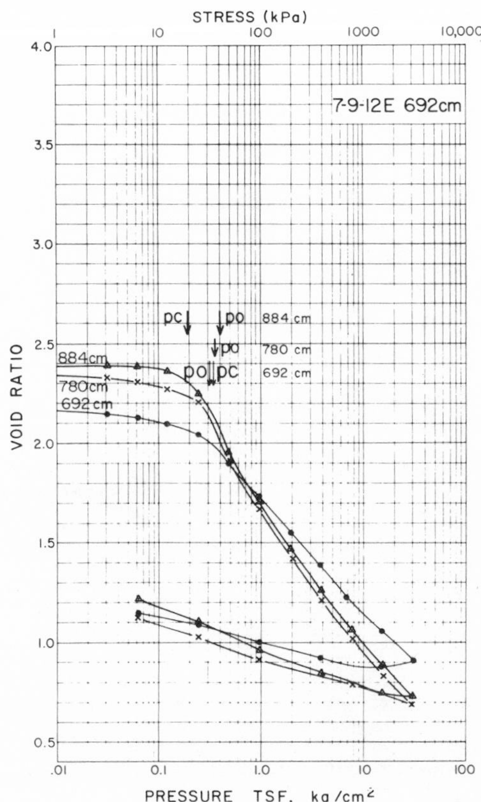

Figure 7. e-log $\bar{p}$ curves of abyssal plain sediments of Core 7-9-12E

sitivity clays, with an initial void ratio greater than about 1.0, behave differently. When presented in the e-log $\bar{p}$ diagram, the virgin compression curve (a) is no longer a straight line. It is concave upward as shown in Figure 11A. As a result of this (1) the existing methods for the determination of the preconsolidation pressure, $\bar{p}_c$, are no longer dependable; (2) the compression index C_c is no longer constant; and (3) the extrapolation of the virgin compression curve to pressure values smaller than $\bar{p}_c$ becomes difficult. The virgin compression curve can be rectified by representing it in another coordinate system consisting of the reduced height, h, as ordinate and the effective pressure, $\bar{p}$, as abscissa, both plotted to logarithmic scales. The reduced height, h, is defined as 1 + e, where e is the void ratio. The reduced height is also the sample's height divided by the height of solids and is therefore dimensionless. When plotted in the log h - log $\bar{p}$ system, the laboratory virgin compression curve of medium to high sensitivity clay soils forms a straight line (Fig. 12). The slope of this line is called the compression exponent, η_v. The virgin compression occurs according to a linear relationship between the true strain ε' and the logarithm of $\bar{p}$. This relationship can be explained by assuming that the distance between clay particles is increasingly proportional to $\bar{p}^{\eta v}$ or:

$$h\ \bar{p}^{\eta v} = \text{constant}$$

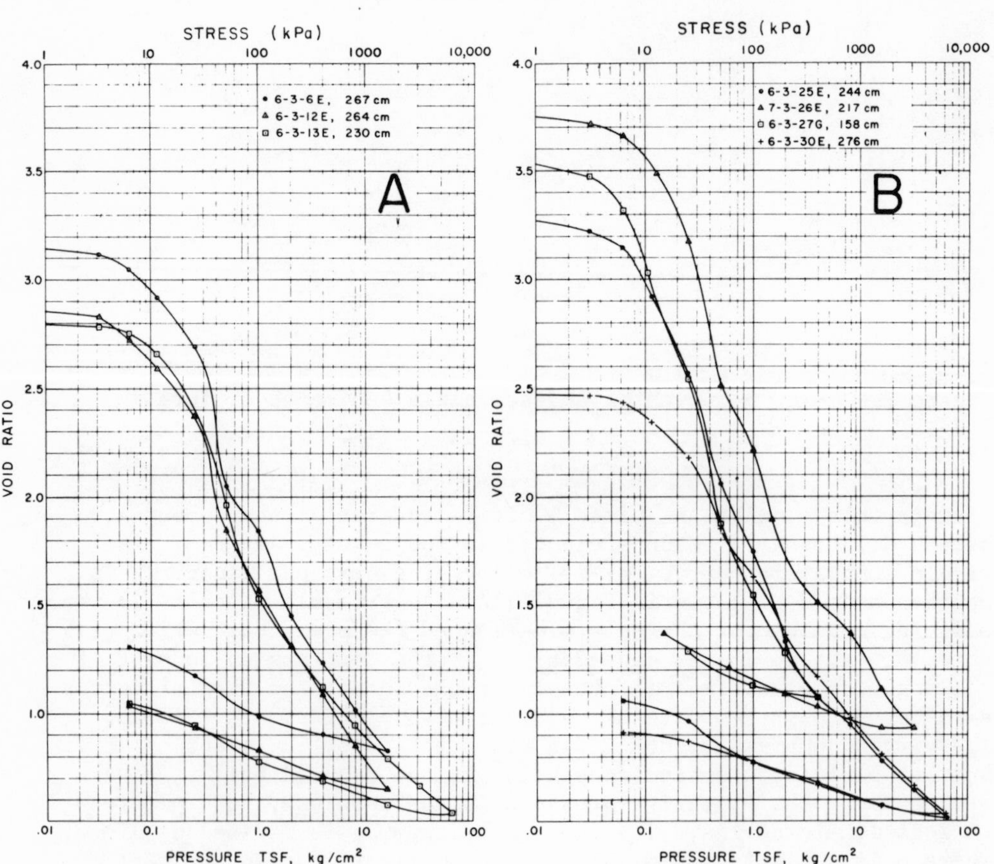

Figures 8a & b. e-log $\bar{p}$ curves of sediments from the Mississippi Fan

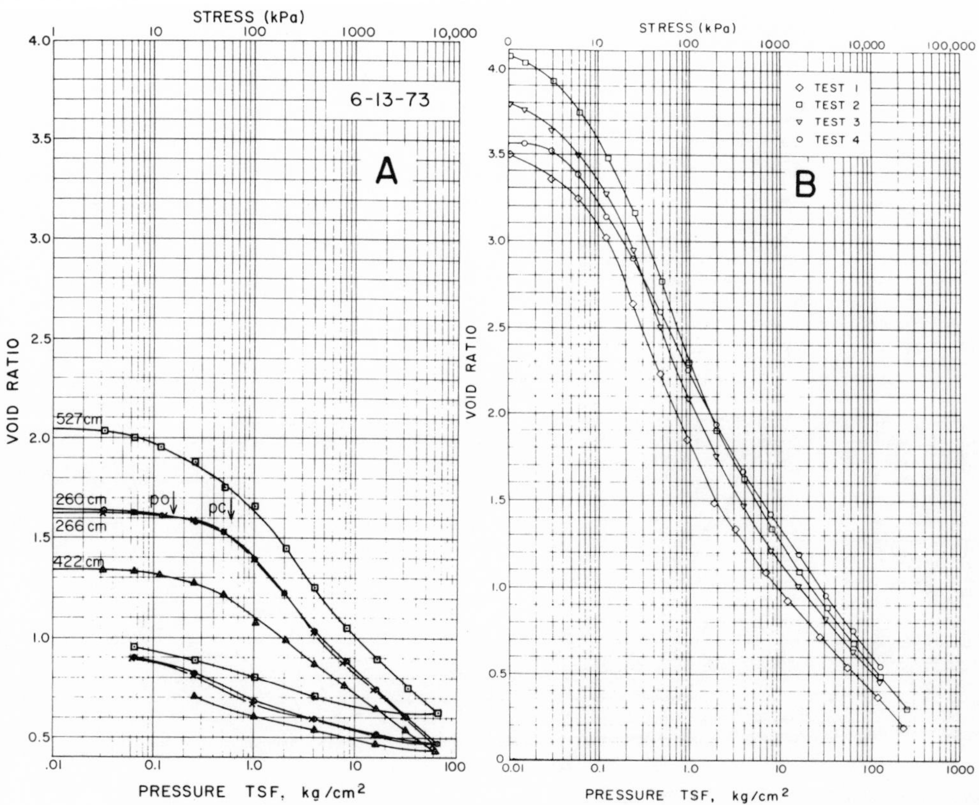

Figure 9a. e-log $\bar{p}$ curves of sediments of Core 6-13-73 from the continental slope

Figure 9b. e-log $\bar{p}$ curves of high void ratio sediments under high normal loads

CONSOLIDATION OF HIGH-VOID RATIO MARINE SEDIMENTS

The conventional representation of the void ratio-pressure relation in the e-log $\bar{p}$ diagram for the 202 cm sample is shown in Figure 11A. It differs from the classical compression curves of clay soils in that (1) the limit of the recompression curve, c, is ill-defined, and (2) the virgin compression portion, a, of the e-log $\bar{p}$ curve is no longer a straight line.

The effective overburden pressure, $\bar{p}_o$, for the 202 cm sample was calculated to be 7.55 kPa (77 g/cm^2) and, as shown in Figure 11A, the preconsolidation pressure was estimated at 12.75 kPa (130

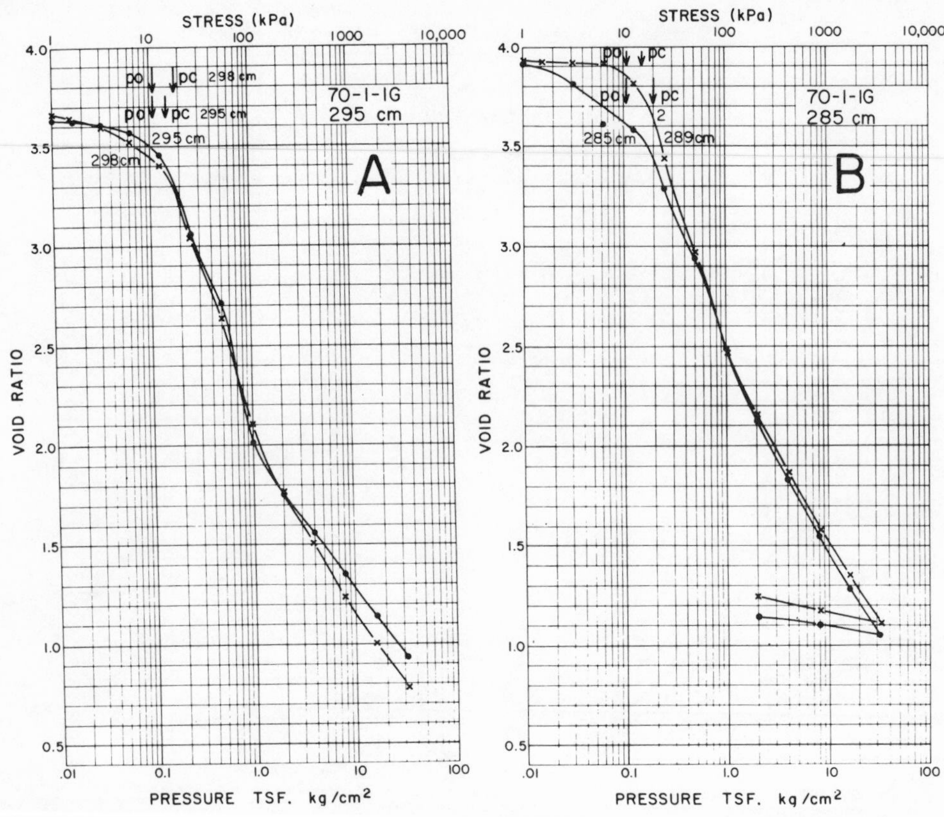

Figures 10a & b. e-log $\bar{p}$ curves of closely spaced sediment samples

g/cm^2). This would indicate that the sample is overconsolidated which is contrary to geological interpretation.

At least two values for the compression index can be obtained: 0.89 and 2.24.

Similarly, as shown in Figure 11B, the effective overburden pressure for the 402 cm sample was calculated at 15.30 kPa (156 g/cm^2), the preconsolidation pressure was estimated at 23.54 kPa (240 g/cm^2) and at least two values for the compression index can be obtained: 0.87 and 4.13.

In both Figures 11A and 11B the compression indices were computed on the field consolidation lines (dashed lines) which represent the e-log $\bar{p}$ relation in the field. These lines can be determined by an extrapolation process from the results of laboratory tests. For these high-void ratio sediments (greater than 2.5), the shape of the laboratory compression curve is somewhat similar to that described by Terzaghi and Peck (1948) for extra-sensitive clays.

As shown in Figure 13 for samples removed from within approximately the upper 5 m below mudline, the laboratory void ratio-pressure relation plotted to arithmetic scales consists of three parts:

1. a linear recompression curve, c_1, for pressure values not in excess of $\bar{p}_c$,

2. an apparent stuctural collapse, a_1, for pressure values greater than $\bar{p}_c$ but smaller than $\bar{p}_b$,

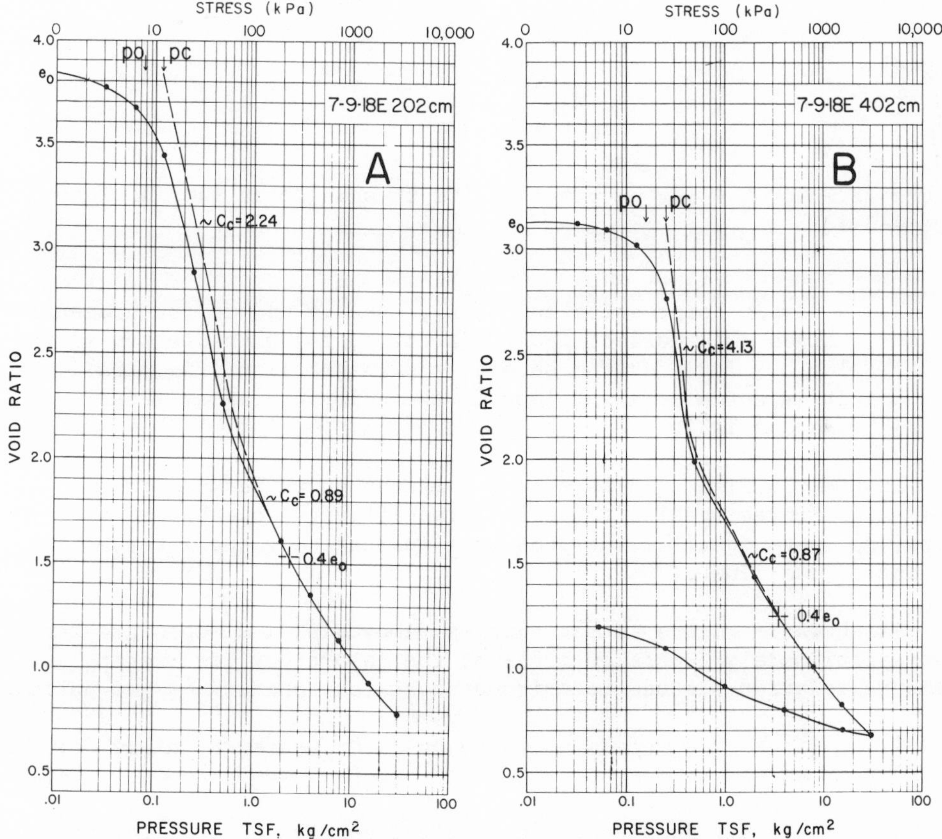

Figure 11a. e-log $\bar{p}$ curve of medium to high sensitivity clays

Figure 11b. e-log $\bar{p}$ curve of 720 cm sample from Core 7-9-18E

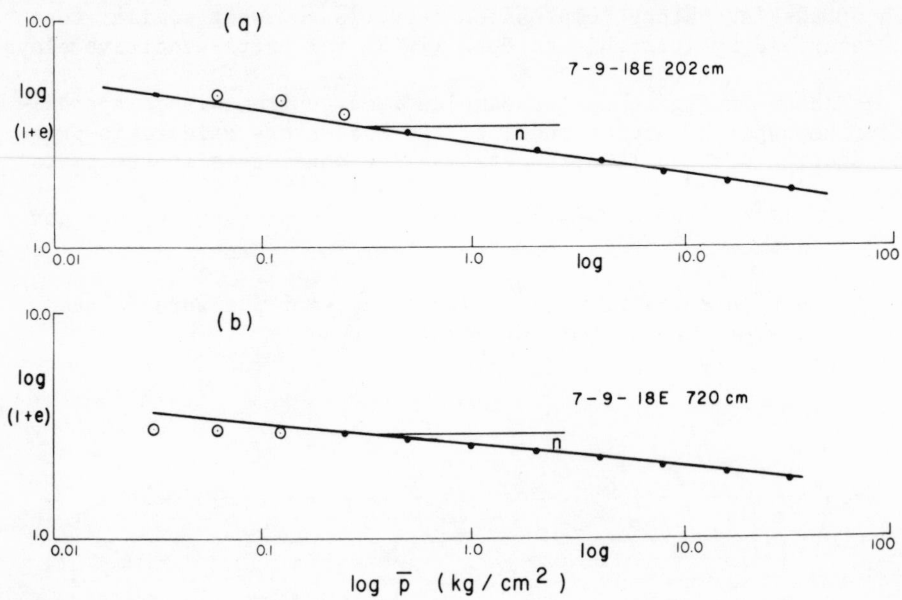

Figure 12. log h - log $\bar{p}$ curves of 202 cm and 720 cm samples from Core 7-8-18E

3. a virgin compression curve, a_2, concave upward for pressure values in excess of $\bar{p}_b$. This curve, a_2, when plotted in the log h - log $\bar{p}$ system becomes a straight line (Fig. 12A). It is not a straight line in the e-log $\bar{p}$ diagram (Figs. 11A and 11B).

Line c_1 in Figure 13 is a recompression curve because in all cases under investigation the calculated effective overburden pressures, $\bar{p}_o$, were equal to the pressures indicated by the intersection of line c_1 and line a_1.

As shown in Figure 14 for samples removed from a depth greater than approximately 5 m below mudline, the laboratory void ratio-pressure relation plotted to arithmetic scales consists of two parts:

1. a linear recompression curve, c_2, for pressure values not in excess of $\bar{p}_c$,

2. the virgin compression curve, a_2, concave upward, for pressure values in excess of $\bar{p}_c$. This curve, a_2, when plotted in the log h - log $\bar{p}$ system becomes a straight line (Fig. 12B).

Two virgin compression curves of a typical high-void ratio

marine clay sediment are represented by line a_1 and curve a_2 in the $e - \bar{p}$ diagram of Figure 15.

The second virgin compression curve, a_2, can be extrapolated to the left (a_2') by transferring to the $e - \bar{p}$ diagram, values found by extending the straight line representing a_2 in the log h - log $\bar{p}$ diagram.

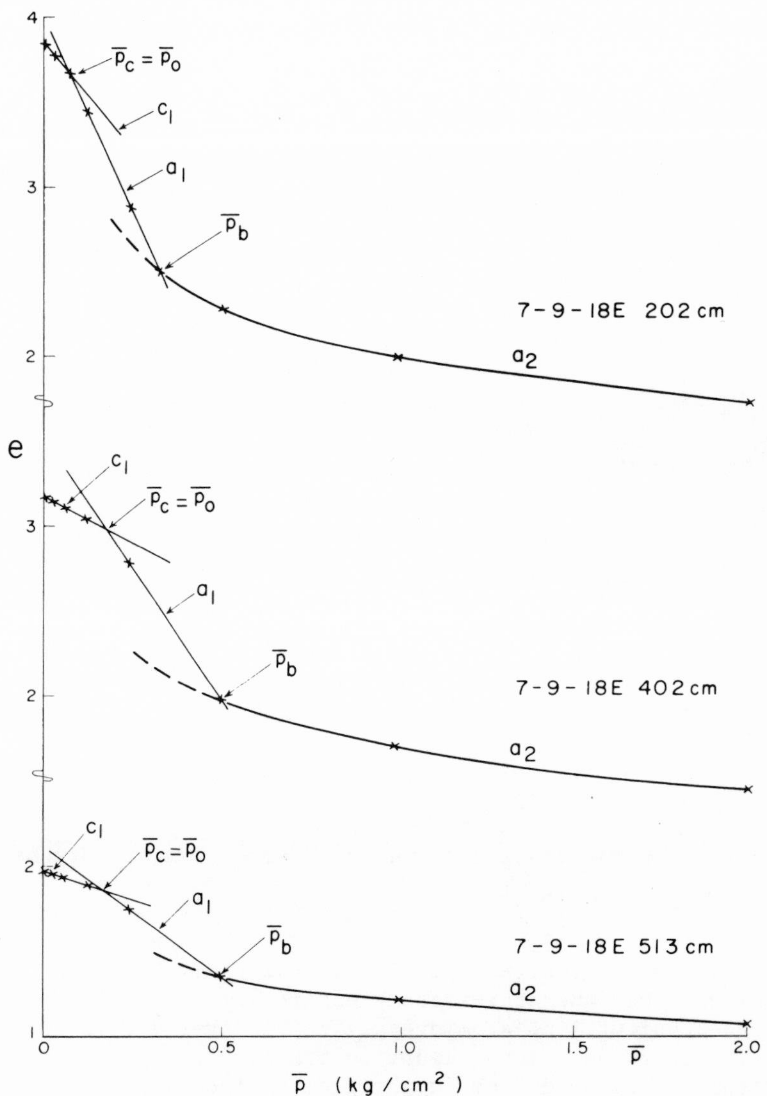

Figure 13. $e - \bar{p}$ compression curves of the upper 5-meter samples

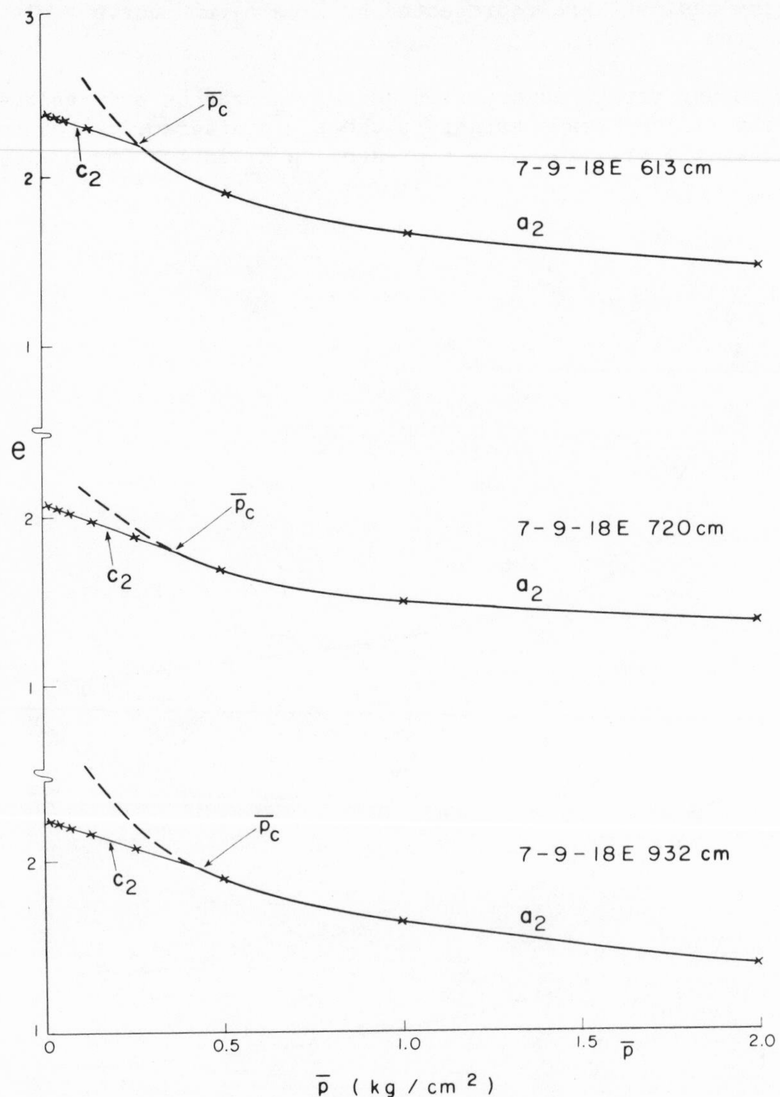

Figure 14. e - $\overline{p}$ compression curves of samples 5 m or more below mudline

For samples removed from within approximately the upper 5 m below mudline, the laboratory recompression line, c_1, with a slope smaller than that of line a_1, intersects the first virgin compression line, a_1, at the preconsolidation pressure $\overline{p}_{c1}$ smaller than p_b. For samples removed from depths greater than approximately 5 m below mudline, the recompression line, c_2, intersects the second virgin compression curve, a_2, at the preconsolidation pressure $\overline{p}_{c2}$.

CONSOLIDATION CHARACTERISTICS OF DEEPLY BURIED MARINE SEDIMENTS

Consolidation tests were performed on samples obtained from the Deep-Sea Drilling Project, Leg 10 (Gulf of Mexico), Leg 16 (Northeastern Equatorial Pacific), and Leg 24 (Western Indian Ocean).

Leg 10 samples were obtained from two sites in the Gulf of Mexico. Site 91 was located on the Sigsbee Abyssal Plain in a water depth of 3763 m and Site 92 from the Sigsbee Scarp, water depth 2573 m.

The e-log $\bar{p}$ curves for Site 91 and 92 are shown in Figure 16. Of the seven samples tested all were determined to be underconsolidated. Although the samples tested did not appear to be excessively disturbed, the high degree of underconsolidation could be attributed to excess pore pressures due to low permeabilities. The age of the sediments tested ranged from Middle Pliocene to Late Pleistocene.

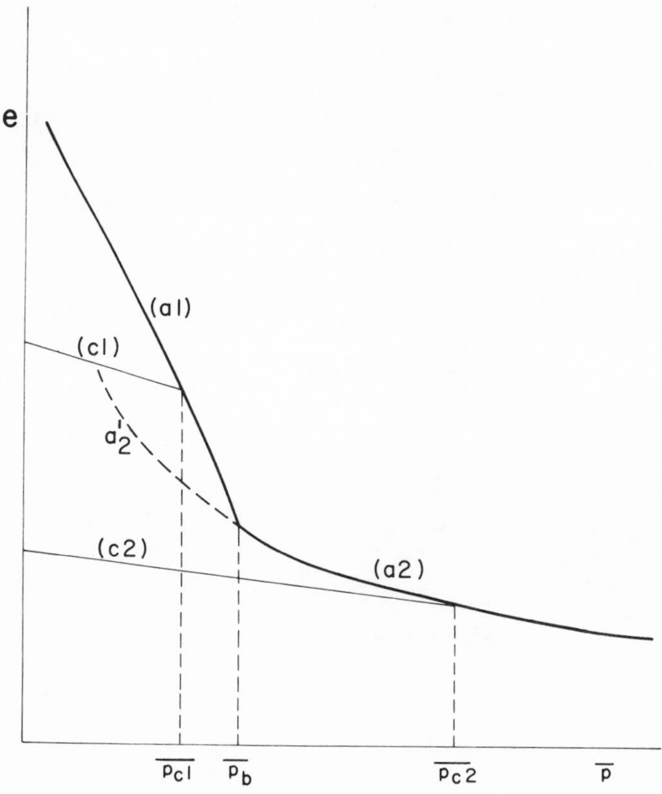

Figure 15. Compression curves of a typical high-void ratio sediment

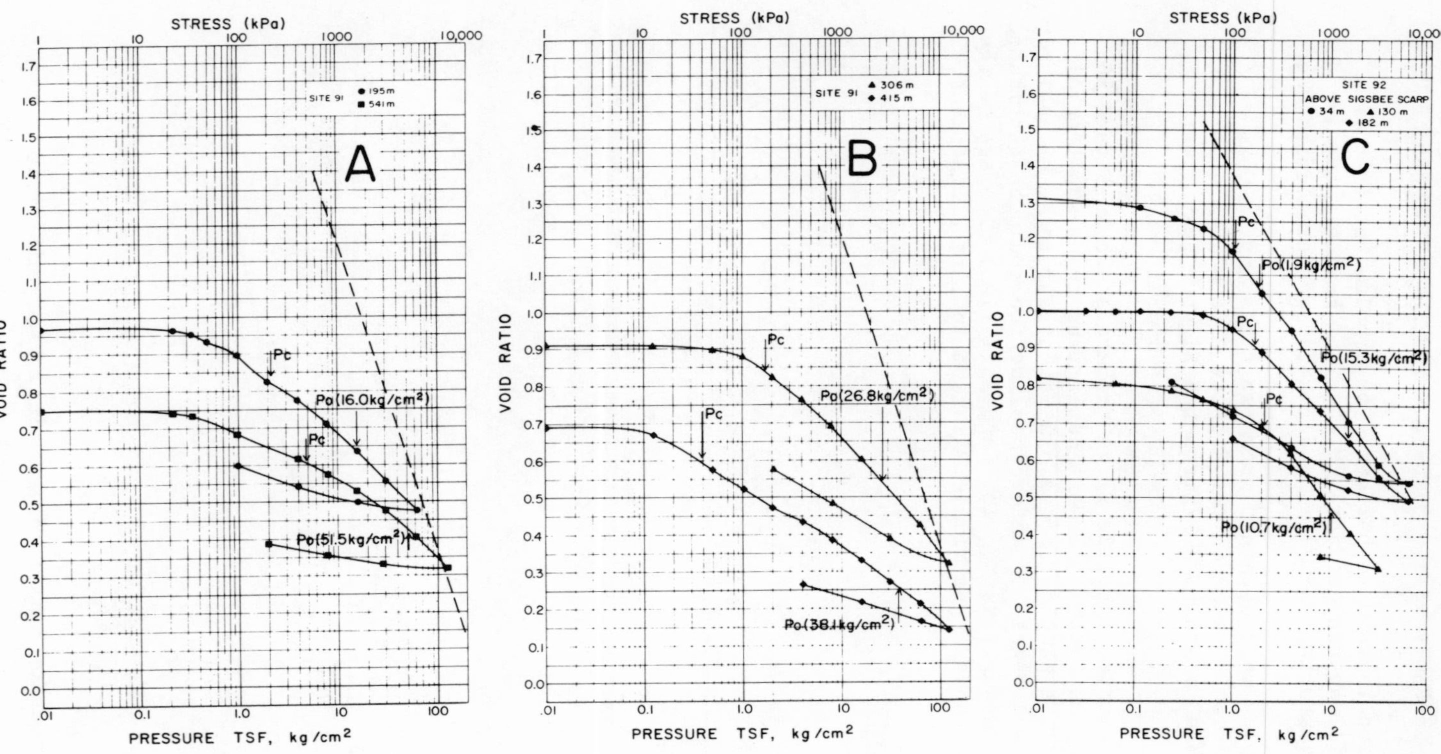

Figures 16a, b & c. e-log $\bar{p}$ curves of samples from D.S.D.P. Sites 91 and 92

The rate of deposition in the area during these epochs averaged 20 cm/10^3 years for the Pleistocene to 4 cm/10^3 years for the Pliocene.

Figure 17 shows the results of four tests on samples from the Northeastern Equatorial Pacific (G. H. Keller and R. H. Bennett, 1973). Samples from Sites 160 and 161 were determined to be underconsolidated but were underconsolidated to a lesser degree than those on Leg 10. The sediment sample from Site 163A was for all practical purposes almost normally consolidated.

Samples taken in the Indian Ocean on Leg 24 are shown in Figure 18 (compliments of Paul Cernock). No additional data is available at this time and the e-log $\bar{p}$ curves are shown for comparison purposes only.

SECONDARY CONSOLIDATION OF MARINE CLAYS

One of the most important factors affecting the consolidation of fine-grained sediments is the rate of loading. The accumulation of marine sediments, within oceanic basins, takes place at rates ranging from a few mm to several cm per 1000 years. The following table shows the equivalent loading rates for such accumulations:

Rate of Sedimentation			g/cm^2 (1 g/cm^2 = 98.0665 Pa)	
cm/1000 yrs	μm/yr	μm/day	load/day	load/year
1	10	0.03	1.35 X 10^{-6}	4.92 X 10^{-4}
10	100	0.37	1.67 X 10^{-5}	6.08 X 10^{-3}
100	1,000	2.71	1.22 X 10^{-4}	4.45 X 10^{-2}
1,000 (extremely high)	10,000	27.08	1.22 X 10^{-3}	.45
60 m (Mississippi Delta)	60 m	.16 cm	.072	26.3

Assuming initial wet density of 1450 kg/m^3 (1.45 g/cm^3)

It is obvious from these data that consolidation within marine sediments will be a very slow process and in no way duplicated in the standard laboratory oedometer test, as the process will be one of "delayed" consolidation (Crawford, 1964). The effects, however, of secondary consolidation may be assessed from long term tests in

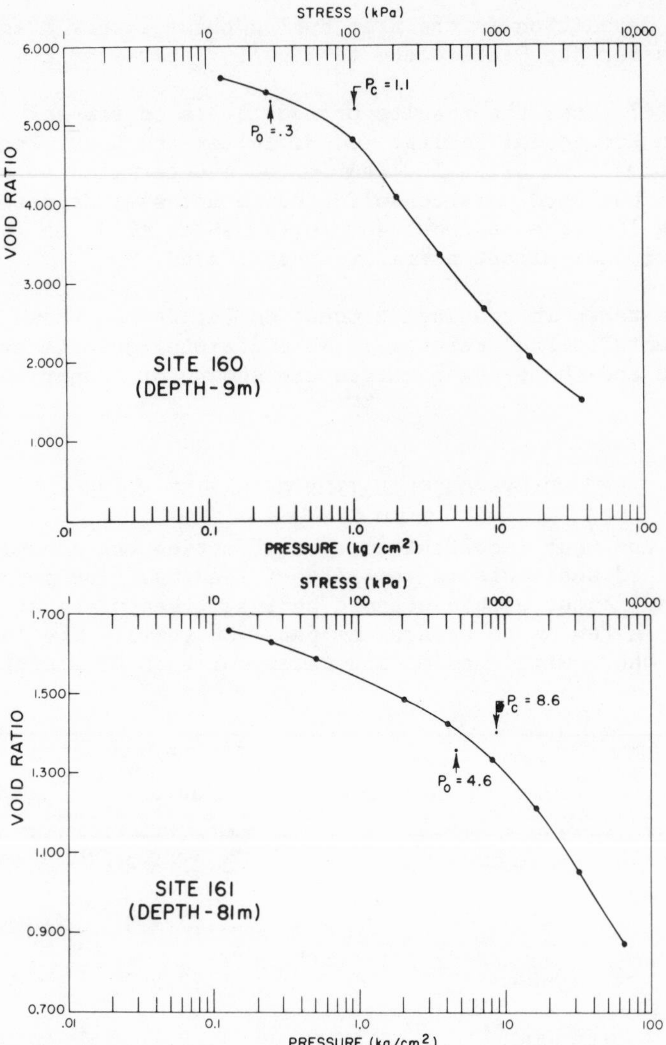

Figures 17a and 17b (opposite). e-log $\bar{p}$ curves of samples from D.S.D.P. Sites 160, 161 and 163A (R. Bennett and G. Keller, 1973)

the laboratory. In the laboratory test, upon dissipation of excess pore pressure a sample under consolidation goes into a "secondary" phenomenon illustrated in the characteristic compression versus log of time curve (Fig. 19). The slope of the second linear portion of the compression-log time curve plot is defined as the rate of secondary compression:

$$C\alpha = \frac{\Delta H/H}{\Delta \log_{10} \tau}$$

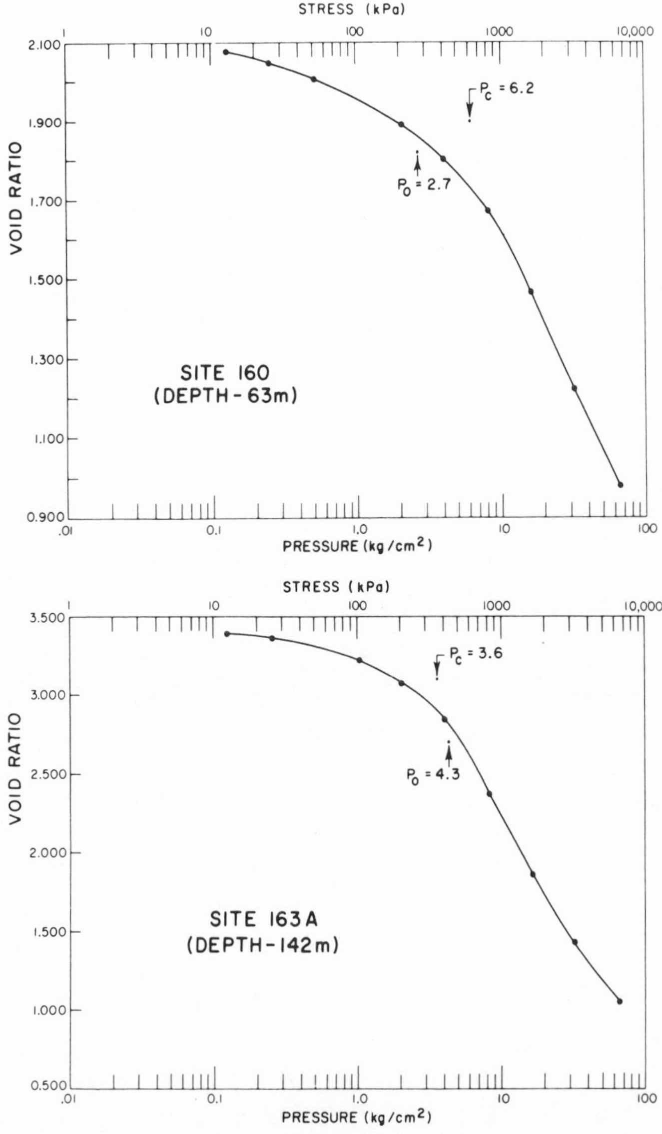

Figure 17b

The laboratory long term consolidation properties for a high void ratio marine sediment were measured on a sample from the fan of the DeSoto Canyon and are illustrated in Figure 19. The scale of the settlement axis is a relative measure only. This work displays the importance of secondary consolidation as it would appear under natural conditions of loading.

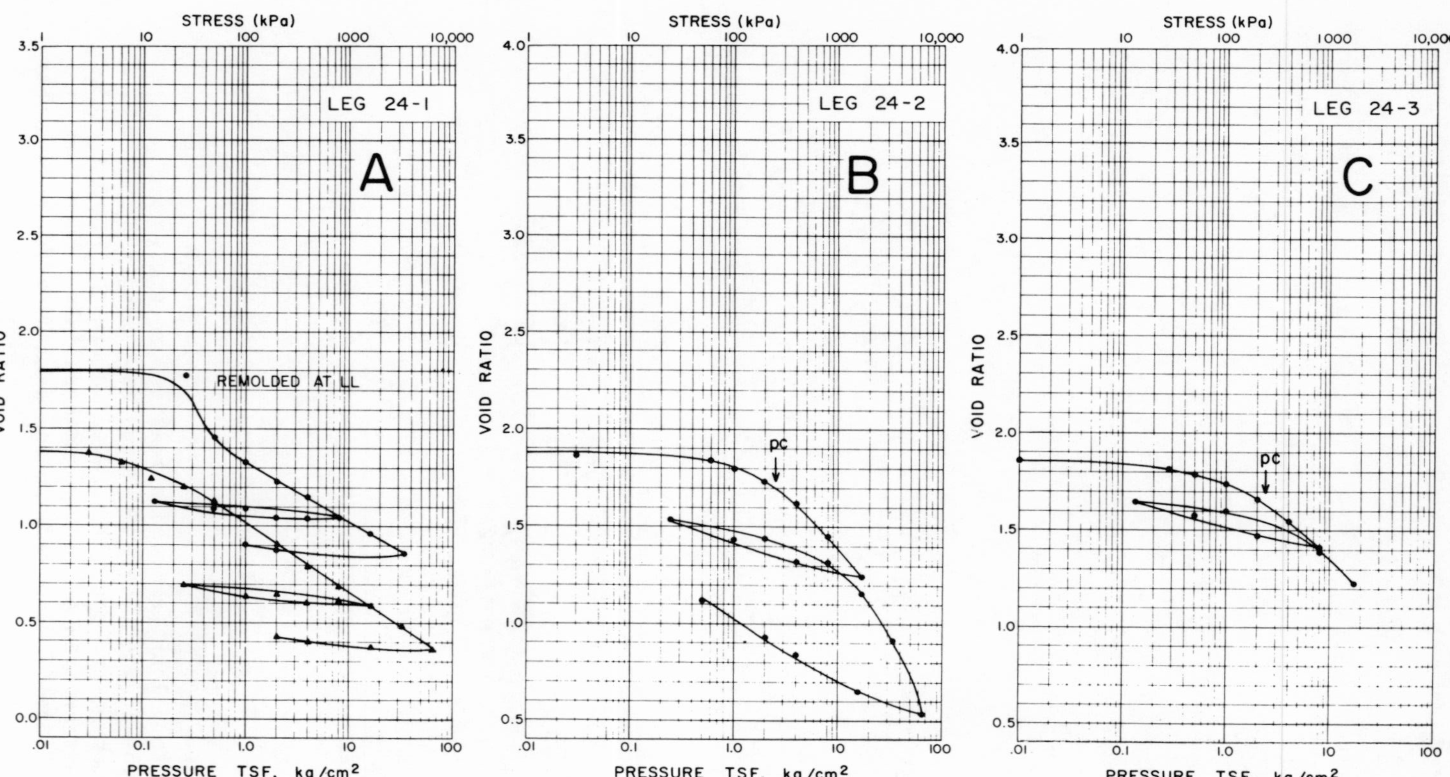

Figures 18a, b & c. e-log $\bar{p}$ curves of samples from D.S.D.P. Leg 24

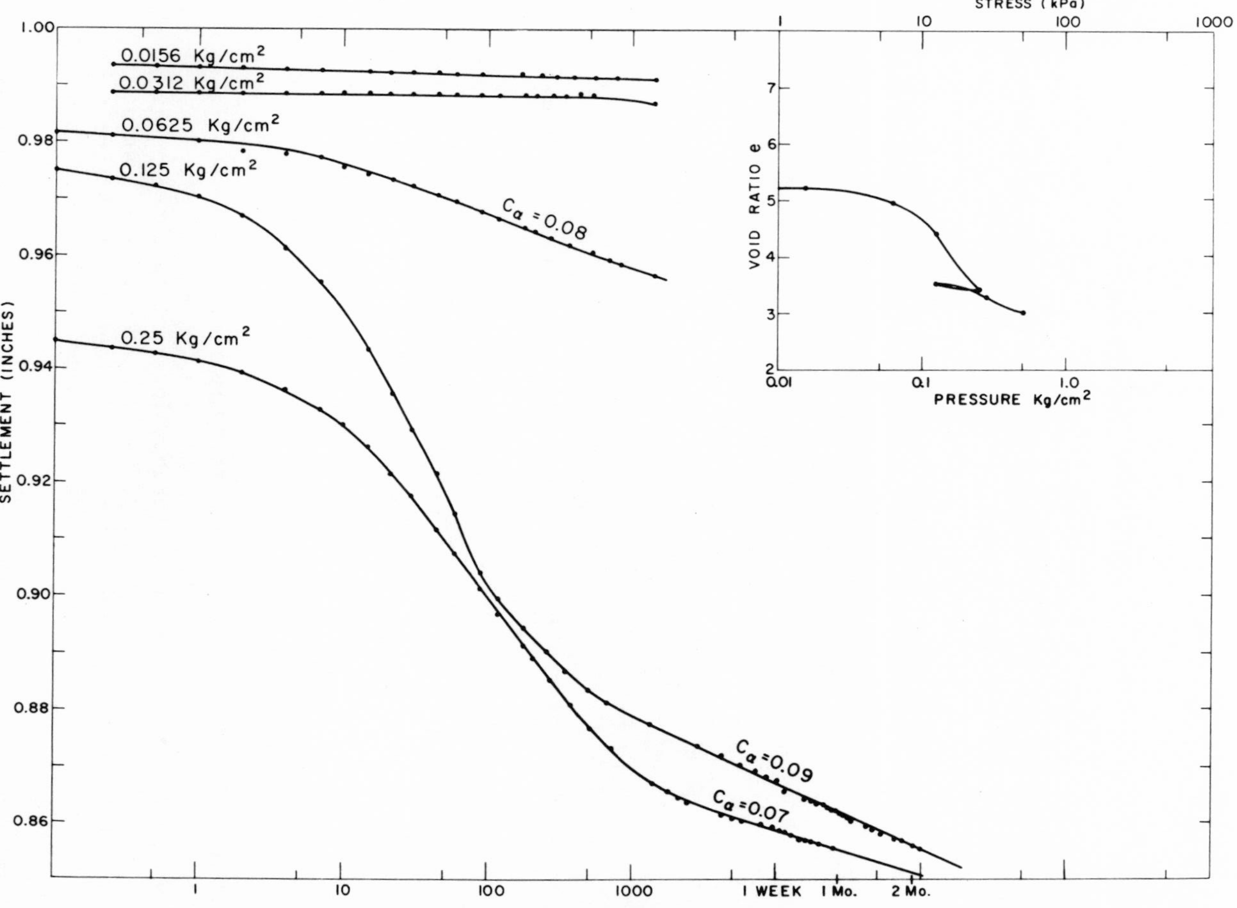

Figure 19. Settlement vs. log time curves of high void ratio sediment. Settlement axis relative scale.

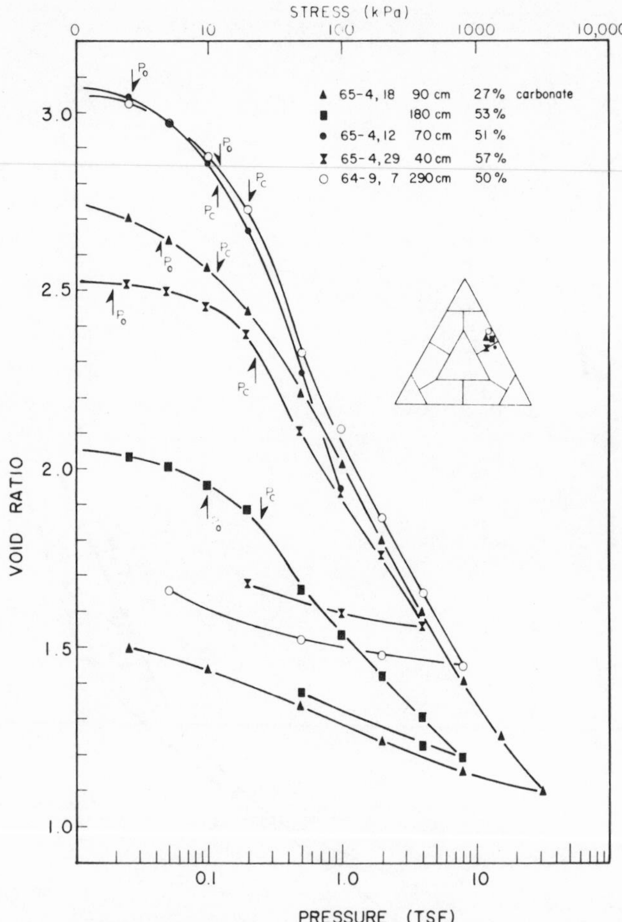

Figure 20. e-log $\bar{p}$ curves of sediments composed of an average of 50 percent carbonate material

CONSOLIDATION OF CARBONATE SEDIMENTS

Although there has been considerable interest in the consolidation of marine carbonate sediments, there has been little actual testing of carbonates. A large number of sediment samples from the Gulf of Mexico, Florida Bay and the Bahama Banks were tested to de-

---

Figure 21a (opposite) e-log $\bar{p}$ curves of high carbonate content sediments from a deep-water environment
Figure 21b (opposite) e-log $\bar{p}$ curves of pure carbonate sediments from Marathon Florida

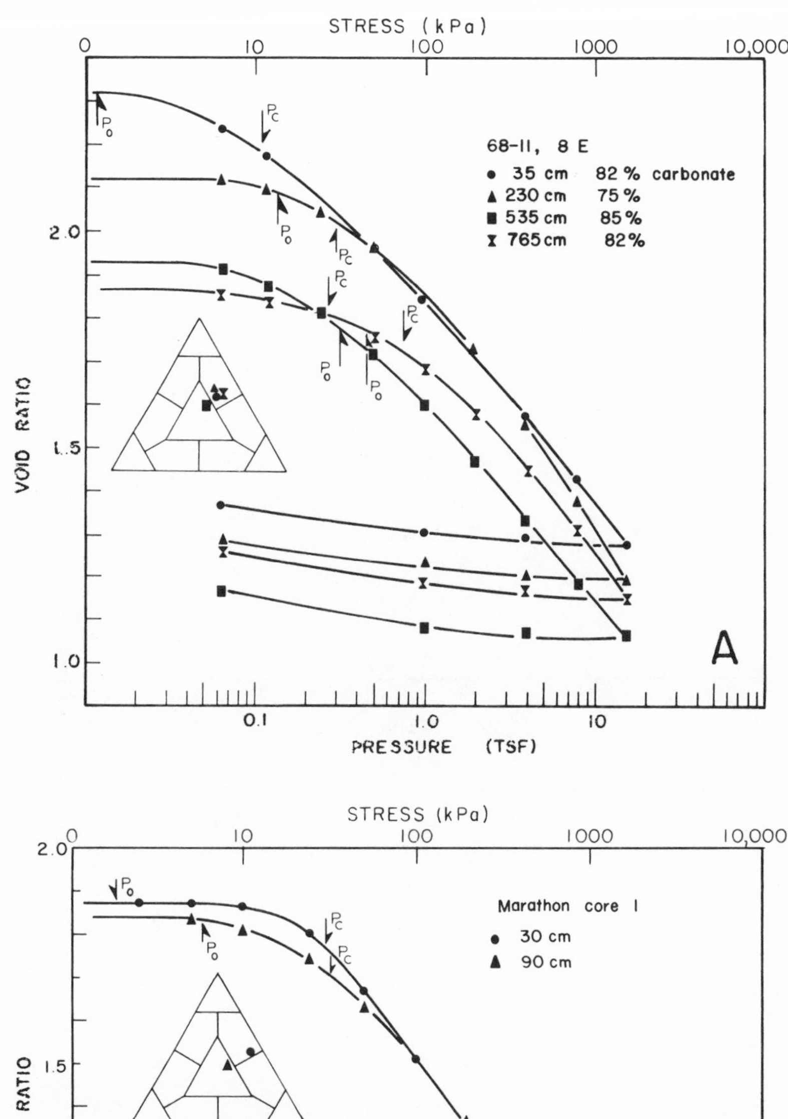
STRESS (kPa)
0
10
100
1000
10,000
68-11, 8 E
● 35 cm 82 % carbonate
▲ 230 cm 75 %
■ 535 cm 85 %
Ⅹ 765 cm 82 %
VOID RATIO
2.0
1.5
1.0
P_o
P_c
A
0.1
1.0
10
PRESSURE (TSF)
STRESS (kPa)
0
10
100
1000
10,000
Marathon core I
● 30 cm
▲ 90 cm
VOID RATIO
2.0
1.5
1.0
P_o
P_c
B
0.1
1.0
10
PRESSURE (TSF)

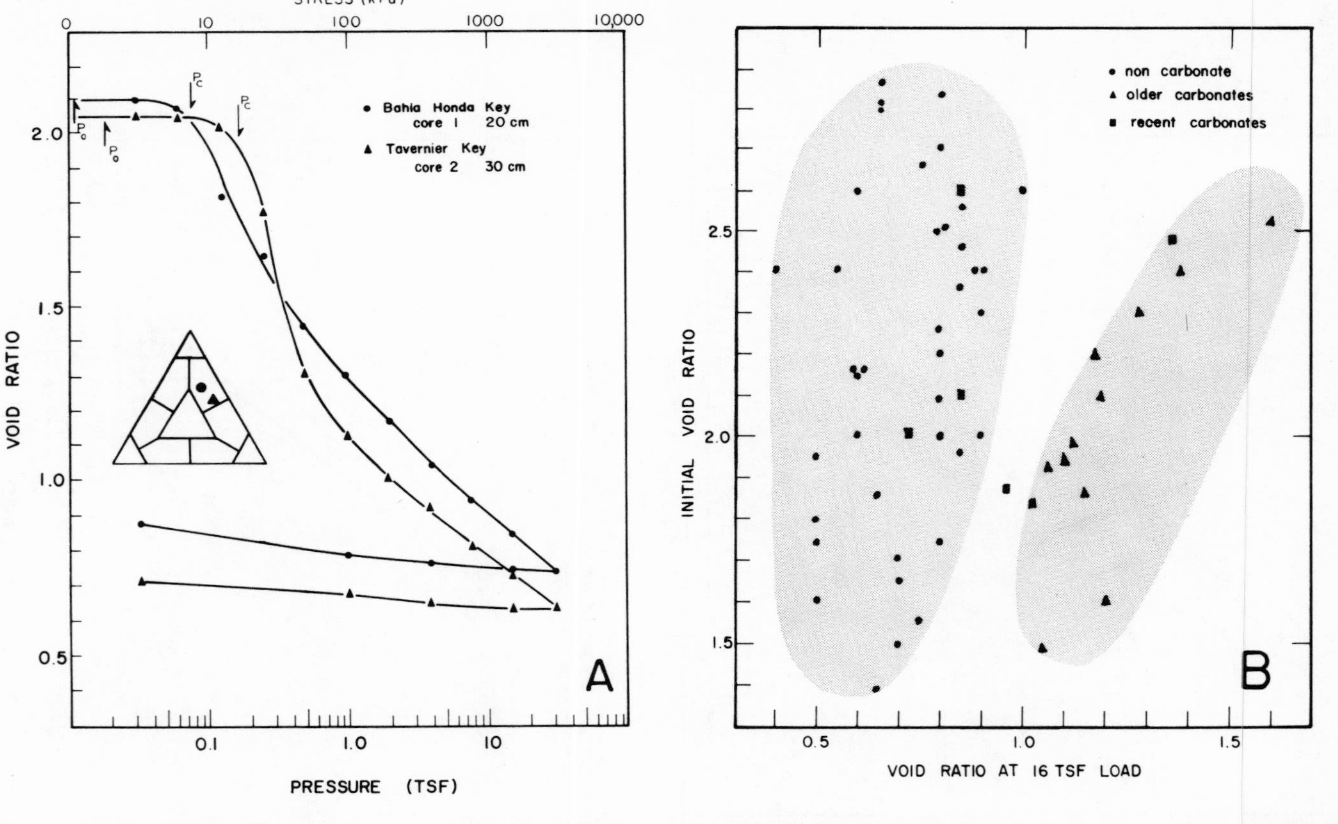

Figure 22a. e-log $\bar{p}$ curves for recent carbonate sediments from the Florida Keys, Tavernier Key and Bahia Honda Key

Figure 22b. Plot of initial void ratio versus void ratio reached under a final normal load of 16 kg/cm^2

termine their consolidation characteristics. Of the samples containing more than 70% carbonate, 12 were of Holocene age, deposited since Wisconsin glaciation, and 11 samples from the continental slope west of Florida were of Pleistocene-Pliocene age.

All samples tested were overconsolidated: that is, there was more structural strength than would be expected from the effect of the present overburden. There was a definite relation between the percentage of fine material present and the resulting consolidation.

In general, the results of consolidation tests were similar to those found by testing noncarbonate silty clay. The main differences observed were between the older carbonate sediments and the noncarbonate or partly carbonate sediments. Under a similar final load, the carbonate sediments did not compact to as low a void ratio as the noncarbonates. Age and incipient cementation must play a part because the Holocene carbonate sediments did not show this characteristic. This conclusion is supported by the results reported by several other workers - that the strength of carbonate sediments increases with age.

The consolidation characteristics of marine sediments containing from 27 to 57% carbonate material is shown in Figure 20. All materials tested fall within the silty clay category as indicated by the triangular diagram for grain-size in each figure. There is no marked departure from the general configuration of the e-log $\bar{p}$ curve for these deep-water carbonate sediments as compared with no or low carbonate sediments. The results of testing sediment samples containing from 75 to 85% carbonate are shown in Figure 21A. These sediments were taken from a material cored in 1850 m of water. The material could be classed as a foram-ooze and ranged in age from Pliocene to Holocene. E-log $\bar{p}$ curves of shallow water carbonate muds whose carbonate content average 97% are shown in Figure 21B and 22A. It is interesting to note that the fine-grain carbonate sediment from Tavernier Key displayed an e-log $\bar{p}$ curve similar to that of a very sensitive clay.

None of the carbonate sediments examined for consolidation show any visible evidence of cementation. Even the older carbonate sediments tested behave quite similarly to noncarbonate silty clays. The main difference between noncarbonate and carbonate sediments of the areas studied was that under a similar load carbonate sediments do not consolidate to as low a void ratio as noncarbonates (Fig. 22B). This may be due to differences in the particle shape and in general structural strength of the individual particles. The nature of the bond between water molecules and particles may also be a factor.

The results of carbonate sediment consolidation testing seem to be valid for engineering purposes and for geological analysis.

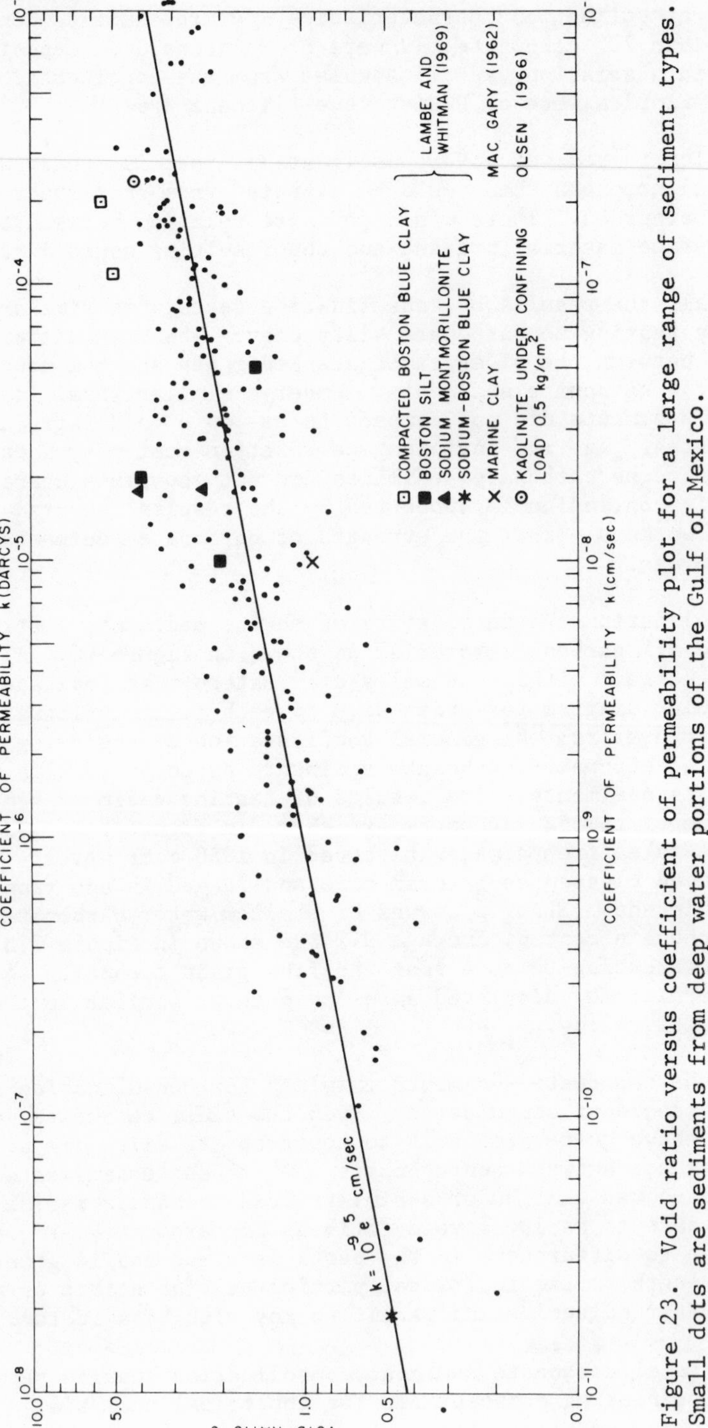

Figure 23. Void ratio versus coefficient of permeability plot for a large range of sediment types. Small dots are sediments from deep water portions of the Gulf of Mexico.

PERMEABILITY OF MARINE SEDIMENTS

During the routine consolidation test the permeability of the sediment may be determined as a function of void ratio. The coefficient of permeability, k, may be determined by the use of the equation for the coefficient of consolidation as derived in Terzaghi's theory where (Taylor, 1948):

$$c_v = \frac{k(1 + e)}{a_v \gamma_w}$$

The method has several advantages over the standard direct methods, which utilize constant or falling head measurements of flow, when dealing with low permeability clays. The sample can be maintained under hydrostatic pressure (back-pressure) while the load (overburden) is varied; these tests were made on Anteus oedometers with a back pressure of 6 atm.

The procedure for obtaining the coefficient of consolidation c_v was based on the square-root-of-time-fitting method where:

$$c_v = \frac{0.848}{t_{90}} \left(\frac{H}{2}\right)^2$$

For double drained specimens the coefficient of permeability was computed from:

$$k = c_v m_v \gamma_{sw} \text{ in cm/sec}$$

where m_v = coefficient of volume compressibility = $\frac{a_v}{1 + e_o}$

a_v = coefficient of compressibility.

The coefficient of compressibility, a_v, is the ratio between the change in void ratio and the change in load for the given increment. The coefficient k is converted to Darcys: by k Darcys = k (1.045 X 10^3). Figure 23 displays the relationship between permeability and void ratio for a large variety of sediment types.

CONCLUSIONS

1. The consolidation of high void ratio ($e > 2$) marine clay sediments exhibits a linear void ratio-pressure relation in contrast to the nonlinear relation as observed in ordinary clay soils.

2. The surficial sediments from the Sigsbee Abyssal Plain, Gulf of Mexico are normally consolidated marine clays. Both the

consolidated tests and geological observation confirm this fact.

3. Deeply buried sediments from the Gulf of Mexico were found to be underconsolidated. The underconsolidation characteristics of these sediments can be explained by the low permeabilities of these fine-grained clays. The rate of deposition exceeds the rate of flow of the interstitial waters of the consolidating sediments.

4. Fine-grained carbonate muds consolidate similar to noncarbonate silty clays. The only difference observed was that under equal loads older carbonate sediments do not consolidate to as low a final void ratio as noncarbonate sediment.

5. The marine clays and silty clays of the Gulf of Mexico become practically impermeable at fairly high void ratios. The relationship between the void ratio and the coefficient of permeability was found to be $k = 10^{-9} e^5$ cm/sec.

ACKNOWLEDGMENT

This work was supported by the Office of Naval Research under Contract N00014-68A-0308-(0002).

NOMENCLATURE

C_c	=	Compression index determined from the e-log $\bar{p}$ diagram
e	=	Void ratio = volume of voids/volume of solids
e_o	=	Original void ratio
h	=	Reduced height = 1 + e
log	=	Logarithm in the base 10
n_v	=	Compression exponent determined from the log h-log $\bar{p}$ diagram
$\bar{p}$	=	Effective pressure
$\bar{p}_b$	=	Pressure at which the characteristics of the virgin compression curves change
$p_c = p_a$	=	Preconsolidation pressure
$p_o = \bar{p}_o$	=	Effective overburden pressure
ε	=	Conventional strain = total compression/original height
ε'	=	True or natural strain = compression true increment/actual height
A_v	=	Coefficient of compressibility
	or	Decrease in unit void ratio per unit increase of pressure
m_v	=	Coefficient of volume decrease
	or	Decrease in unit volume per unit increase of pressure
c_v	=	Coefficient of consolidation
k	=	Coefficient of permeability
w	=	Water content
p	=	Total pressure
u	=	Pore pressure

REFERENCES

Crawford, C. B., Interpretation of the consolidation test, J. Soil Mech. and Fdn. Div., Proc., Am. Soc. Civil Engrs., 87-108, 1964.

Keller, G. H., and R. H. Bennett, Sediment mass physical properties-Panama Basin and Northeastern Equatorial Pacific, in Initial Reports of the Deep-Sea Drilling Project, 16, edited by Tj. H. van Andel, et al., pp. 499-512, U. S. Govt. Printing Office, Washington, D. C., 1973.

Skempton, A. W., The consolidation of clays by gravitational compaction, J. Geol. Soc. London, 125, 373-411, 1970.

Taylor, D. W., Fundamentals of Soil Mechanics, John Wiley and Sons, N. Y., 1948.

Terzaghi, K., Erdbaumechanick, Franz Deuticke, Vienna, 1925.

Terzaghi, K., and R. B. Peck, Soil Mechanics in Engineering Practice, John Wiley and Sons, N. Y., 1948.

DEEP-SEA FOUNDATION AND ANCHOR ENGINEERING

PHILIP J. VALENT

Civil Engineering Laboratory

ABSTRACT

Some aspects of deep-sea foundation and anchor engineering state-of-the-art are presented and discussed. Special emphasis is given to those areas where need exists for research and development. Techniques for predicting bearing capacity are discussed, and their shortcomings outlined. Techniques for estimating the short-term holding capacity of propellant-embedded anchors are discussed, and required information on long-term holding capacity and response to cyclic loading is described. The approaches used to minimize the detrimental effects of scour are reviewed and criticized.

INTRODUCTION

This report considers certain failure modes of bottom-resting and buoyant-suspended structures in the deep ocean, describes the role of sediment engineering properties in designing to prevent such failures, and outlines areas of sediment behavior that are insufficiently understood to permit safe and economical design-problem solution. The purpose of this presentation is to stimulate an exchange between scientist and engineer which will improve the procedures and assist in satisfying the needs of the practicing deep-sea foundation engineer.

First, the catastrophic sinking of a bottom-resting structure into the sediments will be reviewed; this failure mode is termed a bearing capacity failure by foundation engineers. Then the problem will be reversed and pullout failure of an embedment anchor used to tether a buoyant-suspended structure will be reviewed. Lastly, the problem of scour accomodation on the deep-sea floor will be considered.

For each of these three problems, the main body of the paper contains only a brief, qualitative description of the theoretical/empirical relationships used in determining the feasibility, type, and size of a proposed foundation or embedment anchor. The appendices present more detailed, quantitative material for the first two problems, material omitted from the main body of the paper to encourage reading of that portion by non-soils engineering oriented persons.

BEARING CAPACITY

A bearing capacity failure is marked by the near-immediate excessive sinkage of a bottom-resting structure into the supporting sea-floor soil. Bearing capacity failures on the sea floor have not been observed frequently. A mobile jack-up rig applying a pressure of 8.6 kPa (180 psf) to soft Gulf of Mexico soils was involved in a bearing capacity failure and was lost (C. J. Beaupre, personal communication, 1969). A 1.8 m diameter test foundation applying a pressure of 7.2 kPa (150 psf) at a Santa Barbara Channel site in 370 m water depth apparently experienced a bearing capacity failure tilting about 15 degrees between 28 and 56 seconds after initial bottom contact. A bearing capacity analysis shows this foundation to have a calculated factor of safety of 2.8 (Herrmann et al., 1972).

The techniques used in predicting bearing capacity, or ultimate bearing load, are founded in theoretical descriptions of various soil deformation and soil rupture models. However, despite these theoretical foundations, the actual techniques used in the prediction process are quite empirical. Empiricism arises for two reasons: first, exact solutions considering all variables would be far too cumbersome, too time consuming, and might even, for some problems, be impossible; and, second, soil parameters measured by present soil tests would not be adequate for problem solution requiring extensive development of more complicated techniques. (See additional comments by Thompson, 1974). It is for reasons similar to these that much engineering is regarded as an art rather than a science.

Sea-floor foundations will be required to support combined eccentric and inclined loads to a far greater extent than terrestrial foundations. This condition will be common because most sea-floor structures will be emplaced without prior site leveling, causing the foundation to be tilted, and shifting the point of application of the structure dead-load resultant force to one side of the foundation center. Further, significant current drag forces will be applied to the structure, support framework, and any riser lines and buoys. Summation of these forces yields a resultant force which is inclined and eccentric to the footing-bearing surface. Such eccentric and inclined loadings cause the supporting foundation to fail at a lower load than if that load were being applied normal to the footing.

Appendix A outlines techniques used in predicting the bearing capacity of footings on the sea floor.

When designing a foundation, especially one in the deep ocean, the biggest problem by far is the acquisition, or reliable estimation, of sediment-relative density, shear strength, and compressibility properties. This problem of data acquisition, and a problem concerning improper assumptions in the bearing capacity analysis, are expanded below:

1. There exists no good developed technique for evaluating the engineering properties of a sea-floor noncohesive (sand) soil, short of employing a commercial drilling platform or specialized, soil-depth limited, cone penetrometers (Demars and Taylor, 1971; Hirst et al., 1972). One promising technique suggested here for consideration involves the use of acoustic sea-floor reflection equipment with a sediment-grab sampler or a spade-box corer. The acoustic equipment could provide a measure of acoustic energy bottom loss at the water-sediment interface, which in turn shows some promise of correlating well with the sediment porosity (Smith, 1971; Barnes et al., 1971); or a measure of sediment dynamic compressibility could be made and correlated to porosity (Smith, 1971). This second relationship shows the greater promise (D. T. Smith, personal communication, 1973). The soil sample recovered, although remolded, would provide the necessary data, after combination with the acoustic porosity data, to enter a chart such as Figure A-2 and determine the angle of internal friction, ϕ. Unfortunately, the acoustic measurements may involve only a thin surficial sediment layer, and may not be an adequate tool for large sea-floor foundations.

2. There exists no developed technique or hardware to obtain soil engineering properties in all deep-sea sediment from soil depths in excess of about 3.0 m. There are two approaches possible to correct this shortcoming:

a. First, a device should be developed to measure in situ soil engineering parameters to a soil depth approaching 30 m. For instance, consider a propellant-driven projectile to measure dynamic soil parameters on the way into the sea floor, and to measure static parameters while being winched out.

b. Second, existing and proposed sea-floor corers of 15 m soil depth capability should be utilized to develop an understanding of the variation of shear strength and compressibility with soil depth and with core disturbance. Further, deep corers will be needed to obtain good quality samples for the development of correlations with rapid parameter measurement techniques.

3. Existing bearing-capacity relationships should be fundamentally re-examined to determine if some modifications are warranted. Present design techniques supposedly design against local shear failure; but footings placed on the sea floor penetrate for some distance (this penetration is a form of local shear failure). Such inconsistencies between theory and practice should be resolved. As an end product, the factor of safety applied in the bearing capacity analysis might more properly become a measure of the statistical variation of soil strength rather than, as now, a measure of uncertainty with regard to soil deformation and failure mechanisms.

EMBEDMENT ANCHOR-HOLDING CAPACITY

Embedment anchors are flat plate devices buried horizontally in the soil and providing restraint to a tie-down rod or cable. The commonly used form of the embedment anchor is the screw anchor used to anchor guy wires and pipe lines. A typical screw anchor consists of a small-diameter steel shaft with a large-diameter, flat, single flight helix at the bottom. Rotation of the screw anchor via the shaft causes the helix to pull itself into the soil in the manner of an auger. The shaft then transmits the tension or compression load to the plate, which distributes the loading to the soil (Raecke and Migliore, 1971). The newest embedment anchor system employs a flat plate, which is propelled edgewise from 5 m to 15 m into the sea floor, and which then rotates sideways to a horizontal position when load is applied to the attached wire rope. Design load for the anchor system is 89000 N (20,000 lbs) in all types of sea bottoms to 6100 m water depth (Taylor and Beard, 1973).

The techniques used in predicting the holding capacity of an embedment anchor are, like those used for bearing capacity, founded in theory but modified heavily based on test data. The empirical relationships and the procedures used to describe one mode of soil-anchor failure are described in Appendix B. (This mode is termed deep anchor behavior, the difference between modes is described in more detail in the appendix.) Embedment anchors will be called on to resist long-term sustained and cyclic loadings because they tether an object or structure to the sea floor and because that object may be acted upon by significant wind, wave, and/or current drag forces. The response of the soil-anchor system to such loading conditions is imperfectly understood with the following areas in need of extensive examination:

1. Deep embedment anchor-holding capacity in typical, normally-consolidated, cohesive soil is probably increased with time due to

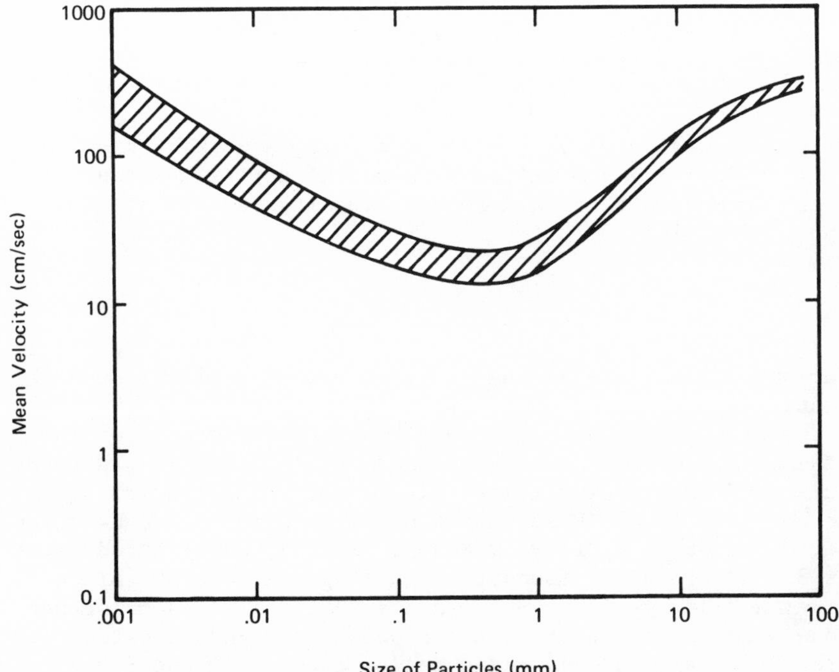

Figure 1. Approximate curves for erosion of uniform material (after Hjulstrom, 1939; Kuenen, 1950).

consolidation and strengthening of the overlying soil and, at the same time the anchor will be creeping upward into weaker soil strata. The relative importance of soil strengthening and creep in different sea-floor soil types and at different loading intensities must be established.

2. Deep embedment anchor holding capacity in noncohesive soil may be reduced from 50 to 90% under repeated or cyclic loading. These loadings will be caused by mooring surface vessels or buoys or by earthquakes. It is important to develop a capability to predict holding capacity reductions under cyclic loadings and typical sea-floor conditions. It is possible that, in certain environments, the necessary high safety factors for embedment-type anchors may render them uneconomical compared to other anchoring systems (but only for a small portion of the sea floor).

3. Short-term load tests of embedment anchors are desirable, but their residual capacity is often masked by the suction effect. Techniques for correcting for the suction effect should receive continuing attention.

4. Penetration of the anchor plate, whether a screw anchor or propellant-driven anchor, will cause some remolding of the overlying soil. The significance of this remolding on the anchor pull-out behavior should be examined.

5. For the design of an embedment anchor, a working knowledge of the soil engineering properties versus soil depth is essential. The earlier comments with regard to deep in situ testing and deep coring apply here also.

SCOUR AND UNDERMINING

Scour and animal undermining are established engineering problems in water depths to 100 m. To this depth, oscillatory bottom currents due to storm waves have considerable influence. For instance, the bottom erosion around offshore oil production platforms as a result of a hurricane attained 2.4 m depth (water depth believed to be about 15 m) (Ralston and Herbich, 1969). The Hydrolab, at a water depth of 15 m off Palm Beach, Florida, experienced scour under its 5.5 by 6.3 m mat foundation in response to a 50 cm/sec current over a 2-day period. Scour had extended along a major portion of one side and a corner of the foundation. The resulting pit, at its greatest, was 1 m deep and extended 1 to 1.5 m under the foundation slab (L. Hallanger, personal communication, 1973). Scour assisted by animal undermining is thought responsible for the unusually large settlements of 1.2 m diameter footings in 37 m water depth in the Santa Barbara Channel area. Aluminum plates, 0.6 m square with 0.15 m skirts completely embedded, were emplaced in the same locale; scour exposed all 0.15 m of the skirts along one side (Muraoka, 1970).

Scour and animal undermining can occur in water depths greater than 100 m, even down into the deep-sea trenches, provided appropriate conditions occur. Sand waves 2 to 6 m high are found at a water depth of 457 m and a manganese pavement is undercut by scour at a depth of 1100 m on the Blake Plateau (Hawkins, 1969). Ripples have been photographed on seamounts in water depths to 3130 m indicating sediment transport and a potential for scour (Einstein and Wiegel, 1970). Fish have been photographed raising sediment into suspension by swimming near the bottom and, in the case of one variety, by plunging into the sediment bottom apparently in search of food (Bowin et al., 1967). Bottom currents in a particular area may be sufficiently strong to transport such suspended sediment, but not strong enough to initiate particle movement (Brundage et al., 1967).

Most parts of the deep-sea floor do not experience water velocities sufficient to erode sediments (Einstein and Wiegel, 1970). The minimum initial erosion velocity is approximately 20 cm/sec (see Fig. 1) (Hjulstrom, 1939; Kuenen, 1950); whereas, examination of the available near-deep sea-floor current measurements collected in Einstein and Wiegel (1970) suggest that current velocities often attain 2 to 5 cm/sec. A high of 16.8 cm/sec was measured on a deep-sea fan during a 46-day deployment period. Exceptions occur about seamounts, ridges, rises, etc., where currents are funneled or otherwise caused to accelerate.

For all sea-floor locations it is recommended that the structure-foundation system display as low and streamlined a profile as possible to these currents because any flow obstruction will cause turbulence and accelerated velocities which can lead to localized scour. In addition, steps must be taken to prevent animal undermining. Undermining has been prevented by installing a mechanical barrier around the foundation perimeter like a skirt or key (Nimomiya et al., 1971), penetrating the sediment to a depth of 0.15 m below the footing base (Muraoka, 1970). This skirt, in addition to functioning as a mechanical barrier, creates an anaerobic environment beneath the footing which will drive away or prove fatal to burrowing sea-floor animals (Muraoka, 1970). Where underscour occurs despite streamlining and skirts, a protective layer (reverse filter) can be employed to form a non-erodable blanket over the immediate area (Posey, 1970).

As indicated earlier, it has been concluded that, in most areas of the deep sea, currents near the bottom are too slow to scour material and put it into suspension (Einstein and Wiegel, 1970). However, parts of a sea-floor structure having a circular or less well-streamlined cross-section will cause local velocities more than double the average (Posey, 1970); thus a greater potential for sediment movement will exist about a sea-floor installation.

With noncohesive sediments, it is possible to reliably predict by model tests the local flow conditions and the resulting scour patterns for a given structure subjected to known uni-directional, oscillatory, and combined flows. However, most deep-sea floor sediments exhibit cohesive behavior for which the modeling art is not nearly so well developed (Einstein and Wiegel, 1970). The background information and technological tools available to the sea-floor structure designer need vast improvement to permit adequate scour potential prediction on cohesive soils. The following specifics are called out:

1. More long-duration current measurements near the deep-sea floor are needed to develop statistical current magnitude/duration models for different sea-floor environments. If this data were available, proper interpretation would provide sufficient input to satisfy most deep-sea floor design requirements.

2. An empirical technique should be developed to predict the

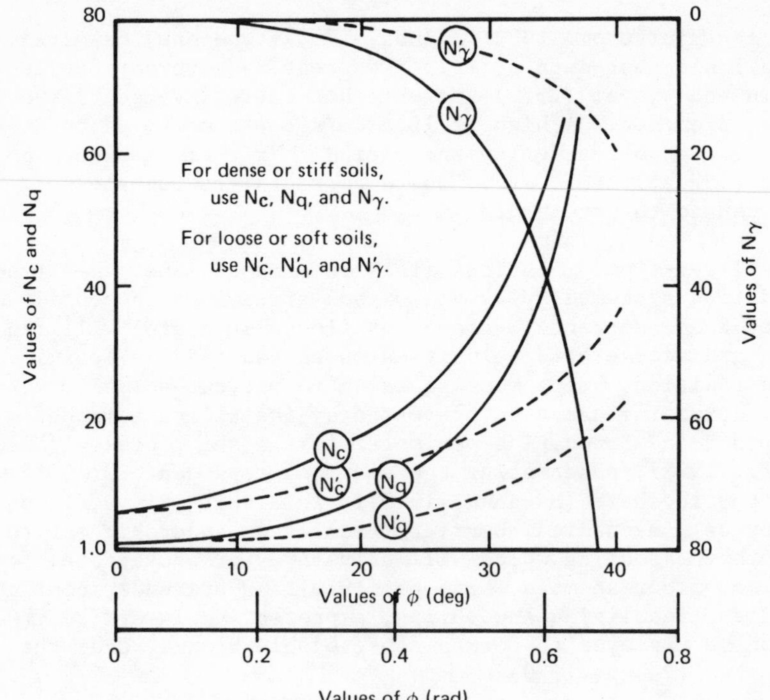

Figure A-1. Bearing capacity factors for the general bearing capacity equation (Hough, 1969). N_c, N_γ, N_q apply for general bearing failure and N'_c, N'_γ, N'_q apply for local failure.

local increase and decrease of the natural flows due to the emplacement of a range of generalized sea-floor structures. Development of this technique would be accomplished through model studies, probably similar in design to those used to study wind forces and distributions around buildings. The study would also provide, as an added benefit, drag coefficient data on the various generalized model configurations for use in predicting the lateral loading. These model studies will require establishment of laws of similitude for these conditions (material and flow) (Einstein and Wiegel, 1970).

3. A technique must be developed to predict the critical erosion velocity of various cohesive sea-floor sediments. Evaluation of the properties and functional relationships controlling the forces resisting erosion in cohesive soils is incomplete and will require considerable research (Einstein and Wiegel, 1970). Flows will be oscillatory, unidirectional, and combinations of both; scour potential will vary with the type of flow (Posey, 1970). Engineering design should optimally seek to prevent the inception of of scour; advanced problems involving the rate of scour and the pit

depth are therefore omitted from consideration (although this limitation may prove too stringent). Scour modeling on cohesive sea-floor soils may require the introduction of some types of distortions, for which no theory is presently available.

CONCLUSIONS

The increasing weight and reliability requirements of sea-floor structures require continuing improvement of deep-sea foundation and embedment anchor engineering capability. This capability includes not only an understanding of soil deformation mechanisms under load but also a knowledge of the in situ sediment properties and of the influence of environmental factors acting upon the structure. The following research needs for deep-sea foundation engineering are considered most important of those discussed in the text:

1. Soil engineering property profile prediction
2. Creep susceptibility of embedment anchors
3. Response of embedment anchors to cyclic loads
4. Near-bottom, deep-sea current data
5. Scour potential about sea-floor structures
6. Sea-floor bearing capacity predictive technique.

APPENDIX A - BEARING CAPACITY ANALYSIS

General Equation

The bearing capacity of a footing is given by the general equation:

$$q_{ult} = K_1 N_c c + K_2 \gamma N_\gamma B + N_q \gamma D \tag{A-1}$$

where q_{ult} = ultimate bearing capacity per unit area

K_1, K_2 = footing shape factors (see Table A-1)

N_c, N_γ, N_q = bearing capacity factors (see Figure A-1)

c = representative cohesive strength

γ = representative submerged density

B = width or diameter of footing

D = depth of footing base below ground level

(modified from Sowers, 1962; Terzaghi and Peck, 1967).

TABLE A-1

SHAPE FACTORS K_1 AND K_2 IN BEARING CAPACITY EQUATION

Shape of Footing	K_1	K_2
Strip	1.0	0.5
Square	1.3	0.4
Circular	1.3	0.3

(adapted from Hough, 1969, and Terzaghi and Peck, 1967)

In equation (A-1), the factor $N_q \gamma D$ may be omitted from further consideration for deep-sea footings because these will, with rare exception, bear on the sea-floor surface, thus D = 0.

Figure A-1 presents two sets of bearing capacity factors: one for general shear failure, which occurs in an ideally plastic material whose volume and strength are unchanged by rupture; and a set of reduced bearing capacity factors applying to a local shear failure condition, where progressive failure and/or large volume changes occur under the footing load. The local shear bearing capacity factors are the result of an arbitrary reduction in ϕ (Terzaghi and Peck, 1967); no rational analysis for the local shear condition is available (Sowers, 1962). In lieu of subsequent data supporting the contrary, it is suggested that local shear factors be used for deep-sea bearing capacity analyses.

In the use of Figure A-1, if the base of the footing should be smooth, for example of plastic or clean metal, then the bearing capacity factors obtained from the figure should be reduced (Ko, 1973) by 10%.

Sands versus Clays

Bearing capacity evaluations of footings are commonly subdivided according to the sediment type supporting the footing, i.e., noncohesive versus cohesive. For noncohesive sediments, c = 0 and equation (A-1) reduces to:

$$\text{For sands, } q_{ult} = K_2 \gamma N_\gamma B. \tag{A-2}$$

For cohesive sediments, the angle of internal friction, ϕ, is assumed equal to 0 (assuming undrained or quick shearing) and from Figure A-1, N_γ and $N'_\gamma = 0$. Then, equation (A-1) becomes:

$$\text{For clays, } q_{ult} = K_1 N_c c. \tag{A-3}$$

In the case of both of these equations, (A-2) and (A-3), assuming a a load acting vertically downward at the center point of a horizontal footing, the ultimate bearing load is:

$$Q_{ult} = A \times q_{ult} \tag{A-4}$$

where A is the bearing area of the footing.

Sands. Foundation design on noncohesive soils requires evaluation of the in situ angle of internal friction, ϕ. This value depends primarily upon the relative density, and, to a lesser extent, on the grain-size distribution, shape of the grains, and hardness of the grains. In terrestrial soils engineering in the U. S., the most expedient procedure for evaluating the relative density is the standard penetration test (Peck et al., 1953; Meigh and Nixon, 1961). This test has been used in continental shelf water depths, but its execution in greater water depths appears uneconomical. Relative densities can be obtained from the data of the static cone penetrometer (Meyerhof, 1956; Meigh and Nixon, 1961); however, only one device is known to be available, requiring specialized ship support with a depth limitation of 2000 m (Demars and Taylor, 1971). A plate-bearing device (Kretschmer and Lee, 1970) could, with some development, obtain the in situ relative density and angle of internal friction, but only for a very limited soil depth (about 0.5 m). Thus, no direct or indirect way is readily available for obtaining the in situ relative density and in situ angle of internal friction of noncohesive sea-floor soils.

In lieu of direct measurements, it appears safe to assume, for clean sea-floor sands, a minimum angle of internal friction, ϕ, of 0.45 rad (26 deg) (see Fig. A-2 adapted from NAVFAC DM-7, 1971). The allowable bearing load is obtained by applying a factor of safety of 3.0 to the calculated bearing capacity, q_{ult}, from equation (A-2). The maximum calculated bearing pressure due to all possible loading combinations must not exceed the allowable bearing load.

Clays. The vast majority of sea-floor footings will be founded on cohesive soils. Knowledge of the soil shear strength profile can be obtained from tests on good quality soil cores. Equipment for obtaining these cores is described elsewhere (Richards and Parker, 1968; Rosfelder and Marshall, 1967). The soil shear strength may be obtained from the core samples via vane shear tests or triaxial consolidated-undrained tests (Lee, 1974). The shear strength profile may also be obtained via in situ vane shear devices; only two of these devices exist having a 3.0 m soil depth capability,

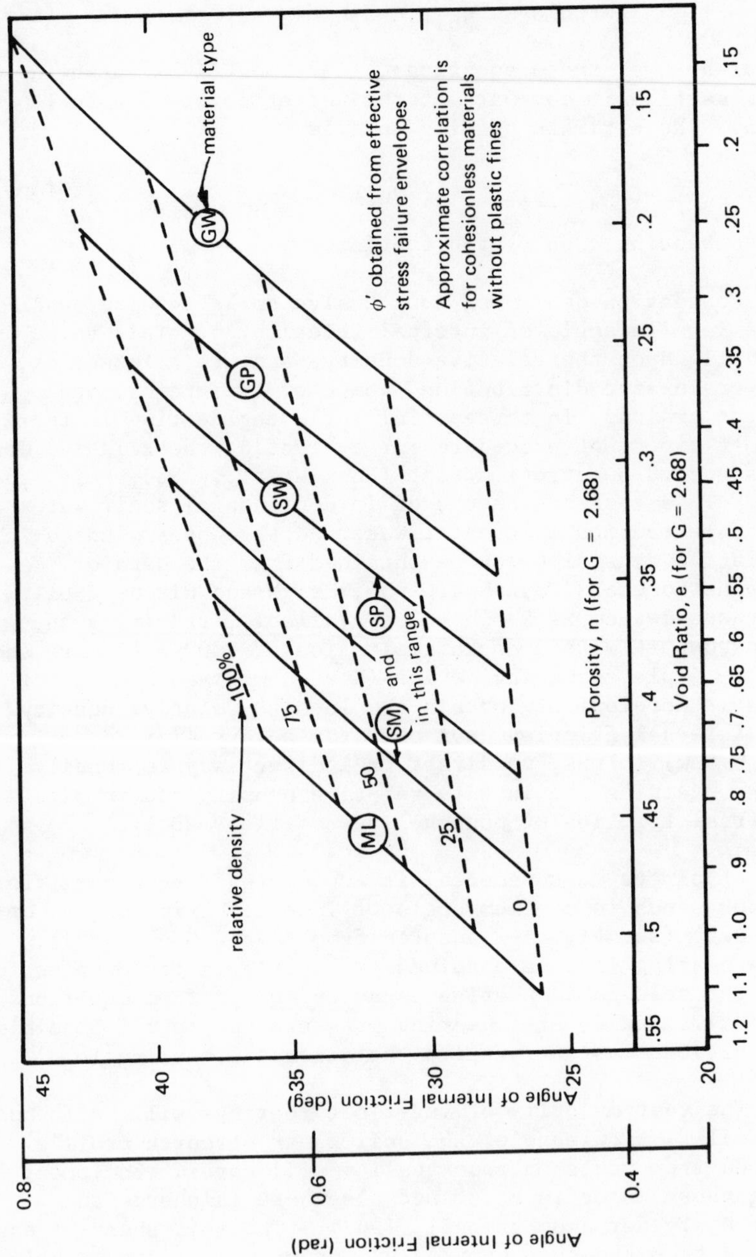

Figure A-2. Angle of internal friction versus porosity and void ratio for noncohesive soils (from NAVFAC DM-7, 1971)

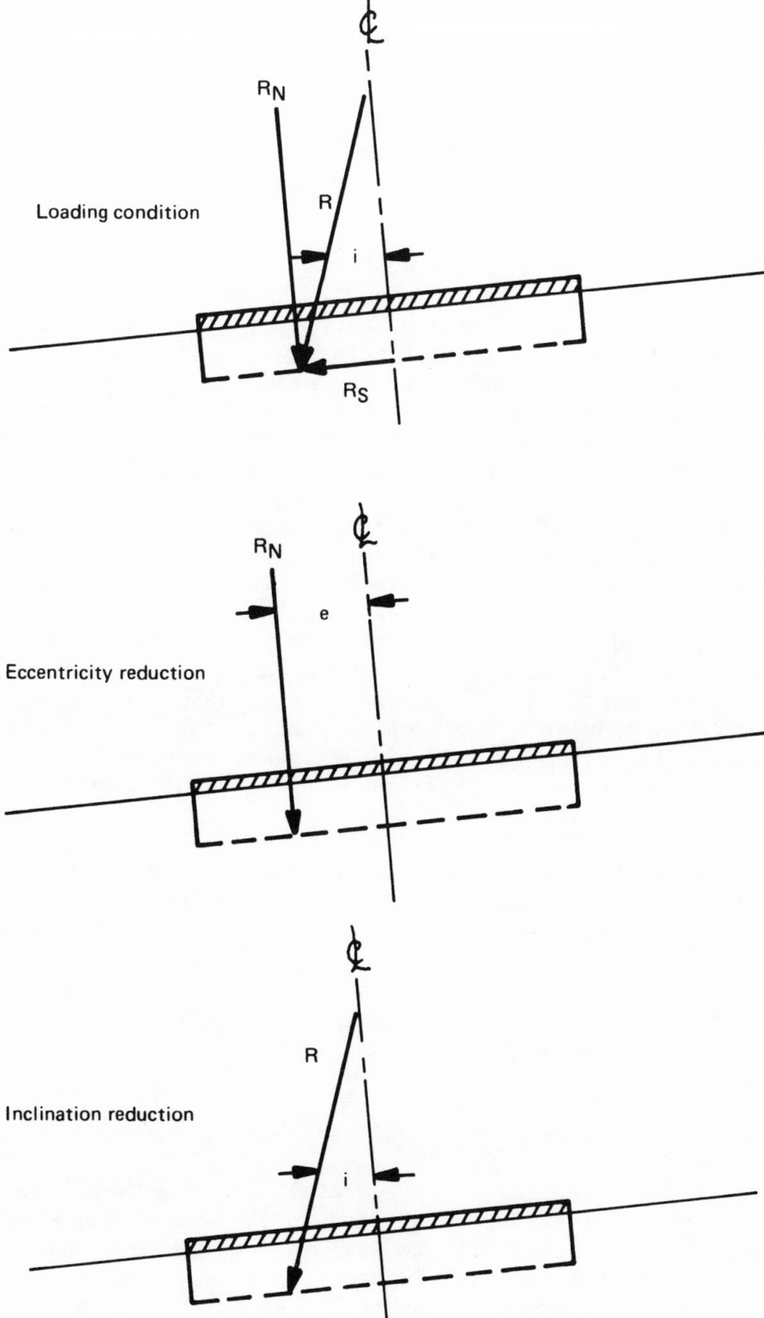

Figure A-3. Pictorial representation of eccentricity and inclination treatments, eccentricity and inclination complementary requiring specialized ship support, and limited to 2000 m and 3600 m water depth, respectively (Demars and Taylor, 1971; Richards et al., 1972).

The limiting condition for bearing capacity on deep-sea cohesive soils is ordinarily the short term, or ϕ = 0 condition, for which equation (A-3) becomes:

$$q_{ult} = 5.70\ K_1 c \qquad \text{(A-5)}$$

Because of the compressible nature of surficial sea-floor soils, it is expected that most sea-floor footings will experience local failure. For this failure condition, the measured soil shear strength, c, will not be realized along the entire failure zone; and, therefore, the analysis should employ a reduced strength value, $c' = 2/3\ c$ (Terzaghi and Peck, 1967; Sowers, 1962).

Equations (A-1), (A-3), and (A-5) assume a uniform soil shear strength, c, with depth in the soil profile to a soil depth B. Such a uniform soil profile is not typical. Investigations of the effect of a variable strength with depth on the bearing capacity factors have resulted in a complex method of prediction (Reddy and Srinivasan, 1967). If, instead of invoking this complex method, the strength at a soil depth of 1/3 x B is selected as c, then the resulting calculated bearing capacities will be 90 to 100% of those obtained via the complex method (H. G. Herrmann, personal communication, 1973). This approach does assume a reasonably consistent variation of strength with depth.

It is recommended that the allowable bearing load be obtained by applying a factor of safety of 4.0 to the calculated bearing capacity, q_{ult}, from equation (A-5). This recommended value of 4.0 is more conservative than the factor of safety of 3.0 recommended for terrestrial structures (Terzaghi and Peck, 1967) because of the writer's concern with regard to certain incompletely explained failures (Herrmann et al., 1972; Keller, 1964).

Eccentric and Inclined Loads

Eccentricity and inclination, separately and together, alter the bearing pressure distribution applied to the supporting soil and cause a footing of given area to fail in bearing at a lower resultant load magnitude than it would under a normal, centered resultant. Analytical treatment is accomplished by separating the effects of eccentricity and inclination (see Fig. A-3). In effect, reduction factors to be applied to the ultimate normal load, Q_{ult}, are calculated individually, first, for eccentricity, and second for inclination. Should the direction of inclination act to push the footing in the direction of eccentricity, as illustrated in

Figure A-3, then the two components act to complement each other and the reduction factors should be applied consecutively to Q_{ult}. If, however, the resultant inclination acts to push the footing in a direction opposite to the eccentricity, as shown in Figure A-4, then the two components act to partially oppose each other, and the net reduction factor is somewhat less than in the previous case (Meyerhof, 1953). However, for this condition, it is recommended for simplicity that the net reduction factor for the Figure A-4 case be calculated in the same way as that for the Figure A-3 case.

Eccentricity. Eccentricity is analytically treated by assuming that only a portion of the bearing area is effective in carrying the perpendicular component of the resultant load. The line of action of the resultant force and its intersection with the effective bearing surface are first determined (see Fig. A-3). Then an effective contact area, A', is determined such that its centroid coincides with the point of action of the resultant (see Fig. A-5) (Sowers, 1962; Meyerhof, 1953). The reduction factor to be applied to the ultimate bearing load (which acts normal to the plane of the footing) is simply the ratio of areas:

$$r_{ec} = A'/A \qquad \text{(A-6)}$$

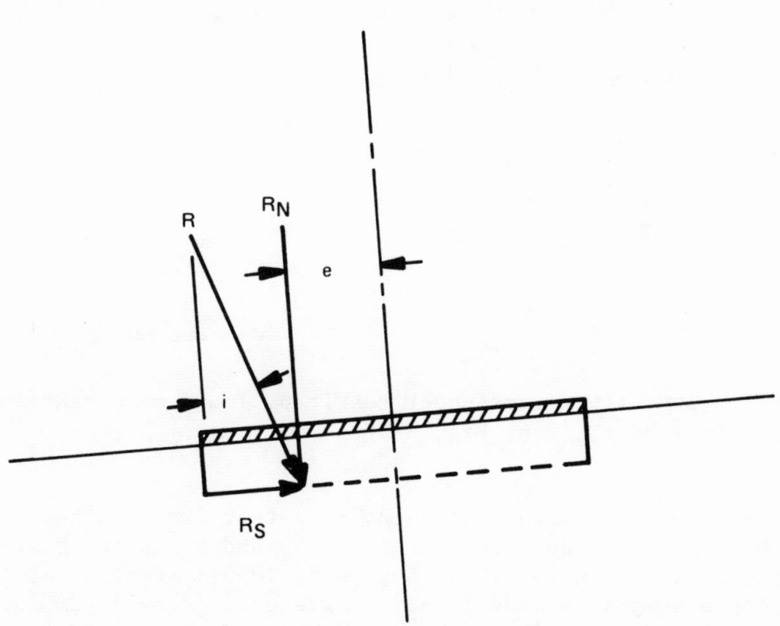

Figure A-4. Eccentricity and inclination opposing

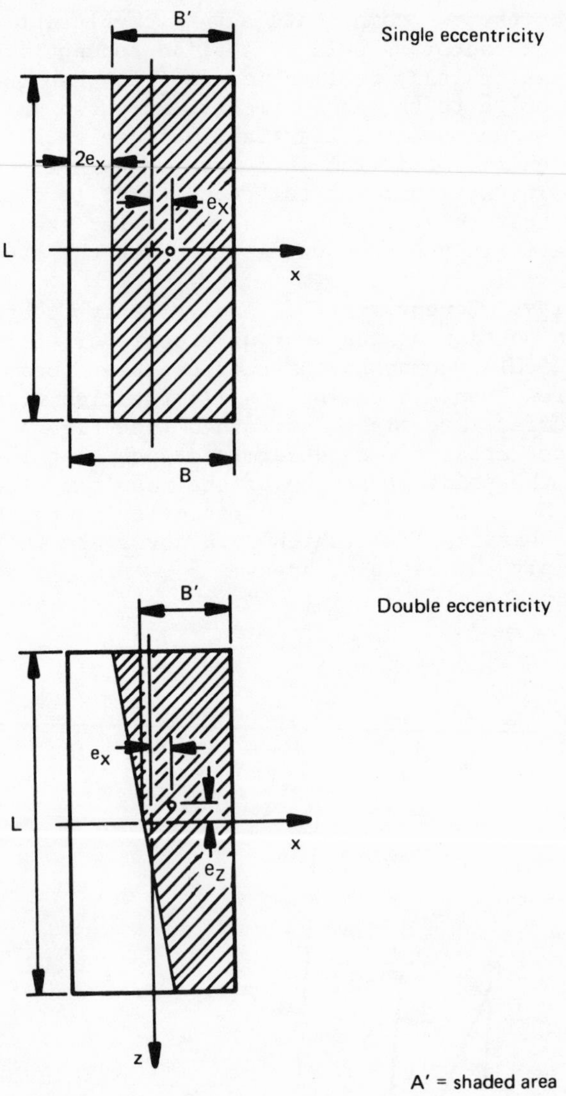

Figure A-5. Effective areas under resultant loads of eccentricity e_x and e_z (from Harr, 1966, Fig. 6-30)

Inclination. Inclination of the resultant load has been treated by applying reduction factors to $N\gamma$ and N_c in the bearing capacity equation (Sowers, 1962; Meyerhof, 1953). Because of the simplifying assumptions made in deriving equation (A-2) for sands and equation (A-4) for clays, these reduction factors too can be directly applied to the ultimate bearing load, Q_{ult}. The reduction factors, r_{in}, are given in Table A-2.

TABLE A-2

REDUCTION FACTORS FOR VERTICAL BEARING CAPACITY OF FOOTINGS UNDER INCLINED LOAD

	Inclination of load with the normal					
degrees/radians	0/0	10/0.17	20/0.35	30/0.52	45/0.79	60/1.05
For sands (equation A-2)	1.0	0.5	0.2	0		
For clays (equation A-3)	1.0	0.8	0.6	0.4	0.25	0.15

(adapted from Sowers, 1962, Table 6-4)

As described earlier, eccentricity and inclination reductions are combined by applying each consecutively to the ultimate normal load, or:

$$Q'_{ult} = Q_{ult} \times r_{ec} \times r_{in}. \quad \text{(A-7)}$$

APPENDIX B - EMBEDMENT ANCHOR-HOLDING CAPACITY ANALYSIS

Embedment anchors in pullout will deform a surrounding soil in either of two modes depending on the depth of the plate beneath the sea-floor surface (see Fig. B-1). If the anchor is quite deep, upward displacement of the anchor will cause soil to move in plastic flow from above the plate to beneath, with no sea-floor surface expression of the action. If the anchor is shallow, upward displacement of the anchor will cause a volume of soil to move with the anchor plate during breakout. The transition from deep to shallow behavior occurs in clays at a D/B of 2 to 5, and in sands at a D/B of 2 to 10 (Vesic, 1969), where:

D = depth of plate below the sea floor,

B = width or diameter of anchor plate.

The pullout resistance of an anchor, even if embedded in an isotropic, homogeneous medium, decreases when the failure mode changes from the deep to the shallow mode. Thus, an efficient design will usually try to emplace the anchor plate at a depth where deep anchor behavior begins.

This paper presents only the deep anchor uplift resistance theory because this situation will be most often encountered and because shallow anchor theory is considerably more complicated and beyond the scope of this paper. (For shallow anchor design recommendations see Taylor and Lee, 1972). As in the evaluation of footing bearing capacity, pullout resistance in sands and clays is treated separately.

In non-cohesive soils, the uplift resistance of an embedded plate may be expressed as follows:

For sands, $Q_u = \gamma' D N_{qu} A\ (0.84 + 0.16\ B/L)$ (B-1)

where Q_u = ultimate uplift capacity

γ' = buoyant unit weight of the soil

N_{qu} = uplift capacity factor from Figure B-2

A = projected area of the plate

L = length of a rectangular plate

(modified from Adams and Klym, 1972; Taylor and Lee, 1972).

This relationship is essentially analogous to the third term of the bearing capacity equation, equation (A-1), with a shape factor added.

In cohesive soils, the uplift resistance expression becomes:

For clays, $Q_u = N_{cu} c A\ (0.84 + 0.16\ B/L)$ (B-2)

where N_{cu} = uplift capacity factor = 9.0 for deep failure mode

c = undrained soil shear strength near the plate elevation

(modified from Adams and Klym, 1972; Taylor and Lee, 1972). This relationship is analogous to the cohesion contribution, the first term, of the bearing capacity equation, equation (A-1).

As in determining the bearing capacity of a footing, predicting the pullout resistance of an anchor requires a knowledge of the soil shear strength profile, specifically the angle of internal friction, ϕ, for sands and the undrained shear strength, c, for clays. The difficulty involved in trying to establish the profile for these properties in sea-floor soil depths to 15 m has been discussed earlier.

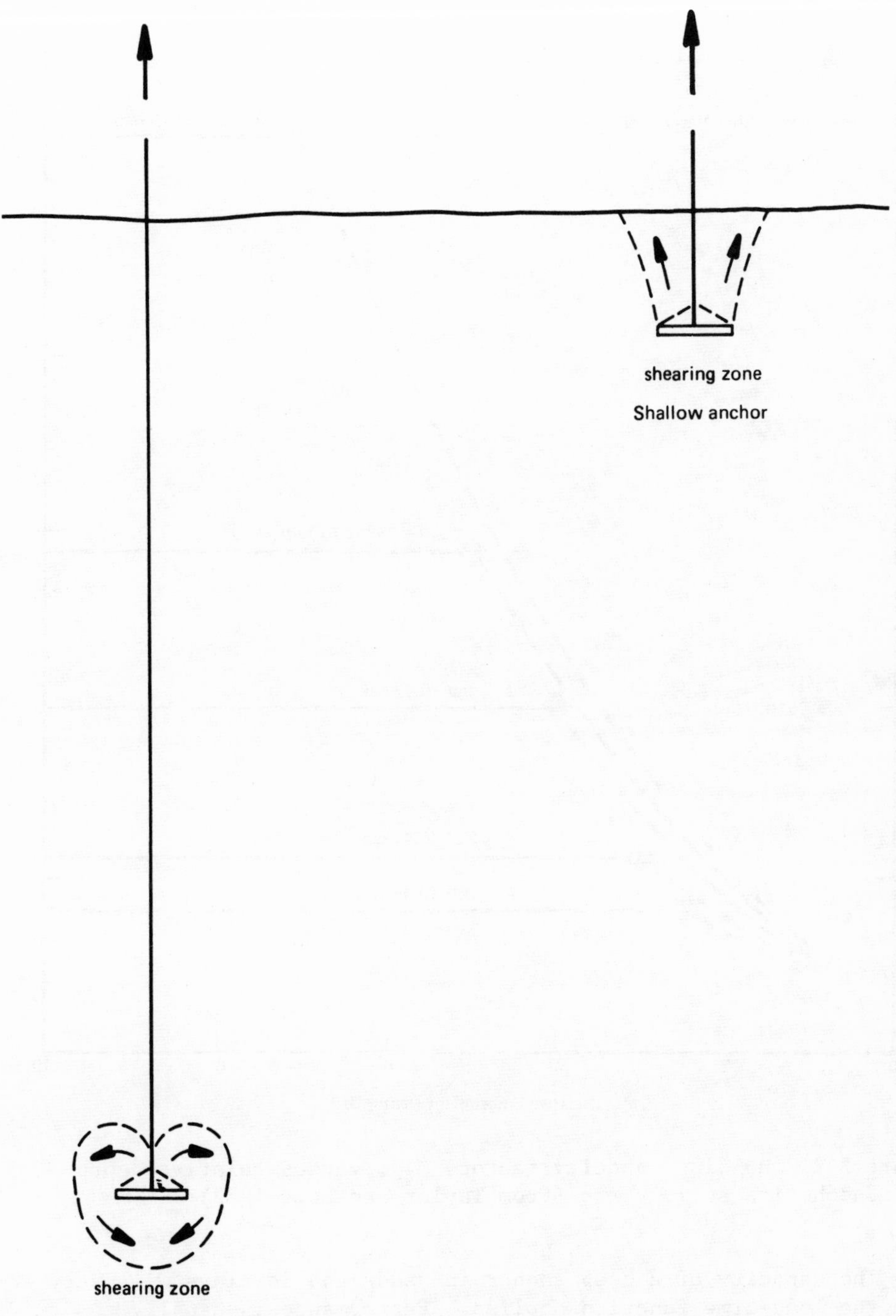

Figure B-1. Failure modes for differing anchor burial depths

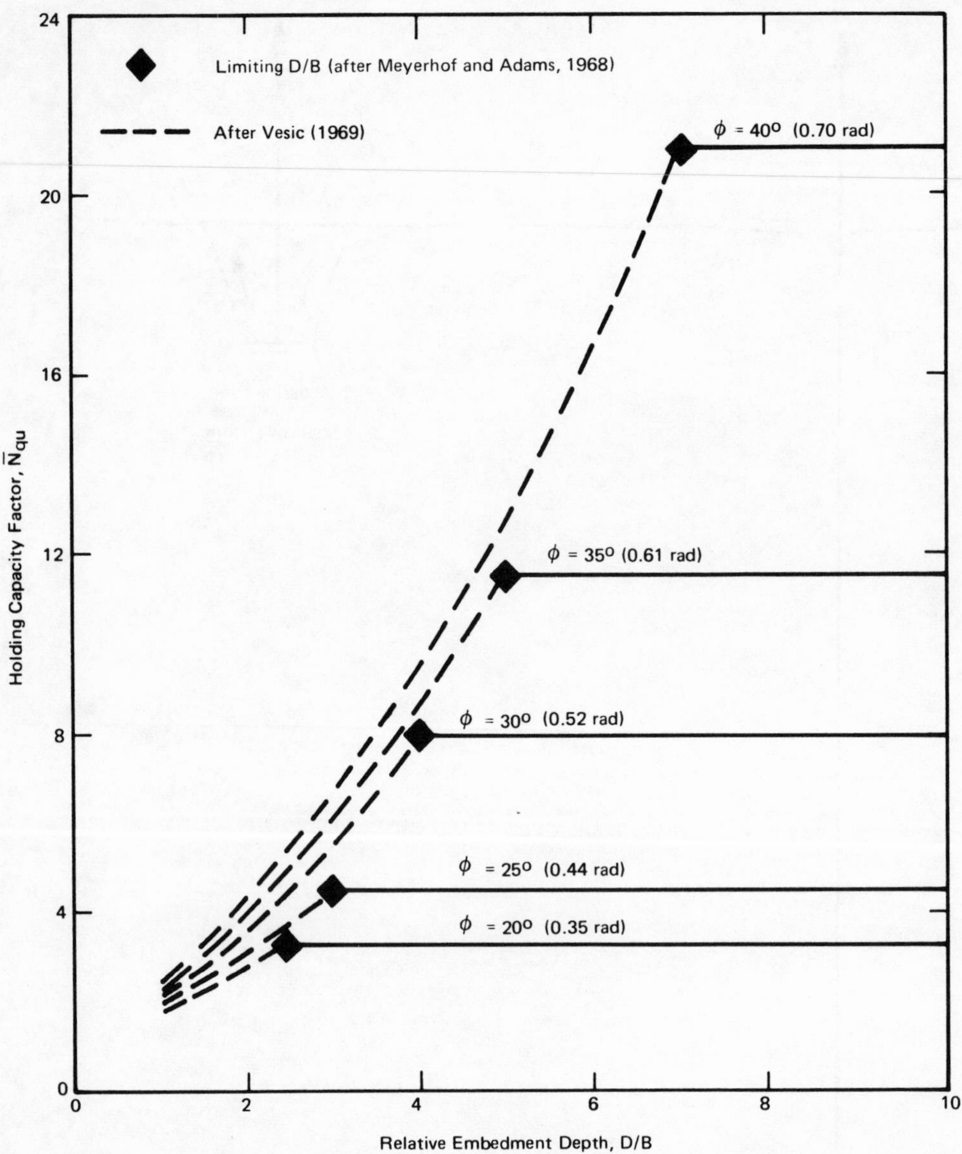

Figure B-2. Holding capacity factor, N_{qu}, versus relative depth for noncohesive soil, c = o (from Taylor and Lee, 1972)

The capacity of a deep anchor in sands and in clays will vary with the load-time function applied. Performance is usually broken down in terms of: (1) short-term or immediate holding capacity, (2) long-term static holding capacity, and (3) the long-term repeated load-holding capacity.

Equations (B-1) and (B-2) supply short-term holding capacities. Care must be taken in evaluating the results of such short-term tests in clays because initially a significant portion of the ultimate pullout load may be due to negative gage pressures (suction) in the water contained by the soil beneath the anchor. Thus, short-term field test results may be unconservatively high. A technique for correcting for the suction effect is presented elsewhere (Taylor and Lee, 1972).

Subjected to long-term static pullout loads, the holding capacity of a deep embedment anchor in noncohesive soil remains constant while the capacity in cohesive soil can be expected to increase (provided that rupture creep does not occur). This capacity increase can be expected because the cohesive soil in the highly stressed soil zone will consolidate with time and its shearing strength will increase (Adams and Hayes, 1967; Taylor and Lee, 1972). An approximate means for estimating the long-term static capacity is available (Meyerhof and Adams, 1968) but questioned for theoretical and practical reasons (Taylor and Lee, 1972). One reason for not utilizing the hypothesized increased holding capacity is that the increased loading may aggravate or accelerate the shear creep of the cohesive medium. Shear creep is a long-term shear straining under which the anchor plate may gradually drift upward under constant load and eventually pull out of the sea floor (creep rupture). In some cohesive soils, creep rupture has occurred at stresses as low as 60% of the measured strength (Singh and Mitchell, 1968). In lieu of information to the contrary, the anchor plate could conceivably creep upward faster than consolidation and soil strengthening could proceed; thus, long-term static anchor design should assume the ultimate uplift capacity to be no greater than 60% of the short-term capacity in cohesive soils. In noncohesive soils the long-term static and short-term static capacities are equal.

A combination of long-term sustained and repeated loads will be applied to embedment anchor systems which are used to moor surface vessels or buoys. These loads vary with the tautness of the system and the nature of wave or tidal action. Subjected to such loadings, in almost all cases pullout will occur at a lower force level. For cohesive soils, the holding capacity reduction will be on the order of 50% of the short-term capacity (Taylor and Lee, 1972; Trafimenkov and Mariupolsuii, 1965). For noncohesive soils, the holding capacity may be reduced by more than 50%; a presentation of design technique for repeated loads involves considering the shallow anchor failure mode and is therefore beyond the scope of this paper (Taylor and Lee, 1972; Kalajian, 1971). For noncohesive soils in the silt-fine sand size range, repetitive load ultimate holding capacities may be as much as 90% less than the short-term static capacities (Taylor and Lee, 1972).

REFERENCES

Adams, J. I., and D. C. Hayes, The uplift capacity of shallow foundations, Ontario Hydro Res. Quart., 19 (1), 1967.

Adams, J. I., and T. W. Klym, A study of anchorages for transmission tower foundations, Canadian Geotech. J., 9 (1), 89-104, 1972.

Barnes, B. B., R. F. Corwin, J. H. Beyer, Jr., and T. G. Hildenbrand, Geologic prediction: developing tools and techniques for the geophysical identification and classification of sea-floor sediments, National Oceanic and Atmospheric Administration Tech. Rept. ERL 224-MMTC 2, 1971.

Bowin, C. O., R. L. Chase, and J. B. Hersey, Geological applications of sea-floor photography, in Deep-Sea Photography, edited by J. B. Hersey, pp. 117-140, John Hopkins Press, Baltimore, 1967.

Brundage, W. L., C. L. Buchanan, and R. B. Patterson, Search and serendipity, in Deep-Sea Photography, edited by J. B. Hersey, p. 81, John Hopkins Press, Baltimore, 1967.

Demars, K. R., and R. J. Taylor, Naval sea-floor soil sampling and in-place test equipment: a performance evaluation, U. S. Naval Civil Engineering Laboratory Tech. Rept. R-730, 1971.

Einstein, H. A., and R. L. Wiegel, A literature review on erosion and deposition of sediment near structures in the ocean, U. S. Naval Civil Engineering Laboratory Contract Rept. CR 70.008, 1970.

Harr, M. E., Foundations of Theoretical Soil Mechanics, McGraw-Hill, N. Y., 1966.

Hawkins, L. K., Visual observations of manganese deposits on the Blake Plateau, J. Geophys. Res., 74 (28), 7009-7017, 1969.

Herrmann, H. G., K. Rocker, Jr., and P. H. Babineau, LOBSTER and FMS: devices for monitoring long-term sea-floor foundation behavior, U. S. Naval Civil Engineering Laboratory Tech. Rept. R-775, 27-28, 1972.

Hirst, T. J., A. F. Richards, and A. L. Inderbitzen, A static cone penetrometer for ocean sediments, in Symp. Underwater Soil Sampling, Testing, and Construction Control, Am. Soc. Test. Mat. Spec. Tech. Publ. 501, pp. 69-80, Phila., 1972.

Hjulstrom, F., Transportation of detritus by moving water, in Rec. Mar. Sediments, edited by P. D. Trask, p. 10, Am. Assoc. of Petrol. Geol., Tulsa, Okla., 1939.

Hough, B. K., Basic Soils Engineering, 2nd edition, Ronald Press, N. Y., 1969.

Kalajian, E. H., The vertical holding capacity of marine anchors in sand subjected to static and cyclic loading, Ph.D. thesis, Univ. of Mass., 1971.

Keller, G. H., Investigation of the application of standard soil mechanics techniques and principles to bay sediments, U. S. Naval Oceanographic Office Informal Manuscript Rept. No. 0-6-64, 1964.

Ko, Hon-Yim, and L. W. Davidson, Bearing capacity of footings in plane strain, J. Soil Mech. Fdns. Div., ASCE, 99 (SM1), 1973.

Kretschmer, T. R., and H. J. Lee, Plate-bearing tests on sea-floor sediments, U. S. Naval Civil Engineering Laboratory Tech. Rept. R-694, 1970.

Kuenen, P. H., Marine Geology, John Wiley and Sons, N. Y., p. 260, 1950.

Lee, H. J., The Role of laboratory testing in the determination of deep-sea sediment engineering properties, (this volume), 1974.

Meigh, A. C., and I. K. Nixon, Comparison on in situ tests for granular soils, Proc. Fifth Intern. Conf. on Soil Mech. and Fdn. Eng., I, 499-507, 1961.

Meyerhof, G. G., The bearing capacity of foundations under eccentric and inclined loads, Proc. Third Intern. Conf. on Soil Mech. and Fdn. Eng., I, 440-445, 1953.

Meyerhof, G. G., Penetration tests and bearing capacity of cohesionless soils, J. Soil Mech. Fdns. Div., ASCE, 82 (SM1), p. 19, 1956.

Meyerhof, G. G., and J. I. Adams, The ultimate uplift capacity of foundations, Canadian Geotech. J., 5, (14), 225-244, 1968.

Muraoka, J. S., Animal undermining of naval sea-floor installations, U. S. Naval Civil Engineering Laboratory Tech. Note N-1124, 1970.

NAVFAC DM-7, Design Manual: Soil Mechanics, Foundations and Earth Structures, Naval Facilities Engineering Command, Washington, D. C., 1971.

Nimomiya, K., K. Tagaya, and Y. Murase, A study on suction breaker and scouring of a submersible offshore structure, 3rd Ann. Offshore Tech. Conf. Preprints, 1971.

Peck, R. B., W. E. Hanson, and T. H. Thornburn, Foundation Engineering, 219-228, John Wiley and Sons, N. Y., 1953.

Posey, C. J., Protection against underscour, 2nd Ann. Offshore Tech. Conf. Preprints, 2, 747-750, 1970.

Raecke, D. A., and J. H. Migliore, Sea-floor pile foundations: state-of-the-art, and deep-ocean emplacement concepts, U. S. Naval Civil Engineering Laboratory Tech. Note N-1182, p. 10, 1971.

Ralston, D. O., and J. B. Herbich, The effects of waves and currents on submerged pipelines (abstract No. 5), Texas A&M University Sea Grant Publ. 101, 1969.

Reddy, S. A., and R. J. Srinivasan, Bearing capacity of footings on layered clays, J. Soil Mech. Fdns. Div., ASCE, 93 (SM2), 83-99, 1967.

Richards, A. F., V. J. McDonald, R. E. Olson, and G. H. Keller, In-place measurement of deep sea soil shear strength, in Symp. Underwater Soil Sampling, Testing, and Construction Control, Am. Soc. Test. Mat. Spec. Tech. Publ. 501, pp. 55-68, Phila., 1972.

Richards, A. F., and H. W. Parker, Surface coring for shear strength measurements, in Civil Eng. in the Oceans, pp. 445-489, ASCE, N. Y., 1968.

Rosfelder, A. M., and N. F. Marshall, Obtaining large, undisturbed, and oriented samples in deep water, in Marine Geotechnique, edited by A. F. Richards, pp. 243-263, Univ. of Ill. Press, Urbana, 1967.

Singh, A., and J. K. Mitchell, General stress-strain-time function for soils, J. Soil Mech. Fdns. Div., ASCE, 94 (SM1), 231-253, 1968.

Smith, D. T., Acoustic and electric techniques for sea-floor sediment identification, in Proc. Intern. Symp. Eng. Properties of Sea-Floor Soils and their Geophys. Ident., pp. 235-267, Univ. of Washington, Seattle, 1971.

Sowers, G. F., Shallow Foundations, Chapter 6 in Foundation Engineering, edited by G. A. Leonards, McGraw-Hill, N. Y., 1962.

Taylor, R. J., and R. M. Beard, Propellant-actuated deep water anchor, 5th Ann. Offshore Tech. Conf. Preprints, I (OTC 1744), 199-208, 1973.

Taylor, R. J., and H. J. Lee, Direct embedment anchor-holding capacity, U. S. Naval Civil Engineering Laboratory Tech. Note N-1245, 1972.

Terzaghi, K., and R. B. Peck, Soil Mechanics in Engineering Practice, 2nd edition, John Wiley and Sons, N. Y., 1967.

Thompson, L. J., Mechanics problems and material properties, (this volume), 1974.

Trofimenkov, J. G., and L. G. Mariupolskii, Screw piles used for mast and tower foundations, Proc. Sixth Intern. Conf. on Soil Mech. and Fdn. Eng., 2, 328-332, 1965.

Vesic, A. S., Breakout resistance of objects embedded in ocean bottom, U. S. Naval Civil Engineering Laboratory Contract Rept. CR 69.031, 1969.

STANDARDIZATION OF MARINE GEOTECHNICS SYMBOLS, DEFINITIONS, UNITS, AND TEST PROCEDURES

ADRIAN F. RICHARDS

Lehigh University

ABSTRACT

The ONR Seminar-Workshop participants approved selected symbols taken from the International Society of Soil Mechanics and Foundation Engineering source book of symbols and definitions. A synopsis of applicable SI units in marine geotechnics is made from the American Society for Testing and Materials (ASTM) and the U. S. National Bureau of Standards guides; relevant conversion formulas from old metric and U. S. customary units to SI units are presented. ASTM tests and recommended test modifications applicable to marine soils are summarized. Commonly used non-ASTM-approved laboratory and field tests are listed, together with selected references to the literature describing each test. Recommendations are made for the effective use of symbols, definitions, units, and test procedures in marine geotechnics.

INTRODUCTION

This paper on the subject of standardization of field and laboratory test procedures, nomenclature, and instrumentation is the result of a request by the organizers of the ONR Seminar-Workshop.

SYMBOLS AND DEFINITIONS

The excellent book, Technical Terms, Symbols and Definitions (hereafter referred to as ISSMFE, 1967), is the accepted international standard. The list of definitions and symbols was prepared by a committee of the International Society of Soil Mechanics and Foundation

TABLE 1

SYMBOLS ADOPTED BY THE SEMINAR-WORKSHOP PARTICIPANTS

Symbol	Definition	NOTES
GENERAL		
t	time	T ASTM preference; footnote 1
V	volume	
W	weight*	
F	factor of safety	
STRESS AND STRAIN		
u	pore pressure	"neutral stress" ASTM
u_w	pore water pressure	"neutral stress" ASTM
σ	normal stress*	
$\bar{\sigma}$	effective normal stress	"effective stress" ASTM
τ	shear stress*	
ε	linear strain*	
SOIL PROPERTIES		
γ	unit weight of soil*	
γ_s	unit weight of solid particles	footnote 2
γ_w	unit weight of water	
γ_{sw}	unit weight of sea water	not defined by ISSMFE or ASTM
γ_{sat}	unit weight of water-saturated soil	not defined by ISSMFE; "saturated unit weight" ASTM
$\bar{\gamma}$	unit weight of submerged soil	γ' in ISSMFE; γ_m ASTM first choice
G_s	specific gravity of solid particles	G ASTM first choice
e	void ratio	
n	porosity	
w	water content	
S_r	degree of saturation	footnote 2
w_L	liquid limit	LL ASTM first choice
w_p	plastic limit	
I_p	plasticity index	
I_L	liquidity index	B ASTM first choice
C_c	compression index	
c_v	coefficient of consolidation	
τ_f	shear strength	s ASTM first choice
$\bar{c}$	effective cohesion intercept	footnote 3
$\bar{\phi}$	effective angle of internal friction	footnote 3
c_u	apparent cohesion intercept	footnote 3
ϕ_u	apparent angle of internal friction	footnote 3
S_t	sensitivity	footnote 4

TABLE 1

(CONTINUED)

OTHER		
K	earth pressure coefficient	p ASTM first choice
K_0	coefficient of earth pressure at rest	P_0 ASTM first choice
N	bearing capacity factor	not defined by ASTM
OCR	over consolidation ratio ($\bar{p}_c$ / $\bar{\sigma}_v$)	ONR seminar adoption; not defined by ISSMFE or ASTM
$\bar{p}_c$	preconsolidation stress (greatest effective stress)	ONR seminar adoption; Pe ASTM preference
$\bar{\sigma}_v$	effective overburden stress	ONR seminar adoption; not defined by ISSMFE or ASTM

* International Organization for Standardization (ISO) approved symbol.

Footnote 1: Also, t is Celsius temperature (and T is Kelvin temperature) in SI units.

Footnote 2: This symbol, believed to be non-controversial, was not presented at the ONR Seminar-Workshop and was not formally adopted by the seminar participants.

Footnote 3: These symbols are not defined in the ASTM (1973) publication. Their use is in conformity with accepted practice.

Footnote 4: The definition (ratio between undrained shear strengths of undisturbed and of remolded soil) given by the ISSMFE (1967) is preferable to the ASTM (1973) definition (the effect of remolding on the consistency of a cohesive soil).

Engineering (ISSMFE). Symbols given in this book are in close conformity to the standard definitions of terms and symbols designated D 653-67 (ASTM, 1973), which were prepared jointly by the American Society of Civil Engineers and the American Society for Testing and Materials (hereafter referred to as ASTM).

Table 1 is a short list of selected symbols chosen to represent those most commonly used in marine geotechnics. All of the symbols are from the ISSMFE (1967) publication, except as noted, where they are fully defined in eight languages. Conformity with the ASTM Standard D 653-67 (ASTM, 1973) definitions is implied unless specifically indicated to the contrary. All symbols in Table 1 were formally adopted by the Seminar-Workshop participants, except for four that were inadvertently omitted from the submitted list: unit weight of solid particles, degree of saturation, coefficient

TABLE 2

SI BASE, SUPPLEMENTARY, AND SELECTED DERIVED AND OTHER UNITS*

Base units		SI Symbol	Formula	Expression in terms of SI base units
length	metre	m		
mass	kilogram	kg		
time	second	s		
electric current	ampere	A		
thermodynamic temperature	kelvin	K		
amount of substance	mole	mol		
luminous intensity	candela	cd		
Supplementary units				
plane angle	radian	rad		
solid angle	steradian	sr		
Derived units (partial list)				
acceleration	metre per second squared	-		m/s^2
angular acceleration	radian per second squared	-		rad/s^2
angular velocity	radian per second	-		rad/s
area	square metre	-		m^2
capacitance	farad	F	C/V	$m^{-2}\cdot kg^{-1}\cdot s^4\cdot A^2$
conductance	siemens	S	A/V	$m^{-2}\cdot kg^{-1}\cdot s^3\cdot A^2$
density	kilogram per cubic metre	-		kg/m^3
electric potential difference, or electromotive force	volt	V	W/A	$m^2\cdot kg\cdot s^{-3}\cdot A^{-1}$
electric resistance	ohm	Ω	V/A	$m^2\cdot kg\cdot s^{-3}\cdot A^{-2}$
energy	joule	J	N/m	$m^2\cdot kg\cdot s^{-2}$
force	newton	N		$m\cdot kg\cdot s^{-2}$
frequency	hertz	Hz		s^{-1}
inductance	henry	H	Wb/A	$m^2\cdot kg\cdot s^{-2}\cdot A^{-2}$
magnetic flux	weber	Wb	V/s	$m^2\cdot kg\cdot s^{-2}\cdot A^{-1}$
permeability	henry per metre	-	H/m	$m\cdot kg\cdot s^{-2}\cdot A^{-2}$
power	watt	W	J/s	$m^2\cdot kg\cdot s^{-3}$
pressure	pascal	Pa	N/m^2	$m^{-1}\cdot kg\cdot s^{-2}$
quantity of electricity	coulomb	C		$s\cdot A$
quantity of heat	joule	J	N·m	$m^2\cdot kg\cdot s^{-2}$
stress	pascal	Pa	N/m^2	$m^{-1}\cdot kg\cdot s^{-2}$
velocity	metre per second	-		m/s
viscosity, dynamic	pascal-second	-	Pa/s	$m^{-1}\cdot kg\cdot s^{-1}$
viscosity, kinematic	square metre per second	-		m^2/s
voltage	volt	V	W/A	$m^2\cdot kg\cdot s^{-3}\cdot A^{-1}$
volume	cubic metre	-		m^3
work	joule	J	N/m	$m^2\cdot kg\cdot s^{-2}$

TABLE 2

(CONTINUED)

Permitted unit		SI Symbol	Formula	Expression in terms of SI base units
temperature	Celsius, degree	°C (degree Celsius)		1°K=273.15 + °C and 1°K=1°C
Units outside the SI, but accepted temporarily				
	bar	bar		1 bar=10^5 Pa
	knot	-		1 knot=(1852/3600) m/s
	international nautical mile	INM		1 nautical mile=1852 m
	standard atmosphere	atm		1 atm=101325 Pa
Units outside the SI, but may be used with SI				
	minute	min		1 min=60 s
	hour	h		1 h=3600 s
	day	d		1 d=86400 s
	degree	°		1°=(π/180) rad
	minute	'		1'=(π/10800) rad
	second	"		1"=(π/648000) rad
	litre	l		1 l=$10^{-3}m^3$
	tonne	t		1 t=10^3kg

* This table was compiled from both the ASTM (1972a) and from Page and Vigoureux (1972).

of earth pressure, and coefficient of earth pressure at rest. The participants agreed to use all of the symbols given in Table 1, except the four above, in future publications. It is hoped that other individuals working in this field will also adopt these symbols. Universal use would greatly simplify problems of effective communication.

UNITS

The definitive description of the modern metric system is Le Système International d'Unités, abbreviated SI in all languages, published in a 1972 revised edition by the International Bureau of Weights and Measures. This document has been independently published in English by the National Bureau of Standards (Page and Vigoureux, 1972). A companion booklet, Standard Metric Practice Guide, was published by the American Society for Testing and Materials (ASTM, 1972a) under designation E 380-72. The contents of this standard also were approved by the American National Standards Institute (ANSI), under designation ANSI Z210.1-1973. The American Society of Civil Engineers (ASCE, 1971), the ASTM, and many other societies require SI units to be used in their publications. The rationale behind the move for U. S. as well as global adoption of modern metric units of measurements has been documented by De Simone (1971).

Table 2 lists the nine basic and supplementary SI units together with derived units and other units of possible concern to marine geotechnologists. It should be noted that the recommended SI English spelling of the basic units of length is metre, not meter. (The corresponding compromise for worldwide uniformity in the English language is the recommended spelling of kilogram, rather than the British kilogramme.) In SI units, the kilogram is the only mass unit and the newton is the only force unit. In combination units including force, such as pressure or stress, the newton should be used rather than kilogram-force; for example, energy (N·m = J), power (N·m/s = W), and stress or pressure (N/m^2 = Pa).

Note that the correct SI symbol for kilogram is kg, not kgm. Also, SI symbols are always written in singular form. Periods are never placed after SI symbols, except at the end of a sentence.

The International Organization for Standardization has recom-

TABLE 3

SI PREFIXES*

Prefix	Symbol	Factor
tera	T	10^{12}
giga	G	10^{9}
mega	M	10^{6}
kilo	k	10^{3}
milli	m	10^{-3}
micro	µ	10^{-6}
nano	n	10^{-9}
pico	p	10^{-12}
femto	f	10^{-15}
atto	a	10^{-18}

* The prefixes hecto (10^{2}), deka (10^{1}), deci (10^{-1}), and centi (10^{-2}) are to be avoided where possible (ASTM, 1972a).

mended that the product of two or more units is best indicated by a dot, unless there is no risk of confusion with another unit symbol: N·m or N m, but not mN (Page and Vigoureux, 1972). Other examples are given in Table 2.

Table 3 gives the ASTM (1972a) recommended prefixes. Note that multiples of 1000 are preferable to smaller steps.

Table 4 is a compilation of conversions believed useful in marine geotechnics that has been taken for the most part from the more extensive list of conversions published in the Standard Metric Practice Guide (ASTM, 1972a). This handy publication is exceedingly useful; no scientist or engineer should be without it. The Guide contains information on accuracy and rounding, on significant digits, and on toleranced dimensions; it also contains an excellent resumé on the development of SI units. (ASTM standard E 380, a SI-U. S. customary units slide rule and a guidelines booklet are available as a kit from the ASTM, under publication code number 12-503801-00, for $5.25 plus shipping and applicable taxes: ASTM, 1916 Race Street, Philadelphia, Pa. 19103.)

The author suggests that most stress or pressure measurements in geotechnics can best be expressed in kilopascals (kPa). The graphical relationship of common U. S. conventional and foreign metric units to kilopascals is shown in Figure 1. Chisholm (1967) presents another source of conversion formulas and tables.

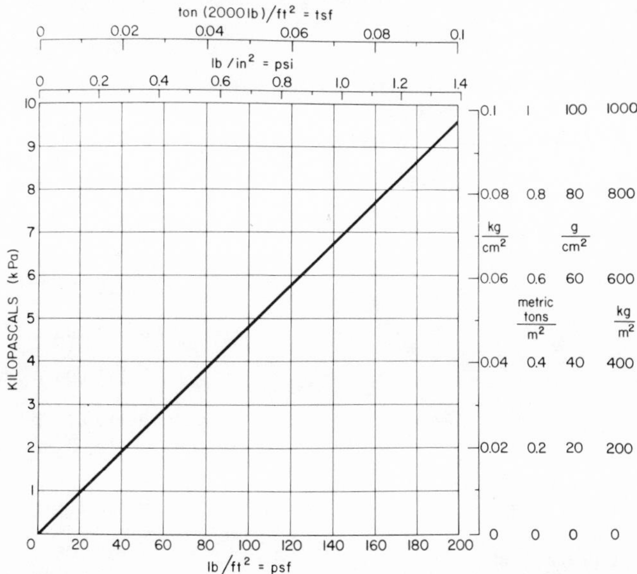

Figure 1. Relationship of stress or pressure in SI units of kPa to old metric and U. S. customary units

TABLE 4

SELECTED CONVERSIONS IN MARINE GEOTECHNICS*

From	To	Multiply by
AREA		
$foot^2$	m^2	9.290 304 E-2
$inch^2$	m^2	6.451 600 E-4
$mile^2$ (U.S. statute)	m^2	2.589 988 E+6
$yard^2$	m^2	8.361 274 E-1
DENSITY OR MASS/VOLUME		
$gram/centimetre^3$	kg/m^3	1.000 000 E+3
ounce (avoirdupois) $(mass)/inch^3$	kg/m^3	1.729 994 E+3
$pound\text{-}mass/foot^3$	kg/m^3	1.601 846 E+1
$pound\text{-}mass/inch^3$	kg/m^3	2.767 990 E+4
ton $(metric)/metre^3$	kg/m^3	1.000 000 E+3*
FORCE		
dyne	N	2.000 000 E-5
kilogram-force	N	9.806 650 E+0
kip	N	4.448 222 E+3
ounce-force (avoirdupois)	N	2.780 139 E-1
pound-force (avoirdupois)	N	4.448 222 E+0
LENGTH		
fathom	m	1.828 800 E+0
foot	m	3.048 000 E-1
inch	m	2.540 000 E-2
microinch	m	2.540 000 E-8
micron	m	1.000 000 E-6
mil	m	2.540 000 E-5
mile (intern. nautical)	m	1.852 000 E+3
mile (U.S. statute)	m	1.609 344 E+3
yard	m	9.144 000 E-1

TABLE 4

(CONTINUED)

From	To	Multiply by
MASS		
gram	kg	1.000 000 E-3
ounce-mass (avoirdupois)	kg	2.834 952 E-2
pound-mass (lbm avoirdupois)	kg	4.535 924 E-1
ton (metric)	kg	1.000 000 E+3
ton (short, 2000 lbm)	kg	9.071 847 E+2
tonne	kg	1.000 000 E+3
PRESSURE OR STRESS (FORCE/AREA)		
atmosphere (normal = 760 torr)	Pa	1.013 25 E+5
atmosphere (technical = 1 kgf/cm^2)	Pa	9.806 650 E+4
bar	Pa	1.000 000 E+5
decibar	Pa	1.000 000 E+4
dyne/centimetre^2	Pa	1.000 000 E-1
gram-force/centimetre^2	Pa	9.806 650 E+1
kilogram-force/centimetre^2	Pa	9.806 650 E+4
kilogram-force/metre^2	Pa	9.806 650 E+0
kip/inch^{2}ksi	Pa	6.894 757 E+6
pound-force/foot^2 (psf)	Pa	4.788 026 E+1
pound-force/inch^2 (psi)	Pa	6.894 757 E+3
ton (metric)/metre^2 (tsm)	Pa	9.806 650 E+3*
ton (short, 2000 lbf)/foot^2 (tsf)	Pa	9.576 056 E+4*
TORQUE		
ounce-force-inch	N·m	7.061 552 E-3
Pound-force-inch	N·m	1.129 848 E-1
pound-force-foot	N·m	1.355 818 E+0

TABLE 4

(CONTINUED)

From	To	Multiply by
VELOCITY AND ANGULAR VELOCITY		
degree/minute	rad/s	2.908 882 E-4*
foot/hour	m/s	8.466 667 E-5
foot/minute	m/s	5.080 000 E-3
foot/second	m/s	3.048 000 E-1
inch/second	m/s	2.540 000 E-2
kilometre/hour	m/s	2.777 778 E-1
knot (international)	m/s	5.144 444 E-1
mile/hour (U.S. statute)	m/s	4.470 400 E-1
VOLUME		
acre-foot	m^3	1.233 484 E+3
$foot^3$	m^3	2.831 685 E-2
gallon (U.S. liquid)	m^3	3.785 412 E-3
litre	m^3	1.000 000 E-3
ounce (U.S. fluid)	m^3	2.957 353 E-5
quart (U.S. fluid)	m^3	9.463 529 E-4
$yard^3$	m^3	7.645 549 E-1
VOLUME/TIME		
$foot^3$/minute	m^3/s	4.719 474 E-4
$foot^3$/second	m^3/s	2.831 685 E-2
$inch^3$/minute	m^3/s	2.731 177 E-7
$yard^3$/minute	m^3/s	1.274 258 E-2

* All conversions from ASTM E 380-72 (ASTM, 1972a) unless followed by an asterisk, in which instance they are by the author.

TEST STANDARDIZATION

Time did not permit the Seminar-Workshop participants to consider the standardization of either laboratory or field tests. This is logically an area of endeavor for Committee D 18 on Soil and Rock for Engineering Purposes of the American Society for Testing and Materials, as well as the comparable committees in standards organizations in other countries.

Table 5 is an assessment of the current ASTM (1973) designations and recommended modifications for marine geotechnical investigations. It should be noted that many of the tests have been approved (date is the last two digits after dash) or re-approved (date given in parenthesis) during the past several years. Old standards may be superseded; they should not be used. The ASTM book of standards is issued annually.

The field vane test (Table 5) and the field and laboratory vane tests (Table 6) require further explanation. The angular speed of vane blade rotation is controversial. The standard speed is 1.7 m·rad/s (0.1°/s). To save time and to ensure that drainage did not occur during the test, which is only valid under undrained conditions, the author and his colleagues (Richards et al., 1972) in 1967 arbitrarily adopted a speed of 23 m·rad/s (1.33°/s). More recently the Naval Civil Engineering Laboratory agreed to adopt 26 m·rad/s (1.5°/s) as standard (H. Gill, personal communication, 1972). This speed was adopted as standard by the JOIDES Committee on Sedimentary Petrology and Mass Properties for the Deep-Sea Drilling Project. The Seminar-Workshop participants were undecided whether to change from the slower to a faster speed, despite the assurance of N. Monney at the Seminar-Workshop that the 23-26 m·rad/s speed appeared satisfactory based on his preliminary test results. The difference between 23 and 26 m·rad/s appears to be minimal.

The D 2573-72 field vane test given in Table 5 is not directly applicable, although the formula for calculating shear strength for a rectangular vane may be considered a standard; in symbols and units previously adopted (Tables 1 and 2):

$$\tau = T/K$$

where τ is the shear stress of the clay (Pa), T is the torque (N·m), and K is a constant depending on the shape of the vane and its dimensions:

$$K = (\pi/10^6)\cdot(D^2H/2)\cdot[1 + (D/3H)]$$

where D is the measured diameter (cm), H is the vane height (cm), and the ratio of vane length to breadth is 2:1.

TABLE 5

ASTM STANDARDIZED TESTS AND MODIFICATIONS FOR MARINE GEOTECHNICS

Test Name	ASTM Designation*	ANSI Designation*	Recommended Modifications+	Notes	Information to be Specified°
Atterberg limits					
Liquid limit	D 423-66 (1972)	None	1. Do not dry nor sieve before test; use D 2217, not D 421, preparation method 2. If water must be added, use sea water of same salinity instead of distilled water	Report results to nearest whole number, according to ASTM	
Plastic limit	D 424-59 (1971)	A37.142-1972	Same as for liquid limit	Same as above	
Bulk density, in situ nuclear methods	D 2922-71	A37.179-1972	See notes	Test probably not applicable to marine soils as stipulated; tests given in Table 6 are preferable	
Color and visual description	D 2488-69	A37.174-1972	Geological Society of America Rock Color Chart is an inexpensive collection of Munsell color chips suitable for marine soil analysis	Descriptive color name usually adequate; use Munsell notation for maximum preciseness	Name of color chart used
Consolidation, one-dimensional	D 2435-70	A37.170-1972	1. Use sea water, if possible, to saturate sample	Test often modified; back-pressured tests also may be made; see Bryant _et al_. (1967), Hermann _et al_. (1972), and Parker and Miller (1970), for examples	

Test Name	ASTM Designation*	ANSI Designation*	Recommended Modifications+	Notes	Information to be Specified°
Grain- or particle-size					
Preparation, dry	D 421-58 (1972)	A37.144	Do not use for water-saturated marine soils	See wet preparation	
Preparation, wet	D 2217-66 (1972)	None	1. Sieving probably not required for most samples	Use test procedure B; do not dry sample prior to test	Procedure B used
Analysis, sieve and hydrometer	D 422-63 (1972)	None	1. Use D 2217, rather than D 421, preparation method 2. Separation of sand- and silt-size using No. 230 sieve (62 μm) probably more common than using No. 200 sieve (75 μm) 3. Separation of silt and clay-size at 2 μm more common than at 5 μm	Pipette and falling-drop tests also used for silts and clays, and settling tube or sedimentation balance tests for sands, among other methods	1. Type of sieve shaker
Separation at 75 μm	D 1140-54 (1971)	A37.143-1972	1. Wet sieving using the No. 230 sieve (62 μm) preferable to using the No. 200 sieve (75 μm) 2. Do not dry sample prior to sieving		
Sampling, thin-wall	D 1587-67	None	1. Alternate use of suitable plastic as well as metal tubes, if bending stresses on sample can be tolerated	See Richards and Parker (1968), and papers in International Group on Soil Sampling (1969, 1971) and ASTM (1972b) for general reviews	1. Inside and outside clearance ratios, area ratio, and tube length and diameter 2. Core recovery, if known 3. Estimated sampler quality

TABLE 5 (CONTINUED)

Test Name	ASTM Designation*	ANSI Designation*	Recommended Modifications+	Notes	Information to be Specified°
Shear strength					
Compressive, unconfined	D 2166-66 (1972)	A37.148-1972		1. It may be necessary to obtain remolded strength using an alternate method if the sample is too soft to stand under its own weight 2. Controlled strain is a more common test than controlled stress 3. Comparison with the vane and fall-cone tests given by Flaate (1965)	
Direct, (consolidated drained)	D 3080-72	None	See notes	Not in common use; author has had insufficient experience to comment	
Triaxial, (unconsolidated, undrained)	D 2850-70	None	See notes	Author has had insufficient experience to comment	
Vane, field	D 2573-72	A37.183-1972	1. Although a vane rotation of 1.7 m•rad/s (6°/min) is the accepted standard, 23-26 m•rad/s (80-90°/min) is recommended	1. Specified vane height to diameter 2:1 2. This test not directly applicable to in situ tests of marine soils 3. See text	1. Rate of vane rotation 2. Vane dimensions 3. Depth of measurement at bottom or midpoint of vane

Test Name	ASTM Designation*	ANSI Designation*	Recommended Modifications+	Notes	Information to be Specified°
Specific gravity	D 854-58 (1972)	A37.145	1. Wash sample to eliminate salts	1. Use of kerosene instead of water is not recommended for marine soils	
Water or moisture content	D 2216-71	A37.141-1972		1. Use of forced-draft type of oven recommended by ASTM and author 2. It is recommended by the author that results be reported only to the nearest whole number 3. Water contents determined by this test represent included salt content, opinion differs on whether or not a correction for salt should be automatically applied.	

* American Society for Testing and Materials designations are listed in ASTM (1973); American National Standards Institute (ANSI) designations are cited in ASTM (1973).

\+ In general usage; recommended by the author.

° In addition to the specifications stipulated in the ASTM standard.

A problem of uncertain magnitude exists in field and laboratory vane testing. The disturbed or remolded strength conventionally determined by a vane test may yield different values of shear stress, depending on whether the soil was carefully remolded by hand to its minimum shear stress at the same water content or whether the soil was retested without hand remolding. At the Marine Geotechnical Laboratory there is an indication that soil remolding by hand, and subsequent retesting by a laboratory vane, yields somewhat lower values of shear stress. In this instance, higher sensitivity values (ratio of "undisturbed" shear stress to disturbed or remolded shear stress at the same water content) result. This clearly is an area requiring further research on different marine soils to determine the magnitude of the difference.

Table 6 was prepared to identify some of the more common laboratory and field tests currently being made on marine soils for which ASTM standards do not exist; it was not practical to include every test in this table. The carbonate and organic carbon tests, commonly made on the same machine using the same test procedure, have almost attained the status of provisional standards. The laboratory vane is most commonly used for measuring the shear stress in cohesive soils. However, until the vane size and the rate of angular rotation of the vane blade is standardized, this test cannot be considered a standard. The other test methods are in varying states of adoption by the profession.

It would be highly desirable for each geotechnical or geological laboratory testing marine soils to publish their test methods. At present, only the SACLANT ASW Research Centre (Kermabon and Blavier, 1967) appears to have accomplished this laudable first step towards permitting an assessment of laboratory and field test methods in marine geotechnics.

SUMMARY OF RECOMMENDATIONS

1. The book, Technical Terms, Symbols and Definitions (ISSMFE, 1967), should be considered the recommended source of terms, symbols, and definitions in marine geotechnics and in the mass physical properties of soils (sediments).

2. Symbols given in Table 1 (in some instances different from those given in the ISSMFE, 1967) are recommended for general adoption. They were formally adopted by the Seminar-Workshop participants.

3. SI units should be adopted immediately by scientists and engineers in all countries. Table 2 lists units believed to be of particular importance in marine geotechnics. Kilopascal (kPa) is recommended as the most useful SI stress or pressure unit in marine geotechnics. Useful conversions are given in Table 3.

TABLE 6

NON-ASTM-STANDARDIZED TESTS FOR MARINE GEOTECHNICS

Test Name	Selected References*	Notes
Bulk density		
nuclear, lab	Evans(1965), Preiss(1968a), Evans and Cottrell(1970)	
nuclear, in situ	Keller(1965), Preiss(1968b), Rose and Roney(1971)	
pycnometer, air-comparison	Hironaka(1966), Kermabon and Blavier(1967)	Helium purge recommended
"weight"/volume	Sutton et al.(1957), Smith and Nunes(1964), Richards(1973)	
Carbonate	Hülsemann(1966), Siesser and Rogers(1971), Boyce and Bode(1972)	Hülsemann test in common use; however, LECO equipment superseding
Grain- or particle-size		
Falling drop	Moum(1965), Pezzetta(1972), Richards(1973)	
Organic carbon	Hironaka(1966), Boyce and Bode(1972), Boyce(1972)	LECO equipment in most common use
Salt content or salinity, pore water+		
Electrical conductivity	Richards(1973)	
Induction salinometer	Friedman et al.(1968), Cernock and Bryant(1969)	
Refractive index	Behrens(1965)	Fastest method; hand-held, direct-reading refractometer (American Optical Co.)
Resistivity	Boyce(1968)	
Titration	Riley(1965), Siever et al.(1965)	
Shear strength		
Fall cone	Moore and Richards(1962), Kessler and Stiles(1968), Richards(1973)	
Penetrometer, field	Taylor and Demars(1970), Hirst et al.(1972)	
Vane, borehole	Fenske(1957), Doyle et al.(1971)	
Vane, miniature lab	Richards(1961), Hironaka(1966), Kravitz(1970)	
Vane, field	Taylor and Demars(1970), Inderbitzen et al.(1971), Richards et al.(1972)	

* Only those references believed representative or containing bibliographies have been cited.

+ All methods require the use of a squeezer: for example, Manheim(1966).

The SI unit of length should be spelled metre, to follow international convention.

4. Whenever possible, American Society for Testing and Materials (ASTM) standards should be followed in North America. Elsewhere the national standards should be followed. Deviations

from standards should be specified in publications. Recommended modifications of ASTM standards for marine geotechnics are given in Table 5.

5. ASTM standards are issued annually. As they may be changed, the most recent standard should be used. The standard designation, which should be cited, indicates the date of approval, or reapproval if the date is in parenthesis (see Table 5 for examples).

6. Many tests commonly used in marine geotechnics have not been formally standardized. It is recommended that ASTM Committee D 18 undertake a review of the tests cited in Table 6, as well as others, with the objective of considering applicable standardization. In the meanwhile, test variables should be specified whenever cited in publications.

7. The early publication of the test methods used at major marine geotechnical and marine geological laboratories is recommended. Only after these methods are known to others can standardization of tests logically occur.

8. It is recommended that comparable tests on similar (identical, if possible) soils be made at two or more laboratories to determine the degree of similarity and difference of test methods. Interlaboratory calibration of tests and test methods is almost unknown in marine geotechnics.

ACKNOWLEDGMENTS

Appreciation is expressed to Michael Perlow, Jr. for his help in reviewing this paper and checking conversions. This paper was prepared in part under Office of Naval Research contract N00014-67-A-0370-005.

REFERENCES

ASCE, Guidelines for the use of SI units of measurement in the publications of the American Society of Civil Engineers, ASCE, N. Y., 1971.

ASTM, Standard Metric Practice Guide, Standard E 380-72, American Society for Testing and Materials, Phila., 1972a; also in 1973 Annual Book of ASTM Standards, Part 30, General test methods, pp. 1188-1221.

ASTM, Underwater soil sampling, testing, and construction control, Am. Soc. Test. Mat. Spec. Tech. Publ. 501, Phila., 1972b.

ASTM, 1973 Annual Book of ASTM Standards, Part II, Bituminous materials for highway construction, waterproofing, and roofing; soil and rock; peats, mosses, and humus; skid resistance, Am. Soc. Test. Mat., Phila., 1973.

Behrens, E. W., Use of the Goldberg refractometer as a salinometer for biological and geological field work, J. Mar. Res., 23, 165-171, 1965.

Boyce, R. E., Electrical resistivity of modern sediments from the Bering Sea, J. Geophys. Res., 73, 4759-4766, 1968.

Boyce, R. E., Carbon and carbonate analyses, leg 11, in Initial Reports of the Deep-Sea Drilling Project, 11, edited by C. D. Hollister, and J. I. Ewing et al., pp. 1059-1071, U. S. Govt. Printing Office, Washington, D. C., 1972.

Boyce, R. E., and G. W. Bode, Carbon and carbonate analyses, leg 9, in Initial Reports of the Deep-Sea Drilling Project, 9, edited by J. D. Hays et al., pp. 797-816, U. S. Govt. Printing Office, Washington, D. C., 1972.

Bryant, W. R., P. Cernock, and J. Morelock, Shear strength and consolidation characteristics of marine sediments from the western Gulf of Mexico, in Marine Geotechnique, edited by A. F. Richards, pp. 41-62, Univ. of Ill. Press, Urbana, 1967.

Cernock, P. J., and W. R. Bryant, Salinities of interstitial waters from selected sediments in the Gulf of Mexico, J. Sediment. Petrol., 39, 1633-1639, 1969.

Chisholm, L. J., Units of weight and measure, international (metric) and U. S. customary, U. S. National Bureau of Standards Misc. Publ. 286, Washington, D. C., 1967.

De Simone, D. V., A metric America: a decision whose time has come, U. S. National Bureau of Standards Spec. Publ. 345, Washington, D. C., 1971.

Doyle, E. H., B. McClelland, and G. H. Ferguson, Wire-line vane probe for deep penetration measurements of ocean sediment strength, 3rd Ann. Offshore Tech. Conf. Preprints, 1, 21-32, 1971.

Evans, H. B., GRAPE-A device for continuous determination of material density and porosity, Soc. Prof. Well Log Analysts, 6th Ann. Symp. Trans., 2B, 1-25, 1965.

Evans, H. B., and C. H. Cotterell, Gamma ray attenuation density scanner, in Initial Reports of the Deep-Sea Drilling Project, 18, edited by M. N. A. Peterson et al., pp. 460-472, U. S. Govt. Printing Office, Washington, D. C., 1970.

Fenske, C. W., Deep vane tests in Gulf of Mexico, in In-Place Shear Testing of Soil by the Vane Method, Am. Soc. Test. Mat. Spec. Tech. Publ. 193, pp. 16-25, Phila., 1957.

Flaate, K., A statistical analysis of some methods for shear strength determination in soil mechanics, Norwegian Geotechnical Institute Publ. 62, 1-8, 1965.

Friedman, G. M., B. P. Fabricand, E. S. Imbimbo, M. E. Brey, and J. E. Sanders, Chemical changes in interstitial waters from continental shelf sediments, J. Sediment. Petrol., 38, 1313-1319, 1968.

Herrmann, H. G., K. Rocker, Jr., and P. H. Babineau, LOBSTER and FMS: devices for monitoring long-term sea-floor foundation behavior, U. S. Naval Civil Engineering Laboratory Tech. Rept. R775, 1972.

Hironaka, M. C., Engineering properties of marine sediments near San Miguel Island, Calif., U. S. Naval Civil Engineering Laboratory Tech. Rept. R503, 1966.

Hirst, T. J., A. F. Richards, and A. L. Inderbitzen, A static cone penetrometer for ocean sediments, in Symp. Underwater Soil Sampling, Testing, and Construction Control, Am. Soc. Test. Mat. Spec. Tech. Publ. 501, pp. 69-80, Phila., 1972.

Hülsemann, J., On the routine analysis of carbonates in unconsolidated sediments, J. Sediment. Petrol., 36, 622-625, 1966.

Inderbitzen, A. L., F. Simpson, and G. Goss, A comparison of in situ and laboratory vane shear measurements, Mar. Tech. Soc. J., 5, 24-34, 1971.

International Group on Soil Sampling, Soil sampling, Proc. Specialty Session No. 1, pp. 1-110, in Proc. 7th Intern. Conf. on Soil Mech. and Fdn. Eng., Mexico, 1969, Intern. Group on Soil Sampling, Melbourne, 1969. (Obtainable from IGOSS, CSIRO, P. O. Box 54, Mount Waverley, Victoria, Australia 3149.)

International Group on Soil Sampling, Quality in Soil Sampling, Proc. Specialty Session, 1 and 2, 4th Asian Regional Conf. of the Intern. Soc. for Soil Mechanics and Fdn. Eng., Intern. Group on Soil Sampling, Melbourne, 1971. (Obtainable from

IGOSS, CSIRO, P. O. Box 54, Mount Waverly, Victoria, Australia 3149.)

ISSMFE, Technical Terms, Symbols and Definitions, 3rd edition, Intern. Soc. Soil Mech. and Fdn. Eng., Zürich, 1967.

Keller, G. H., Nuclear density probe for in-place measurement in deep-sea sediments, Mar. Tech. Soc. and Am. Soc. Limnol. and Ocean., Trans. Joint Conf. Exhibit, 1, 363-372, 1965.

Kermabon, A., and P. Blavier, Principles and methods of core analysis at the SACLANT ASW Research Centre, SACLANT ASW Research Centre Tech. Rept. 71, 1967.

Kessler, R. S., and N. T. Stiles, Comparison of shear strength measurements with the laboratory vane shear and fall-cone devices, U. S. Naval Oceanographic Office Informal Rept. 68-75, 1968.

Kravitz, F. H., Repeatability of three instruments used to determine the undrained shear strength of extremely weak, saturated, cohesive sediments, J. Sediment. Petrol., 40, 1026-1037, 1970.

Manheim, F. T., A hydraulic squeezer for obtaining interstitial water from consolidated and unconsolidated sediments, U. S. Geol. Survey Prof. Paper 550-C, 256-261, 1966.

Moore, D. G., and A. F. Richards, Conversion of "relative shear strength" measurements by Arrhenius on East Pacific deep-sea cores to conventional units of shear strength, Géotechnique, 12, 55-59, 1962.

Moum, J., Falling drop used for grain-size analysis of fine-grained materials, Sedimentology, 5, 343-347, 1965.

Page, C. H., and P. Vigoureux, editors, The International System of Units (SI), U. S. National Bureau of Standards Spec. Publ. 330, 1972.

Parker, H. W., and D. G. Miller, Jr., Operation manual for model A Anteus back pressure consolidometers, Lehigh Univ., Marine Geotechnical Laboratory Internal Rept., 1970. (Obtainable from the Laboratory).

Pezzetta, J. M., Falling-drop technique for silt-clay sediment analysis, University of Wisconsin Sea Grant Program Tech. Rept. 15, WIS-SG-72-215, 1972.

Preiss, K., Non-destructive laboratory measurement of marine sediment density in a core barrel using gamma radiation, Deep-Sea Res., 15, 401-407, 1968a.

Preiss, K., In-situ measurement of marine sediment density by gamma radiation, Deep-Sea Res., 15, 637-641, 1968b.

Richards, A. F., Investigations of deep-sea sediment cores, I. Shear strength, bearing capacity, and consolidation, U. S. Navy Hydrographic Office Tech. Rept. 63, 1961.

Richards, A. F., Geotechnical properties of submarine soils, Oslofjorden and vicinity, Norway, Norwegian Geotechnical Institute Tech. Rept. 13, 1973.

Richards, A. F., and H. W. Parker, Surface coring for shear strength measurements, in Civil Eng. in the Oceans, pp. 445-489, Am. Soc. Civil Engrs., N. Y., 1968.

Richards, A. F., V. J. McDonald, R. E. Olson, and G. H. Keller, In-place measurement of deep-sea soil shear strength, in Symp. Underwater Soil Sampling, Testing, and Construction Control, Am. Soc. Test. Mat. Spec. Tech. Publ. 501, pp. 55-68, Phila., 1972.

Riley, J. P., Analytical chemistry of sea water, in Chemical Oceanography, 2, edited by J. P. Riley and G. Skirrow, pp. 295-424, Academic Press, London, 1965.

Rose, V. C., and J. R. Roney, A nuclear gage for in-place measurement of sediment density, 3rd Ann. Offshore Tech. Conf. Preprints, 1, 43-49, 1971.

Siesser, W. G., and J. Rogers, An investigation of the suitability of four methods used in routine carbonate analysis of marine sediments, Deep-Sea Res., 18, 135-139, 1971.

Siever, R., K. C. Beck, and R. A. Berner, Composition of interstitial waters of modern sediments, J. Geol., 73, 39-73, 1965.

Smith, R. J., and L. Nunes, Undeformed sectioning of plastic core tubing, Deep-Sea Res., 11, 261-262, 1964.

Sutton, G. H., H. Berckhemer, and J. E. Nafe, Physical analysis of deep-sea sediments, Geophysics, 22, 779-812, 1957.

Taylor, R. J., and K. R. Demars, Naval in-place sea-floor soil test equipment: a performance evaluation, U. S. Naval Civil Engineering Laboratory Tech. Note N-1135, 1970.

SOIL MECHANICS CONSIDERATIONS IN THE DESIGN OF DEEP OCEAN MINING EQUIPMENT

JOHN E. HALKYARD AND JOHN T. FULLER

Kennecott Exploration, Inc.

ABSTRACT

Deep ocean mining equipment must be designed to operate on the surficial layer of the sea bottom with a high degree of reliability, efficiency, and controllability. By modeling the nodule collector as an element of a dynamic system, the collector motions may be predicted and its effectiveness evaluated. Such a model depends heavily on accurate formulation of the soil forces acting on the collecting vehicle. The methodology for determining these forces is discussed, and recommendations for future research are presented.

INTRODUCTION

Manganese nodules have long been considered potentially one of the most economically attractive of the ocean resources likely to be seriously pursued in the near future. Recent activities by American and foreign companies have confirmed this belief, and there is little doubt that commercial ocean mining will indeed be a reality within a few years.

The ultimate economic success of these ventures is far from being assured, however. An economically profitable mining system will have to perform at a high rate of production, probably over 9072 metric tons per day of nodules, and will have to maintain a high degree of reliability. These requirements are not easily met, and a great deal of designing and testing must yet be done before the ultimate success of ocean mining is attained.

TABLE 1

TOTAL OCEAN MINING SYSTEM

Subsystem	Mission(s)
Collector	a) Traverse the ocean bottom b) Harvest nodules from the sea floor c) Separate mud and debris from the harvested material d) Inject the collected and screened nodules into a lifting mechanism
Lift	a) Accept the harvested nodules and transport them to the surface b) Provide a connecting link between the surface system and the collector system for propulsion, motion and control of the total system
Surface	a) Traverse the ocean surface in a manner compatible with sea bottom mining operations as performed by subsurface equipment b) Provide support for the operation of subsurface equipment during operations c) Be capable of deploying, retrieving and maintaining sea bottom equipment d) Provide buffer storage for material delivered by the lift system e) Deliver the mined product to a transportation vessel for transport to a shore-based processing plant

These requirements place a heavy burden on deep ocean mining equipment. Such a system may conceptually be divided into three interrelated subsystems: collector, lift, and surface systems. Table 1 describes the missions of each of these subsystems.

Of all the subsystems, the collector represents the most uncertain piece of hardware, and the most critical for successful operations. It must be able to traverse a predictable and controllable path across a sometimes rather rough and obstacle-ridden sea

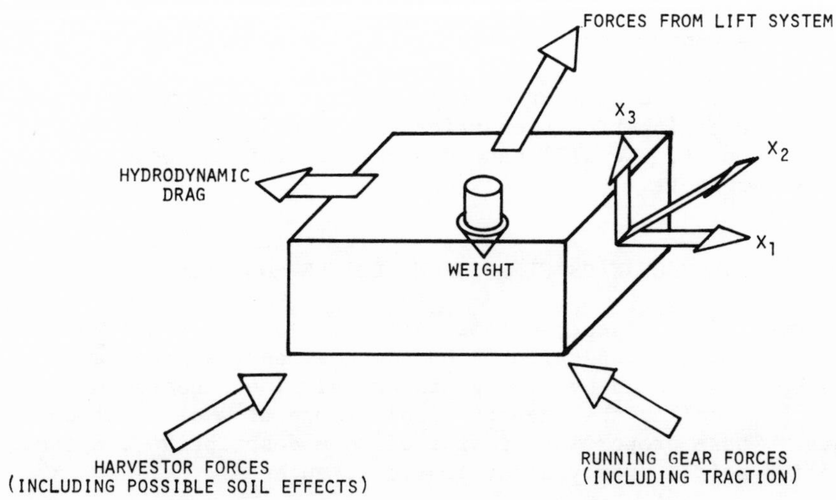

Figure 1. Conceptual drawing of the collector as a dynamic system

bottom. It must maintain a close affinity for the upper few inches of the bottom where the nodules lie, lest its harvesting function be severely handicapped.

This paper will present, from an engineering point of view, the critical problem of predicting and confirming the collector's motion over the sea floor, particularly as it is affected by the sediment properties. It will be impossible to cover all aspects of the soil/collector interaction in this short time. No discussion of the harvesting function is included.

NODULE COLLECTOR AS A DYNAMIC SYSTEM

In order to predict the motions of a nodule collector, all the forces acting on it must be considered. To do this, the collector must be analyzed as a dynamic system. A conceptual representation of the collector in this sense is shown in Figure 1.

Once all the external forces acting on the collector are known, as well as its inertial properties, the motion of the collector may be defined. Unfortunately, these forces represent a combination of complex physical effects which are not readily amenable to analysis. Of all these forces, however, the soil related forces are by far the most uncertain, and the hardest to predict. The collector's inertial properties, its drag properties and its weight are all straightforward results of calculation or model tests. Also, the interaction of the collector with the lift system, while quite complex, is solvable by means of a structural analysis of whatever mechanical systems are under consideration.

It is within the context of this overall dynamic model, therefore, that the engineering properties of the sediments must be taken into account. We will next discuss how this is being accomplished by Kennecott.

SOIL MECHANICS EFFECTS ON VEHICLE MOBILITY

A survey of pertinent soil mechanics research reveals at least two distinct perspectives. On the one hand, a great deal of work has been conducted by geologists and civil engineers into what have been called the "geotechnical" properties of sea-floor sediments. These properties consist of some descriptive and some quantitative soil property determinations, such as

Shear strength
Compressibility
Sensitivity
Atterberg Limits
Activity
Classification by size
Porosity

The primary application of such information is with foundation problems or such subjects as anchoring, erosion and other similar questions.

The problem of designing a nodule harvesting vehicle, however, is more akin to the problems being addressed by off-road mobility researchers, such as those at the U. S. Army Corps of Engineers Waterways Experiment Station in Vicksburg, Mississippi. Mobility research, while based on fundamental soil mechanics principles, necessarily requires ad hoc measurements and analysis to evaluate soil/machine interactions for each case under investigation.

The evaluation of the dynamic forces on a harvesting vehicle actually requires consideration of both points of view.

The nodule collector may be thought of as a vehicle which is supported by certain running gear. These may be wheels, tracks, skis, chains, or some other form, depending on the weight of the vehicle and the amount of traction required. The selection of the proper running gear is certainly a critical design decision, and requires geotechnical information.

Wheels, for example, function very well for traction and flotation in cohesionless soils, while their relatively small contact areas and consequent high ground pressures and sinkages would be a serious handicap for operations in the very soft soils (ooze) of the abyssal plains (Wiendieck, 1970).

Of special significance, however, is the quantitative determination of the intrinsic soil reaction forces on a particular running gear, or on a particular vehicle operating on a set of running gear.

THE "MICRO" APPROACH

There are two methods of approaching this problem. The first method, which the authors choose to refer to as the "micro" approach, utilizes measured soil properties. In this method, the particular running gear is divided into a finite number of surface elements. At any specific time, as the vehicle is moving through or over the soil, those elements which are in contact with the sediment experience forces normal and tangential to the surface. The direction of the tangential component must be opposite to the direction in which the sediment is being sheared, or, opposite the direction of travel of the element in question. Quantitatively, these elemental forces arise from the shear strength and bearing capacities of the sediment, as well as adhesion characteristics of the soil/metal (if the running gear is metal) interface.

Using this "micro" approach, it is postulated that the force on each element may be computed by some adaptation of Coulomb's law and bearing capacity formulas from Bekker (1969). The gross soil loadings may then be predicted from a mathematical integration of these loads over all the elements of the running gear in contact with the sediment.

An important consideration in this approach is the estimation of which running gear elements, at any particular time, are in contact with the mud, and what normal loads are present. Typically, this requires a numerical integration of the vehicle's motion from some initial conditions. Because of the non-linearity of the soil forces, this requires a non-linear numerical integration scheme for the vehicle motions, and must take into account the coupling of vehicle motions and topography with the soil reactions on the running gear.

This method of predicting soil forces may prove successful, provided sufficient empirical data are available to upgrade the mathematical model. To accomplish this, the authors have made modifications to the usual soil force coefficients to account for high strains and strain rates. A modified Coulomb formula becomes:

$$\tau' = c' + p \tan \phi$$

where τ' = modified shear strength (Pa)

$c' = \alpha(c + c_v v)$

α = adhesion coefficient

c = nominal soil cohesion at zero strain rate

c_v = velocity dependent cohesion

v = velocity of element with respect to soil

p = normal pressure (see below)

ϕ = internal angle of friction.

A modified normal force formula is also proposed to take into account both the "springiness" of the sediment and strain rate effects:

$$p = p_o + (a + bv)Z^n$$

where p = unit normal load (Pa)

p_o = static normal load

Z = penetration of running gear element

a,b,n = soil coefficients.

A preliminary assessment of the coefficients α, c, ϕ, c_v, p_o, a, b, and n may be obtained by laboratory tests. Indeed, c, ϕ, and p_o are standard soil mechanic properties and need no elucidation here. The adhesion coefficient α, and c_v, will be estimated from shear force measurements on annular rings or plates shearing the soil at various rates in KEI developed tests. a, b, and n may be estimated by special plate sinkage tests.

Once these coefficients are evaluated on a small (micro) scale, it remains to validate their utility in the prediction of gross vehicle loads. Unfortunately, this requires physical modeling of the vehicle, preferably on a large scale since little is known of soil/metal interactions.

Whatever the running gear configuration, it is necessary to validate the dynamic predictions based on preliminary soil coefficients by means of a test/comparison/update procedure, such as that shown schematically in Figure 2.

This procedure naturally leads to some modification of the original soil coefficients. The nominal soil cohesion and angle of internal function measured by an unconfined compression test may differ slightly from vane shear or annular ring tests (Bekker, 1969). The closer the lab scale measurements simulate full-scale effects, of course, the better.

There is, unfortunately, very little experience to draw upon for this type of comparison. Land mobility research has shown this method of predicting vehicle performance to be highly variable in its success, depending on the particular type of soil and running gear involved (Erlich, 1972). Practically all mobility research is conducted in frictional or frictional-cohesive soils. Most nodule deposits of economic interest, on the other hand, are located in areas of the equatorial Pacific characterized by brown clay or siliceous ooze (Horn et al., 1973). These are cohesive sediments, often with sensitivities unusual on land (up to 10 in some cases) (Noorany, 1970). Only a healthy bit of data, of the type generated by the methodology represented in Figure 2, can clear up the question of predicting vehicle motions on the sea floor. This requires:

a) measurement of "micro" properties of the sediment,

b) mathematical modeling of the intrinsic behavior of the vehicle,

c) testing of a model vehicle, at least half scale,

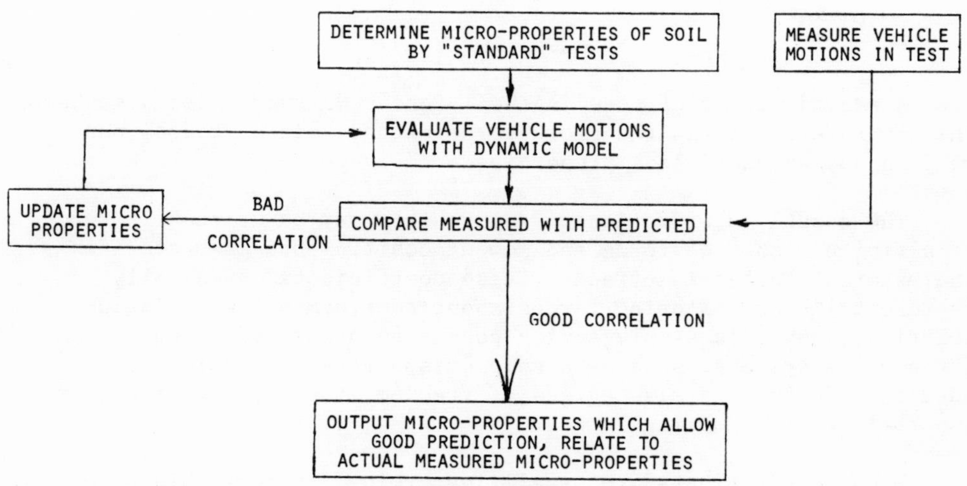

Figure 2. Methodology for validation of soil model

DISPLACEMENT OF VEHICLE C.G. = $\vec{X} = (X_1, X_2, X_3)$

ROTATION OF VEHICLE ABOUT C.G. = $\vec{\theta} = (\theta_1, \theta_2, \theta_3)$

FORCES ON VEHICLE = $\vec{F} = (F_1, F_2, F_3)$

MOMENTS ON VEHICLE = $\vec{M} = (M_1, M_2, M_3)$

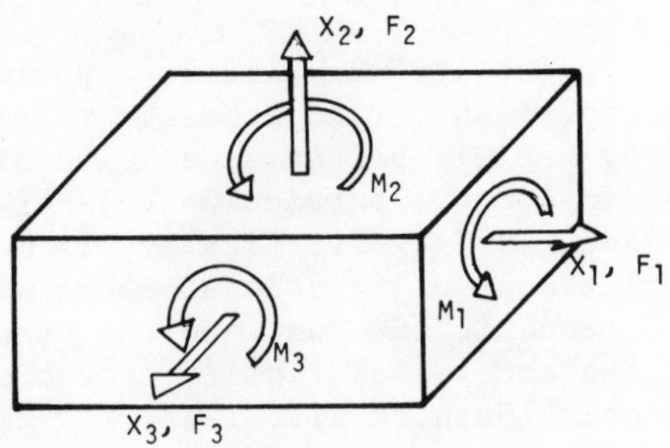

REPRESENTATION OF NON-INERTIAL FORCES:

$$(F_1, M) = [A]\,(X - X_0, \theta - \theta_0) + [B]\,(\dot{X} - \dot{X}_0, \dot{\theta} - \dot{\theta}_0)$$

X_0 AND θ_0 REPRESENT THE EQUILIBRIUM POSITIONS OF THE VEHICLE

Figure 3. Generalized forces and moments acting on vehicle

d) comparison of test results with predicted results, and

e) iteration to determine "correct" soil coefficients for the model.

"MACRO" FORCES

A second method of modeling soil forces warrants some discussion. This is referred to as the macro force method, since it does not rely on any standard soil properties.

The equations of motion of any vehicle may be written in terms of a finite number of force and moment coefficients characterizing the external forces in effect. These coefficients may usually represent linear restoring forces about some nominal equilibrium position, such as a simple spring constant in a mass/spring system. For complex systems, such as a ship or spacecraft, consideration must be made for the six degrees of freedom, their derivatives, and important cross coupling forces.

These coefficients are usually determined by perturbing the vehicle from a nominal equilibrium position, and measuring the resulting loads in the six degrees of freedom. A mathematical characterization of this is given in Figure 3.

These coefficients, represented by the A and B matrices, are determined from a vast number of tests on a model vehicle, probably, as mentioned earlier, at least one-half scale. To determine each coefficient, at least four tests must be conducted. A_{11}, for example, is determined by plotting F_1 vs $x_1 - x_{10}$ (surge) and evaluating $\partial F_1/\partial x_1$ at the equilibrium position. For the general case where A and B are both sixth order matrices, up to 2x36x4=288 individual tests would be required to determine the force coefficients for one soil and one vehicle!

This method is mentioned here only because of its widespread use in naval architecture and aerodynamics. For problems of soil vehicle dynamics, its effectiveness is severely limited.

CONCLUSIONS AND RECOMMENDATIONS

Soil properties of importance to deep ocean mining are those which affect vehicle mobility. These are primarily soil strength parameters, including dynamic damping parameters. For a particular soil, a model of the soil forces acting on a vehicle may be developed with eight soil coefficients determined from lab scale tests. The validity of these coefficients in the dynamic model of the system has to be determined through large-scale testing, and comparison of measured with predicted vehicle motions. More studies of this type, particularly of vehicles traversing the sensitive cohesive sediments of the equatorial Pacific, need to be carried out to evaluate this methodology. The need to perform this research is fostered by the necessity to build mining machines which can traverse the ocean bottom uniformly, and which may be controlled with a minimum of power and a maximum of reliability.

REFERENCES

Bekker, M. G., Introduction to Terrain-Vehicle Systems, Univ. of Mich. Press, Ann Arbor, Mich., 1969.

Erlich, I. R., and C. J. Nuttall, Preliminary notes on submerged soil strengths, Stevens Institute of Technology Letter Rept. SIT-UL-72-1581, Hoboken, N. J., 1972 (Confidential).

Horn, D. B., B. M. Horn, and M. N. Delach, Ocean manganese nodules metals values and mining sites, National Science Foundation Tech. Rept., No. 4, NSF-X-GX33616, 1973 (unpublished manuscript).

Noorany, I., Engineering properties of submarine calcerous soils from the Pacific, in Proc. Intern. Symp. Eng. Properties of Sea-Floor Soils and their Geophys. Ident., Univ. of Washington, Seattle, 1971.

Wiendieck, K. W., A preliminary study of sea-floor trafficability and its prediction, U. S. Army Engineers Waterways Experiment Station Tech. Rept. M-70-8, 1970.

INITIAL PENETRATION AND SETTLEMENT OF CONCRETE BLOCKS INTO DEEP-OCEAN SEDIMENTS

FRANK SIMPSON, ANTON L. INDERBITZEN, AND ATWAR SINGH

Lockheed Missiles & Space Company, Inc.; University of Delaware; and Lockwood-Singh & Associates

ABSTRACT

Two sets of concrete blocks of various shapes (a cube, a rectangular parallelepiped and a cylinder) of unit loads 3.78 kN/m^2 - 7.31 kN/m^2 (38.6 - 74.5 g/cm^2) were placed on the floor of the San Diego Trough at a water depth of 1240 m to observe their short-term and long-term settlement behavior. Emplacement of the blocks and periodic observations of their settlement during a two-year period were accomplished using Lockheed's submersible DEEP QUEST. In situ vane shear tests to a sediment depth of 122 cm and sediment cores varying in length from 93-120 cm also were obtained at the emplacement sites with DEEP QUEST's vane shear device and coring unit.

From the in situ tests and laboratory tests on cores, engineering and mass physical properties of the sea-floor sediments were determined and used in conjunction with conventional soil mechanics methods to predict the settlement of the blocks at the two test sites. Settlement analyses included consideration of initial settlement, primary consolidation, secondary compression, and shear failure phenomena.

Results of settlement predictions approximate the total actual settlement observed. Differences between the predicted settlement and the observed settlement for the concrete blocks entering the sediment in an upright position varied from 0 to 2.4 cm with the maximum difference representing approximately 28% of the total settlement predicted. The amount of agreement between predicted and actual settlement values demonstrates the degree of applicability of terrestrial soil mechanics methods for analyzing fine-grained saturated sea-floor sediments.

INTRODUCTION

The objective of this project was to obtain controlled empirical data on the short- and long-term settlement behavior of various shaped objects on the sea floor, and compare the actual settlement to that predicted by conventioanl soil mechanics approach. Data of this nature are essential for:

1. evaluating the reliability of standard soil mechanic theories for predicting total settlement and penetration rate of objects into the sea floor, and, if necessary,

2. for determining a more accurate method of predicting these phenomena in the marine environment.

Such information is needed when designing any structure to be placed on the sea floor or when trying to search for an object embedded in the ocean bottom. Traditionally, the settlement, or sinking, of an object into a saturated fine-grained sediment (clay and silt) is considered to occur in three stages: initial, primary consolidation, and secondary compression settlement.

Initial settlement occurs immediately upon loading the sediment and is caused by elastic deformation of the soil below and around the emplaced object without dissipation of excess pore water pressures. In this study the initial settlement phase also includes any shear failure which may occur if the unit loads of the objects exceed the bearing capacity of the sediment.

During the primary consolidation phase of settlement, the material underlying the loaded area consolidates as the pore water flows out of the soil matrix and the load is transferred directly to the soil particles. Settlement occurs at a much slower rate than in the initial stage and continues until the excess pore pressure generated by the loading is dissipated and hydrostatic equilibrium is achieved.

Settlement due to secondary compression is caused by plastic deformation and creep of the sediment particles. It occurs at very slow rates, and is time-dependent. Secondary compression is the least understood of the three stages of settlement. By observing the blocks over an extended period of time, and performing extensive tests on sediment samples from the sites, it was hoped to gain a better understanding of this long-term settlement and its importance in offshore engineering problems.

Very few empirical data have been gathered on the settlement behavior of objects emplaced on the sea floor or on the bearing capacity characteristics of the surficial layers of marine sediment. Keller (1964) compared the amounts of initial penetration of concrete blocks into soft bay sediments to that predicted by standard soil mechanics formulas.

For his study, the amounts of predicted penetration were less than the actual penetration that occurred. Penetration of the blocks into the sediment, measured during Keller's study, was believed due to shear failure of the surface sediments and no measurements were made of the time-dependent settlement behavior of the blocks. Rucker et al., (1967) made a study of predicted versus actual dynamic penetration of various shaped objects into the sea floor. Predicted depths of penetration, based upon the kinetic energy of the object on impact, were 1.1 to 3.2 times greater than actual values.

Hironaka and Smith (1968) investigated the foundation characteristics of sea-floor sediments at several locations off the Southern California coast prior to emplacement of a Submersible Test Unit II series. No serious foundation engineering problems were encountered as the sea-floor material at all locations was composed primarily of sand and the bearing loads applied were much lower than the bearing capacity of the sediments. Predicted settlement values agreed closely with observed penetration markings on the test units. More recently, Carlmark (1971) performed a comprehensive investigation of penetration of free-falling objects into deep-sea sediments and developed a method for predicting the expected penetration.

Plate-loading devices have been designed and tested (Kretschmer, 1967; Harrison & Richardson, 1967) for in situ determination of the short-term bearing capacity and settlement response of ocean floor sediments. Although not depth restricted, use of plate-loading devices is limited to measurement of short-term settlement only (initial settlement). Kretschmer and Lee (1970) performed a series of in situ plate-bearing tests off the Southern California coast in cohesive sediments and developed a technique for predicting the initial settlement of sea-floor footings using elastic and bearing capacity theory solutions.

Herrmann et al., (1972) described the development and evaluation of two devices for monitoring long-term and differential settlement behavior of sea-floor sediment. Results of his primary consolidation predictions generally overestimated settlements by 25%, while their methods for predicting initial and secondary compression were less accurate. Other related publications include those of Anderson and Herrmann (1971), Herrmann (1972), and Raecke (1973) which summarize sea-floor foundation analysis of numerous case histories and/or describe guidelines for sea-floor installations.

The present study performed by LMSC is the first known to the

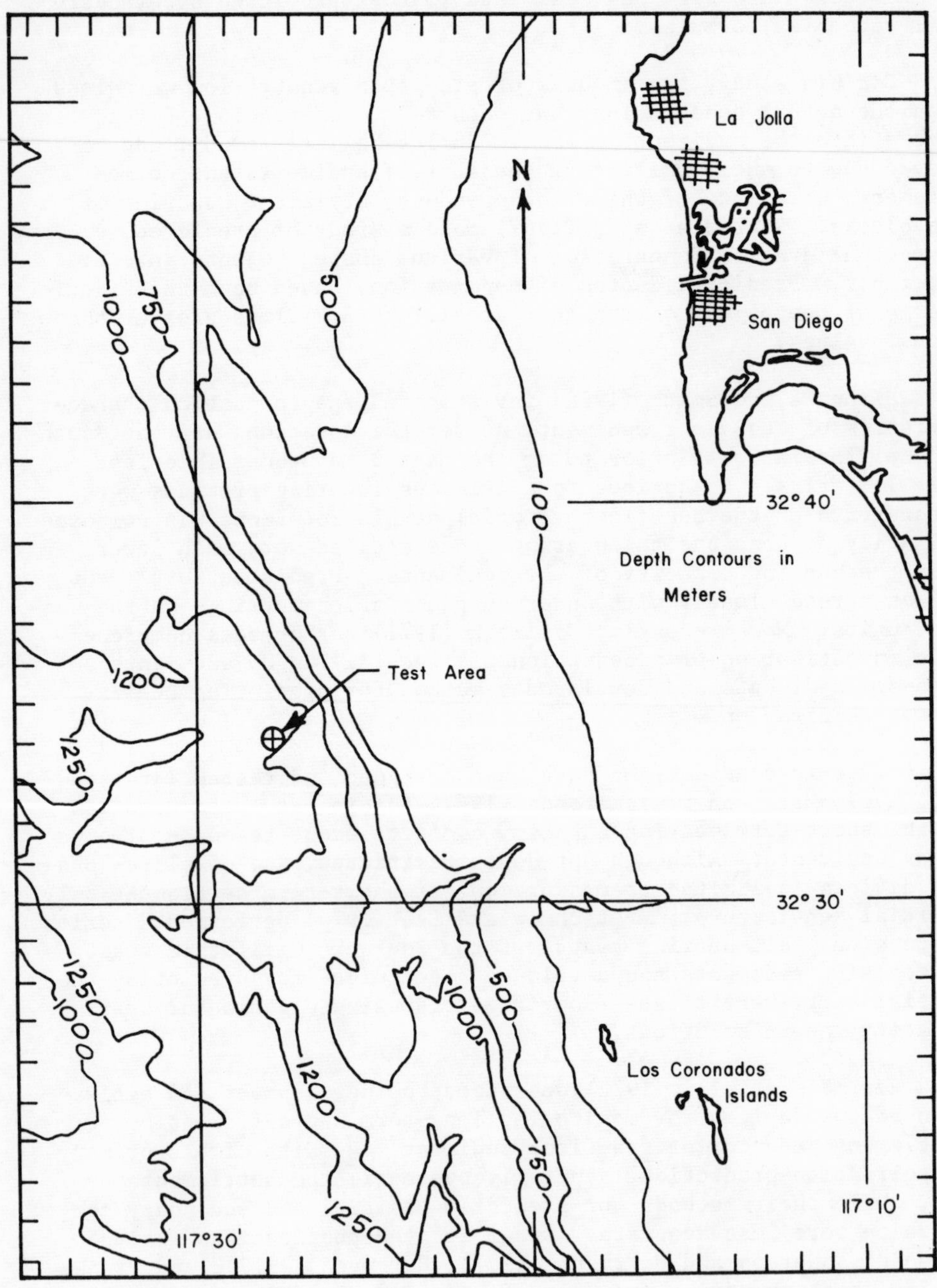

Figure 1. Location of settlement tests

authors where:

1. concrete blocks of various shapes were emplaced gently on the sea floor;

2. the necessary data on sediment properties were obtained by a combination of in situ measurements and coring adjacent to the blocks, and

3. the blocks were periodically monitored to measure their continuing settlement during a two-year period.

All these tasks were performed with the aid of the submersible DEEP QUEST. In this manner, accurate, controlled data were obtained on all three phases of the settlement. Only by studying the entire settlement phenomena directly on the ocean bottom can one begin to obtain the data necessary to develop accurate relationships for predicting total settlement of objects into the sea floor.

APPROACH

Two test sites were established on the floor of the San Diego Trough at a water depth of 1240 m. Locations of Sites 1 and 2 are 32°33.9'N Lat., 117°27.3'W Long., and 32°34.1'N Lat., 117°28.0'W Long., respectively. These sites are approximately 1040 m apart and are located about 24 km SW of Point Loma, San Diego, California (Fig. 1).

Extensive observations from DEEP QUEST in the San Diego Trough area indicate that the sea floor is practically flat except for microtopographic relief features. Bottom currents are very weak, 10 cm/sec or less, as measured by the Savonious rotor current meter mounted on DEEP QUEST. Additional evidence of weak currents is the presence of holothurians on the sea floor and detailed tracks created by bottom organisms. Based on numerous grain size analyses, the bottom sediments are classified as clayey silt (Shepard's 1954 nomenclature), and are homogenous in both lateral and vertical directions.

Concrete blocks used for this study consisted of a cube, a rectangular parallelepiped, and a cylinder. Characteristics of the blocks are as follows:

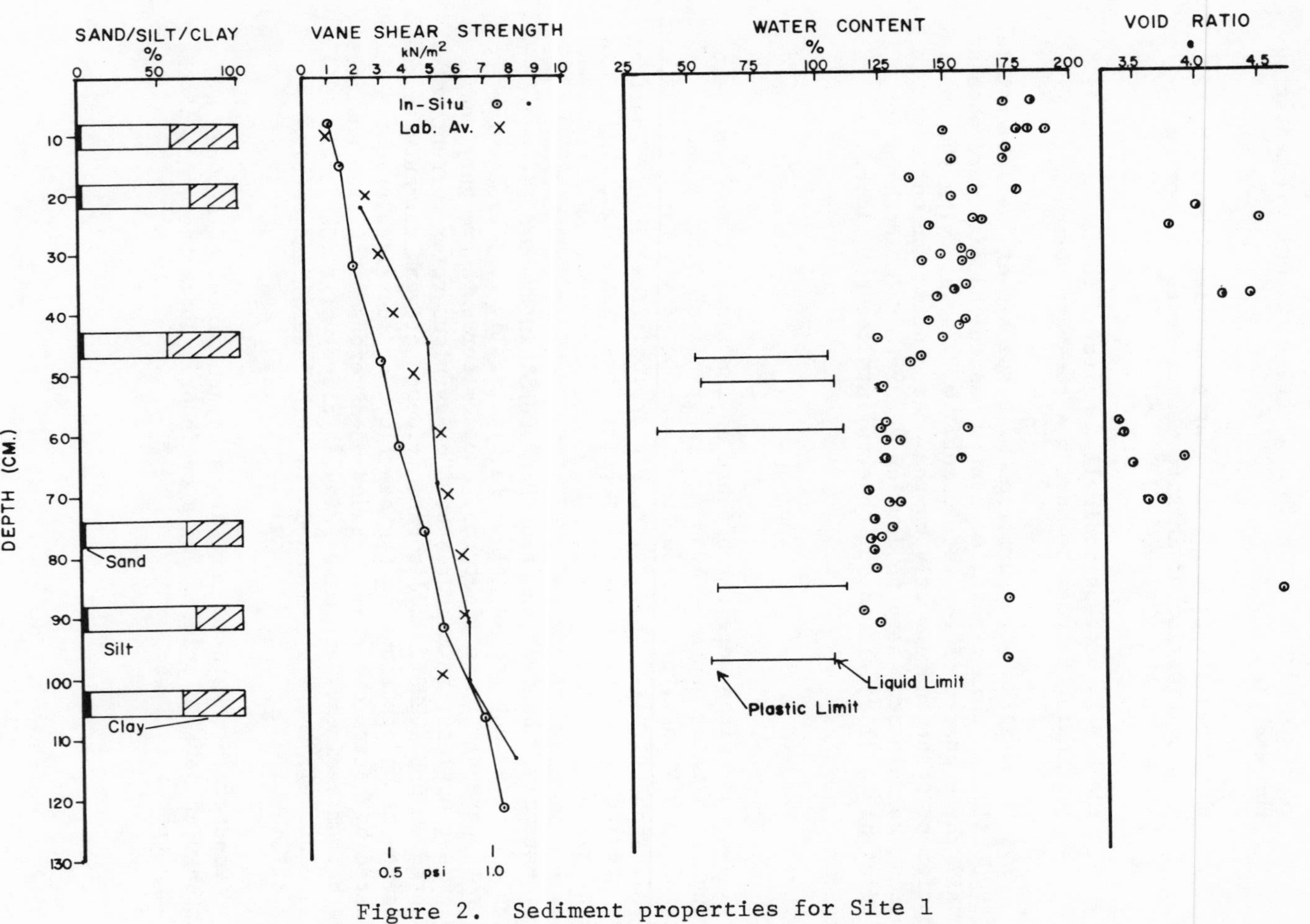

Figure 2. Sediment properties for Site 1

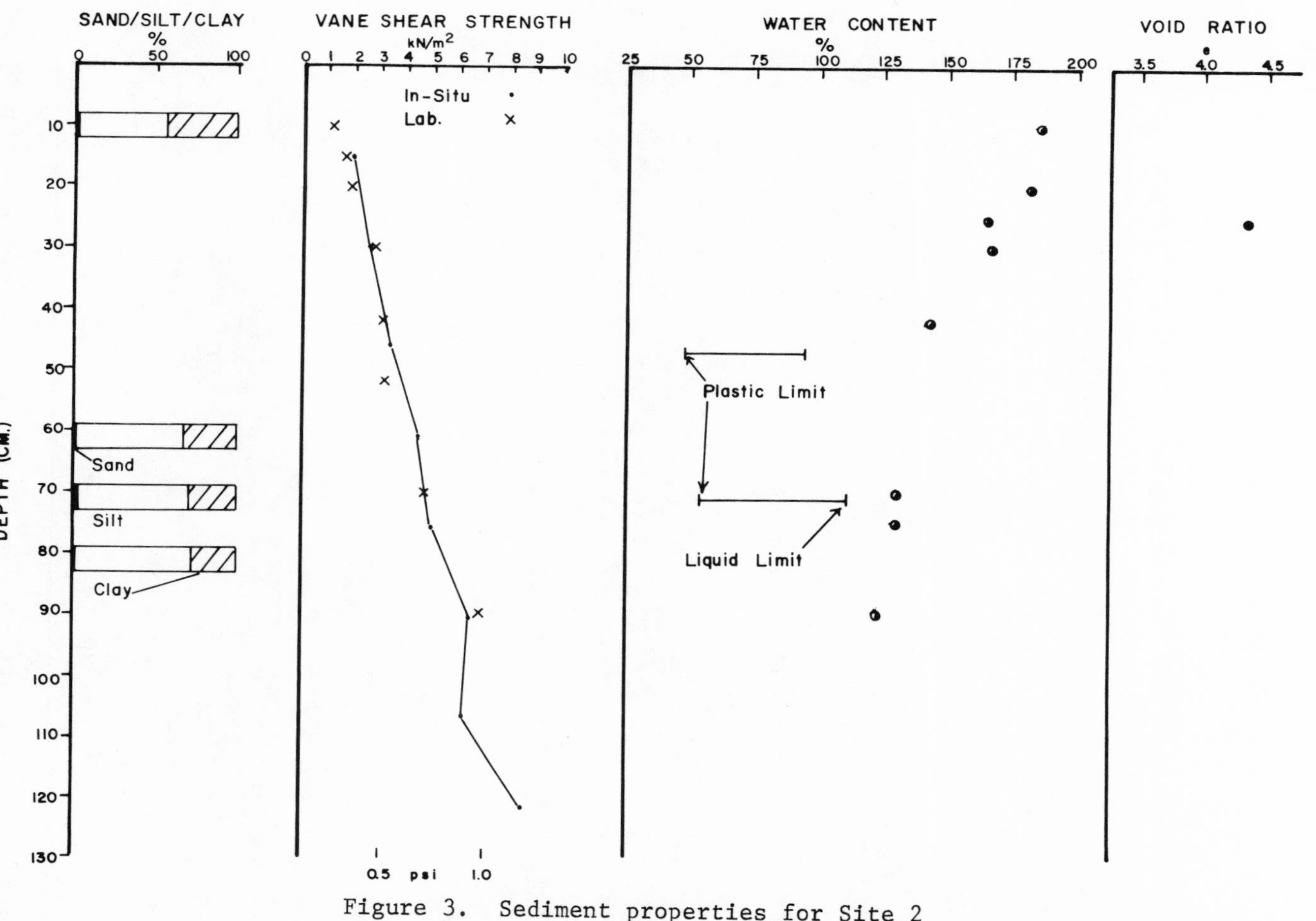

Figure 3. Sediment properties for Site 2

TABLE 1

OBSERVED SETTLEMENT OF CONCRETE BLOCKS

(cm)

SITE 1		July 1970	Nov. 1970	Oct. 1970	June 1972
Cube No. 10					
	Corner 1	11	14	13	13
	Corner 2	0	9	N/A[2]	N/A
	Corner 3	5	8	N/A	2
	Corner 4	11	11	9	5
	Average	7	11	11	7
Rectangular Block No. 8					
	Corner 1	4	13	13	13
	Corner 2	3	10	13	13
	Corner 3	N/A	18	N/A	N/A
	Corner 4	20	25	23	23
Vertical Settlement		14	19	19	19

SITE 2		July 1970	Oct. 1970	Oct./Nov. 1971	June 1972
Cube No. 12					
	Corner 1	0	9	6	6
	Corner 2	N/A	8	8	6
	Corner 3	9	9	8	N/A
	Corner 4	9	9	8	9
	Average	6	9	7	7
Rectangular Block No. 5					
	Corner 1	5	N/A	8	10
	Corner 2	5	9	13	10
	Corner 3	5	9	N/A	N/A
	Corner 4	3	9	10	5
	Average	4	9	10	9

[1] Block tilted excessively during emplacement thus the settlement is is computed geometrically and represents the maximum vertical distance of penetration.

[2] N/A - corner could not be observed - data not available.

Shape	Dimensions	Av. Saturated Buoyant Wt.	Av. Buoyant Unit Load
Cube	30.5 x 30.5 x 30.5 cm	37.5 kg	3.78 kN/m^2
Rectangular Block	15.2 x 61.0 cm base x 30.5 cm ht.	35.7 kg	3.78 kN/m^2
Cylinder	61 cm length 20.3 cm diameter	24.2 kg	1.31 kN/m^2

One block of each shape was emplaced gently on the sea floor at each site in July 1970 by use of DEEP QUEST's manipulator arms. Initial penetration of the blocks into the sea floor was observed through the viewport of DEEP QUEST and recorded. Amounts of settlement were measured by means of reference lines painted on the sides of the blocks. Two reference lines are painted on each side of the blocks with cross-lines every 1.27 cm. In this manner, differential, as well as total settlement, could be noted.

In situ shear strength measurements were made at each site to sediment depths of 122 cm using Lockheed's vane shear device mounted on DEEP QUEST (Inderbitzen et al., 1971) and the results used in bearing capacity computations. Sediment cores varying in length from 93 to 120 cm also were obtained at each site for laboratory measurement of mass physical properties and consolidation tests. A full description of the coring device is available in Inderbitzen and Simpson (1972). Consolidation tests were necessary in order to compute the expected amount of settlement beneath each block due to primary consolidation and secondary compression.

After the initial emplacement of the blocks in July 1970, settlement observations were made at each site in October and November 1970, October 1971, and June 1972. Additional cores and in situ vane tests also were obtained to verify the earlier measurements. Results of shear strength and mass physical property measurements for sediments at Sites 1 and 2 are presented in Figures 2 amd 3. Laboratory test procedures for DEEP QUEST cores are in close agreement with ASTM standard procedures.

Presently, the blocks are still on the sea floor and their locations have been marked with acoustic pingers to enable additional observations during future submersible dives in this area.

RESULTS OF THE FIELD OBSERVATIONS

Table 1 lists the average settlement observed for each of the blocks during the two-year program. Average settlement computations are based on the settlement observed at the corners of each block. Settlement values for the cylinders have been omitted as these objects could not be emplaced in an upright position. All attempts to gently set the cylinders upright on the sea floor resulted in the cylinders sinking 10 cm or more into the sediment, tilting, and falling over. Cylinder No. 1 at Site 1 was finally left on its side and Cylinder No. 4 at Site 2 was forced into the bottom to determine how deep it had to be pushed before it would remain upright. A penetration of approximately 30 cm was required before it would remain upright.

July 1970 Measurements

The July 1970 readings are the settlement observations recorded immediately upon emplacement of the blocks on the sea floor. The readings represent the settlement within the first 5-15 minutes. At both sites, the blocks tended to enter the sediment in an uneven or canted position. Greater tilting of the blocks occurred at Site 1. As expected by their shape, the rectangular blocks tilted more than the cubic blocks, and, as mentioned earlier, the cylinders were the least stable.

Average initial penetration or settlement for the cubes at Sites 1 and 2 was 7 and 6 cm respectively. For rectangular block No. 8 at Site 1, settlement was very uneven (tilt about 42° from vertical). Therefore, the settlement values for this block were not obtained by averaging the penetration depth observed at each corner, but by calculating the maximum vertical distance the block had penetrated into the sediment. This distance was determined to be 14 cm. Values for block No. 8 are given for general information only. Because of the excessive tilting, this block was eliminated from the later computations. At Site 2, the average initial settlement of rectangular block No. 5 was only 4 cm with the settlement at each corner being very uniform.

Analyses of the in situ and laboratory core data (grain size, water content and shear strength) did not demonstrate a significant difference between the characteristics of sediments at Sites 1 and 2. Thus, the different settlement behavior of the two rectangular blocks is not completely understood but may be due to some small local variations in the mechanical properties of the sediment.

November 1970 Measurements

Observations made in November 1970 at Sites 1 and 2 indicated that the corners or sides of each block, which initially penetrated the deepest, had settled very little compared to the remainder of the block. Except for rectangular block No. 8 at Site 1, which tilted even farther (about 51°), the rest of the blocks had attained a nearly upright position. Average total settlement for all the blocks is shown on Table 1.

Sea life observed around the concrete blocks were small brittle stars (ophiuroids) and some star fish (asteroids). Burrowing by fish or other benthic organisms and scouring was not apparent. The blocks were in a surprisingly good, clean condition. No biological growth or accumulation of sediments was present on their upper parts.

October/November 1971 Measurements

Observations of the blocks at Sites 1 and 2 during the October/November 1971 DEEP QUEST dives indicated that, in general, no additional settlement had occurred since the 1970 observations. Rectangular block No. 5 at Site 2 had an average increase in settlement of 1 cm, and rectangular block No. 8 had tilted farther (about 57°). The rest of the blocks maintained their previous penetration level or had some minor amounts of the surrounding sediment removed. Agents responsible for the removal of sediment could include biological activity or possibly water movements caused by DEEP QUEST's thrustors. As in the 1970 observations, the blocks appeared to be in excellent condition.

June 1972 Measurements

Results of the settlement measurements in June 1972 were quite similar to those of November 1971. All blocks had remained at their previous penetration level or had small amounts of surrounding sediment eroded by biological activity or DEEP QUEST's thrustors. Markings on the blocks were still clearly visible and little sediment had accumulated on the top surface of the blocks.

SETTLEMENT ANALYSIS

Methods investigated for predicting the initial settlement included elastic and bearing capacity theory solutions. Predictions for settlement due to primary consolidation are based on Terzaghi's theory of consolidation. Secondary compression predictions were determined from coefficients of secondary compression obtained in the laboratory.

A review of the theory and testing procedures for initial and primary consolidation settlement is presented in Seed (1965). Lee (1970) discusses in detail the use of elastic and consolidation theories and secondary consolidation methods for settlement calculations of shallow foundations on clay soils. All the above methods are also presented in Lambe and Whitman (1969).

Initial Settlement

Because of the observed semi-liquid nature of the surface sediments in the San Diego Trough, initial shear failure of sediment below the blocks was expected. This was verified by the observed behavior of the blocks immediately after emplacement. Settlement observations within the first 5-15 minutes were considered to include shear failure, plastic deformation and to a lesser degree settlement due to primary consolidation.

The approach taken to analyze the initial settlement consisted of determining the expected settlement attributable to both elastic and rupture phenomena and comparing the sum of these components with the observed initial settlement. Any discrepancies remaining between the observed and anticipated settlement values were attributable to primary consolidation.

Elastic Theory Solution. Predictions of initial settlement caused by elastic deformation are based on the integration of Boussinesq's equation for a point load on the surface of a homogeneous, isotropic, linearly-elastic half space (Terzaghi, 1943). The equation for calculating the settlement at the center of the rectangular loaded areas are obtained from Lambe and Whitman (1969, p. 215):

for the cube
$$S_i = p\frac{B}{E}\ 1.12\ (1 - \mu^2) \qquad (1)$$

for the rectangular block
$$S_i = p\frac{B}{E}\ 1.8\ (1 - \mu^2) \qquad (2)$$

where S_i = initial settlement (cm)

p = stress applied to the loaded area (kN/m^2)

B = width of the loaded area (smallest dimension of base in cm)

E = modulus of elasticity (kN/m^2)

μ = Poisson's ratio

In order to use these equations, a value of the secant modulus, E, was obtained from results of triaxial undrained tests conducted on DEEP QUEST cores from the San Diego Trough area. The value of E derived for a vertical stress approximately equal to the unit load of the concrete blocks was 225.7 kN/m^2. A Poisson ratio of 0.5 was assumed for the sediments at Sites 1 and 2 (Kretschmer and Lee

1970). Initial settlement predictions by this approach are presented in Table 2. Settlement due to elastic deformation was estimated to be approximately 0.5 cm for all concrete blocks.

The significant difference between the elastic theory settlement prediction and the observed settlement indicates that a large portion of the initial settlement apparently was due to shear failure and possibly primary consolidation. In order to determine the settlement predicted from shear failure, the bearing capacity approach was investigated.

Bearing Capacity Solution. An assumption made by the authors, regarding the shear failure and rupture phenomena, is that an object will initially penetrate into the sediment until it reaches the depth at which the bearing capacity of the soil is sufficient to support the imposed load. A similar assumption was made by Kretschmer and Lee (1970). The approach used to determine initial settlement is based on Skempton's bearing capacity equation for an embedded rectangular footing of width B and length L (Terzaghi and Peck, 1948). The equation is as follows:

$$p = 5c\left(1 + 0.2\frac{S}{B}\right) \cdot \left(1 + 0.2\frac{B}{L}\right) + \gamma_s S\,(9.806 \times 10^{-2}) \qquad (3)$$

where p = ultimate bearing capacity (kN/m^2)

c = shear strength of the soil (kN/m^2)

S = depth of embedment or settlement of concrete blocks (cm)

γ_s = submerged unit weight of the soil (g/cm^3)

B and L = rectangular dimension of the base of the blocks (cm).

This equation assumes a saturated clay under undrained shear conditions having an angle of internal friction, ϕ, equal to zero. A shear strength or cohesion, c, value for local failure was used for our calculations. Therefore, c becomes c' which is equal to 2/3 the shear strength value, S_u, measured by the vane tests (Terzaghi and Peck, 1948, pp. 169-173).

In order to determine c, a least-square line was fitted to the vane data from Sites 1 and 2 (Figs. 2 and 3), which resulted in the following line equations for S_u in kN/m^2:

$$S_u = .054\,Z + .883 \qquad \text{Site 1} \qquad (4)$$

$$S_u = .056\,Z + .785 \qquad \text{Site 2} \qquad (5)$$

where Z = depth below sediment surface in cm.

TABLE 2

INITIAL SETTLEMENT OBSERVED AND PREDICTED

	Settlement Observed[1] (cm)	Elastic Theory Prediction (cm)		Bearing Capacity Theory Prediction (cm)		Total Initial Settlement Predicted (cm)	Error Based on Prediction
SITE 1							
Cube No. 10	7	≃.5	+	4.1	=	≃4.6	52%
*Rectangular Block No. 8[2]	14	≃.5	+	4.5	=	≃5.0	--
SITE 2							
Cube No. 12	6	≃.5	+	4.1	=	≃4.6	30%
Rectangular Block No. 5	4	≃.5	+	4.5	=	≃5.0	20%

(1) Settlement observed after emplacement of blocks (July 1970 readings).

(2) Direct comparison of settlement predictions with observed settlement for rectangular block No. 8 are not valid, as the block tilted excessively upon emplacement.

Since equations (4) and (5) are almost identical, an average least-square line was adopted for Sites 1 and 2 combined (Fig. 4).

One important aspect of the shear strength profile shown in Figure 4 is the change in slope at the 5 cm depth. Results of laboratory and in situ vane tests conducted at 3 cm and 6.5 cm

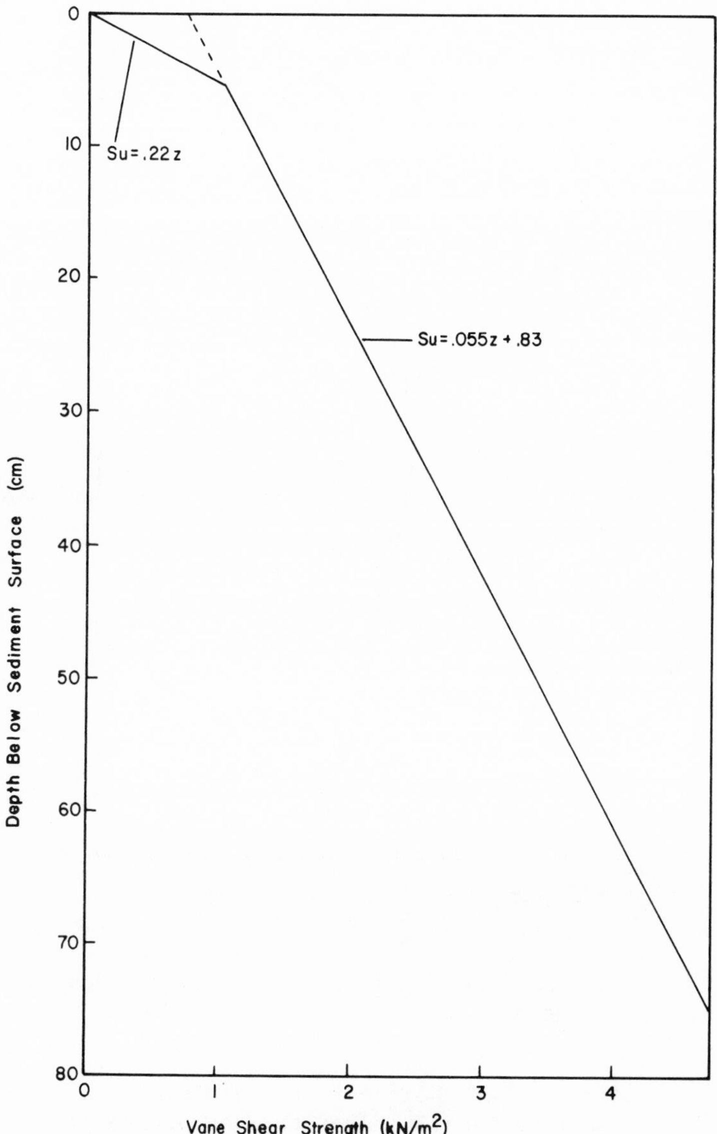

Figure 4. Least-square line fitted to vane strength data for Sites 1 and 2

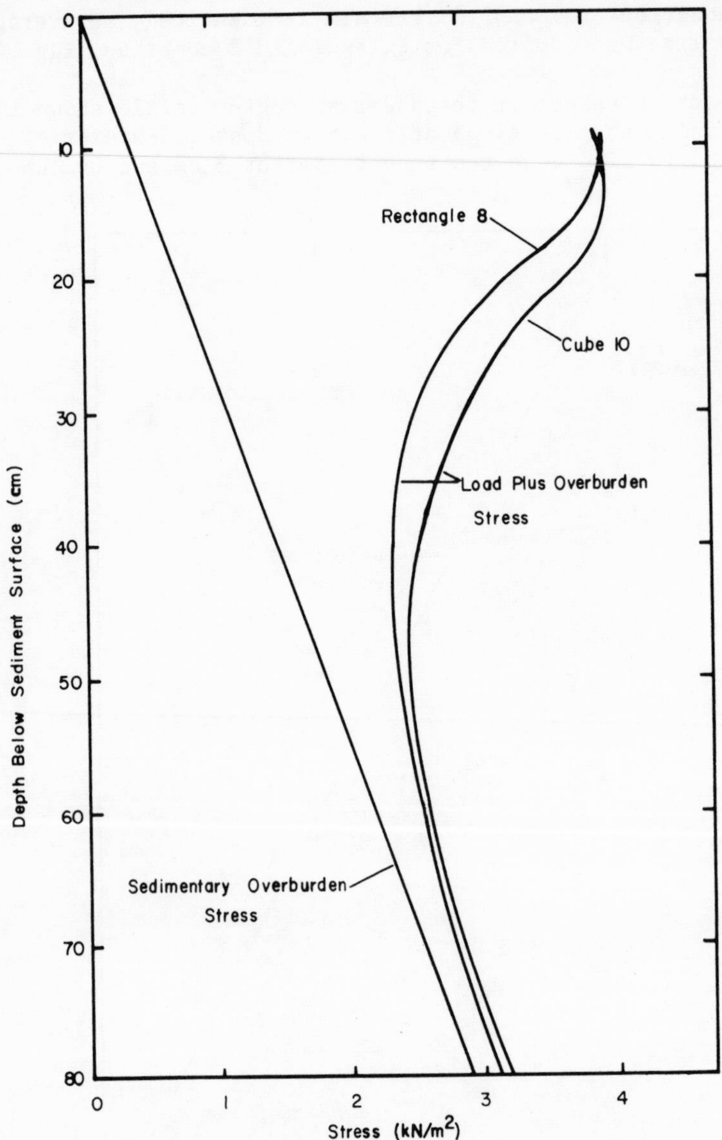

Figure 5. Effective overburden pressure and vertical pressure under concrete blocks at Site 1

sediment depths indicate that the shear strength values are very small near the sediment surface. Therefore, use of the single least-square line derived for Sites 1 and 2 vane data is considered inappropriate for the 0-5 cm depth region.

The following equations, derived from available test data, represent the best estimate of the shear strength profile for the upper 100 cm of sediment depth at Sites 1 and 2:

$$\text{for } Z < 5 \text{ cm:} \quad S_u = 0.216Z \text{ (in kN/m}^2) \tag{6}$$

$$\text{for } Z > 5 \text{ cm:} \quad S_u = 0.055Z + .834 \text{ (in kN/m}^2). \tag{7}$$

By multiplying the above equations by 2/3, the local failure conditions are obtained. The resulting equations are then substituted in equation (3) for the value of c in order to solve for the settlement of the blocks.

Predicted settlements, by Skempton's equation, are 4.1 cm for cubes No. 10 and 12, and 4.5 cm for rectangular blocks No. 8 and 5. When these settlement values are added to the initial settlement predicted by elastic theory, the total values for predicted settlement are 20 to 52% below the observed settlement of cubes No. 10 and 12 and rectangular block No. 5 (see Table 2).

Consolidation Settlement

Settlement due to consolidation of the sediment beneath the concrete blocks was determined from soil properties measured in the laboratory and Terzaghi's theory of consolidation. Predicted values of settlement were obtained by the formula (Hough, 1957, pp. 360-361):

$$\Delta H = \frac{HC_c}{1+e_o} \log \left(1 + \frac{\Delta p}{p}\right) \tag{8}$$

where H = original height of the sediment increment being considered (cm)

C_c = compression index (slope of the straight line portion of e vs log p curve)

e_o = original void ratio (ratio of volume of voids/ volume of solids)

Δp = change in effective stress or stress due to concrete block (kN/m^2)

p = effective overburden stress or stress (kN/m^2).

The equation is solved for incremental layers of sediment beneath each block to a depth where the stress induced by the concrete blocks is assumed negligible. This is the depth at which the load-plus-overburden stress curve becomes asymptotic to the overburden stress curve. Figure 5 illustrates these curves as computed for the rectangular and cubic block at Site 1, beneath the center of the blocks. The sediment depth at which the curves become asymptotic to the overburden stress curve is about 60 cm. Settlement computations were performed beyond this point to a depth of 78 cm to be sure that all stresses induced by the blocks were taken into account. The curves were of approximately the same shape and magnitude for the blocks at Site 2. Computations to determine ΔH for blocks at both sites were made for six 5-cm thick increments and then for four 10-cm thick increments of sediment below each block. All incremental ΔH values for each block were then added to determine total settlement below that block. A value of 1.15 was used for the compression index, C_c, in the computations, as determined by laboratory tests.

These settlement calculations assume that initial depth of penetration of the blocks is equal to the observed initial settlement (Table 2). Overburden stress, p, was calculated on the basis of the submerged unit weight of the sediment. Void ratios also were determined by laboratory tests on cores from the two sites. Changes in stress, Δp, induced in the sediment by each block were computed by use of Newmark's stress charts (Newmark, 1942). The entire computational procedure for determining settlement due to consolidation has been outlined by Richards (1961, pp. 49-53). Time necessary for the settlement of the concrete blocks due to primary consolidation (90% consolidation) was determined to be 96 hours based on interpretation of the laboratory consolidation tests. Computations for the cylinders were not made because depths of settlement cannot be observed accurately for the cylinder on its side and the other cylinder was forced into the sediment.

Table 3 lists the predicted values of settlement due to consolidation alone for each of the four blocks and the additional settlement observed during the October 1970 dives. Observations made in October 1970 are believed to represent settlement values after the completion of the settlement phase due to primary consolidation. This assumption is based upon the determination of time necessary for 90% consolidation as extrapolated from the laboratory tests. Predicted consolidation settlement approximates the settlement observed in the field. The predicted settlement values based on Terzaghi's consolidation theory and shown in the second column of Table 3, are believed to include the settlement due to elastic deformation (Seed, 1965). However, the settlement attributed to elastic deformation is very minor, as shown in the previous section of this paper.

Settlement Due to Secondary Compression

Observations of the concrete blocks in October/November 1971 and June 1972 indicated that, in general, no additional settle-

TABLE 3

PRIMARY CONSOLIDATION SETTLEMENT OBSERVED AND PREDICTED

	Settlement Observed[1] (cm)	Predicted Settlement Due to Consolidation[2] (cm)	Error Based on Prediction
SITE 1			
Cube No. 10	4	4.5	11%
Rectangular Block No. 8[3]	5	4.0	---
SITE 2			
Cube No. 12	3	5.0	40%
Rectangular Block No. 5	5	4.5	11%

(1) Settlement values obtained by subtracting the July 1970 settlement readings from the Oct./Nov. 1970 readings.

(2) Predicted settlement due to primary consolidation includes elastic deformation (Seed 1965).

(3) Direct comparison of settlement predictions with observed settlement for rectangular block No. 8 are not valid, as the block tilted excessively upon emplacement.

ment had occurred since the 1970 measurements (see Table 1). It appears that secondary compression phenomena has been negligible for the loads used in this study. For comparison purposes, secondary compression predictions were made using coefficients of secondary compression, C_α, determined from laboratory consolidation tests. A description of the method used is presented in Lee (1970). Results of tests conducted by Singh and Yang (1971), on DEEP QUEST cores from the San Diego Trough demonstrated that the rate of secondary compression, C_α, for deep-ocean sediments is stress-dependent. Based on their laboratory tests, C_α values were found to vary from

TABLE 4

TOTAL SETTLEMENT OBSERVED VERSUS TOTAL PREDICTED SETTLEMENT

	Settlement Observed[1] (cm)	Total Settlement Predicted[2] (cm)	% Error
SITE 1			
Cube No. 10	11	≃ 8.6	28%
Rectangular Block No. 8[3]	19	≃ 8.5	---
SITE 2			
Cube No. 12	9	≃ 9.1	1.1%
Rectangular Block No. 5	9	≃ 9.0	0%

(1) Total settlement observed during Oct./Nov. 1970 DEEP QUEST dives. Later observations were the same or slightly less due to biological disturbance.

(2) Total settlement predictions include settlement calculated by bearing capacity theory and by Terzaghi's consolidation theory.

(3) Direct comparison of settlement predictions with observed settlement for rectangular block No. 8 are not valid, as the block tilted excessively upon emplacement.

a low of 0.5% at a stress of 3.43 kN/m^2 to a high 3.0% at a stress of 55.11 kN/m^2.

Inasmuch as the applied stresses used in this study were low (≃3.82 kN/m^2), only minor settlement was anticipated due to secondary compression. For the cubes, the settlement predicted for a two-year period was determined to be approximately .5 cm. In the case of the rectangular blocks, the computed settlement due to secondary compression was even less. Apparently, secondary compression settlement is insignificant compared to the other settlement mechanisms present.

Total Settlement

Table 4 presents the resulting values of total predicted settlement and the total observed settlement for each block. The predicted values for total settlement were obtained by summing the initial settlement calculated from bearing capacity theory (Table 2, column 3) and the settlement due to primary consolidation as obtained by Terzaghi's consolidation theory (Table 3, column 2). Predicted settlement computations based on Terzaghi's theory appear to include settlement due to elastic deformation (Seed, 1965). Therefore, the elastic settlement predicted on the basis of the Boussinesq equation ($\simeq$.5 cm) has not been added in separately to obtain the total predicted settlement values shown in Table 4. Secondary compression settlement was ignored as it was determined to be insignificant for the unit loads used in this study.

Results indicate that, in general, the total predicted settlement better approximates the total actual settlement than any of the partial settlements. The greatest difference observed between predicted and actual settlement was 2.4 cm or 28% (excluding block No. 8). Yet, the agreement is still below what one should anticipate if the assumptions are valid.

SUMMARY AND CONCLUSIONS

Results of this study indicate that the settlement of rectangular footings of low bearing pressures on the sea floor can probably be predicted to within 30% by conventional soil mechanics theory. The amount of agreement between predicted and actual settlement values demonstrates the degree of applicability of terrestrial soil mechanics methods for analyzing fine grained saturated sea floor sediments. It is assumed that even less agreement would have been indicated had it not been for the quality of the geotechnical data obtained at the test sites.

The study has demonstrated the following:

1. The strength profile is very critical in the upper 5-10 cm of sediment for proper prediction of initial penetration by shear failure and rupture using the bearing capacity approach. Use of shear strength values derived from the least-square line fitted to vane shear data obtained at test depths greater than 5 cm was found to seriously distort the predicted initial settlement.

2. Terzaghi's theory for primary consolidation settlement approximates the observed settlement and, in this study, accounts for about 50% of the total settlement observed.

3. Settlement due to secondary compression was both predicted and observed to be insignificant for the range of unit loads used in this study.

4. Large amounts of initial settlement can be expected if the footings tilt excessively upon emplacement on the sea floor.

5. A high ratio of slenderness (ratio of the height of the block to its minimum base dimension) may cause eccentric loading and extreme differential settlement.

6. Initial settlement caused by elastic deformation is considered minor for the unit loads tested and for practical purposes should be considered part of the settlement due to primary consolidation.

ACKNOWLEDGMENTS

The authors wish to thank the entire crew and staff of DEEP QUEST and TRANSQUEST for their help during the at-sea operations. In particular, special thanks go to R. K. R. Worthington, Director of DEEP QUEST operations and to Don Saner and Larry Shumaker, pilots of DEEP QUEST for their many suggestions throughout the project. Thanks are also given to C. S. Wallin and J. G. Wilder, III, for their assistance in the laboratory testing of cores and in the reduction of in situ vane shear strength data acquired with DEEP QUEST. Laboratory consolidation tests were performed at the University of California at Los Angeles, under the direction of A. Singh, School of Engineering and Applied Science. Triaxial undrained tests on DEEP QUEST cores were performed by Woodward-Gizienski & Associates of San Diego, California, as part of a joint research project with Lockheed Ocean Laboratory. Conception and development of the project, acquisition of the initial data, and data analyses were under the direction of A. L. Inderbitzen. Final data acquisition, reduction, analyses and report preparation were under the direction of F. Simpson.

The work discussed in this report was initiated in 1970 as part of Lockheed's Sea floor Geotechnical studies. The study was funded by Lockheed Missiles & Space Company as part of their 1970-1972 Independent Research Program at Lockheed Ocean Laboratory.

REFERENCES

Anderson, D. G., and H. G. Herrmann, Sea-floor foundations: analysis of case histories, U. S. Naval Civil Engineering Laboratory Tech. Rept. R-731, 1971.

Carlmark, J. W., Penetration of free-falling objects into deep-sea sediments, Masters thesis, Naval Postgraduate School, Monterey, Calif., 1971.

Harrison, W., and A. M. Richardson, Jr., Plate-load tests on sandy marine sediments, lower Chesapeake Bay, in Marine Geotechnique, edited by A. F. Richards, pp. 274-290, Univ. of Ill. Press., Chicago, Ill., 1967.

Herrmann, H. G., Foundations for small sea-floor installations, U. S. Naval Civil Engineering Laboratory Tech. Note N-1246, 1972.

Herrmann, H. G., K. Rocker, and P. H. Babineau, LOBSTER and FMS: devices for monitoring long-term sea floor foundation behavior, U. S. Naval Civil Engineering Laboratory Tech. Rept. R-775, 1972.

Hironaka, M. C., and R. J. Smith, Foundation study for materials test structure, in Civil Eng. in the Oceans, pp. 489-530, Am. Soc. Civil Engrs., 1968.

Hough, B. K., Basic Soils Engineering, Ronald Press, New York, 1957.

Inderbitzen, A. L., F. Simpson, and G. Goss, A comparison of in situ and laboratory vane shear measurements, Mar. Tech. Soc. J., 5 (4), 24-34, 1971.

Inderbitzen, A. L., and F. Simpson, A study of the strength characteristics of marine sediments utilizing a submersible, in Symp. Underwater Soil Sampling, Testing, and Construction Control, Am. Soc. Test. Mat. Spec. Tech. Publ. 501, pp. 204-215, Phila., 1972.

Keller, G. H., Investigations of the application of standard soil mechanics techniques and principles to bay sediments, U. S. Naval Oceanographic Office Informal Manuscript Rept. No. 0-4-64, 1964.

Kretschmer, T. R., In situ sea-floor plate bearing device: a performance evaluation, U. S. Naval Civil Engineering Laboratory Tech. Rept. TR-537, 1967.

Kretschmer, T. R., and H. J. Lee, Plate bearing tests of sea-floor sediments, U. S. Naval Civil Engineering Laboratory Tech. Rept. R-694, 1970.

Lambe, T. W., and R. V. Whitman, Soil Mechanics, Wiley, N. Y., 1969.

Lee, K. L., Shallow foundations of clay soil - settlement, lecture notes from: Recent Advances in Soil Mechanics - Design and Shallow Foundations, Univ. Ext., Univ. of Calif., Los Angeles, March 23-27, 1970.

Newmark, N. M., Influence charts for computation of stresses in elastic foundations, Univ. of Ill. Bull., 40 (12), Univ. of Ill., Urbana, Ill., 1942.

Raecke, D. A., Supplement to sea-floor foundations: analysis of case histories, U. S. Naval Civil Engineering Laboratory Tech. Rept. R-7315, 1973.

Richards, A. F., Investigations of deep-sea sediment cores, I. Shear strength, bearing capacity, and consolidation, U. S. Navy Hydrographic Office Tech. Rept. TR-63, 1961.

Rucker, J. B., N. T. Stiles, and R. F. Busby, Sea-floor strength observations from the DRV ALVIN in the Tongue of the Ocean, Bahamas, South-Eastern Geol., 8, 1-8, 1967.

Seed, H. B., Settlement analyses, a review of theory and testing procedures, J. Soil Mech. and Fdn. Div., ASCE, 91 (SM2), 1965.

Shepard, F. P., Nomenclature based on sand-silt-clay ratios, J. Sediment. Petrol., 24, 151-158, 1954.

Singh, A., and Zen Yang, Secondary compression characteristics of a deep ocean sediment, in Proc. Intern. Symp. Eng. Properties of Sea-Floor Soils and their Geophys. Ident., pp. 121-129, Univ. of Washington, Seattle, 1971.

Terzaghi, K., Theoretical Soil Mechanics, Wiley, N. Y., 1943.

Terzaghi, K., and R. B. Peck, Soil Mechanics in Engineering Practice, Wiley, N. Y., 1948.

IN SITU MEASUREMENT OF SEDIMENT ACOUSTIC PROPERTIES DURING CORING

AUBREY L. ANDERSON AND LOYD D. HAMPTON

The University of Texas at Austin

ABSTRACT

The acoustical properties of liquid-saturated and gas-bearing sediments are discussed and related to other sediment properties. A system is described which has been developed for attachment to sediment corers in order to obtain an in situ sound-speed profile during a coring operation. The system uses two electroacoustic transducers mounted in the cutting head of the corer and associated electronic circuitry to measure the travel time of an acoustic pulse traversing the diameter of the sediment core. Results of laboratory and field tests are presented. There is an ongoing study of the feasibility of expanding the sound speed measurement system capabilities to include a measure of sediment acoustic attenuation and internal volume scattering. Each of these measured acoustical parameters is useful for sediment description and gas assessment.

BACKGROUND

Knowledge of the acoustical properties of sediments is important in its own right. Sediments form one of the boundaries which control long range underwater sound transmission. Acoustical tools such as sub-bottom profilers or high resolution sonar for buried pipeline, mineral deposit, or archeological artifact location use the sediments as an acoustical propagation medium.

Knowledge of the relationships between acoustical and other physical properties of sediments allows the use of acoustical tools for determining the physical and engineering properties of sediments. Several of the speakers at this symposium are discussing the relationship between acoustic and other properties of liquid-saturated sediments. There is a need to develop a similar collection of information about interrelationships among properties of gas-bearing sediments.

ACOUSTICAL AND OTHER PHYSICAL PROPERTIES OF SEDIMENTS

Saturated Sediments

Attenuation of sound in saturated sediments depends on frequency of the propagating sound and on properties of the sediment such as grain size and porosity. Figure 1 summarizes measured attenuation values reported by several investigators for the frequency range from 1 to 1000 kHz. These data are given to indicate the values of attenuation to be expected for a saturated mud sediment and to allow later comparison with attenuation values for gassy sediments. Figure 1 includes attenuation data from both laboratory and field measurements of natural and artificial sediments.

Sound speed in saturated sediments is a function of porosity. The normal incidence interface reflection coefficient is strongly correlated with sediment porosity (Faas, 1969; Hamilton, 1970a; Akal, 1972). High porosity, small grain size clays and silts are poor reflectors. Acoustic reflectivity increases with increasing sand content.

Gas-Bearing Sediments

Gas bubbles, when present in sediments in even small amounts, can dominate the characteristics of the sediment. Acoustic attenuation, propagation speed, and sediment reflective characteristics are different for liquid-saturated and partially saturated (gas-bearing) sediments.

Gas bubbles in sediments have been recognized as important in production of vent and collapse features (Maxson, 1940; Cloud, 1960; Zangerl and Richardson, 1963; Zangerl et al., 1969). Reduced stability and consequent slumping of sediments results from sediment dilation as gas bubbles form when bacterial decomposition of organic matter proceeds to supersaturation of interstitial waters with gaseous products (Monroe, 1969).

Attenuation of sound propagating in bubbly sediments has been the subject of only a few measurement programs. Muir (1972) reports attenuations from 25 to 90 dB/m over the frequency range from 40 to

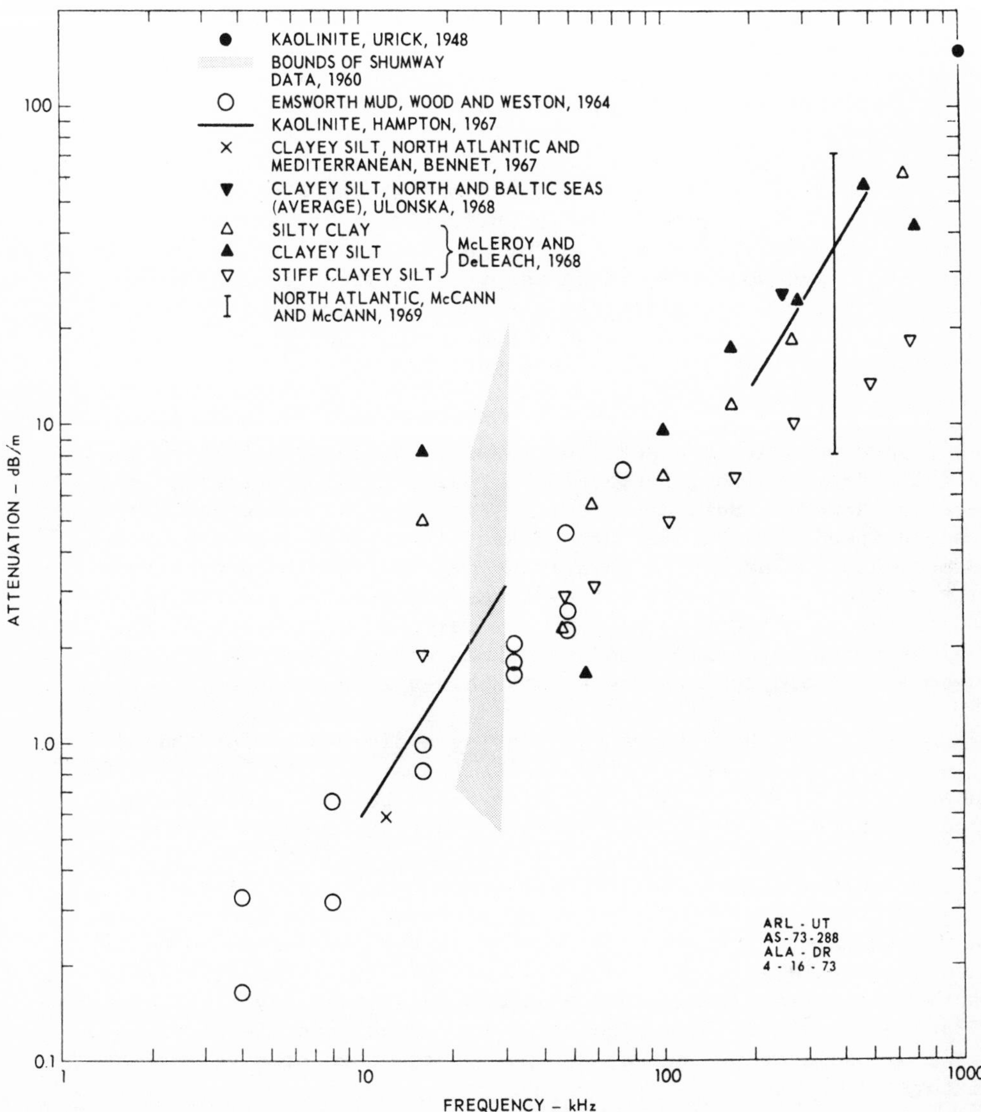

Figure 1. Acoustical attenuation vs. frequency in clays and silts (less than 1% sand)

80 kHz in the gassy mud bottom of the Lake Travis, Texas. Anderson et al., report attenuations of 13 dB/m at 40 kHz in gassy Lake Austin mud (1971) and attenuations in excess of 130 dB/m at 300 kHz in the gassy bottom of the lower reaches of the Brazos River (1968). In referring to Figure 1, it can be seen that these values for sediment with gas are only slightly larger than saturated-sediment values.

For the most part, attenuation due to gas bubbles in sediments has been reported as an interference with measurements in saturated sediments. For example, Nyborg, Rudnick, and Schilling (1950) report prohibitive attenuations, as high as 2,600 dB/m at 10 kHz to 6,400 dB/m at 30 kHz in their partially saturated soil samples. Wood and Weston (1964) report attenuations of 1,750 dB/m at 8 kHz and of 2,400 dB/m at 14 kHz in samples of natural bay sediments. Again referring to Figure 1, it is obvious that these values are orders of magnitude greater than for saturated sediments. Both Nyborg et al., and Wood and Weston report significantly decreased values for attenuation after the gas had been drawn from the sediments by placing them in a vacuum chamber (see the data of Wood and Weston for saturated sediments in Figure 1).

Many authors report a greatly decreased sound speed when gas bubbles are present in a sediment (Jones, Leslie, and Barton, 1958 and 1964; Brandt, 1960; Levin, 1962; Lewis, 1966; Hochstein, 1970). There is, however, some disagreement reported: an observed increase of sound speed when bubbles are present (Brutsaert and Luthin, 1964, discuss this). The key to this issue may well be the degree to which bubbles in sediments act as a resonant system and the relation of the acoustic frequency to the resonance frequency of the bubble-sediment system. Sound-speed measurements versus frequency can be a valuable tool for in situ bubble-size assessment.

Many studies have shown gas-bearing sediments to form highly reflective pressure release boundaries (Bobber, 1959; Leslie, 1960; Grubnik, 1961; Levin, 1962; Ruff, 1967; Schubel, in press). These measurements are all for frequencies less than 10 kHz.

The examples given have been for gas-bearing sediments in shallow water. Very little published information is available on the acoustical properties of sediments in the deep ocean which are known to contain gas. A notable exception (Stoll et al., 1971) describes anomalously high sound speeds in a gassy layer of sediment in 3600 m of water. This high value for sound speed is attributed to the probable occurrence of the gas in the form of a gas hydrate (clathrate).

MEASUREMENT OF SOUND SPEED WHILE CORING

Sediment acoustical and mechanical properties, which are measured in the laboratory for either constructed sediments or cores, require correction to determine their in situ values. Limits to the capability of obtaining undisturbed samples for such measurements contribute to uncertainty about actual in situ values. Hamilton (1971) has given a technique for extrapolating laboratory sound-speed measurements to in situ values. His method is based on laboratory and in situ measurements of surficial sediments of the North Pacific. Extending this technique to other regions or to sediments deeper than the surface layer remains to be tested. Furthermore, the presence of gas either as bubbles or as a hydrate (clathrate) complicates any attempt to extrapolate to sea-floor conditions. In situ measurements of sound speed by probing have previously been limited to surficial sediments (Hamilton, 1963; Bennin and Clay, 1967; Lewis et al., 1970).

Sediment cores are commonly taken to depths of 20 m and more in the bottom. Applied Research Laboratories (ARL), the University of Texas at Austin (UT), has developed a system, for attachment to

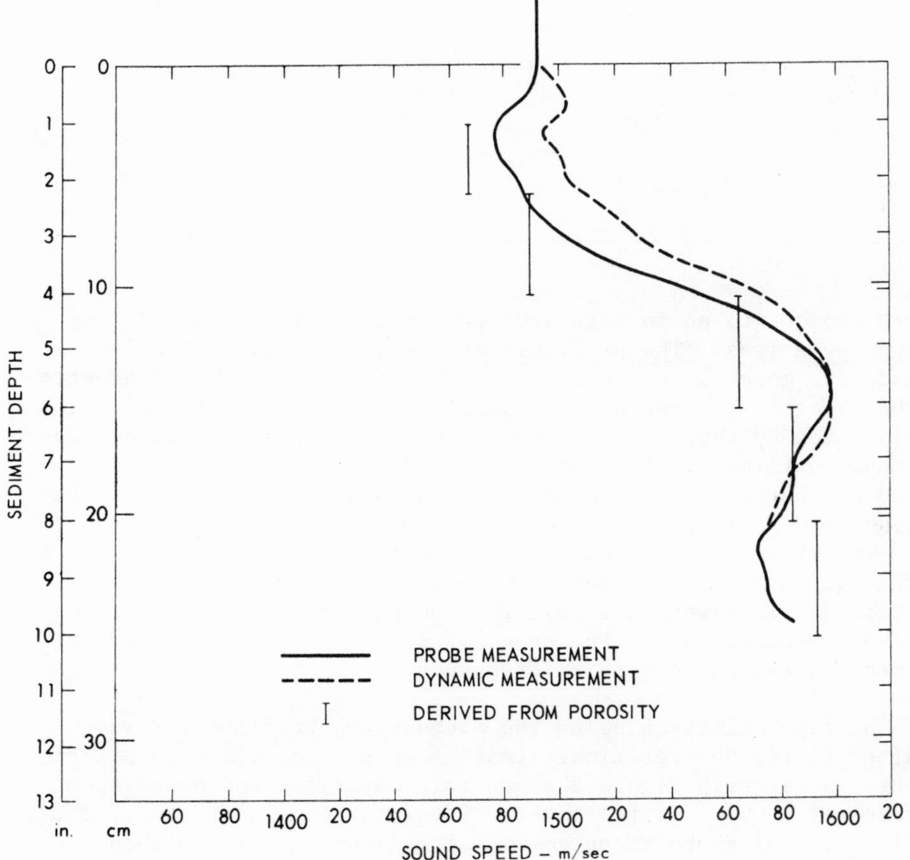

Figure 2. Sound-speed in graduated sediment
T = 23° C

ARL - UT
AS-71-799
LDH - RFO
7 - 8 - 71

TABLE I

WATER CONTENT, POROSITY, AND SOUND-SPEED PROFILE OF LABORATORY SEDIMENT

Sediment Depth Interval (cm)	Water Content (% of Dry Weight)	Porosity (%)	Sound Speed from Akal (1972) (m/sec)
2.5 - 5.7	102.2	73.0	1466
5.7 - 10	62.6	62.4	1490
10 - 15	37.3	49.7	1564
15 - 20	33.6	47.1	1584
20 - 25	32.2	46.0	1593

corers, to obtain an in situ sediment sound-speed profile during coring operations. The profilometer system uses two piezoelectric transducers mounted in the cutting head of the corer and connected to the electronics housing by coaxial cables. An acoustical pulse is generated by the transmit transducer. This pulse traverses the sediment contained in the cutting head at that instant and is received by the second transducer which is diametrically across the cutting head from the projector. The time delay resulting from the pulse traversing the sediment in the cutter head is converted to a voltage output which is presently sent to the surface by a cable and recorded on magnetic tape. This output voltage is calibrated to give a measurement of the speed of propagation of sound in the sediment at the pulse-carrier frequency.

The feasibility of using the sound-speed profilometer was first confirmed by preliminary testing under controlled conditions in the laboratory. Figure 2 shows for comparison two sound-speed profiles of a 0.3 m (1 ft) layer of sediment in a test tank. The profile labeled probe measurement was made using three probes (one projector and two receivers at different distances) lowered incrementally into the sediment. Time delay between the two received pulses was measured with an interval timer and the sound speed calculated. The profile labeled dynamic measurement is a plot of the sound speed calculated from the voltage output of the electronic circuitry as the coring head was penetrating the sediment after free fall from a height of 1.2 m (4 ft).

The sediment for which sound speed is shown in Figure 2 was sampled by coring with a 3.2 cm (1 1/4 in) diam. tube. Subsamples of these core samples were analyzed for porosity. The results are summarized in Table 1. Akal (1972) has summarized the relationship between porosity and relative sound speed (ratio of sediment sound speed to water sound speed) which is shown by more than 450 marine sediment cores, based on data drawn from numerous authors. He fitted a second order equation to these data and presents a graph of the resulting curve. This curve was used to determine the values of sound speed shown in Table 1 for the measured porosities. These values are shown in Figure 2 for comparison with the measured sound speeds. The absolute values of the measured sound speeds are well within the bounds of the observed sound speeds reported by Akal (about ± 75 m/sec).

Figure 3 shows a sound-speed profile obtained in about 55 m (30 fathoms) of water aboard RV ALAMINOS on station off the Mississippi delta in the Gulf of Mexico. The top trace is the unintegrated output of an accelerometer attached to the coring tool and is used as indication of corer entry into the sediment and end of corer motion. The uniform (unlayered) nature of this soft sediment is shown by the sound-speed profile. Analysis of the cored material confirmed the unlayered nature of the sediment. Porosity of the sediment, when compared with Akal's (1972) best fit curve relating porosity and sound speed, predicts a sound speed in the sediment about 15 m/sec lower than in the water with variations along the core of about ± 15 m/sec. The values are in good agreement with the measured profile in Figure 3.

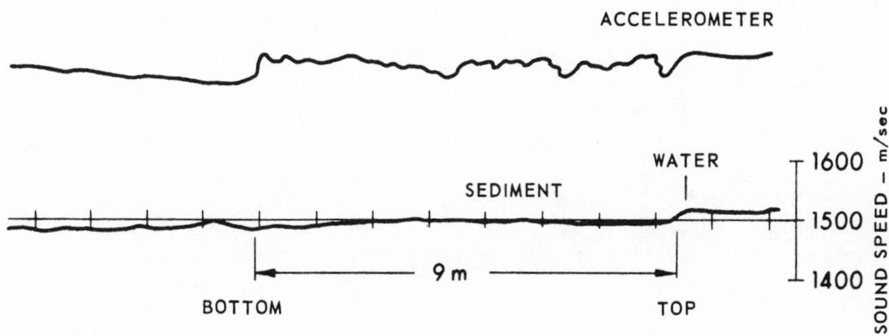

Figure 3. Sound-speed profile, Mississippi Delta, Gulf of Mexico

ARL - UT
AS-73-296
ALA - DR
4 - 16 - 73

Cores obtained in cooperation with the UT Marine Sciences Institute used a shallow water coring rig designed by Dr. E. W. Behrens. This device pushed the 7.6 cm (3 in) diam. core into the bottom using block and tackle rigging on a shallow draft platform. Using this equipment, cores were obtained in Baffin Bay and Redfish Bay on the Texas Gulf Coast near Corpus Christi, Texas.

Figure 4 shows sound-speed profiles made in a single core taken in Baffin Bay (August 1972). The top profile labeled "in situ" was made with the transducers mounted on the corer. The other three profiles were made using transducers that are designed to slide on the core liner. The profile labeled "on deck" was made approximately 3 hours after the core was taken. The two lowest profiles were made the first and second days after removing the cores to the laboratory.

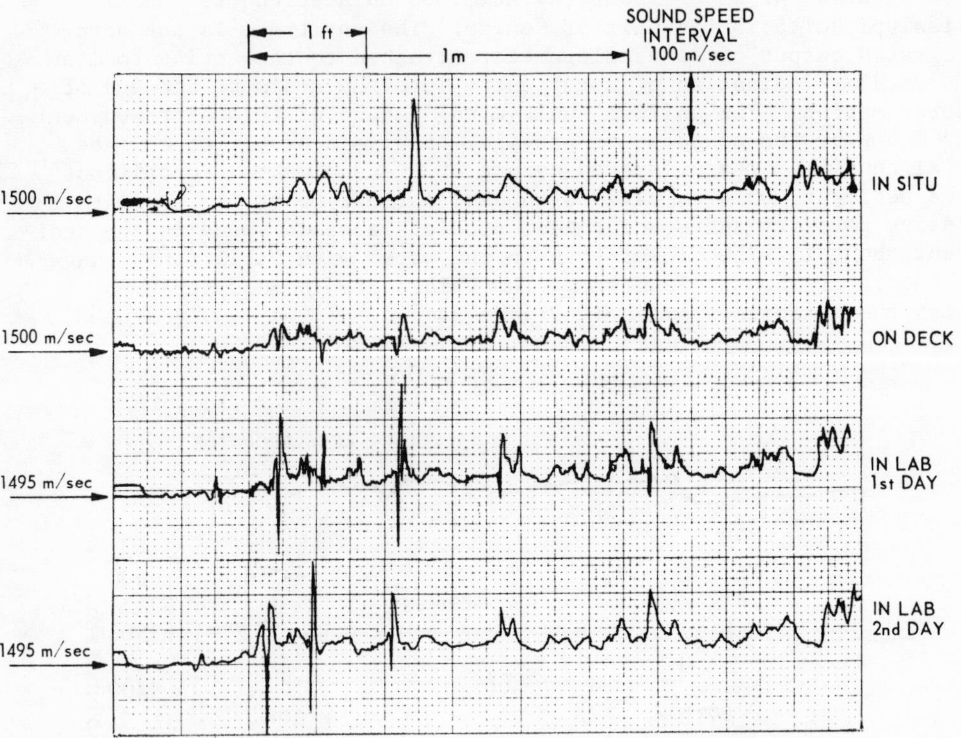

Figure 4. Sound-speed profiles of Core 1. From Baffin Bay, Texas, August 1972.

ARL - UT
AS-72-1045-S
DJS - RFO
8 - 28 - 72

Notice that the absolute value of sound-speed changes somewhat from one determination to the next, and the major highlights show some variation in each profile.

The large excursions in the bottom two profiles are the result of large attenuation of the signal beyond the capabilities of the AGC system. This attenuation was the result of either gas generation in these layers or physical separation of the material of the core during transportation or possibly both. Profiles taken later than these showed even greater deterioration.

In Figure 5, the in situ sound-speed profile from Figure 4 is compared with the core lithology, as are two other profiles from this region of Baffin Bay. The three cores were taken in a line and spaced approximately 0.6 m (2 ft) apart making a maximum separation between cores of only 1.2 m (4 ft). However, both the sound-speed profiles and lithology diagrams show a large variation in layering between the cores. Some layers even show a variation across the 7.6 cm (3 in) diam. of the core. Lines on Core 3, Figure 5, indicate the degree of correlation between the lithology and sound-speed record. Cores 1, 2, and 3 are representative of the cores taken from this area in that they consist of alternating layers of mud, shell, and fine sand with intermixing of the three types to some degree in each layer.

In Figure 5 the water-sediment interface is located at the top of the figure and is marked as zero on the depth scale. The top 15 to 20 cm (6 to 8 in) was a light fluffy mud containing gas and it showed a sound-speed lower than that of the overlying water. At about 0.3 m (1 ft) depth, the first high sound-speed layer was encountered and it consisted of a mixture of mud, sand, and shell. These top shell and sand layers as well as those deeper are seen from Figure 5 to be uneven, discontinuous, and quite often hard to distinguish. However, there is a good correlation between the amount of sand and shell in the cores and the value of sound speed. Sound speed increases with increasing percentage of sand and shell, with sand content the more dominant feature. The next distinctive layer encountered varied in depth from 0.5 m (1 3/4 ft) in Core 3 to 0.6 m (2 ft) in Core 1 and consisted of a very sandy layer about 5 to 7.5 cm (2 to 3 in) thick. This layer seemingly slopes downward from Core 3 to Core 1. The layer is distinguished mainly by a sound speed that is higher than that of the other layers in the cores. This is especially true for Core 1. For the next 0.6 to 0.9 m (2 to 3 ft), the layering consisted of sandy, shelly layers, which correlate with high sound speed, interspersed with muddy layers having a sound speed only slightly higher than that of water. Many of the sandy layers are too thin to have been detected by the sound-speed profile and so do not show up on the profile. Finally, at a depth of 1.5 to 1.8 m (5 to 6 ft), a layer consisting of almost pure shell was encountered which could not be penetrated by the

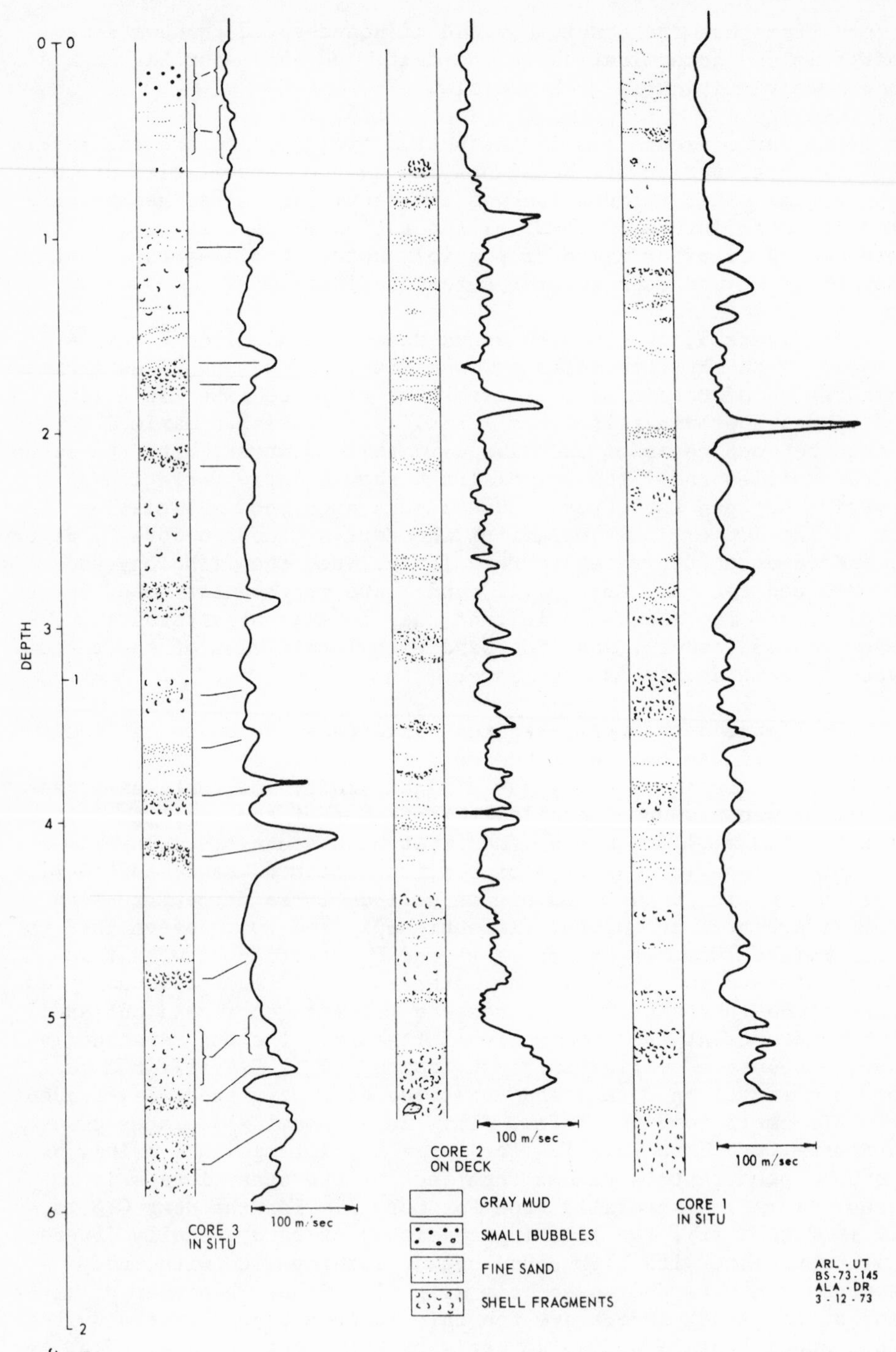

Figure 5. Correlation of lithologies and sound-speed profiles of cores from Baffin Bay, Texas. August 1972.

corer. This bottom layering was distinctive for all the cores taken in the area and shows up on the sound-speed profiles as a broad, medium high speed peak with a dip in the middle. This distinctive layer occurred in all 15 profiles taken in this region as shown in Figure 6.

The 15 cores were taken in groups with approximately 0.3 to 0.6 m (1 to 2 ft) separation between cores within a group. All cores were taken within an area of radius of approximately 9 m (30 ft). The groupings were: Cores 1, 3, and 4; 5 and 6; 7, 8 and 9; 10, 11, and 12; 13, 14, 15, and 16.

It is possible from the sound-speed profiles to identify the distinct bottom layer as continuous throughout the sampled region. Note also that conclusions about complexity of the area, dimensions of features, and extent of layering in the sediment can be drawn from these profiles without supplemental information.

For the most part the complexity of the core lithology precluded meaningful sampling of portions of the core for sound-speed prediction from lithology; however, the high speed layer at about 0.6 m (2 ft) in Core 1 was almost pure sand. Grain size distribution was determined for this material by sieve analysis. Geologically, the layer is described as poorly sorted muddy very fine sand with a mean grain diameter of 0.094 mm (3.4 ϕ). Hamilton (1970b) gives a graph of sound speed versus mean grain diameter which has been measured for sediments of the North Pacific. The measured sound speed of 1660 m/sec for the sand layer of Core 1 (Figure 5) is well within the bounds of 1640 to 1700 m/sec reported by Hamilton for material of the same mean grain size. Significantly, the value measured in the lab, through the core liner (Figure 4), for this same layer is 1550 m/sec, which is lower than the Hamilton values. Thus, the in situ profilometer measurement provides a better value than the in-lab measurement.

A core sound-speed profile which is of different character than either the homogeneous cores from the Gulf of Mexico (Figure 3) or the complex Baffin Bay cores (Figures 4, 5, and 6) is given in Figure 7. This core was taken in Redfish Bay, Texas, on 9 May 1972. The sound-speed profile shown is representative of measurements on 7 cores taken at this site. The major character of the Redfish Bay cores is slow changes of sound speed with depth, in particular a uniformly high sound speed in the first 0.45 m (1 1/2 ft) of depth and gradual sound-speed decrease from 0.45 m (1 1/2 ft) to the bottom of the core. Many large shells are scattered throughout the core, explaining the presence of fine structure in addition to the broad features.

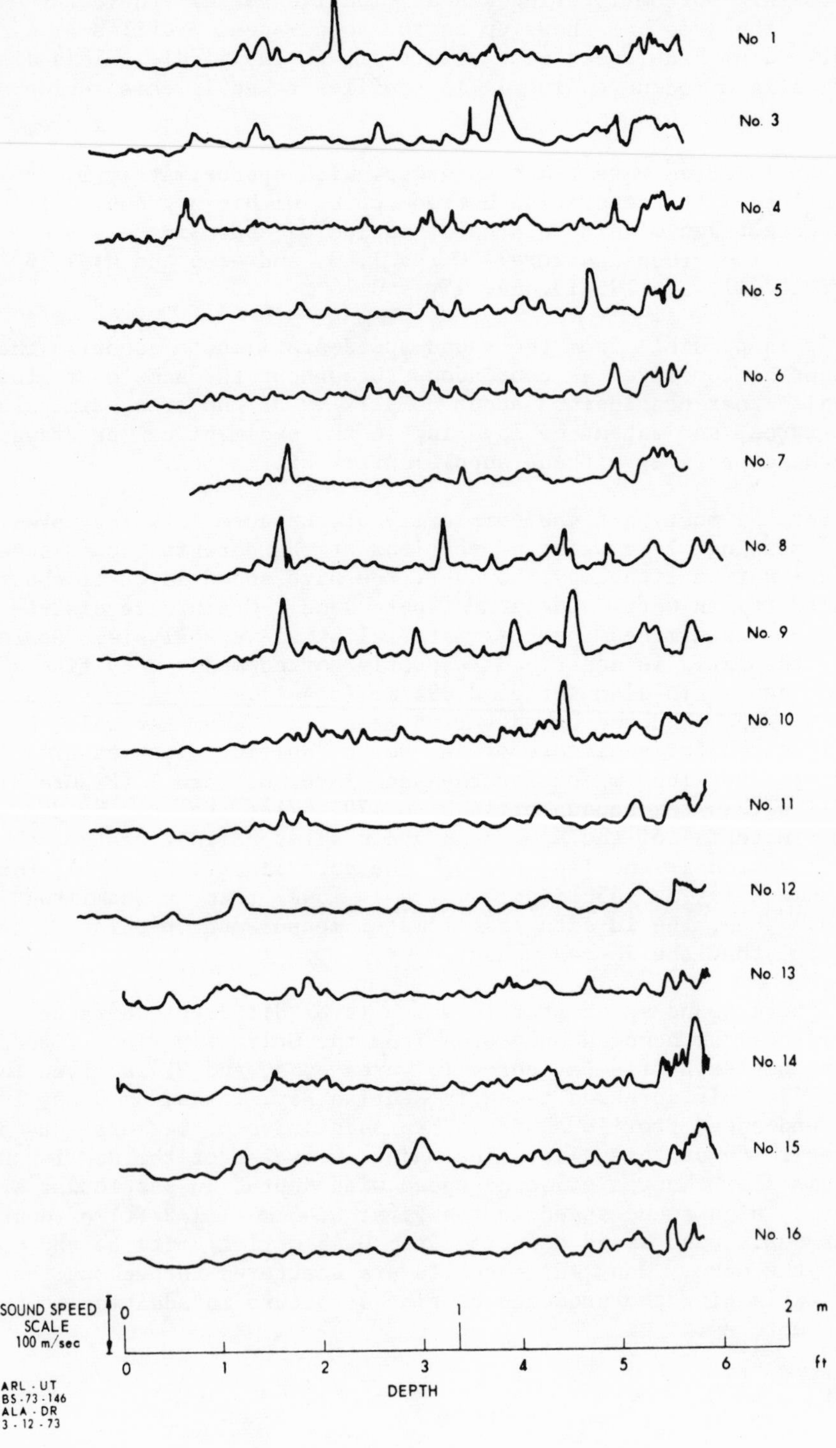
No 1
No. 3
No. 4
No. 5
No. 6
No. 7
No. 8
No. 9
No. 10
No. 11
No. 12
No. 13
No. 14
No. 15
No. 16
SOUND SPEED
SCALE
100 m/sec
0
1
2
m
0
1
2
3
4
5
6
ft
DEPTH
ARL - UT
BS - 73 - 146
ALA - DR
3 - 12 - 73

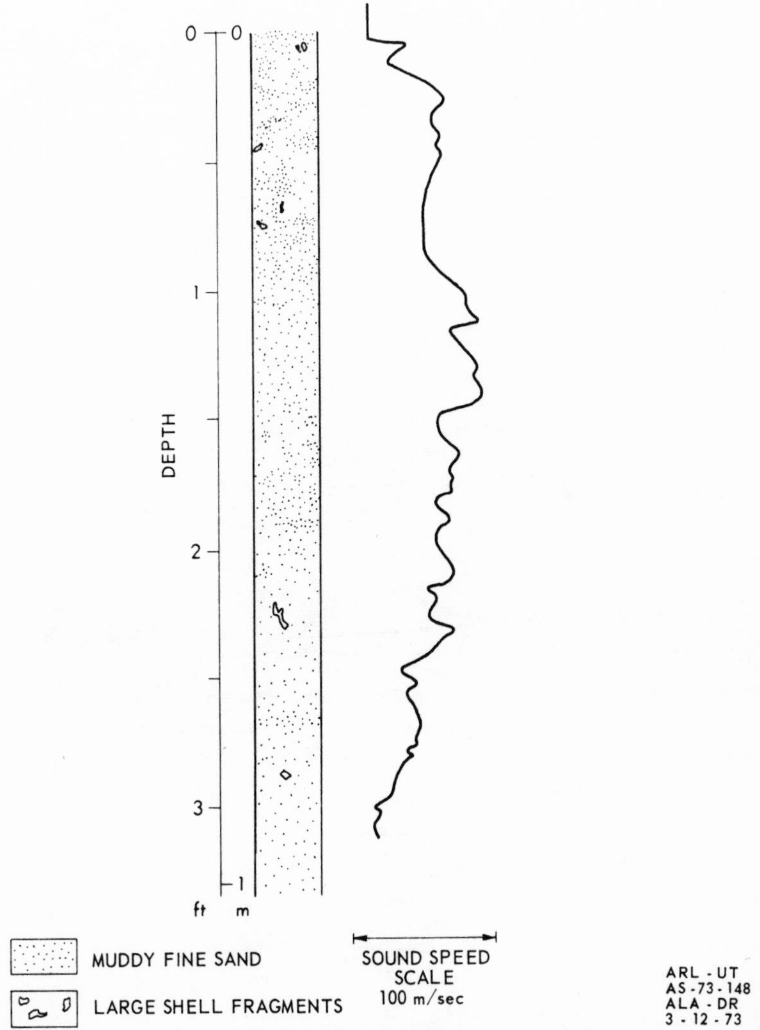

Figure 7. Correlation of lithology and sound-speed profile of Core 2 from Redfish Bay, Texas. May 1972.

Figure 6 (opposite). In situ sound-speed profiles for fifteen cores in sequence from Baffin Bay, Texas. August 1972.

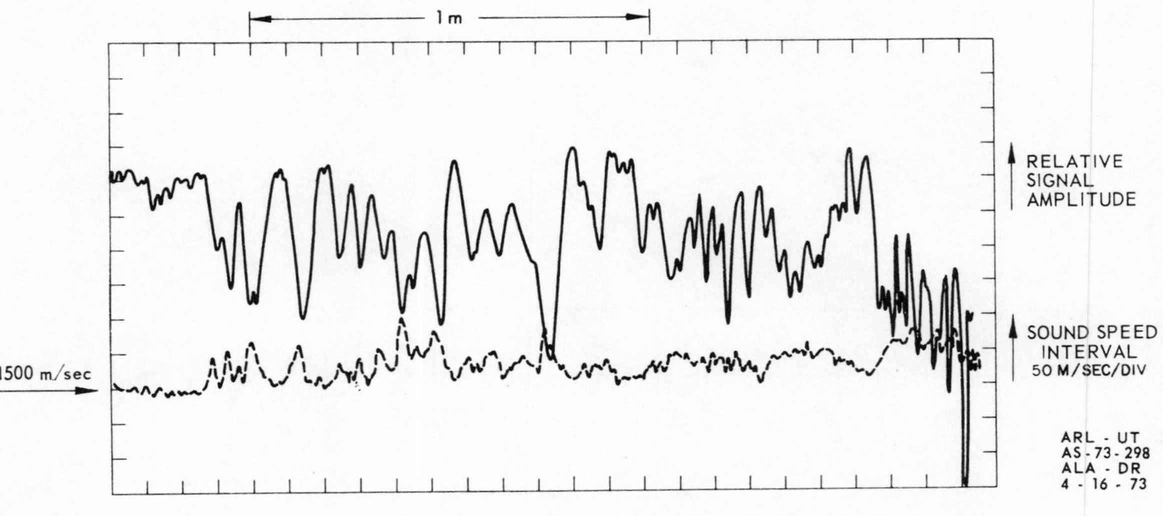

Figure 8. Relative signal amplitude and sound-speed profiles for a core from Baffin Bay, Texas.

EXPANSION OF PROFILOMETER CAPABILITIES

At ARL, there is an ongoing study of the feasibility of expanding the profilometer capabilities to include a measure of sediment acoustic attenuation and internal volume scattering.

The automatic gain control feedback voltage is a measure of the insertion loss of the material between the profilometer transducers at any instant. This voltage has been separately recorded in a limited number of coring tests. Figure 8 compares profiles of sound speed and relative received signal amplitude (as measured by the AGC feedback voltage) for a single core from Baffin Bay, Texas. Increased sound speed in these cores is well correlated with increased sand content. As Figure 8 demonstrates, the increased sound-speed layers measured are also well correlated with decreases in the received signal amplitude. Thus, though work has not yet been devoted to reducing this information to a numerical measure of acoustical attenuation, the expected increase of attenuation with sand content is observed.

The internal acoustical volume scattering strength of sediments is not documented in the literature. The acoustical scattering cross section of gas bubbles in water is very large at resonance, more than one thousand times the physical bubble cross section. Bubble scattering cross section in sediments is not known, but should be large at resonance. To test the feasibility of adding measurements of this parameter to the profilometer, a third transducer was added to the cutter head. This transducer was placed so that its acoustical axis intersected at 90° the primary acoustical path joining the sound-speed measurement transducer pair. Using the third transducer as a receiver, acoustical energy which is side scattered out of the principal transmission path in the sediment would be measured.

The acoustical feedover to the volume scatter transducer in clear water was measured to be about 14 dB below the direct path signal. Insertion of the cutter head into saturated sand sediments and muddy sand sediments in the laboratory indicated that the volume scattered levels were below the acoustical feedover levels. This measurement will therefore require greater isolation between the two acoustic paths.

An encouraging indication was obtained from a series of measurements with the 3-transducer arrangement in water. Gas bubbles, generated by hydrolysis, were allowed to rise into the region which is common to both acoustical paths. Side-scattered acoustical levels were above the acoustical feedover for some bubble sizes and concentrations. Thus, a sufficient reduction of the feedover level by change of frequency, different transducer arrangement or acoustical baffling should allow measurement of volume scattering in sediments. Also, the high levels of scattering from bubbles in water suggest the possibility of internal volume scattering measurements as useful indications of gas bubbles in the sediment.

ACKNOWLEDGMENTS

Much of the development of, and experimentation with, the profilometer was carried out by Mr. D. J. Shirley. The work was sponsored by Office of Naval Research, Code 480.

REFERENCES

Akal, T., The relationship between the physical properties of underwater sediments that affect bottom reflection, Mar. Geol., 13, 251-266, 1972.

Anderson, A. L., T. G. Muir, Jr., R. S. Adair, and W. H. Tolbert, A geoacoustic survey of the Brazos River, part I: environmental studies, Defense Research Laboratory Acoustical Rept. No. 294, DRL-A-294, Applied Research Laboratories, Univ. of Texas, Austin, 1968.

Anderson, A. L., R. J. Harwood, and R. T. Lovelace, Investigation of gas in bottom sediments, Applied Research Laboratories Tech. Rept. No. 70-28, ARL-TR-70-28, Univ. of Texas, Austin, 1971.

Bennett, L. C., Jr., In situ measurements of acoustic absorption in unconsolidated sediments (abstract), Trans. Am. Geophys. Union, 48, 144, 1967.

Bennin, R. S., and C. S. Clay, Development of an in situ sediment velocimeter, Tech. Rept. No. 131, Hudson Laboratories, Columbia Univ., N. Y., 1967.

Bobber, R. J., Acoustic characteristics of a Florida lake bottom, J. Acoust. Soc. Am., 31, 250-251, 1959.

Brandt, H., Factors affecting compressional wave velocity in unconsolidated marine sand sediments, J. Acoust. Soc. Am., 32, 171-179, 1960.

Brutsaert, W., and J. N. Luthin, The velocity of sound in soils near the surface as a function of the moisture content, J. Geophys. Res., 69, 643-652, 1964.

Cloud, P. E., Gas as a sedimentary and diagenetic agent, Am. J. Sci., 258-A, 34-35, 1960.

Faas, R. W., Analysis of the relationship between acoustic reflectivity and sediment porosity, Geophysics., 34, 546-553, 1969.

Grubnik, N. A., Investigation of the acoustic properties of underwater soil at high acoustic frequencies, Sov. Phys. Acoust., 6, 447-454, 1961.

Hamilton, E. L., Sediment sound velocity measured in situ from bathyscaph TRIESTE, J. Geophys. Res., 68, 5991-5994, 1963.

Hamilton, E. L., Reflection coefficients and bottom losses at normal incidence computed from Pacific sediment properties, Geophysics, 35, 995-1004, 1970a.

Hamilton, E. L., Sound velocity and related properties of marine sediments, North Pacific, J. Geophys. Res., 75, 4423-4446, 1970b.

Hamilton, E. L., Prediction of in situ acoustic and elastic properties of marine sediments, Geophysics, 36, 266-284, 1971.

Hampton, L. D., Acoustic properties of sediments, J. Acoust. Soc. Amer., 42, 882-890, 1967.

Hochstein, M. P., Seismic measurements in Suva Harbour (Fiji), New Zealand, J. Geol. Geophys., 13, 269-281, 1970.

Jones, J. L., C. B. Leslie, and L. E. Barton, Acoustic characteristics of a lake bottom, J. Acoust. Soc. Am., 30, 142-145, 1958.

Jones, J. L., C. B. Leslie, and L. E. Barton, Acoustic characteristics of underwater bottoms, J. Acoust. Soc. Am., 36, 154-157, 1964.

Leslie, C. B., Normal incidence measurement of acoustic bottom constants, U. S. Naval Ordnance Laboratory Rept. No. 6832, 1960.

Levin, F. K., The seismic properties of Lake Maracaibo, Geophysics, 27, 35-47, 1962.

Lewis, L. F., Speed of sound in unconsolidated sediments of Boston Harbor, Mass., Masters thesis, Mass. Inst. Technol., Cambridge, 1966.

Lewis, L. F., V. A. Nacci, and J. J. Gallagher, In situ marine sediment probe and coring assembly, U. S. Naval Underwater Sound Laboratory Rept. No. 1094, 1970.

Maxson, J. H., Gas pits in nonmarine sediments, J. Sediment Petrol., 10, 142-145, 1940.

McCann, C., and D. M. McCann, The attenuation of compressional waves in marine sediments, Geophysics, 34, 882-892, 1969.

McLeroy, E. G., and A. DeLeach, Sound speed and attenuation, from 15 to 1500 kHz, measured in natural sea-floor sediments, J. Acoust. Soc. Am., 44, 1148-1150, 1968.

Monroe, J. N., Slumping structures caused by organically derived gases in sediments, Science, 164, 1394-1395, 1969.

Muir, T. G., Experimental capabilities of the ARL sediment tank facility in the study of buried object detection, Applied Research Laboratories Tech. Memo. No. 72-32, ARL-TM-72-32, Univ. of Texas, Austin, 1972.

Nyborg, W. L., I. Rudnick, and H. K. Schilling, Experiments on acoustic absorption in sand and soil, J. Acoust. Soc. Am., 22, 422-425, 1950.

Ruff, G. A., Acoustic characteristics of Black Moshannon Lake bottom, J. Acoust. Soc. Am., 42, 524-525, 1967.

Schubel, J. R., Gas bubbles and the acoustically impenetrable, or turbid, character of some estuarine sediments, Proc. Conf. Natural Gases in Mar. Sediments, Lake Arrowhead, Calif., (in press).

Shumway, G., Sound speed and absorption studies of marine sediments by a resonance method, Part I, Geophysics, 25, 451-467, 1960.

Stoll, R. D., J. Ewing, and G. M. Bryan, Anomalous wave velocities in sediments containing gas hydrates, J. Geophys. Res., 76, 2090-2094, 1971.

Ulonska, A., Versuche zer Messung der Schallgeschwindig keit und Schalldampfung im Sediment in situ, Deutsche Hydrographische Zeitschrift, 21 (2), 49-58, 1968.

Urick, R. J., The absorption of sound in suspensions of irregular particles, J. Acoust. Soc. Am., 20, 283-289, 1948.

Wood, A. B., and D. E. Weston, The propagation of sound in mud, Acustica, 14, 156-162, 1964.

Zangerl, R., and E. Richardson, The paleoecological history of two Pennsylvanian black shales, Fieldiana: Geol. Mem., 4, Field Museum of Natural History, Chicago, 1963.

Zangerl, R., B. G. Woodland, E. G. Richardson, and D. L. Zachry, Early diagenetic phenomena in the Fayetteville black shale (Mississippian) of Arkansas, Sediment. Geol., 3, 87-119, 1969.

SEA-FLOOR SOIL MECHANICS AND TRAFFICABILITY MEASUREMENTS WITH THE TRACKED VEHICLE "RUM"

DANIEL K. GIBSON AND VICTOR C. ANDERSON

Marine Physical Laboratory of the Scripps Institution of Oceanography

ABSTRACT

A brief description of the ORB-RUM Sea Floor Work System, program history, philosophy and direction is given. An instrumentation suite developed for the use of RUM in a soil mechanics and trafficability study program is also described. Highlights of over 400 operating hours on the sea floor at depths ranging from 40 to 1880 meters illustrates the problems encountered and the experience that has been obtained in this program. A limited amount of data is presented.

BACKGROUND

In 1957 a program was undertaken at the Marine Physical Laboratory, under the sponsorship of the Office of Naval Research, to develop an experimental remotely manned sea-floor work vehicle, named "RUM" for Remote Underwater Manipulator. The program was carried through to the completion of some shallow water operational tests, near shore, operating off the beach near Scripps Institution at La Jolla, California in 1959. At that point, a change in direction in the sponsoring program shifted the priorities to other Laboratory projects of more urgent interest to the Navy, and the RUM was placed in storage.

Nearly ten years later there was a renewed interest in RUM. The remotely controlled, cable-tethered, bottom-crawling vehicle appeared to offer a significant advantage over a manned submersible in the performance of some sea-floor tasks; for other tasks the manned submersible appeared to offer more advantage. The factors where advantages occur are in economics, operator safety, unlimited mission time, platform stability for manipulator, extended payload capability, and the ability to exert large reaction forces in drilling, excavating, etc. In order to investigate these possible advantages, work on the RUM was resumed in 1968. Operational tests at sea were begun in 1970.

THE RUM/ORB SYSTEM

RUM as presently configured is shown in Figure 1. The basic chassis, tracks and suspension system are from a surplus Marine Corps combat vehicle called "ONTOS." All else is specially designed and developed for this application. The vehicle mounts a 15-year old electro-mechanical manipulator on a derrick which pivots about a king post at the rear. Sensors include three TV cameras, two scanning sonars, listening hydrophones, an acoustic navigation system and a host of additional sensors for monitoring the equipment and the surrounding environment, as an aid to the operators in the control and operation of the vehicle. In addition, there is an instrumentation interface providing a number of control signals and power for operation of external mission orientated instruments, tools or devices. A sampling 8-bit A to D converter provides for telemetry of up to 8 sensor outputs from this instrumentation interface to the surface for real time analysis, recording or observation.

RUM is operated from a surface support platform called "ORB" for Oceanographic Research Buoy. ORB, shown in Figure 2, was also designed and developed at the Marine Physical Laboratory under the sponsorship of the Office of Naval Research. ORB is a 21.34 m long by 13.7 m wide, rectangular box-like barge with a 4.57 by 6.1 m center well from which RUM is launched and recovered. ORB has no propulsion capability of its own so must be towed to and from the operating area where it is usually moored in a multi-point taut moor over the work site. ORB is equipped with living accommodations for 14, spacious laboratory and work space, 240 kW of electrical power, and the capability to remain at sea for 30 to 40 days.

The RUM launch or recovery is usually completed in 10 to 15 minutes. RUM descends and ascends at approximately 20 m per minute.

The RUM system is illustrated in Figure 3. With RUM on the bottom the operator sits in relative comfort and safety at the control console on board ORB. All controls for the vehicle and its equipment, including the manipulator, along with readouts from the many sensors, are grouped according to function, on the desk top and panels before him.

The operator control console is connected to the vehicle via a single coaxial cable 3000 m in length. Twenty-five kilowatts of power to the vehicle, along with all television, sonar and telem-

Figure 1. "RUM" vehicle

Figure 2. "ORB"

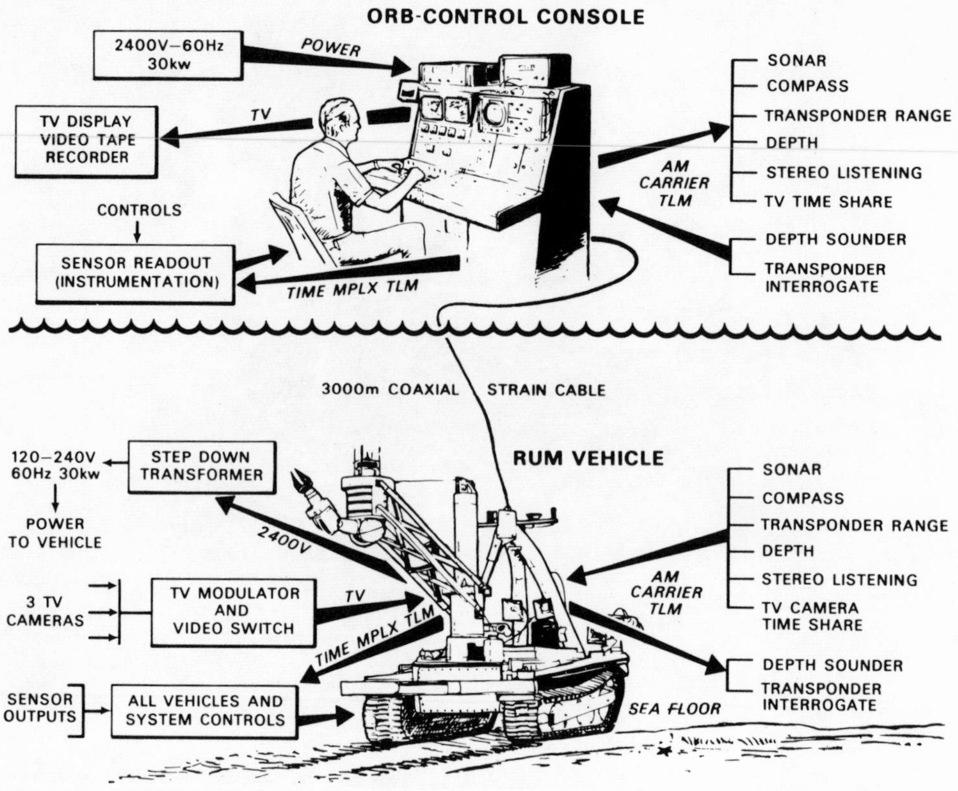

Figure 3. "RUM" system drawing

etry signals, for control and instrumentation of the vehicle, its related sensors and instrumentation suite, are time and frequency multiplexed onto this single shielded electrical conductor. Strain members, wrapped around the coaxial transmission line, support the vehicle weight.

THE SEA-FLOOR WORK SYSTEMS PROJECT

In late 1970 a project was undertaken at the Marine Physical Laboratory to use the ORB-RUM Sea Floor Work System operational experience, as a model, in an attempt to determine the work performance effectiveness and to develop the economics of a remotely controlled cable-tethered sea-floor work vehicle in the performance of a variety of underwater research tasks. This study has been carried out under the sponsorship of the Sea Grant Office of the National Oceanic and Atmospheric Agency.

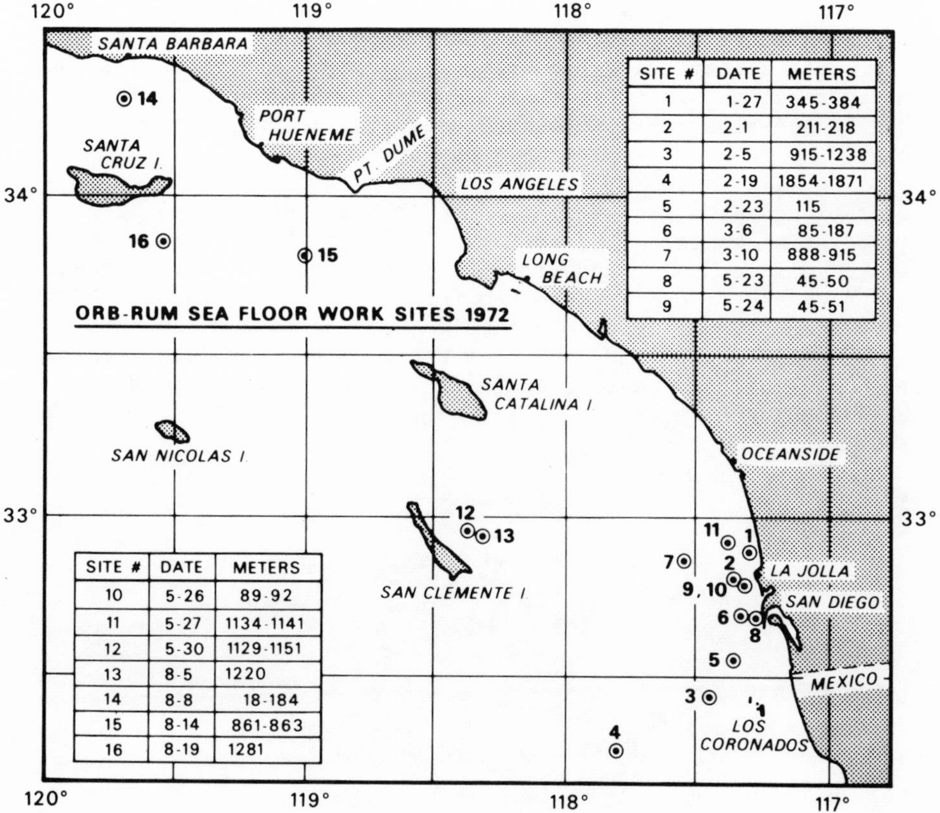

SITE #	DATE	METERS
1	1-27	345-384
2	2-1	211-218
3	2-5	915-1238
4	2-19	1854-1871
5	2-23	115
6	3-6	85-187
7	3-10	888-915
8	5-23	45-50
9	5-24	45-51

SITE #	DATE	METERS
10	5-26	89-92
11	5-27	1134-1141
12	5-30	1129-1151
13	8-5	1220
14	8-8	18-184
15	8-14	861-863
16	8-19	1281

Figure 4. "ORB-RUM" sea-floor work site 1972

In order to provide a specific set of repetitive useful work tasks from which work performance data could be obtained, a suite of instruments specifically designed for handling by the RUM manipulator and interfacing with the RUM telemetry have been developed under this Sea Grant Project. This suite is described in detail by Anderson et al. (1972) and Clinton (1972). The purpose of the instrumentation suite is to make in situ measurements of sediment strengths, collect relatively undisturbed sediment samples for later laboratory analysis and relate these to vehicle trafficability. The instrument suite consists of a vane shear meter, a cone penetrometer, a set of core samplers, an anchor-winch-tensiometer system for measuring vehicle drawbar pull capability, and a track depression profiler.

OPERATIONS

By 1972 the ORB-RUM system including the newly developed instrumentation suite was declared tested, proven and operationally ready for data collection. From January through August of 1972 the system was taken on six trips to sea, during which ORB was moored over the 16 sea-floor work sites indicated in the chart shown in Figure 4.

Operating depths ranged from 48 to 1880 m. RUM logged 237.5 actual operating hours on the bottom. Of these, 88.5 hours were devoted to soil mechanics and trafficability studies. During these studies 34 traverses over the bottom for a total distance of 1680 m were completed, 49 sediment core samples were collected, 29 sets of vane shear measurements made, 67 cone penetromoter profiles recorded, 60 traces made of track depression profiles and 13 drawbar pull measurements completed.

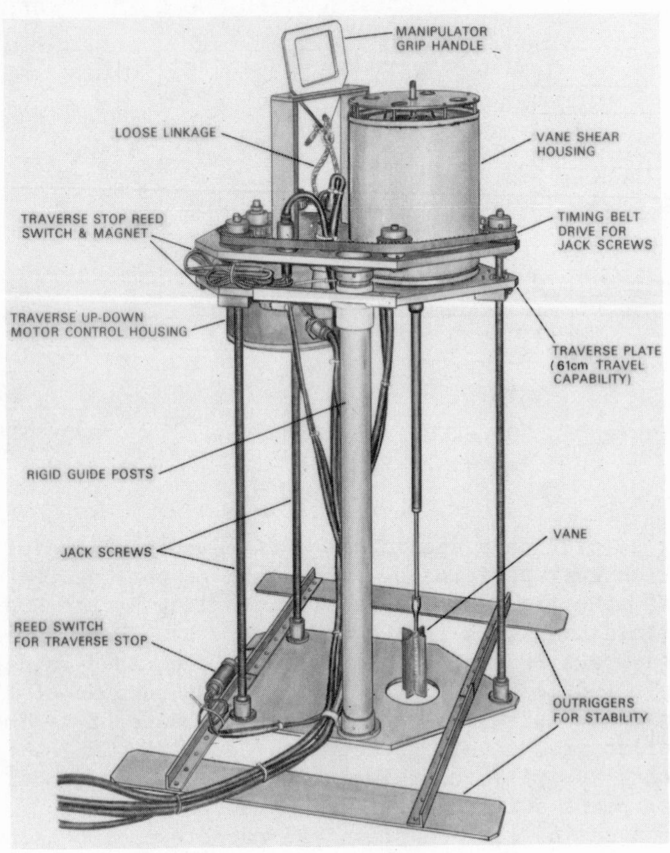

Figure 5. Vane shear meter assembly

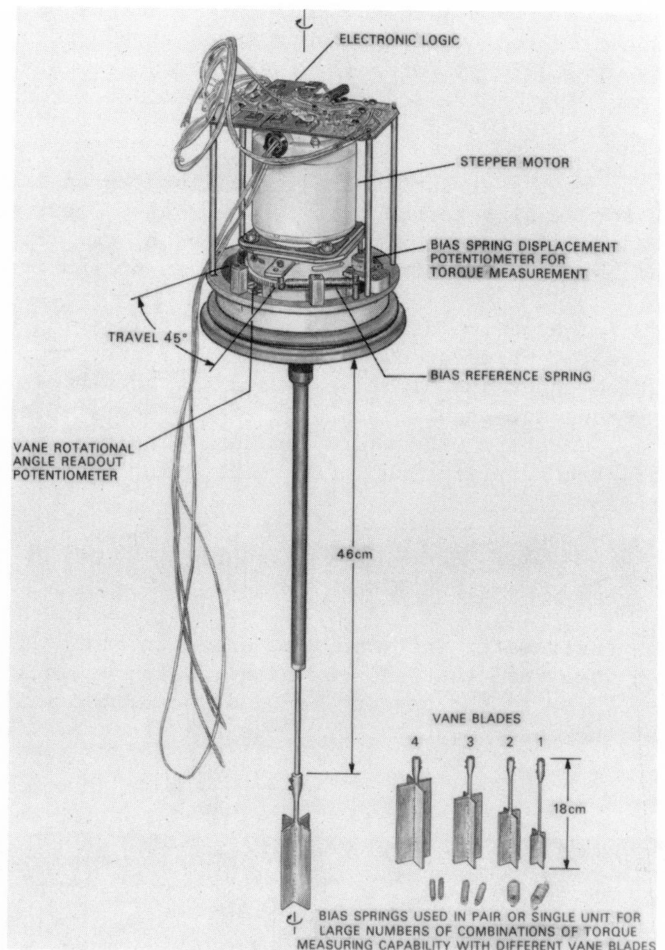

Figure 6. Vane shear meter internal

Other activities utilizing the remaining two-thirds or 149 of the operating hours, on the bottom, included search, inspection, object recovery, photography, biological studies, general visual observations, sample collection and system test and evaluation.

These operations ranged from 1.5 to 65 nautical miles off shore and from near the Mexican border on the south to the Santa Cruz Island area on the north.

VANE SHEAR METER

The instrument shown in Figure 5 is the vane shear meter. The four bladed vane, which has a height to diameter ratio of 2.5

to 1, can be slowly driven into the sediment to any depth up to 60 cm. The motor driven shaft is then rotated at a constant angular velocity of 0.025 rpm for soil strength measurements and 0.8 rpm for remolding of the sediment. The remolded strength is then measured using the 0.025 rpm speed again.

Figure 6 shows the arrangement using calibrated springs and potentiometer for sensing torque on the vane shaft. Measurements are usually made at four depths, 15, 30, 46 and 61 cm. Four vane sizes are available for shifting operating range of the instrument.

In Figure 7 the manipulator is viewed as it is being used to set the vane shear instrument out onto the sea floor, away from any disturbance which otherwise might be caused by the vehicle. Depth of the vane, torque and angular position are telemetered to the control console, on board ORB, where they are displayed on a multichannel graphic recorder, for real time monitoring of the experimental measurement.

CONE PENETROMETER

The cone penetrometer instrument is shown in Figure 8. The cone is driven down into the sediment at a constant velocity of about 20 cm/sec. As in the use of the vane shear, the manipulator is used to set the instrument out onto the sea floor away from the

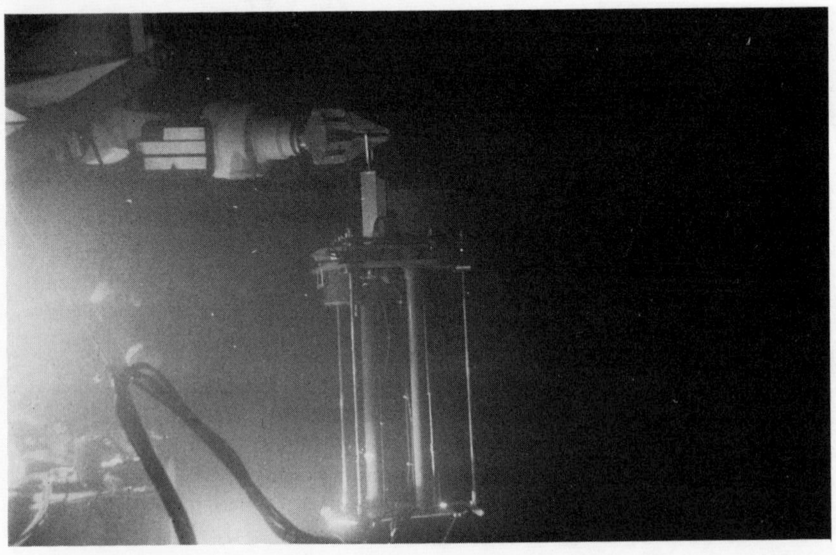

Figure 7. Manipulator deploys vane shear

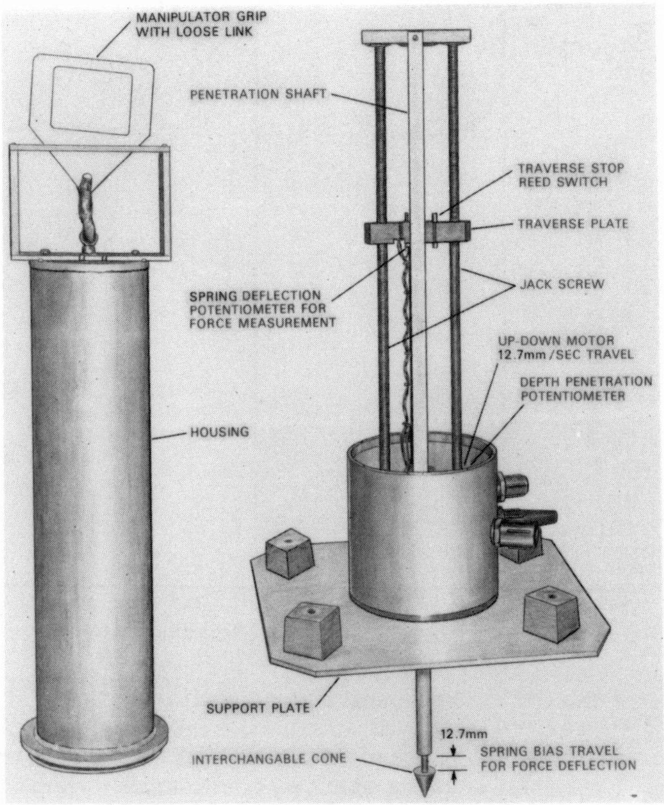

Figure 8. Cone penetrometer

vehicle. Depth of cone and force on the cone are telemetered to the console where, as with the vane shear instrument, they are displayed on a multi-channel graphic recorder.

Although it is generally agreed that the vane shear meter, compared to the cone pentrometer, appears to obtain results which are more easily related to soil shearing strengths, the cone penetrometer has two apparent advantages: the time to complete a set of measurements is at least an order of magnitude less than with the vane shear, and fine-grained structure, such as thin layering or an abrupt change in soil properties with depth, is much more readily detected with the cone penetrometer.

This fine-grained structure is dramatically illustrated in Figure 9. The three sets of data shown were all taken in the same general area in the San Diego Trough, off the mouth of the La Jolla Canyon. All three sets were taken within a 100-meter radius. The first set on the left shows four cone penetrometer profiles made with less than 1-meter horizontal separation between profiles. The

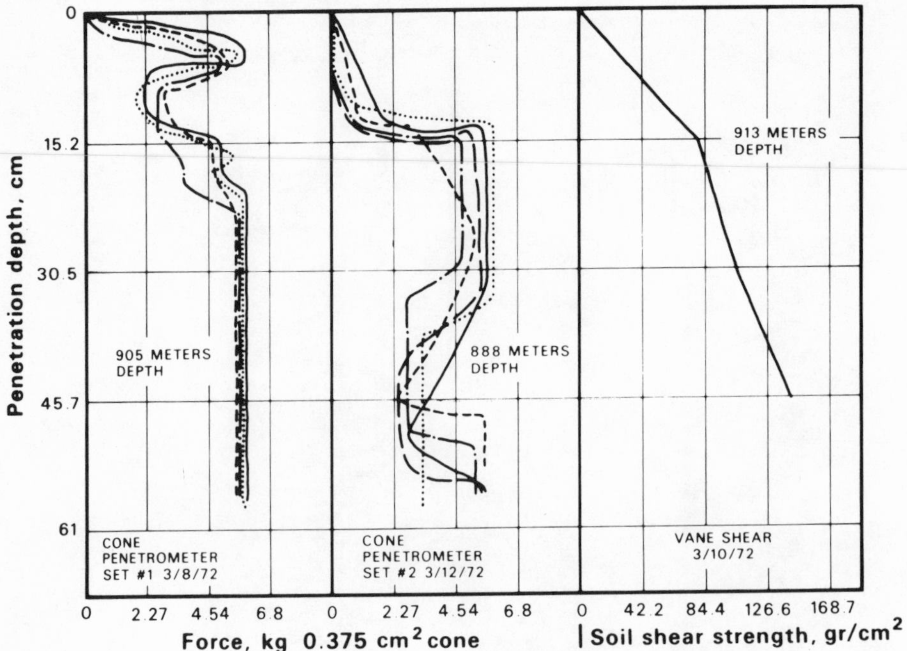

Figure 9. Cone-vane data comparison

second set, taken some 150 m away, contains five cone penetrometer profiles taken less than a meter apart. Considerable variation is observed between individual profiles within the sets. The average profile differs greatly between the two sets. The third profile obtained by taking vane shear measurements at 15, 30 and 46 cm depths, not far from the site of the first cone penetrometer set shows nothing of the layering observed in the cone data. In addition to the yield of detailed structual information, the cone has the advantage that it takes only 2 minutes to complete a profile with the cone penetrometer compared to 35 minutes for a profile with the vane shear instrument. Or stated another way, during 1972, 5.9 hours of on-bottom time were devoted to making 67 cone penetrometer profiles while 45.5 hours were devoted to making 29 sets of vane shear measurements.

This experience would indicate that it would probably be most efficient to survey with the cone penetrometer, bringing the vane shear into play for reference measurements at appropriate sampling points or whenever an anomaly is noted.

The RUM vehicle, as presently configured, cannot accommodate both instruments at the same time, so it must be returned to the surface for a change of instruments. This is a deficiency which can be rectified as soon as time and resources permit.

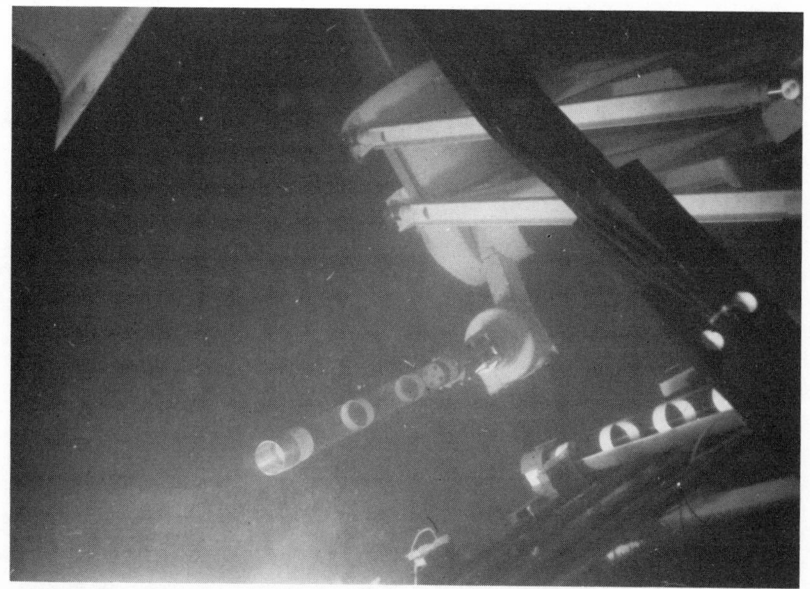

Figure 10. Manipulator deploys core barrel

Figure 11. Winch tensiometer system

CORE SAMPLING

In order to provide the opportunity for laboratory analysis on sediment samples taken from the same locations as the in-situ measurements, the RUM has been equipped to take core samples as shown in Figure 10. The coring tubes, which are fitted with a special handle designed to fit the manipulator grip, are carried in a storage rack on the vehicle. They are removed, deployed, inserted into and withdrawn from the sediment, and returned to the rack by the manipulator. The core tubes are 61 cm in length to match the depth range of the soil strength measurements. The core samples obtained have all been analyzed by Dr. I. Noorany at California State University, San Diego. His results are presented in Noorany and Luke (1972) and Noorany and Zinser (in press).

DRAWBAR PULL TESTS

The anchor-winch-tensiometer system for making drawbar pull measurements is shown in Figure 11. An anchor is deployed by the manipulator. Wire is played out as the vehicle is driven ahead. The winch is then locked, the anchor set, and drawbar pull force, developed by the vehicle as it is driven ahead, is measured as a function of track pressure. In the past, the anchor deployment scheme used permitted only one set of drawbar pull measurements to be made per vehicle lowering. Improvements have been incorporated which will make multiple sets of measurements possible in the future.

During the first drawbar pull measurements, it became apparent that silting up or clogging of the track cleats greatly reduced drawbar pull capability. In the highly plastic silt of the canyon and trough bottoms, off San Diego, this appears to have a more profound effect on vehicle drawbar pull capability than does track pressure. This phenomenon is illustrated in Figure 12.

Data from four sets of measurements are shown. Drawbar pull, on the left and drawbar coefficient, which is the pull divided by total track area, at the right, are plotted as a function of track pressure. In graphs numbered 1 and 2, the tests were conducted starting almost at the point of set down on the bottom with very little prior maneuvering. The initial pull test with clean tracks is over 55% higher than successive tests, in both cases. In number 3 where a small amount of maneuvering was carried out prior to the test the initial pull was 47% higher than the one to follow. In the last case where the track cleats were badly clogged from extensive traverse prior to the test, the initial pull test and the one to follow are identical and less than 24% higher than the lowest of the set. The RUM is driven ahead one vehicle length at the proper track pressure just prior to each pull test, including the initial test, in order to avoid the influence of any sediment disturbance caused by vehicle impact on set down, or previous pull tests. The sediments from core samples taken at the four test sites were all classified as clay of high plasticity by Dr. I. Noorany following laboratory analysis (Noorany and Luke, 1972).

A jet pump system for cleaning the tracks during drawbar pull tests was installed on the vehicle but problems with the pump forestalled its use.

Numerous problems have been encountered in attempts to make drawbar pull measurements, not the least of which were related to deployment and setting of the anchor and tending the wire. The anchor handling problem now appears to be solved. However, there is still a need to implement a more accurate means of determining bottom slope and the horizontal force component in the taut tether

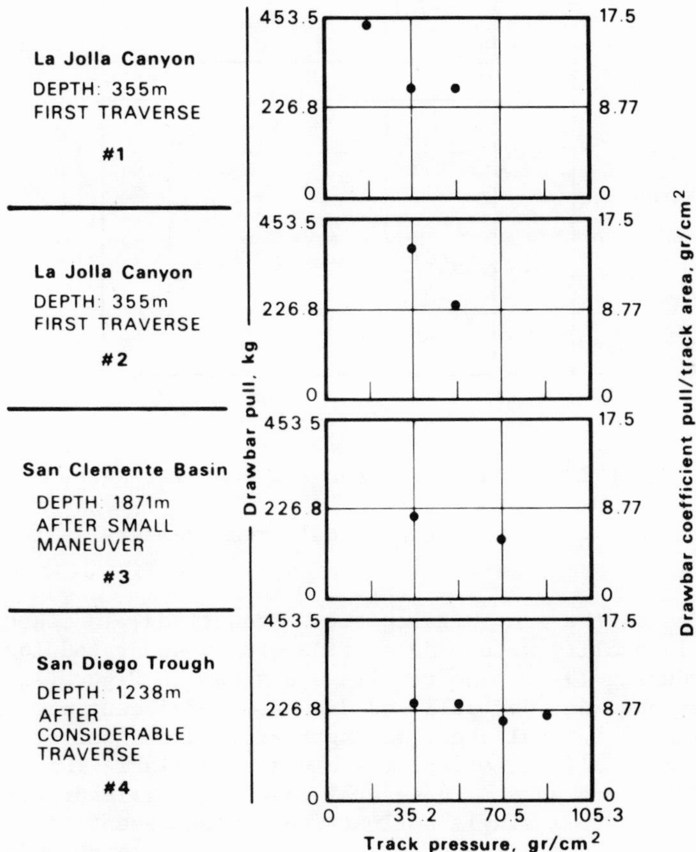

Figure 12. Drawbar versus track pressure

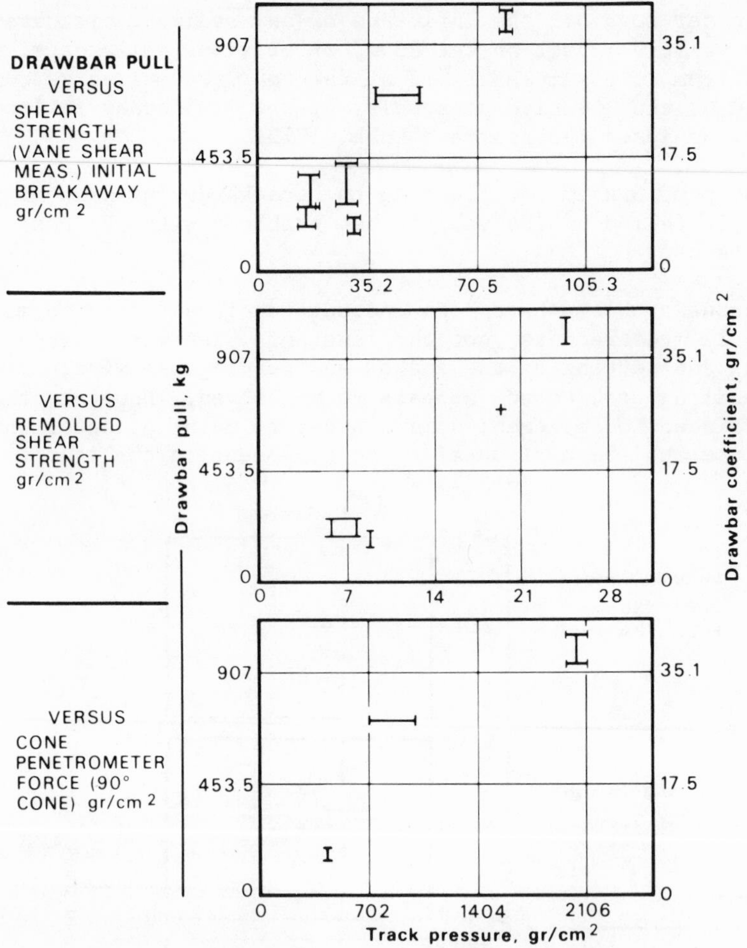

Figure 13. Drawbar versus sediment strength

cable if the quality of the drawbar pull measurements is to be improved.

In spite of the many difficulties encountered, 13 measurements of acceptable quality were made at five widely separated locations and depths during 1972. The results are shown in Figure 13. In the three graphs, drawbar pull and drawbar coefficient are plotted as a function of initial shear strength, remolded shear strength and cone force. All three factors were not available for each of the drawbar measurements. In several cases, multiple drawbar measurements relate to a single soil strength measurement, in others, multiply soil strengths relate to single or multiple drawbar measurements. This is the cause for the spread observed. As one would expect, even with this limited amount of data, drawbar coefficient appears to correlate best with remolded shear strength. This data would further indicate that drawbar coefficient is approximately equal to remolded shear strength at the low end of the range and at least 50% higher at the high end. This may be attributable to the effectiveness of the track cleats in the stronger sediments where there was less tendency for the tracks to become clogged.

TRACK DEPRESSION PROFILER

The track depression profiler, shown in Figure 14, is another instrument developed under this program as an aid to trafficability study. It is a 3.5 MHz short range high resolution scanning sonar. The profiler is mounted at the rear of the RUM, just behind one of the tracks, where it scans across the track depression left in the sediment by the vehicle traverse. A profile of the track depression is developed which has a 0.6 cm vertical resolution and a 2.5 cm horizontal resolution. Currently, for reasons of expediency, the sonar scans in polar coordinates and is displayed in rectangular coordinates. A correction must therefore be applied to reproduce the true track depression. Although implementation of a corrected display should be a relatively simple task, the lack of time,

Figure 14. Track depression profiler

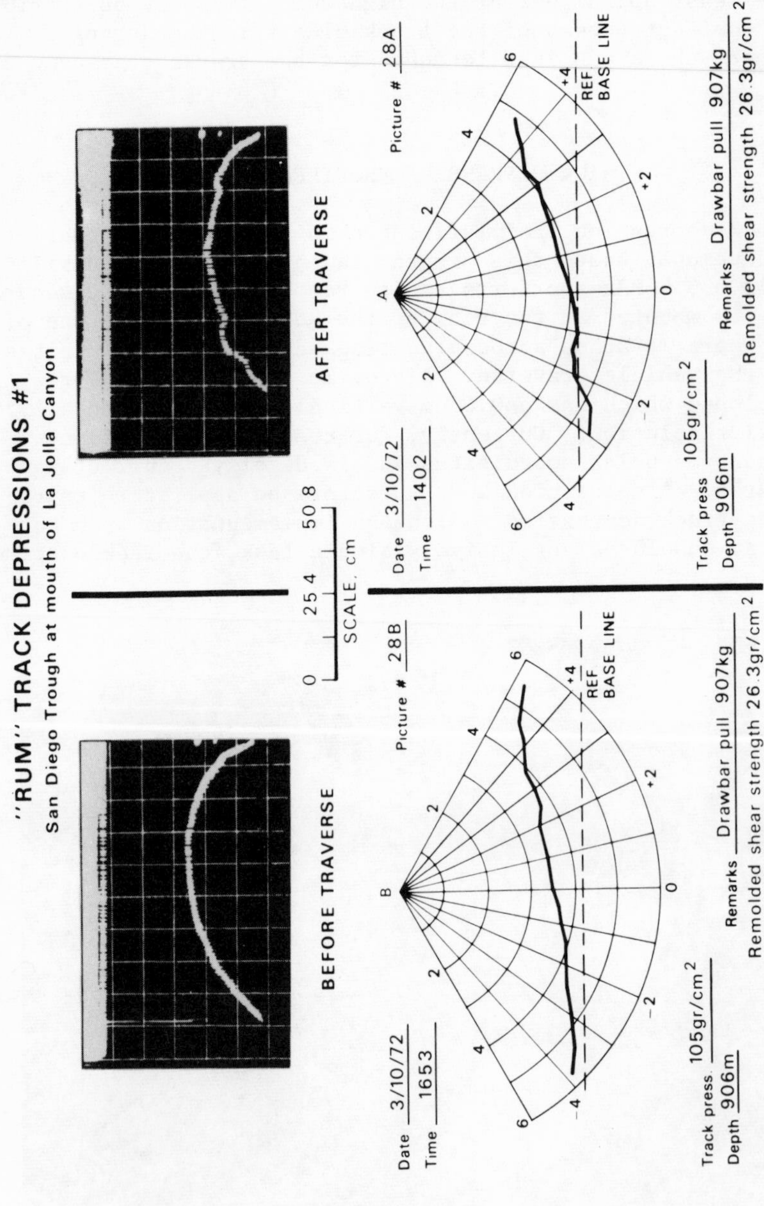

Figure 15. "RUM" track depressions #1 oscillographic display and replotted profiles

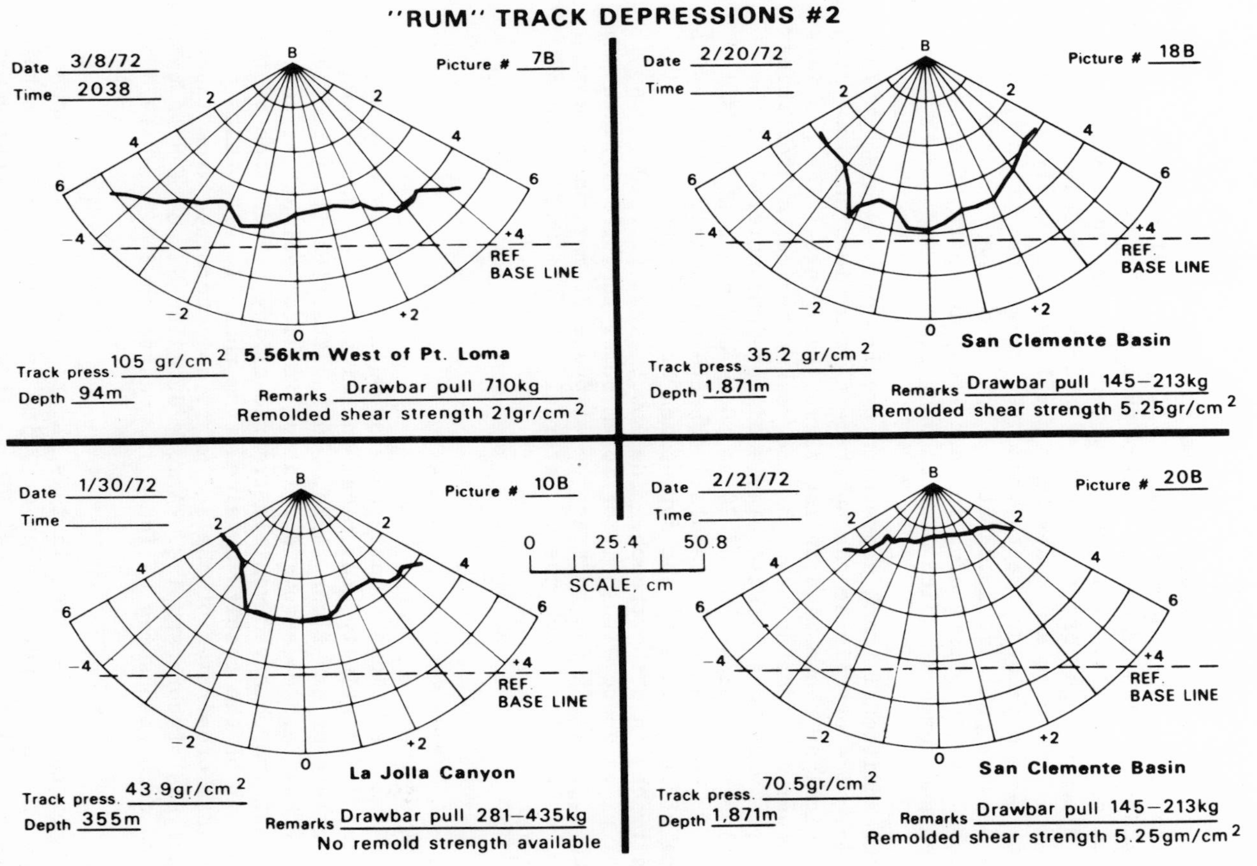

Figure 16. "RUM" track depressions #2 replotted profiles

resources and the pressures of other priorities have thus far prevented it. At present photographs of the display are read and replotted to obtain corrected data.

The replotting is illustrated in Figure 15. The two pictures at the top are of the uncorrected oscillographic display of profiles, before and after RUM traverse. Below them are the corrected replotted track depressions. This traverse was made in the same area of the San Diego Trough, off the mouth of the La Jolla Canyon, where the cone penetrometer and vane shear profiles, previously shown in Figure 8, were obtained. Depth, measured vane shear strength and vehicle drawbar pull are noted at the bottom of the illustration. Trafficability was outstanding in this area, in spite of a number of ridges and hummocks which were encountered.

Four additional track depression profile plots are shown in Figure 16. The profile at the upper left is a clean shallow track print typical of those made in the firmer sediments. The profiles at the lower left and upper right are rather typical of footprints made in the very soft plastic clay sediments encountered on the canyon floors. Depressions of 20 to 30 cm depth are not at all uncommon. The humps and irregularities are due mainly to material sluffing off the sides of the depression or from material falling off the tracks into the trench. The lower right profile shows material piled up to within 15 or 18 cm below the profiler scanning transducer. This is due to traction slippage at maximum drawbar pull. The slipping track pushes material out from under into a pile at the rear.

SUMMARY

The sea operations to date have shown RUM to be a very useful, versatile and relatively economical facility for the performance of a wide variety of sea-floor tasks. In the trafficability studies a beginning has been made toward relating in situ sediment strength measurements to observed vehicle trafficability and to look at the relationship of the results obtained from these in situ tests to the physical properties of the sediments as determined from laboratory analysis of core samples. A more detailed discussion of the tests and comparisons may be found in Anderson and Gibson (1970 and 1972).

As pointed out earlier in this paper, experience has shown that the vane shear meter is a very time consuming means of obtaining in situ sediment strengths. Thus, it would be to great advantage if the cone penetrometer could be used as the principal survey instrument.

FUTURE PLANS

Looking toward the future, it may even be feasible to conduct surveys with acoustic devices if sufficient data base can be accumulated to establish the correlation between the engineering properties and the acoustic properties of sea-floor sediments. Work currently being carried out at the Naval Undersea Center and the Applied Research Laboratory at the University of Texas, should contribute greatly to the investigation of this approach.

The RUM is currently being outfitted with a rock drill capable of obtaining 3 m long 5 cm diameter documented, oriented core samples. This drill should become operational before the end of 1973.

It is hoped that sufficient operational support for this unique RUM/ORB facility can be found to provide for its cooperative use by a variety of researchers having needs for the performance of sustained work upon the sea-floor. Two such cooperative studies have already been planned for 1973. During the first week in July, the Marine Physical Laboratory and the Lockheed Ocean Laboratory have planned an intensive trafficability study to be carried out in the San Diego Trough, at about 1200 m depth, in an area where a good deal of base line data is available regarding the engineering properties of the sediments. In October and November it is planned to use RUM to carry out a highly sophisticated benthic biological study which will involve a team of noted researchers from a number of institutions. This study also will be carried out in the San Diego Trough at approximately 1200 m depth.

REFERENCES

Anderson, V. C., R. Clinton, D. K. Gibson, and O. H. Kirsten, Instrumenting RUM for in situ sub-sea soil surveys, in Symp. Underwater Soil Sampling, Testing, and Construction Control, Am. Soc. Test. Mat. Spec. Tech. Publ. 501, pp. 216-231, Phila., 1972.

Anderson, V. C., and D. K. Gibson, RUM-ORB operations January 1970, Scripps Institution of Oceanography Sea Floor Tech. Rept. No. 1, Ref. No. 70-13, 1970.

Anderson, V. C., and D. K. Gibson, RUM-ORB operations March through August 1970, Scripps Institution of Oceanography Sea Floor Tech. Rept. No. 2, Ref. No. 71-29, 1972.

Clinton, J. R., Soil mechanics with the RUM-ORB sea floor work system, Scripps Institution of Oceanography Sea Floor Tech. Rept. No. 4, Ref. No. 72-63, 1972.

Noorany, I., and G. L. Luke, Engineering properties of sea-floor sediments off the coast of San Diego, Dept. of Civil Eng., Calif. State Univ., San Diego, Res. Rept., 1972.

Noorany, I., and R. A. Zinser, Engineering properties of sea-floor sediments from La Jolla Canyon, in Proc. 13th Intern. Conf. Coastal Eng., Vancouver, (in press).

Relationships Between Physical, Mechanical and Geologic Properties

ERODIBILITY OF FINE ABYSSAL SEDIMENT

JOHN B. SOUTHARD

Massachussetts Institute of Technology

ABSTRACT

A fundamental approach to erodibility of fine abyssal sediment must take into account the great complexity of forces acting on surface sediment particles exposed to bottom currents. Important controls on these forces involve the dynamics of turbulent boundary-layer flow, the physical and chemical nature of the sediment particles, and the past history of the sediment, including effects of flocculation and the activity of benthic organisms. Time scale also determines, at least in part, the way we view the problem of erodibility. Experiments on erodibility have largely been aimed at threshold velocity or boundary shear stress for incipient erosion. Many such experiments have been made, but comparison is difficult because of differing sediment composition and state of the sediment bed. The only general conclusion that can be drawn is that, for a given sediment, threshold velocity or shear stress varies strongly with water content but depends also on method of sediment preparation and time since deposition. Mode of erosion also varies with water content.

INTRODUCTION

The deep ocean floor is a repository for terrigenous and biogenous sediment and must therefore be largely depositional; in fact, until not too many years ago it was viewed as a totally depositional and very dull place. But we now have good evidence that substantial areas of the deep ocean floor have been sites of protracted erosion over geologic time scales (e. g., Johnson, 1972, and various of the Initial Reports of the Deep-Sea Drilling Project). Thus, although it many seem more important to study controls on modes and rates of deposition rather than erosion, erodibility of abyssal sediment is a geologically significant problem. Also, of course, aside from its geological significance, erodibility is of interest to those involved in locally altering the normal bottom-current regime for whatever scientific, technological, or economic reason. In many respects, related both to fluid dynamics and to the nature of fine sediments, the two problems of erosion and deposition call for different approaches; in this paper we will focus mainly on erosion.

Erosion of fine sediment involves the action of fluid forces on surficial particles or aggregates thereof, of certain sizes, shapes, and compositions, in overcoming both particle weight and interparticle cohesive forces that tend to keep the particles in contact with, or attached to, underlying material. In the following section we review some important aspects of the problem: (1) what we mean by erodibility, (2) the nature of fine abyssal sediments and of erosive or resisting forces; and (3) how the erosion of fine sediment differs from that of coarser sediment. In the latter part of the paper we review briefly some experimental results on erosion of fine sediments.

REVIEW

Erodibility

By erodibility we mean the behavior of a sediment surface when subjected to an erosive current. In one respect this is not a very satisfactory definition, because only in the case of very fine sediment is an entrained particle carried a great distance away; grains of coarse sediment are carried only short distances even if put into temporary suspension by the current, so that both erosion and deposition are equally involved at any point on the sediment surface. There are four principal aspects of erodibility: (1) the lowest velocity, termed the critical velocity or threshold velocity, which causes erosion, (2) rates of erosion at given velocities greater than the threshold velocity, (3) size, nature, and location of particles or larger solid units removed from the sediment surface, and (4) form or geometry of the eroded surface. Of these, most attention has been given to threshold velocities. A disadvantage in using velocity to characterize the strength of the current acting on a sediment surface is that time-average velocity must increase away from the sediment-fluid interface from zero to some maximum or free-stream value well away from the sediment surface, and the nature of the velocity distribution depends upon both the strength of the flow and the detailed geometry of the sediment surface. An alternative is to specify the average shear stress exerted by the flow on the sediment surface, except that more about both flow and fluid must be specified in order to characterize the spatial and temporal variation of the boundary shear stress.

Sediments

For the purposes of a general discussion of erodibility, most fine abyssal sediments can be viewed as mixtures of three components: clay-mineral particles, skeletal remains of planktonic plants and animals, and terrigenous silt, volcanic ash, and extraterrestrial debris carried by the wind. Clay is derived either from the continents or from in-place alteration of volcanic ash. The distribution of most of the clay seems to be controlled by continental sources (Biscaye, 1965; Griffin et al., 1968). All the major compositional groups are well represented in the oceans. Individual clay-mineral particles are mainly in the size range of small fractions of a micron to a few microns. Flocculation state when deposited is not well known. Calcareous and siliceous hard parts of planktonic plants and animals form a significant part of much of abyssal sediment; their distribution is controlled both by surface productivity and by abyssal solution. Particle size varies greatly, depending on the organism, but is typically between fine silt and fine sand. Atmospheric transport accounts for the presence of quartz, feldspar, some clay minerals (mainly in the coarsest clay and finest silt sizes), and meteoritic dust. These components, ubiquitous but nowhere as abundant as clay minerals, commonly form up to several percent of abyssal sediments. Since it is largely clay-mineral content that gives fine sediments their distinctive characteristics, a highly generalized but satisfactory view of abyssal sediment, from the standpoint of erodibility, is that of clay-mineral material everywhere diluted by at least small percentages of coarser and more equant material characterized by much weaker interparticle effects. Where percentages of coarser material are small, or perhaps even moderate, the sediment must behave much like pure clay sediment. Where content of coarse material is very high, the controlling erosional effects must be those of the more familiar coarse sediments.

Why are the erodibility characteristics of clay sediment so distinctive? Most clay particles have the shape of plates or flakes whose largest dimension is one or two microns or less. Particles of this size form colloidal systems, in that they are much larger than atomic dimensions but so small that forces acting between particles completely overshadow the gravity forces acting on individual particles. Mainly because of ionic substitutions in the aluminosilicate crystal lattice, clay particles typically carry negative charges on their faces. When the particles are dispersed in water, these charges attract a swarm of cations to the surface of the particle to form what is known as the electric double layer. Because of the thermal motion of both the cations and the surrounding water molecules, and the shielding effect of cations close to the particle surface, the density of positive charge decreases continuously outward for some distance from the surface of the particle. Because of this surrounding cloud of positive charges, individual particles tend to repel one another.

In sea water, with its high concentration of cations, the double layer is effectively thinner than in pure water, and the repulsive forces are too weak to prevent flocculation caused by attractive forces of an atomic nature between particles independent of the nature of the solution. The kinetics and dynamics of flocculation depend not only upon the magnitude and distribution of charges around the sediment particles but also upon the magnitude and distribution of local rate of fluid shear in the medium.

Flocculation is especially complicated for clays, partly because of particle shape but perhaps more importantly because the edges of clay-mineral particles tend to carry a different charge than the faces, so that edge-to-face flocculation as well as face-to-face flocculation can be important (van Olphen, 1963). We should thus expect that when condensed into a deposit the individual clay particles will be connected by a complex system of electrostatic forces. Interparticle packing arrangement and forces near the surface of an actual clay deposit must depend very strongly on such things as grain size and composition of the clay, state of flocculation of the sediment upon settling to the interface, and degree of subsequent rearrangement attendant upon disturbance by bottom-dwelling organisms and compaction by loading. Given our present state of knowledge, there seems to be little hope of formulating quantitative theories of erosion based on the details of interparticle forces at or near the sediment-water interface.

Hydrodynamics

Oceanographers commonly characterize the strength of bottom currents by measuring current velocity at a single level above the bottom, usually at a height of something like a meter or two. Only an indirect estimate of boundary shear stress can be obtained, even given velocities at several positions above the bottom (Sternberg, 1970). Experimentalists more commonly work with boundary shear stress than velocity, because it is easy to measure in uniform open-channel flow.

A great many measurements of current velocity near the deep-sea floor in the past few decades have shown that current speeds of 5 to 15 cm/sec are common, but not nearly as common as weaker currents of only a few centimeters per second (Heezen and Hollister, 1972). Much higher current speeds of a few tens of centimeters per second occur but are very uncommon in areas of fine sediment. Typical magnitudes are ultimately governed by the strength of the general oceanic circulation, which, being dependent on both the climatic state of the earth and the configuration of continents and oceans, must have varied substantially in the geological past. There seems to be some truth in the view that nature has presented us with an oceanic circulation that works with just such an intensity as to put abyssal bottom-current velocities largely in a midrange between nonerosive and strongly erosive. If velocities were typically less by a factor of two or three there would be little erosion, and if they were greater by the same factor the ocean bottom would be a much more active place than it is.

Even for bottom-current velocities far too low to erode sediment, flow structure near the bed is that of a turbulent boundary layer. Provided that bottom roughness elements are not too large, such flow is characterized by a thin layer adjacent to the boundary, called the viscous sublayer. Here, in contrast to the rest of the boundary layer, vertical exchange of fluid momentum by turbulent eddies is unimportant, because vertical components of fluid velocity are weak, and the shear stress in the sheared fluid is developed dominantly by molecular motions. Since momentum exchange is less effective by molecular than by turbulent motions, the velocity gradient in this zone is very steep. The flow is not truly laminar, however, because random fluctuations in velocity (and therefore also in instantaneous fluid forces on exposed sediment particles), though not the dominant control on flow structure, are substantial. The viscous sublayer passes outward into a thin zone of small-scale turbulence which grades in turn into the main part of the boundary layer, characterized by large but weak eddies. Since bottom irregularities on the scale of sediment particles or potentially erodible aggregates (whether on flat sediment surfaces or located on the large roughness elements commonly observed in abyssal bottom photographs) are certainly always well within the local viscous sublayer, consideration of forces and motions in the viscous sublayer is essential in studying erodibility of fine sediment. This is an active area of research in fluid dynamics (Kline et al., 1967; Corino and Brodkey, 1969) and certainly represents one of the most fruitful avenues for fundamental advance in our understanding of erosive processes.

A fluid flowing over a sediment particle exerts both normal and tangential forces at all points of the surface. If we knew how to find the distribution of these forces we could integrate to find the magnitude, direction, and line of action of the resultant force on the particle. Unfortunately this is difficult or impossible even for the simplest of geometries and flow conditions. Since we must deal with irregular particle geometry, irregular surface grain packing, and a complexly unsteady and nonuniform local flow pattern, there is no hope of obtaining exact solutions. There have been a number of attempts at rational analysis of the conditions for initiation of grain motion; these have usually involved a balance of opposing forces and moments on representative surface grains (ASCE, 1966). Because of the inherent irregularity and complexity of the problem, this sort of approach can do nothing more than provide a general framework, even in the simpler case of noncohesive sediments. Intelligent empiricism has been the only recourse in dealing with actual sedimentological or engineering problems.

Shields (1936) developed the first rational approach to graphical representation of initiation of motion. This approach, used extensively by both engineers and geoscientists, can be summarized by assuming that the variables important in governing the state of sediment movement are boundary shear stress, τ_o, grain size and submerged specific weight, d and γ_s, and fluid density and viscosity, ρ and μ. Dimensional analysis then supplies two independent variables, usually the roughness Reynolds number:

$$R_b = \tau_o^{1/2}\rho^{-1/2}d/\mu$$

and the Shields parameter:

$$S = \tau_o/d\gamma_s$$

(though R_b and ρ^2gd^3/μ^2 is perhaps a better set, because then τ_o is not plotted against itself). A graph of these two dimensionless variables showing a curve separating states of sediment movement from those of no movement is called the Shields diagram. There is considerable scatter of experimental points around this curve, which can be partly eliminated in the analysis by replacing γ_s by the two variables sediment density ρ_s and acceleration of gravity g to take into account relative-density effects as well as simply particle weight, and replacing τ_o by flow depth h and flow velocity U to better characterize the state of the flow.

In the part of the Shields curve for which grain size is large and interparticle forces are negligible compared to particle weight, past history of the deposit has no important effect on sediment transport, and a single curve suffices to describe erodibility. In the fine-sediment part of the diagram, however, various interparticle effects are important, especially in clay-rich sediments. Such partially interrelated things as time since deposition, water content, and grain arrangement, as well as grain size distribution and mineral composition then become very important in governing erodibility, and the Shields curve for a given sediment can lie anywhere within a wide envelope. The Shields diagram is thus not nearly as useful for fine sediments as for coarser sediments. The graph of grain diameter versus current velocity for erosion, transportation, and deposition of bottom sediments developed by Heezen and Hollister (1964, 1972) from various theoretical and experimental sources brings out this same effect.

Time Scales

Although we concentrate on the experimental approach in this paper, flume studies of fine-sediment erosion are immediately applicable only to problems of man-induced local erosional events, and may be irrelevant to problems of long-term erosion on the sea floor. We can discuss erosion of abyssal sediment on two entirely different time scales: laboratory time (hours, days, or, at the very most, months), or geological time (tens or hundreds of thousands of years, or even millions of years). Bottom photographs show good evidence of erosional features (Heezen and Hollister, 1971); since these are not obliterated by bottom-dwelling organisms, erosion must be on relatively short time scales. We also have good evidence for, but no way of directly observing, erosion on geological time scales at rates that, on the average, are orders of magnitude lower than the lowest rates that can be measured reliably (or possibly even produced) in the laboratory. The relation between processes at these two time scales is not at all clear. It is possible that geologically slow erosion has been governed not by steady erosion of the sort we observe in the laboratory but instead by either unusual erosional events that happen a large number of times over a long time span, or by the long-continued activity of bottom-dwelling animals. Little can be said about the former possibility; as for the latter possibility, the effect of animals on abyssal erosion has received little attention but must be a significant if not predominant factor. Except in anaerobic environments, even at the greatest depths the sea floor is populated by large numbers of such animals as worms, mollusks, and crustaceans that ingest sediment in order to extract food. In typical abyssal areas with low rates of deposition, each sediment particle must be cycled through an animal many times before it becomes buried so deeply that it is out of reach. Erodibility of the surficial sediment, and other physical and mechanical properties as well, must therefore be governed very strongly by the bottom fauna. Moreover, sediment must to some extent be resuspended directly by animals rather than by currents alone; in this way there could be slow net erosion in an area affected by only weak bottom currents.

EXPERIMENTAL WORK

Most experimental work on erosion of cohesive sediments has been done by civil engineers, who have been interested mainly in design of stable channels in fine cohesive soils but also in estuarial silting problems. There has been some, but much less, work by geologists and sedimentologists, from the standpoint of both erosive activity of turbidity currents on soft mud bottoms and the behavior of the sediment surface in the oceans. As noted above, theory is intractable, and only a few general things can be said; only Partheniades (1965) has ventured the attempt to work out phenomenological theories in this area, with limited success. In this section we review briefly some of the experiments and conclusions most relevant to erosion of abyssal sediments.

What are some of the important questions we might hope to answer from laboratory experiments on erosion of fine sediment? By means of well designed experiments we ought to be able to get some idea about two of the more obvious questions: what current velocity is needed to erode a given sediment, and how does rate of erosion depend on current velocity? Much of the experimental work that could be reviewed has been concerned with these questions. Behind these questions is another that is far less tractable, though it too has been the concern of many experimental studies: what properties of a sediment determine or characterize its erodibility? A broader question which has received relatively little attention is: what bearing do laboratory experiments on erodibility have on abyssal erosion problems? One aspect of this last problem which severely limits the value of flume studies is that it has so far been impossible to transfer even a small segment of sediment-water interface from nature to the laboratory without some disturbance of the sediment, and any kind of reconstitution, remolding, or redeposition of the sediment introduces major uncertainties about comparability of results, because erodibility is so much a function of past history rather than simply grain size and mineral composition. An obvious way to get around this problem is to take the flume to the ocean bottom rather than the other way around; little work seems to have been done along this line. Another aspect of this same problem, discussed in the last section, is what bearing short-term laboratory experiments have on geologically slow erosion.

A large number of experimental studies have been made to determine either critical erosion velocity or rate of erosion for clay-rich sediment or soil. These have varied widely in various aspects of experimental materials or technique: in experimental apparatus (straight flumes, circular flumes, scouring jet), erosive medium (fresh water or salt water), sediment composition (natural clay-rich sediments, pure clays, artificial mixtures), and sediment state (natural, remolded, resedimented, artificially consolidated). In accordance with their goals, most of these studies have involved fresh-water systems and have dealt with sediment having a much higher degree of consolidation than is typical of abyssal sediments; these studies are well summarized by the American Society of Civil Engineers (ASCE, 1968). Less attention has been given to mode of erosion and to geometry of the eroded surface. We will discuss here only those studies dealing with erosion of natural sediment by sea-water flows.

Threshold velocity or bed shear stress for erosion of clay-rich sediment by sea-water flows has been studied by Postma (1967), Migniot (1968), Southard et al. (1971), and Lonsdale and Southard

(1973). The most extensive experiments are those of Migniot (1968); six muds differing considerably in both grain size and mineral composition were used, though only one set of experiments in sea water was reported. Postma (1967) used estuarine mud, Southard et al. (1971) used calcareous ooze, and Lonsdale and Southard (1973) used abyssal clay. Postma's experiments were made in a small circular flume; the others were made in recirculating straight flumes.

Direct comparison of results of such experiments is difficult. In studies with coarse sediment, only sorting and grain shape (aside from differences in flow system) stand in the way of comparability, and it is usually tacitly assumed that these effects are not so important as to preclude direct comparison of different flume studies. But fine sediments almost invariably have several compositional components, each with a potentially substantial effect on erodibility, and other properties of the sediment, such as water content and

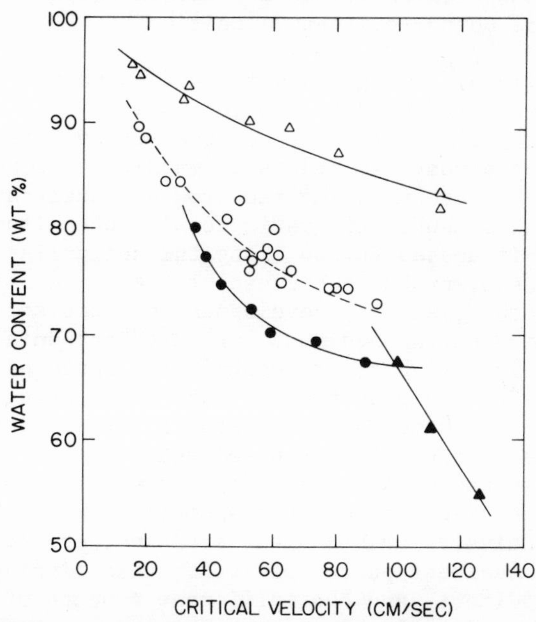

Figure 1. Data on weight-percent water content (weight of water in sample divided by total wet weight of sample) vs. threshold velocity for experimental sea-water erosion of calcareous ooze (Southard et al., 1971). Runs were made in a recirculating rectangular channel 6 m long and 17 cm wide with a uniform flow about 6 cm deep. Measured flow velocities were approximately corrected for effects of flow depth and water temperature; velocities shown on the graph are equivalent to velocities that would be measured 1 m above the ocean bottom. Time after preparation of the sediment as a slurry in the flume is shown by the symbols: open circles, 1 hr; solid circles, 1-2 hr; open triangles, 2-10 hr; solid triangles, 10 hr.

grain arrangement, which are only partly related to mineral composition, are of even greater importance in determining erodibility. Each study of this sort therefore largely stands by itself, and can be valuable not so much for specific values obtained but for any general trends revealed. There has not yet been any extensive and systematic study of the effect of grain size and clay-mineral composition on erodibility under controlled conditions of sediment preparation.

Figures 1 and 2 show results from these four studies. The only general conclusion that can be drawn is that, for a given mode of sediment preparation, threshold velocity or shear stress varies strongly with water content. For a range of water content substantially higher than is typical of abyssal sediments, erosion velocities are not greatly different from typical bottom-current velocities, but for values of water content more nearly representative of oceanic conditions erosion velocities are an order of magnitude greater than typical bottom-current velocities.

Partheniades (1965), using sea water and estuarine mud in a large recirculating flume, studied erosion rate as a function of bed shear stress for two different states of the sediment that differed in shear strength by almost two orders of magnitude: remolded but at natural water content, and resedimented from a concentrated flowing suspension. Despite somewhat enigmatic results occasioned by localization of erosion in a deep furrow along the centerline of the flume, for some reason, probably because of the effect of sidewalls on turbulence structure, and the development of some sort of stabilized crust elsewhere on the sediment surface, Partheniades' experiments demonstrate clearly that, at least in the range of shear strengths used, there is a wide range of bed shear stress above the threshold condition for which erosion rate is largely independent of the internal shear strength of the sediment, but that above a certain shear stress the erosion rate depends strongly on shear strength. Though no visual observations could be made, this suggests that, at relatively low bed shear stresses, erosion involves removal of surface sediment particles rather than ripping up of larger chunks of sediment, and that different aspects of particle arrangement and interparticle forces are involved in such removal of surface particles than those which come into play in resistance to internal shearing in the sediment. Unfortunately, the water content of the resedimented bed was not reported, so there is no way of knowing whether water content was similar for these two states of the sediment. The occurrence of two different modes of erosion has also been noted by Migniot (1968) and Lonsdale and Southard (1973). In both cases, a sharp change in the curve of threshold velocity or shear stress vs. water content coincides with this change in mode of erosion as observed in the flume.

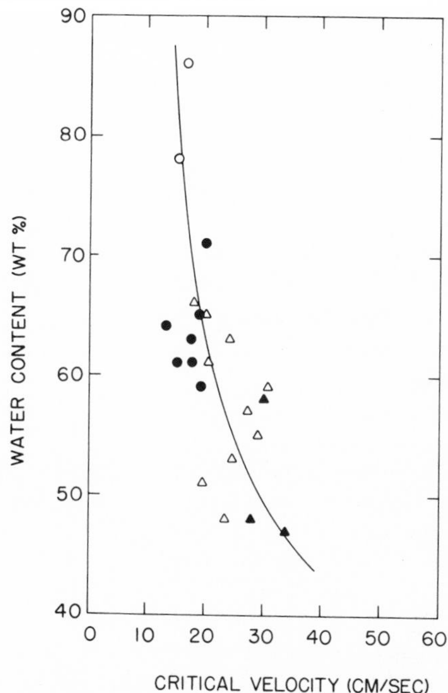

Figure 2. Data on weight-percent water content (weight of water in sample divided by total wet weight of sample) vs. threshold velocity for experimental sea-water erosion of nonmarine (?) clay (open triangles; Migniot, 1968), estuarine mud (open circles; Postma, 1967), and abyssal brown clay from the North Pacific (Lonsdale and Southard, 1973). Migniot's runs were made in uniform flows 10 cm deep in a rectangular recirculating flume 12 m long and 40 cm wide; velocities shown are mean flow velocity in the flume. The runs by Lonsdale and Southard were made in uniform flows about 7 cm deep in a rectangular recirculating flume 6 m long and 17 cm wide. Velocities shown are corrected to be equivalent to those that would be observed 1 m above the ocean bottom; actual mean velocities in the flume can be obtained by multiplying by 0.56. Postma's experiments were made in a small annular paddle-driven flume; velocities are approximately those of the main body of flow in the tank, but are difficult to compare with the other set of experiments.

REFERENCES

ASCE, Task Committee on Preparation of Sediment Manual, Sediment transport mechanics: initiation of motion, Am. Soc. Civil Engrs., Proc., J. Hydraulics Div., 92, 291-314, 1966.

ASCE, Task Committee on Erosion of Cohesive Sediment, Erosion of cohesive sediments, Am. Soc. Civil Engrs., Proc. J. Hydraulics Div., 94, 1017-1049, 1968.

Biscaye, P. E., Mineralogy and sedimentation of recent deep-sea clay in the Atlantic Ocean and adjacent seas and oceans, Bull. Geol. Soc. Am., 76, 803-832, 1965.

Corino, E. R., and R. S. Brodkey, A visual investigation of the wall region in turbulent flow, J. Fluid Mech., 37, 1-30, 1969.

Griffin, J. J., H. Windom, and E. D. Goldberg, The distribution of clay minerals in the World Ocean, Deep-Sea Res., 15, 433-460, 1968.

Heezen, B. C., and C. D. Hollister, Deep-sea current evidence from abyssal sediments, Mar. Geol., 1, 141-174, 1964.

Heezen, B. C., and C. D. Hollister, The Face of the Deep, Oxford Univ. Press, Inc., N. Y., 1971.

Heezen, B. C., and C. D. Hollister, Geologic effects of ocean bottom currents: western North Atlantic, in Studies in Physical Oceanography; A Tribute to Georg Wüst on his 80th Birthday, 2, edited by A. L. Gordon, pp. 37-66, Gordon and Breach, N. Y., 1972.

Johnson, D. A., Ocean-floor erosion in the Equatorial Pacific, Bull. Geol. Soc. Am., 83, 3121-3144, 1972.

Kline, S. J., W. C. Reynolds, F. A. Schraub, and P. W. Runstadler, The structure of turbulent boundary layers, J. Fluid Mech., 30, 741-773, 1967.

Lonsdale, P. F., and J. B. Southard, Experimental erosion of abyssal clay (manuscript), 1973.

Migniot, C., Étude des propriétés physiques de différents sédiments très fins et leur comportements sous des actions hydrodynamiques, La Houille Blanche, 591-620, 1968.

Partheniades, E., Erosion and deposition of cohesive soils, Am. Soc. Civil Engrs., Proc. J. Hydraulics Div., 91, 105-139, 1965.

Postma, H., Sediment transport and sedimentation in the estuarine environment, _in_ Estuaries, edited by G. H. Lauff, pp. 158-179, Am. Assoc. Adv. Sci., Publ. 83, Washington, D. C., 1967.

Shields, A., Anwendung der Ähnlichkeitsmechanik und der Turbulenzforschung auf die Geschiebebewegung, Preussische Versuchanstalt für Wasserbau und Schiffbau, Berlin, Mitteilungen, _26_, 1936.

Southard, J. B., R. A. Young, and C. D. Hollister, Experimental erosion of calcareous ooze, J. Geophys. Res., _76_, 5903-5909, 1971.

Sternberg, T. E., Field measurements of the hydrodynamic roughness of the deep-sea boundary, Deep-Sea Res., _17_, 413-420, 1970.

van Olphen, H., An Introduction to Clay Colloid Chemistry, Interscience, N. Y., 1963.

THE EFFECTS OF THE BENTHIC FAUNA ON THE PHYSICAL PROPERTIES OF DEEP-SEA SEDIMENTS

GILBERT T. ROWE

Woods Hole Oceanographic Institution

ABSTRACT

Benthic invertebrates are known to rework and modify sediments mechanically through such activities as burrowing, tube building and deposit feeding. The results are a variety of identifiable structures which are abundant both at the sediment-water interface and preserved in deeper layers. These "Lebensspuren" are always common in sediments, except where sedimentation rates or other physical processes have obscured them.

In the deep sea, invertebrate assemblages are depauperate in both abundance and biomass, and although the diversity of the communities is relatively high, metabolism appears to be extremely slow. These structural and functional characters of the communities might imply that faunal effects on sediments would be proportionately lower in deep water, but no evidence supports this. In deep depauperate regions where sedimentation rate and organic carbon are low, the fauna appears to have accentuated effects. Bioturbation may be the catalyst for most deep-sea sediment erosion. In anomalous deep deposits where organic-rich material has accumulated, either naturally or from man, animal abundance and activities are measurably increased. Time-lapse photography indicates the bottom fauna has pronounced catalytic effects on sediment erosion by currents at the head of Hudson Submarine Canyon.

INTRODUCTION

The sediment-water interface of the ocean is inhabited by a diverse and abundant invertebrate fauna. Many species markedly modify sediments through a variety of feeding activities and the construction of protective habitation. Animal activities often are manifest in the mass physical properties, erodibility, chemical composition and continued dispersion of sediments. In shallow water, many sedimentary products of extant animals are easily recognized in the fossil record. Although some information is available on the modifications attributable to shallow-living species, the effects of the deep-sea biota on sediments must be confined to conjecture. For the purpose of this paper, I have attempted to review, briefly, the history of biological sedimentology in shallow water, adding, where appropriate, information which might be important in the deep sea. I follow that with my own conjecture, based mostly on new knowledge of biological energy cycling, and some suggestions on how sediment-animal interactions could best be investigated in the deep ocean.

Ichnology is the study of the physical manifestations of benthic biota in sediments (Seilacher, 1953). A classification of sediment structures, based on ethology (animal behaviour) (Seilacher, 1953), includes: 1) resting traces, which are the imprints within the sediments of motile animals not in motion; 2) crawling traces, which are records of movement through sediment usually characterized by a "bow-wave"; 3) feeding structures, formed by deposit feeders in the process of engulfing whole sediment; 4) grazing traces, which are delicate surficial markings on the sediment water interface; and 5) dwellings, which are habitats of hemisessile suspension feeders, carnivores or scavengers. These different phenomena have different effects on the physical character of the sediments. According to Frey and Howard (1972), crawling is "transient" and results in considerable bioturbation (mixing) and destruction of any record of physical conditions during sedimentation; resting and feeding traces are "semipermanent" and therefore result in an intermediate degree of bioturbation, and dwellings are "permanent" and result more in "burrow mottling" of deposits rather than mixing. Frey and Howard worked with shallow-water species in aquaria, but all the above forms are common also in the deep-sea. "Grazing traces" are not common in shallow water, but they are one of the most common features of sediment surfaces in the deep-sea; this is probably because of the relative physical stability of the bottom and the fact that most of the biodegradable material is confined to the top millimeter or so (Heezen and Hollister, 1971).

The "Lebensspuren", or "records of life activity imparted to a substratum" (Hertweck, 1972), are said to have exaggerated environmental significance if the burrows are more common than the animals themselves. This is more usual for deep burrowing species in shallow dynamic environments. Deep burrows appear to be absent from the deep sea, but too little is known of lebensspuren frequency or population densities to confirm this.

The most prominent shallow water sediment modifications are

the tunnels of several families of modest-sized decapod crustaceans. These are dendritic excavations which may be as much as a meter deep and several centimeters in diameter. They are most common on intertidal mud or sand flats and have been described by numerous authors (Verrill, 1873; Pearse, 1912; Dembowski, 1926a, b; Stevens, 1928; Verwey, 1929; MacGinitie, 1934; Crane, 1941, 1943, 1947, 1958; MacGinitie and MacGinitie, 1949; Teal, 1959; Crichton, 1960; Bennett, 1968; MacNae, 1968; Phillips, 1971). In early studies tunnels were recreated by pouring plaster down the entrances at low tide (Stevens, 1928), but recently epoxy and polyester resins, which can harden under water (Burger et al., 1969), have allowed casting of sublittoral burrows by SCUBA-diving investigators (Shinn, 1968; Farrow, 1971; Dybern and Hoiseter, 1965; Chapman and Rice, 1971; Rice and Chapman, 1971). Similar experiments have not been attempted in deep water.

Polychaete annelid worms often make tubes, either of naturally occurring elements in the sediments, of some organic material they secrete, or a combination of both, rather than large excavated galleries. As a result, the worms' tubes can be more easily recognized as the products of particular species. The onuphid *Diopatra cuprea* embeds shell fragments in a membranous tube (Myers, 1970) whereas *Chaetopterus variopedatus* secretes a tube (Barnes, 1965). Both polychaetes are common and easy to recognize on intertidal mud flats. The polychaete *Pectinaria* cements quartz sand grains together into a precise, cone-shaped tube (Reineck, 1960; Fager, 1964; Gordon, 1966). Polychaetes of the family Serpulidae construct coiled calcium carbonate tubes (Neff, 1969), but these are usually confined to a hard substratum such as rock. Another polychaete tube builder, *Sabellaria alveolata*, can build massive reefs of small sand tubes (Richter, 1920, 1921, 1927; Galeine and Houlbert, 1922). No such structure would be likely in deep water because of the high supply of energy required to support such an abundant population.

Burrow builders or tube dwellers increase the stability of naturally thixotropic sediments and prevent them from collapsing. This modification ranges from agglutinating the linings of the galleries of the crustaceans with secretions (Weimer and Hoyt, 1964; Shinn, 1968; Farrow, 1971) to elaborate compartmentalization of chitinous tubes by highly specialized polychaetes. Pogonophora also live in septate membranous tubes, and they are usually only about 0.1 mm in diameter. These modifications of the physical structure of sediment add a resilience (Kennedy, 1967) often manifested in a fossil record, but continued maintenance is often required for habitation of the simpler burrows or they will collapse within several days (Rice and Chapman, 1971).

Movement through the sediments themselves has been most successfully mastered by bivalve molluscs and annelid polychaetes. The bivalve taxa which are good burrowers are able to work their muscular foot out into the sediments, anchor it by distorting its shape, and then pull their shells up to it (Ropes and Merrill, 1966; Pfitzenmeyer and Drobeck, 1967; Trueman, 1967, 1968a, b). For most annelids on the other hand, burrowing is accomplished by whip-like side-to-side action or peristaltic movements progressing down the length of the worm (Trueman, 1966; Wells, 1969). Burrowing has been investigated also in anemones (Ansell and Trueman, 1968), amphipod crustaceans (Engle, 1966; Sameoto, 1969) and scaphopod molluscs (Trueman, 1968c). Protobranchs, the most common bivalve order in deep water, are also the most accomplished burrowers in soft sediments.

Frequency of burrowing has been used to indicate historical benthic activity (Clarke, 1968). By dividing the number of burrows in a layer of sediment by its age, he found that burrowing maxima occur at 50,000, 90,000 and 160,000 years BP.

Another salient biogenous feature of sediments is invertebrate fecal material, usually in the form of castings or pellets and common in fossil as well as Recent deposits (Kuenen, 1961; Young, 1971). As the products of specialized feeding behaviors, some feces can be attributed by size and shape to particular species. Non-selective deposit feeders engulf sediments with little or no regard for its contents. They rely on their digestive tracts to remove organic nutrient, whether it is detritus or the microbiota living on the detritus. Castings of long coils or loops are attributed to this behavior, and it can be expected that the concentration of organic matter in these casts will be lower than in uningested sediments. Selective feeders also produce castings or pellets, but the remaining organic content cannot easily be related to the concentrations in uningested sediments.

Pseudofeces are materials not passed through the digestive system, but they are a byproduct of the feeding activity of filter feeders. If particulate matter in the water is too concentrated or of unsuitable composition, the material is shunted back into the environment (Moore, 1939; Haven and Morales-Alamo, 1966). As most animals are deposit feeders in the deep sea, pseudofeces will be uncommon. Beach deposits are often mottled by pellet-like material (Chakrabarti, 1972). The deposits are good indicators of paleo-beach deposits (Emery, 1953), but as deep burrows are not common in the deep sea, these deposits would not be expected there.

Bioturbation is sediment mixing by benthic animal activity. Its most salient result is the mixing of surface sediments with concomittant destruction both of precise strata of pelagic sediments and the structures constructed by the benthic fauna. This homogenation is absent only from regions where bottom water oxygen has been low enough to prohibit survival of benthos or where sedimentation rates have been extremely high. Quantification of the degree or rate of this reworking is difficult. Berger and Heath (1968) have attempted to quantitatively describe vertical mixing of sediments. Sedimented particles of a particular kind are mixed down through a homogeneous layer of continuously mixed surface sediments having a particular thickness. As this layer migrates upward with continued sediment deposition in time, particles of this particular species are left in what has become the historical layers below, which are no longer subject to mixing. Likewise, after the extinction of a species, it will be mixed upward too. With a species' first appearance, its concentration, because of mixing by the fauna, increases rapidly through the mixed layer. With extinction, the concentration decreases gradually, and significant contamination can extend a distance of about 6 mixed-layer-thicknesses above the level at which it became extinct.

Macrofauna are not the only modifiers of sediments. Controlling faunal composition by removing macrofauna, Cullen (1973) found that smaller species, called meiofauna and less than 1.0 mm in size, quickly destroy large lebensspuren made by macrofauna.

The pelleting of the sediments by deposit-feeding bivalves and polychaetes appears to increase resuspension by decreasing the physical stability of the bottom (Rhoads, 1973). The sediments of organic-rich shallow muds are probably reingested numerous times in the form of fecal material of some sort (Young, 1971). Coprophagy is common among benthic invertebrates (Johannes and Satomi, 1966; Frankenberg and Smith, 1967). This reuse and pelletization as well as continuous bacterial degradation accounts for the decrease in organic carbon and nitrogen at depth in the sediment column (Rittenberg et al, 1955; Trask, 1939; Hobson and Menzel, 1969).

Although macrofaunal intertebrates are the predominant sediment modifiers, other organisms have significant effects. Endolithic algae and fungi contribute to the degradation of carbonate deposits (Nadsen, 1927a, b; Swinchatt, 1969; Perkins and Halsey, 1971). Dr. Ronald Perkins has an experiment in place now at the WHOI deep-ocean station which he hopes will elucidate $CaCO_3$ breakdown rates in the deep ocean.

Rapid sedimentation due to flooding of river basins can bury benthos and cause mass mortalities (Brongersma-Sanders, 1957). A turbidity current could do the same thing in deep water. Such deposits, where recognized in the sediment, often are characterized by successful or unsuccessful escape attempts upward through the deposit. These burrows are not related to feeding or habitation and they result from unpredictable phenomenon (Schafer, 1962). Whether or not these catastrophes have had selective influences is unknown (Rowe, 1971b).

FAUNAL EFFECTS ON DEEP-SEA SEDIMENTS

Although little is known of the effects of animals on the sediments of the deep ocean, let us attempt, given our knowledge of shallow sediment-biota interactions and some generalizations about the deep-sea biota, to predict what might be occurring in the deep basins.

The exponential decrease in the standing stock of macrobenthos (Rowe, 1971a; Rowe et al., in press) and the depression in rates of life processes (Jannasch and Wirsen, 1973; Smith and Teal, 1973) imply that biological effects on the sediments will be much lower than in shallow deposits. The high diversity of the communities (Sanders, 1968; Hessler and Sanders, 1967) suggests that single-species dominance of an assemblage of animals is unlikely, and as most of the species are considered deposit feeders (Dayton and Hessler, 1972), relating most effects to particular species will be very difficult.

Most organic matter produced in the upper ocean is remineralized before it sinks below the permanent thermocline (Menzel, 1967) several hundred meters below the surface. Most organic matter at greater depths is refractory or is broken down only slowly by bacteria. Because of this sparseness of organic energy, we can infer that tubes, shells, etc., constructed of organic materials by benthic animals will be very rare. If they do occur, when they are abandoned by their maker, they will be used as a source of energy for the other organisms of the community. That is, organic remains are not to be expected in oxic deep-sea sediments.

But are these predictions warranted? Perhaps not. First of all, scattered investigations by geologists indicate that sediment reworking occurs everywhere except in anoxic basins. Pools of sediment in the median rift valley at the Mid-Atlantic Ridge contain numerous tracks and trails where, by our predictions, little should be found. The paucity of organic matter might mean that more mud must be worked on to support a given amount of biomass. There are other numerous exceptional areas which also seem to defy our expectations that little happens in deep water. Opportunists, such as giant amphipods (Hessler et al., 1972) and wood-boring bivalves (Turner, 1973) are extremely active when organic substrates are added to the deep sea.

The Cascadia deep-sea channel appears to funnel organic-rich sediment into the deep sea. The canyon fauna and sediment modifications differ from adjacent depths (Griggs et al., 1969). Large asteroids burrow into the soft mud of the levees of United States east coast canyons and continental rise valleys (Rowe, 1971b, 1972). Biological activity has been reported as the major initiator of erosion in east coast canyons (based on bottom photographs, Stanley, 1971a, b; Dillon and Zimmerman, 1970). This has been confirmed at the head of Hudson Canyon using time-lapse photography (Rowe et al., in press). Rock burrowers cause erosion in the California Scripps Canyon (Warme et al., 1973).

Several reviews are useful for inferring the effects of animals on sediments in the deep sea. Trench biology has been covered in detail (Belyaev, 1966). Menzies et al. (1973) discussed zonation at length and illustrated many deep-sea animals. Hersey et al. (1968) presented numerous bottom photographs of animals and tracks and Heezen and Hollister (1971) made some conjecture about sediment reworking, based on comparisons with shallow species.

John Southard, in reviewing the erodibility of abyssal sediments, has stressed elsewhere in this volume the potential for catalysis of erosion by the benthic fauna. In the ensuing discussion, he suggested than an organism's energy requirements might be insightful. That is, can the amount of sediment moved through the organisms be estimated, given rough estimates of organic energy supply and demand. Estimates of community oxygen uptake at 1800 m amount to about 0.5 ml $m^{-2}hr^{-1}$ (Smith and Teal, 1973). This is equivalent to 2.42 calories of energy, or about 0.61 mg organic carbon, if it is assumed that about 1.22 mg organic carbon is oxidized by 1 ml of oxygen. This is equivalent to 14.6 mg per day or 5.3 grams per year.

There are approximately 1000-2000 macrofaunal (>0.42 mm sieve mesh) animals per square meter at 1800 m south of Cape Cod (Sanders et al., 1965; Rowe, 1971a). If these account for half the metabolism measured (assuming the other half is smaller fauna and bacteria), then they consume about 2.5 g of organic carbon per year. The average organic carbon content is about 0.5% dry weight of sediment (Sanders et al., 1965); and the animals must pass therefore at least 500 g of whole sediment through their guts per year. This assumes that all organic matter ingested is assimilated, which is probably a gross overestimate. As the organic matter may be highly refractory and require incorporation into bacterial cells first, this estimate may be as much as an order of magnitude or more too low. This means that as much as 5000 g would have to be moved through animal guts per square meter per year, which appears large, but one individual animal would consume only 3.33 g per year at most. (This was calculated by dividing mean density of 1500 animals/m^2 by 5000 g of sediment.)

Another useful approach has been to make estimates of feces production based on bottom photographs (Heezen and Hollister, 1971). Castings, strings or coils (10 cm long by 2 or 3 cm in diameter) have been seen to be defecated from large elasipode holothurians which are assumed to ingest only the top millimeter of mud. *Scotoplanes sp.* has been seen to move 40 cm per hour (Eugene Lafond, fidé

Heezen and Hollister, 1971). If feces are egested twice a day (Crozier, 1918) and every ten meters, then the sea cucumber would move about 50 cm per hour, a comparable figure. If the feces have a density of 2 g cm^{-3} and the feces have a volume of 25 cm^3, 100 g of sediment would pass through a single animal per day.

If the animals of this size occur only about every 100 m^2 (Rowe, 1971b; Rowe and Menzies, 1969), then they account for only one gram of sediment turned over per square meter per day. Crozier (1918), working in shallow water with *Stichopus*, estimates that 2.7 to 2.9 g are passed per day, a number comparable to Heezen and Hollister's. Patches of densely spaced individuals may have more significant local effects. For instance, in Hatteras Canyon, the holothurian *Peniagone* was observed to have an average density of about 30 m^{-2} (Rowe, 1971b) and reworking under these clumps might be much more significant. Why they clump is unknown.

Bottom photographs are useful for gaining inferences about faunal effects on surface sediments (Bourne and Heezen, 1965), as are cores for subsurface remains or deeply-buried tubes and burrows, but these samples isolate the bottom at a single point in time. Rates of modification are therefore unknown. Short-term time-lapse photography near the head of Hudson Canyon indicated that a small patch of bottom was influenced by numerous species moving around over undetermined distances. The tracks they made were surprisingly ephemeral, lasting for only several hours before they became markedly smoothed. Animals disturbance of surface materials acted as a catalyst for sediment movements by periodic bottom currents (Rowe et al., in press; Keller et al., 1973).

Submersibles are useful tools for investigating mechanisms or rates of surface-sediment modification (Emery and Ross, 1969). Charles Hollister (personal communication) has suggested that the large red crab *Geryon quinquedins* makes burrows in the upper Hudson Canyon and this is gradually eating away the steep walls. The materials fall to the axis and are kept in suspension by the activity of the hake (fish) *Phycis*. Once suspended, the sediments can move by periodic up-and-down axis flushing (Keller et al., 1973). Several of us (Rowe and Keller) have seen similar large burrows of the lobster *Homarus americanus*, sometimes collapsed, in the central basins of the Gulf of Maine.

CONTINUED INVESTIGATION

How best can deep-sea sediment modifications by the biota be investigated? I suggest we classify the kinds of modifications in shallow water, and then make estimates of the possible rates of formation and occurrence of modifications in the deep sea, based on our knowledge of the decreases in life processes and rates and abundances with depth. The classification based on ethology (Seilacher, 1953) might not be as useful to sedimentologists as one based on the actual physical modifications. Table 1 therefore attempts to separate and define modifications on the basis of their effects on the mass physical or "geotechnical" properties of the sediment. This is incomplete and cannot be quantified, but may be a starting point for accounting for some of the variance in standardly measured physical properties. All the categories are based on research referenced from shallow water although whether a modification will strengthen or weaken the category is based on my intuition rather than measurement. The initial division separates

TABLE 1

THE EFFECTS OF THE BENTHOS ON SEDIMENTS

				EXAMPLES	OCCURRENCE IN THE DEEP SEA
B. Weaken, disperse or unconsolidate	Feeding	Non-selective	Deposit	Ingest sediments whole, removes organic matter, produces castings	Common*
		Selective	Filter	Filters or sieves bottom water, produces both pellets and pseudofeces	Uncommon*
			Deposit	Removes biota or detritus from surface (10 cm) sediments, produces pellets and castings	Common(?)*
	Locomotion		Swimming	Fishes, some holothurians and crustaceans, smooth surface	Common
			Crawling	Crustaceans and echinoderms, may either smooth or cut sediments	Common
			Burrowing	Bivalves and polychaetes, infaunal worms (*Glycera* et al.) and mollusks (*Mya* et al.), accounts for most bioturbation.	Common
				Homarus in Gulf of Maine, *Geryon* in Hudson Canyon	Less common, but often in canyons or steep topography
	Habitation		Burrow Construction	Mostly decapod crustaceans *Nephrops*, *Goneplax*, *Uca*, etc. in shallow sand and mud	
	Escape			Response to catastrophic burial - success depends on the taxon, not common	Unknown
A. Strengthen or consolidate	Tube construction	Inorganic		Calcium carbonate tubes - Polychaetes	Uncommon
				Onuphids (polychaete) use of *Globigerina* tests	Common above $CaCO_3$ compensation depth
				Pectinaria (polychaete) quartz sand tubes	Uncommon
		Organic		Pogonophoran tubes	Common
				Cerianthis anemone tubes	Common at slope depths
				Chitinous polychaete tubes, *Hyalinoecia*, chaetopterids, onuphids, etc.	Patchy, remains remineralized.

*Although much work has been done on deep-sea feeding (Sokolova, 1968), how and on what animals feed is still an open question (Dayton and Hessler, 1972).

what will weaken versus what will strengthen a deposit. This is not as straight-forward as I have made it, for some structures may weaken an extant interface of mud and seawater but strengthen a fossil deposit and vice versa. Likewise, something that might increase shear strength might also increase porosity. The diversity of habits that weaken, disperse or mix is far greater than those that solidify. The relative absence of organic remains produced on the bottom in the deep sea is worth noting as this is based on experience of a qualitative nature. The next step for the biologist is to quantify the various categories of Table 1 in terms of conventionally measured sediment properties. There is no question that this would be valuable, but how easy it would be is open to dispute.

The initial separation of dispersal and stabilization is easy to conceptualize, but difficult to measure. Adrian Richards pointed out to me that the most salient result of biotic modifications in sediments is an increase in the variance in geotechnical properties. For example, suppose shear strength is being measured every 10 cm down the length of a core, with three measurements being taken at each increment. The range of the three values will be greater at any level with worm tubes in it than at levels without them, regardless of whether the tubes have increased or decreased sediment strength.

In shallow water of Buzzards Bay where Rhoads (1973) and Young (1971) have jointly worked on benthos and sediments, we have used an in situ diver-held vane shear apparatus. Rhoads and Young attribute most sediment characters to the numerically dominant bivalve *Nucula* and to polychaetes, but another common larger animal is the burrowing anemone *Cerianthus* which constructs a thick membranous tube. While the smaller species form pellets of the soft sediment and make it susceptible to current erosion, the anemone increases sediment shear strength. Several occur per square meter, and at distances of greater than 20 cm from a tube, shear strength in the surface 5 cm is about .98 kPa (10 g/cm^2). But from 20 cm away, up to the tube, sediment shear strength gradually increases to about 1.83 kPa (18.7 g/cm^2).

A species of *Cerianthus* also occurs at the WHOI deep-ocean station at a depth of 1800 m on the continental slope off Cape Cod. Repeated visits with ALVIN have been used to determine the abundance of *Cerianthus*, as well as numerous other species. It is much sparser than its shallow congener, as are all the invertebrates. The most common species are echinoderms. The brittle star *Ophiomusium lymani* quickly smooths the sharp corners of ALVIN tracks made in the mud, but the depressions remain from one year to the next. A small cage was left last year over one track to determine if preventing ophiuroid movement over it will eliminate the smoothing altogether.

Several species of urchins are also abundant on the continental slope. Echinus affinus appears often, and when unusual materials are introduced, it clumps around or on them. Ruth Turner has put down blocks of wood which in three months were attacked by boring pelecypods. Echinus climbs all over these blocks. A two-ton bale of shredded, compressed refuse, encapsulated in 6-ml plastic and set on a wooden palette, was deployed at the bottom station to determine how it might affect the biota and environment there. The information of Jannasch and Wirsen (1973) suggests that bacterial decay of the refuse should be only minor. But after one month, observation of the bale found Echinus in a dense aggregation all around it. How and why they are attracted to it is unknown.

A functional shift may occur in communities within very predictable environments such as the deep sea. Bacterial metabolism may decrease, thereby increasing the importance of macrofaunal energy cycling relative to total energy flow through the community. Would an ALVIN sandwich be eaten be brittle stars and urchins? This meager experiment is essentially what Isaacs et al. have been doing with their imaginative bait experiments in the Pacific and based on their findings the answer would be emphatically yes.

MAN'S EFFECTS

The ocean floor is for the most part a desert, depauperate of organisms because it is so distant from a source of food. We have interest in the physical properties of sediments because the space of the sea floor will be exploited and in doing so we will introduce change. Experimentally therefore let us introduce change to see how the fauna and sediments respond. Inundate an area of the deep sea with azoic dredge spoils and see if the fauna can escape burial. What is the successional pattern of repopulation? Fertilize with organic matter. Does the abundance of the fauna increase? How fast does this bottom return to normal? How far are animals attracted to such things? If an opportunistic population explodes under such conditions, what will the effects be on the sediments?

Experiments investigating these questions are underway by J. Frederick Grassle at the WHOI 1800 m deep-ocean station using DSRV ALVIN for monitoring environmental changes from year to year. There may be an effect of scale with extrapolation to large-scale, long-term phenomena.

There are potential experiments serendipitously initiated already, however. Experimentalists are not capable of introducing a large-scale change onto the sea floor, but full advantage should be taken to find out the effects of artifacts man has already introduced. A specific example is the work of Drs. Robert Menzies and Robert George with the U. S. Navy ship R/V MIZAR studying the

CHASE II munitions dump at 2200 m just west of the WHOI ALVIN bottom station at 1800 m. With ALVIN's new depth (12000 ft or 3800 m) capabilities, CHASE II could be used to determine how the imposition of foreign materials affects the deep sea.

CONCLUSIONS

1) Animals can either consolidate and strengthen or disperse and weaken sediments.

2) The overall decreases in animal abundance and metabolic processes in the deep sea imply that the total amount of sediment reworking by animals will be much less than in shallow water, but this has not been demonstrated. It is counteracted by the vastly slower rates of sediment accumulation. Faunal effects therefore are probably accentuated in deep water because sedimentation rate and the supplies of organic matter are so low.

3) Some areas of the deep sea, such as the continental slope, submarine canyons and trenches, contain flourishing faunas; bioturbation is probably proportional to that on the continental shelf.

4) As exploitation of the deep sea will undoubtedly increase in coming years, experiments on the rates and modes of sediment modification should be initiated using conventional surface vessels and submersibles in areas where we may have already inadvertently imposed on the sea floor.

ACKNOWLEDGMENTS

Contribution No. 3169 from the Woods Hole Oceanographic Institution, Woods Hole, Massachusetts 02543. This work was supported by Office of Naval Research Contract N00014-66-C0241. I would like to thank K. L. Smith, Jr., Charles Hollister, John Milliman and Robert S. Carney for enlightening conversations and criticisms of this paper.

REFERENCES

Ansell, A., and E. Trueman, The mechanisms of burrowing in the anemone, Peachia hastata Gosse, J. Exp. Mar. Bio. Ecol., 2, 124-134, 1968.

Barnes, R. D., Tube-building and feeding in chaetopterid polychaetes, Biol. Bull., 129, 217-233, 1965.

Belyaev, G. M., Bottom fauna of the ultra-abyssal depths of the world ocean, Akad. Nauk SSSR Tr. Inst. Okeanol., 591, 1-248, 1966.

Bennett, I., The mud lobster, Aust. Mus. Mag. (Nat. Hist.), 16, 22-25, 1968.

Berger, W. H., and G. R. Heath, Vertical mixing in pelagic sediments, J. Mar. Res., 26, 134-143, 1968.

Bourne, D. W., and B. C. Heezen, A wandering enteropneust from the abyssal Pacific, and the distribution of "spiral" tracks on the sea floor, Science, 150, 60-63, 1965.

Brongersma-Sanders, M., Mass mortality in the sea, in Treatise in Mar. Ecol. and Paleoecol., edited by J. Hedgpeth, pp. 941-1010, Memoir 67, U. S. Geol. Soc., 1957.

Burger, J. A., G. deV. Klein, and J. E. Sanders, A field technique for making epoxy relief-peels in sandy sediments saturated with salt water, J. Sediment. Petrol., 39, 338-341, 1969.

Chakrabarti, A., Beach structures produced by crab pellets, Sedimentology, 18, 129-134, 1972.

Chapman, C. J., and A. L. Rice, Some direct observations on the ecology and behavior of the Norway lobster Nephrops norvegicus, Mar. Biol., 10, 321-329, 1971.

Clarke, R. H., Burrow frequency in abyssal sediments, Deep-Sea Res., 15, 397-400, 1968.

Crane, J., Eastern Pacific expeditions of the New York Zoological Society, XXVI. Crabs of the genus Uca from the west coast of Central America, Zoologica, 26, 145-207, 1941.

Crane, J., Crabs of the genus Uca from Venezuela, Zoologica, 28, 33-44, 1943.

Crane, J., Eastern Pacific expeditions of the New York Zoological

Society. XXXVIII. Intertidal brachygnathus crabs from the west coast of tropical America, Zoologica, 32, 69-95, 1947.

Crane, J., Aspects of social behavior in fiddler crabs, with special reference to *Uca maracoani* (Latrille), Zoologica, 43, 113-130, 1958.

Crichton, O. W., Marsh crab, Estuar. Bull., 5, 3-10, 1960.

Crozier, W. J., The amount of bottom material ingested by holothurians (Stichopus), J. Exp. Zool., 26, 379-389, 1918.

Cullen, D., Bioturbation of superficial marine sediments by interstitial meiobenthos, Nature, 242, 323-324, 1973.

Dayton, P., and R. Hessler, Role of biological disturbance in maintaining diversity in the deep sea, Deep-Sea Res., 19, 199-208, 1972.

Dembowski, J. B., Notes on the behavior of the fiddler crab, Biol. Bull, 50, 179-201, 1926a.

Dembowski, W. S., Study on the habits of the crab *Dromia vulgaris* m.e., Biol. Bull., 50, 179-201, 1926b.

Dillon, W. P., and H. B. Zimmerman, Erosion by biological activity in two New England submarine canyons, J. Sediment. Petrol., 40, 542-547, 1970.

Dybern, B. I., and T. Hoiseter, On the burrowing behavior of *Nephrops norvegicus* (L), Sarsia, 21, 49-55, 1965.

Emery, K. O., Some surface features of marine sediment made by animals, J. Sediment. Petrol., 23, 202-204, 1953.

Emery, K. O., and D. A. Ross, Topography and sediments of a small area of the continental slope south of Martha's Vineyard, Deep-Sea Res., 15, 415-422, 1969.

Engle, R., An account of the burrowing behavior of the amphipod *Corophium arenarium* Crawford, Ann. Mag. Natur. Hist., 9, 309-317, 1966.

Fager, E. W., Marine sediment: effects of a tube-building polychaete, Science, 143, 356-359, 1964.

Farrow, G. E., Back-reef and lagoonal environments of Aldabra Atoll distinguished by their crustacean burrows, *in* Regional Variation in Indian Ocean Coral Reefs, edited by D. R. Stoddart and C. M. Yonge, pp. 455-500, Symp. Zool. Soc. London, 1971.

Frankenberg, D., and K. L. Smith, Jr., Coprophagy in marine animals, Limnol. Oceanogr., 12, 443-450, 1967.

Frey, R. W., and J. D. Howard, Georgia Coast Region, Sapelo Island, U. S. A.: Sedimentology and Biology. VI. Radiographic study of sedimentary structures made by beach and offshore animals in aquaria, Senckenberg. marit. 4, 169-182, 1972.

Galeine, C., and C. Houlbert, Les recifes d'Hermelles et l'assechement de la baie du Mont-Saint Michael. Bull Soc. geol. et min. de Bretagne, 2, 319-324, 1922.

Gordon, D., The effects of the deposit-feeding polychaete Pectinaria gouldii on the intertidal sediments of Barnstable Harbor, Limnol. Ocean., 11, 327-332, 1966.

Griggs, G. B., A. G. Carey, and L. D. Kuln, Deep-sea sedimentation and sediment-fauna interaction in Cascadia Channel and Cascadia Abyssal Plain, Deep-Sea Res., 16, 157-170, 1969.

Haven, D. S., and R. Morales-Alamo, Aspects of biodeposition by oysters and other invertebrate filter feeders, Limnol. Ocean., 11, 487-498, 1966.

Heezen, B. C., and C. D. Hollister, The Face of the Deep, Oxford Univ. Press, N. Y., 1971.

Hersey, J. B. (ed.), Deep-sea photography, Johns Hopkins Press, Baltimore, 1968.

Hertweck, G., Georgia Coastal Region, Sapelo Island, U. S. A.: Sedimentology and Biology. V. Distribution and environmental significance of lebensspuren and in situ skeletal remains, Senckenberg. marit., 4, 125-167, 1972.

Hessler, R. R., J. D. Isaacs, and E. L. Mills, Giant amphipod from the abyssal Pacific Ocean, Science, 175, 636-637, 1972.

Hessler, R. R., and H. L. Sanders, Faunal diversity in the deep sea, Deep-Sea Res., 14, 65-78, 1967.

Hobson, L. A., and D. W. Menzel, The distribution and chemical composition of organic particulate matter in the sea and sediments off the east coast of South America, Limnol. Ocean., 14, 159-163, 1969.

Jannasch, H. W., and C. O. Wirsen, Deep-sea microorganisms: in situ response to nutrient enrichment, Science, 180, 641-643, 1973.

Johannes, R. E., and M. Satomi, Composition and nutritive value of fecal pellets of a marine crustacean, Limnol. Ocean., 11, 191-197, 1966.

Keller, G. H., D. L. Lambert, G. T. Rowe, and N. Staresinic, Bottom currents in the Hudson Canyon, Science, 180, 181-183, 1973.

Kennedy, W. J., Burrows and surface traces from the lower chalk of southern England, Bull. Brit. Mus. Nat. Hist. (Geol.), 15, 127-167, 1967.

Kuenen, H., Some arched and spiral structures in sediments, Geol. en Mignbouw 40 Jaargang, 71-74, 1961.

MacGinitie, G. E., The natural history of Callianassa californiensis Dana, Am. Midl. Nat., 15, 166-177, 1934.

MacGinitie, H., and G. E. MacGinitie, Natural History of Marine Animals, McGraw Hill Book Co., N. Y., 1949.

MacNae, W., A general account of the fauna and flora of mangrove swamps and forests in the Indo West Pacific region, Adv. Mar. Biol., 6, 73-270, 1968.

Menzel, D. W., Particulate organic carbon in the deep sea, Deep-Sea Res., 14, 229-238, 1967.

Menzies, R. J., R. Y. George, and G. T. Rowe, Abyssal Environment and Ecology of the World Oceans, John Wiley and Sons, N. Y., 1973.

Moore, H. B., Faecal pellets in relation to marine deposits, in Rec. Mar. Sediments, pp. 516-524, 1939.

Myers, A. C., Some palaeoichnological observations on the tube of Diopatra cuprea (Bosc); Polychaeta, Onuphidea, in Trace Fossils, edited by T. P. Crimes and J. C. Harper, pp. 331-334, Geol. J., 4, 1970.

Nadson, M. G., Les algues perforantes de la Mer Noire, Acad. Sci. Compté Rendus, 184, 896-989, 1927a.

Nadson, M. G., Les algues perforantes, leur distribution et leur role dans la nature, Acad. Sci. Compté Rendus, 184, 1015-1017, 1927b.

Neff, J. M., Mineral regeneration by serpulid polychaete worms, Biol. Bull., 136, 76-90, 1969.9

Pearse, A. S., The habits of fiddler-crabs, Phillipine J. Sci., 7, 113-134, 1912.

Perkins, R. D., and S. D. Halsey, Geologic significance of microboring fungi and algae in Carolina shelf sediments, J. Sediment. Petrol., 41, 843-853, 1971.

Pfitzenmeyer, H. T., and K. G. Drobeck, Some factors influencing reburrowing activity of softshell clam, Mya arenaria, Chesapeake Sci., 8, 193-199, 1967.

Phillips, P. J., Observations on the biology of mudshrimps of the genus Callianassa (Anomura: Thalassinidea) in Mississippi Sound, Gulf Res. Repts., 3, 165-196, 1971.

Reineck, H. E., Uber eingeregelte und verschachtelte Rohren des goldkocher-Wurmes (Pectinaria koreni), Natur. u. Volk, 90, 334-337, 1960.

Rhoads, D. C., The influence of deposit-feeding benthos on water turbidity and nutrient recycling, Am. J. Sci., 273, 1-22, 1973.

Rice, A. L., and C. J. Chapman, Observations on the burrows and burrowing behavior of two mud-dwelling decapod crustaceans, Nephrops norvegicus and Goneplax rhomboides, Mar. Biol., 10, 330-342, 1971.

Richter, R., Ein devonischer Pfeifen quartzit, Senckenbergiana, 2, 215-235, 1920.

Richter, R., Scolithus, Sabellarifex and Geflechtquarzite, Senckenbergiana, 3, 49-52, 1921.

Richter, R., Sand korallen - Riffe inder Nordsee, Natur und Museum 57. Bericht der Senckenberg. Gesellschaft, Frankfurt a. M., 2, 49-62, 1927.

Rittenberg, S. C., K. O. Emery, and W. L. Orr, Regeneration of nutrients in sediments of marine basins, Deep-Sea Res., 3, 214-228, 1955.

Ropes, T., and A. Merrill, The burrowing activities of the surf clam, clam, Underwater Naturalist, 3, 11-17, 1966.

Rowe, G. T., Benthic biomass and surface productivity, in Fertility of the Sea, 2, edited by J. Costlow, pp. 441-454, 1971a.

Rowe, G. T., Observations on bottom currents and epibenthic populations, Deep-Sea Res., 18, 569-581, 1971b.

Rowe, G. T., The exploration of submarine canyons and their benthic faunal assemblages, Proc. Roy. Soc. Edinburgh, 73, 159-169, 1972.

Rowe, G. T., and R. J. Menzies, Zonation of large benthic invertebrates in the deep sea off the Carolinas, Deep-Sea Res., 16, 531-537, 1969.

Rowe, G. T., G. Keller, H. Edgerton, N. Staresinic, and J. MacIlvaine, Time-lapse photograph of the biological reworking of sediments in Hudson Submarine Canyon, J. Sediment. Petrol., in press.

Rowe, G. T., P. T. Polloni, and S. Horner, Benthic biomass and abundance estimates from the northwestern Atlantic Ocean and the northern Gulf of Mexico, Deep-Sea Res., in press.

Sameoto, D., Comparative ecology, life histories, and behavior of inter-tidal sand burrowing amphipods at Cape Cod, J. Fish. Res. Bd. Canada, 26, 361-388, 1969.

Sanders, H. L., Benthic marine diversity and the stability-time hypothesis, in Brookhaven Symp. Biol. (22), pp. 71-81, 1968.

Sanders, H. L., R. R. Hessler, and G. R. Hampson, An introduction to the study of deep-sea benthic faunal assemblages along the Gay Head-Bermuda transect, Deep-Sea Res., 12, 845-867, 1965.

Schafer, W., Aktuo-Palaontologie Nach studien in der Nordsee, Waldemar Kramer, Frankfurt, 1962.

Seilacher, A., Studien zur palichnologie. I. Uber die methoden der palichnologie, Nues Jahrbuch der Geol. Paläontol., 96, 421-451, 1953.

Shinn, E. A., Burrowing in recent lime sediments of Florida and the Bahamas, J. Paleontol., 42, 879-894, 1968.

Smith, K. L., Jr., and J. M. Teal, Deep-sea benthic community respiration: an in situ study at 1850 m, Science, 179, 282-283, 1973.

Sokolova, M. N., Relationships between feeding groups of bathypelagic macrobenthos and the composition of bottom sediments, Oceanology, 8, 141-151, 1968.

Stanley, D. T., Fish-produced markings on the Atlantic outer continental margin off North-Central United States, J. Sediment. Petrol., 41, 159-170, 1971a.

Stanley, D. T., Bioturbation and sediment failure in some submarine canyons, Vie et Milieu. Suppl., 22, 541-555, 1971b.

Stevens, B. A., Callianassidae from the West Coast of North America Publ. Puget Sound Mar. Biol. Sta., 315-369, 1928.

Swinchatt, J. P., Algal boring: a possible depth indicator in carbonate rock and sediments, Bull. Geol. Soc. Am., 80, 1391-1396, 1969.

Teal, J. M., Respiration of crabs in Georgia salt marshes and its relation to their ecology, Physiol. Zool., 32, 1-14, 1959.

Trask, P. D., Organic content of recent marine sediments, in Rec. Mar. Sediments, edited by P. O. Trask, pp. 428-433, Am. Assoc. Petrol. Geologists, Tulsa, Oklahoma, 1939.

Trueman, E. R., The mechanism of burrowing in the polychaete Arenicola marina (L), Biol. Bull., 131, 369-377, 1966.

Trueman, E. R., The dynamics of burrowing in Ensis (bivalvia), Proc. Roy. Soc. Ser. B. Biol. Sci., 166, 459-476, 1967.

Trueman, E. R., The burrowing activities of bivalves, Symp. Zool. Soc. London, No. 22, 167-186, 1968a.

Trueman, E. R., A comparative account of the burrowing process of species of Mactra and other bivalves, Proc. Malacol. Soc. London, 38, 139-151, 1968b.

Trueman, E. R., The burrowing process of Dentalium (Staphopod), J. Zool., 154, 19-27, 1968c.

Turner, R., Wood boring bivalves, opportunistic species in the deep sea, Science, 180, 1377-1379, 1973.

Verrill, A. E., Results of recent dredging expeditions on the coast of New England, Am. J. Sci. & Arts, 5, 1-14, 1973.

Verwey, J., Einiges aus der Biologie von Talitrus saltator (Mont.) Congres Internatl. X[e] Zool., Budapest, pt. 2, 1156-1162, 1929.

Warme, J. E., T. B. Scanland, and N. F. Marshall, Submarine canyon erosion: contribution of marine rock burrowers, Science, 173, 1127-1129, 1973.

Weimer, R. J., and J. H. Hoyt, Burrows of Callianassa major Say, geologic indicators of littoral and shallow neritic environments, J. Paleont., 38, 761-767, 1964.

Wells, G. P., Mechanisms of movement in worms, Proc. Challanger Soc., 4, 36-50, 1969.

Young, D. K., Effects of infauna on the sediment and seston of a subtidal environment, Vie et Milieu, 22, 557-572, 1971.

SEDIMENTATION IN THE INDIAN OCEAN

KOLLA VENKATARATHNAM AND JAMES D. HAYES

Lamont-Doherty Geological Observatory

ABSTRACT

Based on information from more than 600 sediment cores, dredges, and available literature, several facies are distinguished in the Indian Ocean sediments: (a) terrigenous turbidites in the Ganges and Indus Cones and the Somali Abyssal Plain; (b) siliceous clay and ooze in equatorial and Antarctic regions; (c) pelagic brown and red clay in deep and central areas remote from land; (d) calcareous ooze in relatively shallow areas. In addition, we have mapped consolidated sediments on the elevated plateaus, igneous rock-outcrops on the Mid-Oceanic Ridge and manganese nodules occurring in a variety of physiographic regions. The nature and distribution of these sedimentary facies are the result of a number of factors: proximity of continents which have varying topography and climate, volcanism, oceanic productivity, water depth, physiography and various sediment dispersal mechanisms. In certain areas, the facies are modified by diagenesis through zeolitization and chertification, by ferromanganese accretion, or by carbonate cementation.

Available data suggest that the distribution and sources of the sediments characterizing some of these facies changed markedly in the geological past. For example, the siliceous clay facies of the equatorial region was either very much restricted in distribution or not present during pre-Miocene times. In pre-Miocene times the prime source for the non-biogenic detritus throughout much of the eastern Indian Ocean was in situ basalt volcanics in contrast to the varying sources since Miocene.

Mapping of sedimentary facies with varying mineralogical and physical properties is a first step towards understanding the acoustical and engineering properties of the Indian Ocean floor.

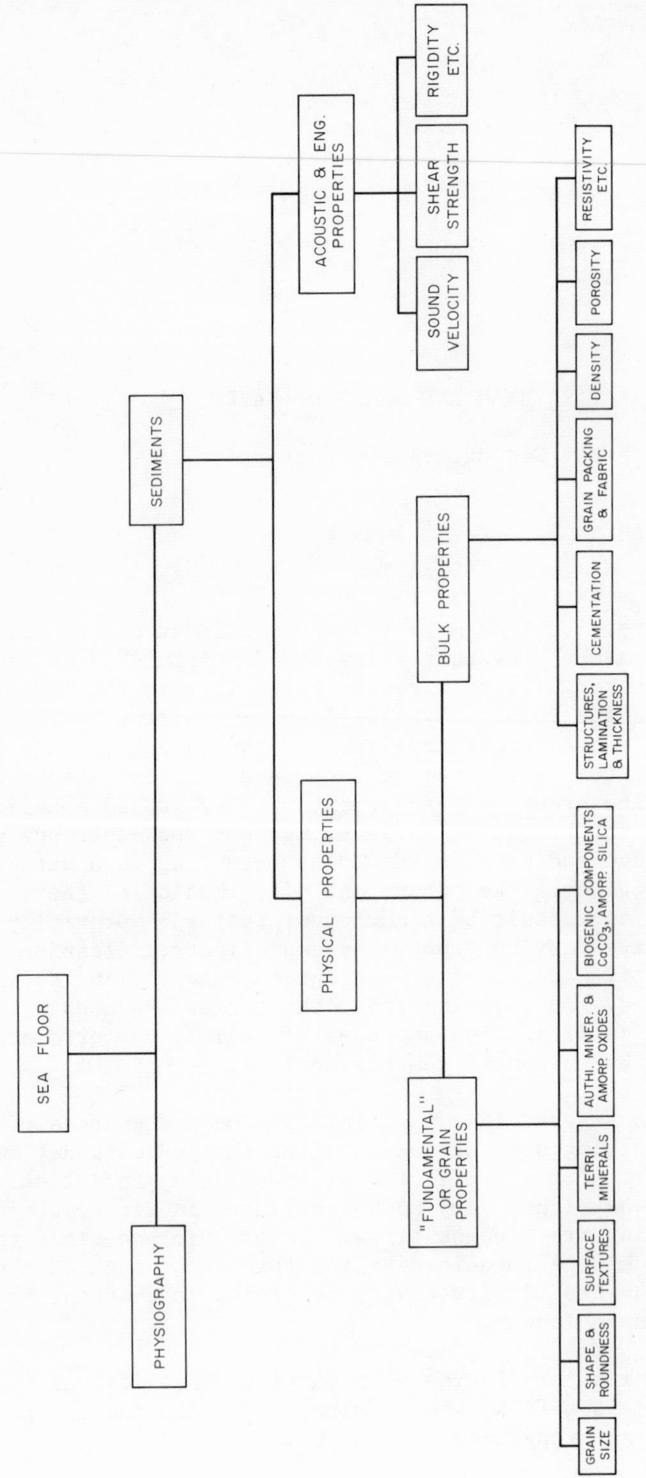

Figure 1. Flow diagram showing the various physical properties of sediments

INTRODUCTION

Broadly speaking, the sea floor may be considered a product of physiography and sediments, which in turn are influenced by a number of factors and processes. An understanding of both physiography and sediments of an ocean basin is essential to the effective use of the sea floor for acoustic and engineering purposes, to gain a better understanding of certain aspects of physical and biological oceanography (e.g., surface and bottom water circulation, organic productivity) and climatology of the area, and to trace the dispersal routes of solids to and within the ocean basin.

The different types of sediment properties are shown in the flow diagram (Fig. 1). The bulk and engineering properties depend upon the attributes of the individual sediment components as well as grain packing, cementation and structures. To understand thoroughly the sedimentary phenomena on the sea floor, one should measure various sediment properties and relate them to water properties, physiography, etc. This paper's objective is to briefly describe the available information on Indian Ocean sediment properties and discuss its significance. The status of the investigations on the physical properties of the Indian Ocean sediments is then discussed and some suggestions are made for future work.

PHYSIOGRAPHY AND SEDIMENT THICKNESS

The most prominant physiographic feature in the Indian Ocean is the seismically-active, rugged, inverted Y-shaped Mid-Indian Ridge, which is cut by numerous north-northeast trending fracture zones (Heezen and Tharp, 1964). Microcontinents and meridional aseismic ridges are unique features of the Indian Ocean (Fig. 2, Heezen and Tharp, 1964). These features are massive but are relatively smooth-surfaced blocks contrasting markedly with the rugged relief of the Mid-Indian Ridge. As in other oceans, there are a number of deep basins in between the topographic highs. Recent geophysical investigations show that the Indian Ocean is the most complex of the three major oceans in its history of formation and hence its physiography (McKenzie and Sclator, 1971).

Sediment cones and abyssal plains with sediment thickness exceeding 1.5 to 2.5 km are present in the Bay of Bengal, Arabian Sea, Somali Basin and Mozambique Basin (Fig. 3, Ewing et al., 1969). In parts of the Ganges cone, sediment thickness exceeds 12 km (Curray and Moore, 1971). In the high southern latitudes near the polar front, smooth "swale" topography with moderate sediment thickness appears as a result of higher productivity of the Antarctic Seas. In other areas, especially topographic highs, the sediment is very thin.

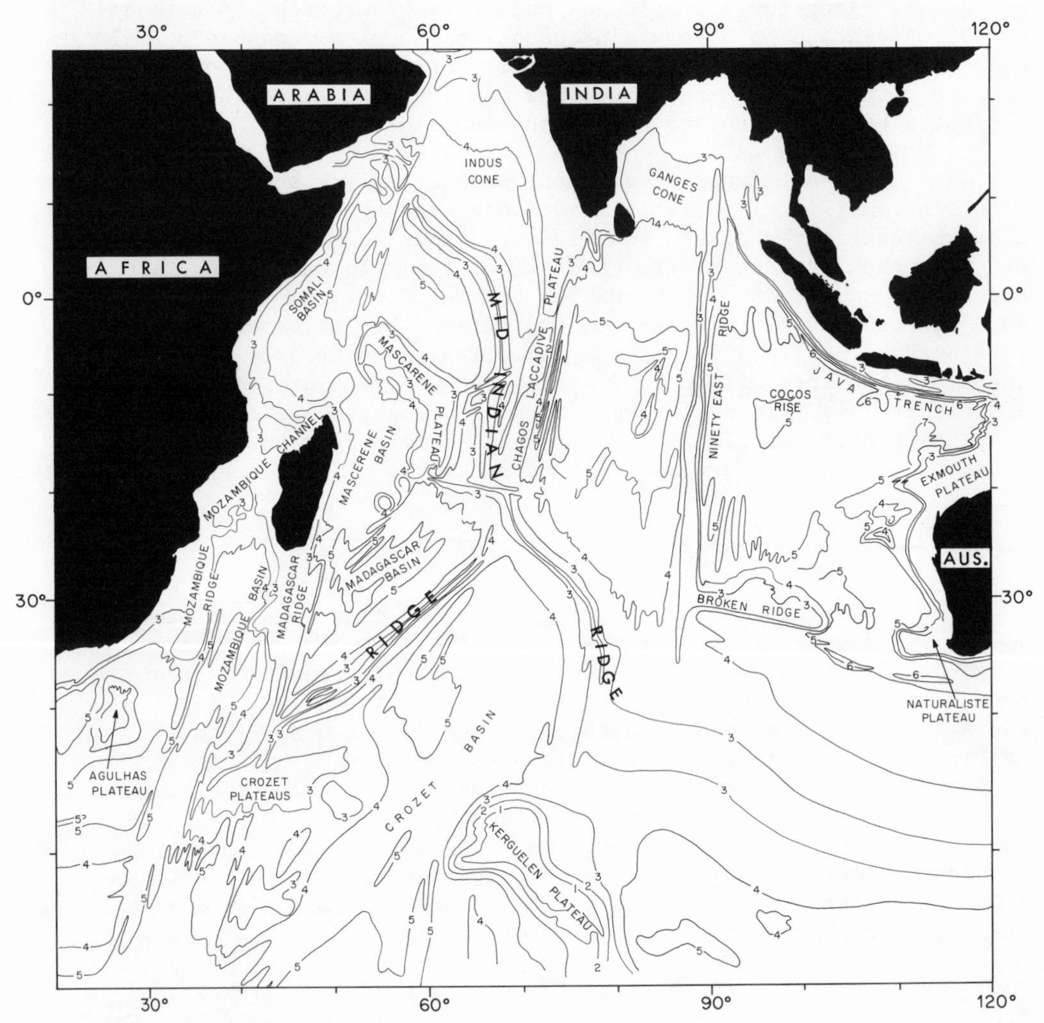

Figure 2. Physiographic provinces of the Indian Ocean from Heezen and Tharp, 1964; Copyright, Bruce C. Heezen. Published by the GSA; reproduced by permission.

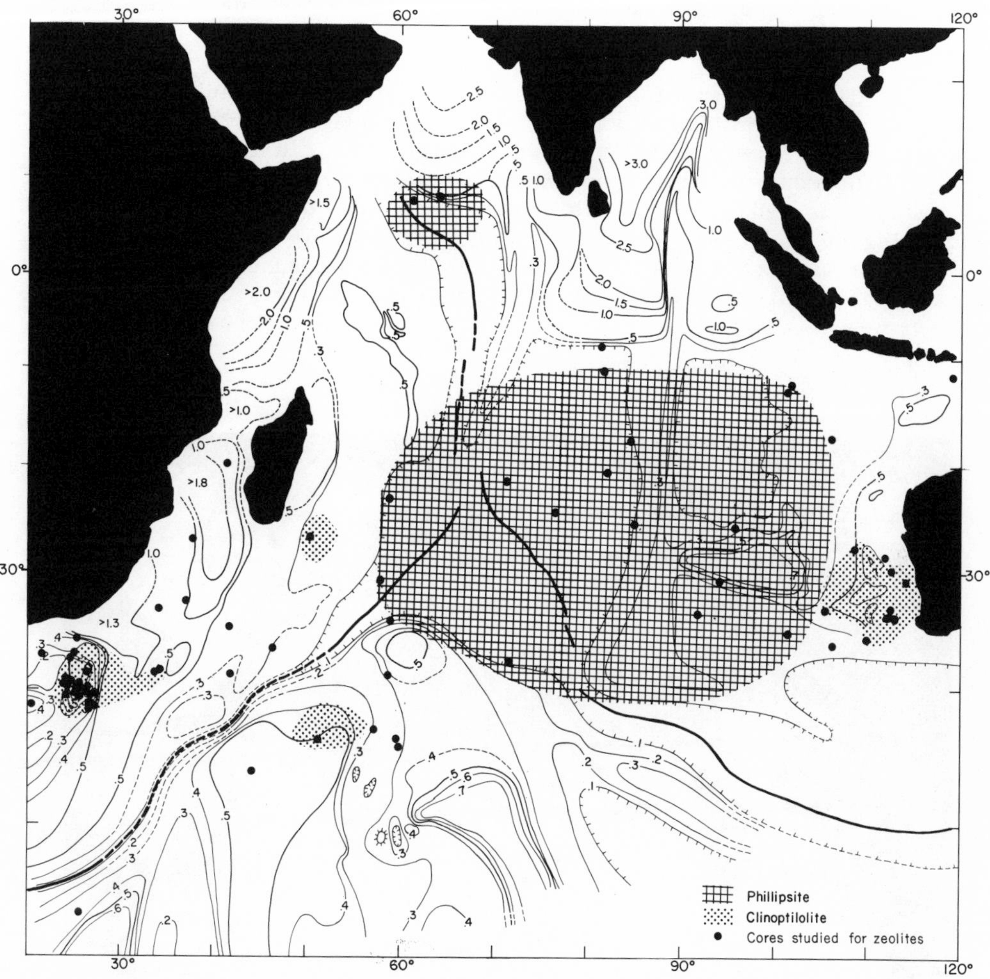

Figure 3. Isopach map of unconsolidated sediments contoured in two-way reflection time in seconds. 1 sec. is equivalent to approximately 1000 m (from Ewing et al., 1969). Also shown in the figure, the distribution of zeolites (from Venkatarathnam and Biscaye, 1973a). ● Cores studied for zeolites.

SEDIMENTARY FACIES

The complexity of the physiography and the asymmetrical distribution of the surrounding continents and their varying geology and climate would result in a complex nature of sediments in the Indian Ocean. The northward closure of the Indian Ocean prevents it from extending into the colder northern latitudes. This factor, together with monsoonal climate, would have considerable effect on the fauna and other sediment characteristics in the northern part of the ocean.

The distribution of broad sedimentary facies, based on purely qualitative information from more than 600 sediment cores, dredges, and available literature, is shown in Figure 4. In terms of areal extent, the most extensive sedimentary facies in the Indian Ocean is calcareous sediments (which covers about 54% of the ocean floor according to Lisitzin, 1971). It occupies the Mid-Indian Ridge and other more shallow areas not influenced by terrigenous influx nor by the influx of siliceous skeletal remains. The critical depth at

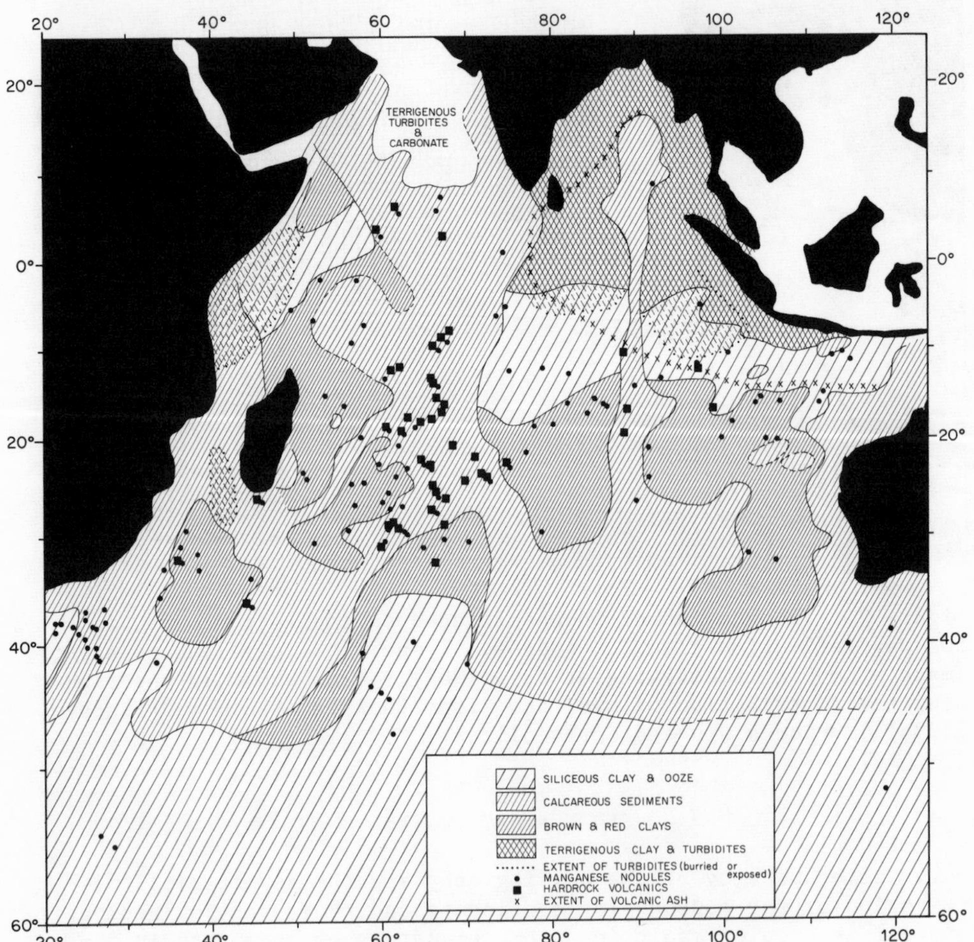

Figure 4. Sedimentary facies in the Indian Ocean. (Areas of buried turbidites are indicated by dashed lines.)

which calcium carbonate becomes a minor component (<10%) varies with latitude (Lisitizin, 1971). The other sedimentary facies are: terrigenous clay adjacent to the major river basins; siliceous clay and ooze in the equatorial and southern latitudes influenced by high biological productivity; brown and red clays in deep (exceeding about 5,000 m) areas outside the productivity belts and areas of terrigenous influx. Although the areal extent of terrigenous sediment is relatively small, because of its enormous thickness (Curray and Moore, 1971), its volume exceeds 70% of the total sediments of the Indian Ocean (Ewing et al., 1969). It may be relevant to note that about 34 X 10^{11} kg of suspended sediment are supplied by rivers draining into the Indian Ocean (see Lisitzin, 1972; Holeman, 1968). Within the above broad facies, the following materials are mapped: (1) hard rocks (volcanics and limestones) on topographically high areas, especially on the Mid-Indian Ridge, (2) Mn-nodules in the deep basins and Agulhas plateau far removed from terrigenous sediment supply, and (3) silicic volcanic ash in the eastern Indian Ocean adjacent to the Indonesian Archipelago.

Little or no information is available on the physical properties of these sediments such as porosity, density, and sound velocity. Speculations on these properties are made on the basis of information gathered by various workers in other oceanic areas (e.g., Hamilton, 1970; Horn et al., 1968) and basic sediment data presented here. In general, the sand-predominant turbidite sediments of Indus and Ganges cones may be expected to have high sound velocity, high wet density and low porosity (Horn et al., 1968). The areas with brown-red clay, siliceous clay-ooze, and calcareous sediments will have relatively low sound velocity and low wet density. The siliceous sediments will be the most porous.

The character of echograms may also reflect the nature of the sediments. Hollister (1967) and others made detailed studies in this respect in the Atlantic Ocean. In the case of the Indian Ocean, one can guess that coherent strong echoes with sub-bottom reflectors will be produced in the areas of sandy abyssal plains. Areas with brown clays and biogenic oozes may give out indistinct echoes. The bottom with hard rocks (like the Mid-Indian Ridge) will have echogram character with abundant irregular hyperbolas. One might expect distinct sub-bottom reflectors in the areas affected by the dispersal of Indonesian silicic volcanic ash. On seismic profiles, the presence of Mn-nodules, sandy horizons, and the consolidated and lithified sediments (to be described below) in the deep sea would show up as good reflectors because they produce sharp contrasts of acoustic impedance. Calcareous and siliceous oozes would show up as poor reflectors.

Bottom currents have been shown to be important in sediment transport in parts of the Atlantic and therefore in influencing the sedimentary facies (e.g., Heezen and Hollister, 1966). Except for the recent work of Burckle et al., (1973) and Ewing et al. (1968), no information is available on the importance of bottom currents in sediment transport in the Indian Ocean. A preliminary examination of some bottom photographs show well defined ripples and lineation marks indicative of strong bottom currents on Agulhas Plateau, Mozambique Ridge, Mascerene Basin and Crozet Basin. In the last area, the echograms depict regular hyperbolas suggesting sediment dune-like features.

MINERAL COMPOSITION

In certain areas, sediments become indurated due to carbonate cementation (discussed by both Milliman, and Rezak in this volume), chertification, zeolitization and precipitation of Fe-Mn oxides. In the Indian Ocean, two types of zeolites--clinoptilolite and phillipsite are observed (Fig. 3, Venkatarathnam and Biscaye, 1973a). Clinoptilolite occurs in the indurated chalks of Cretaceous-Eocene age on Agulhas and Naturalist plateaus. Chert is often observed in these sediments. Phillipsite, on the other hand, occurs in the central Indian Ocean in Miocene to Quaternary age sediments. Some of the phillipsite-rich sediments are also indurated. Alteration of volcanic material is the primary cause for zeolite genesis. The type of parent volcanic material and the extent of mobilization of silica are believed to be the factors controlling zeolite type. The induration of sediments on Agulhas and Naturalist plateaus and elsewhere in the Indian Ocean are due to cementation by zeolites, chert and carbonates. Overburden pressure on sediments appears to be not important in the process of consolidation in the deep sea. The degree of induration and consolidation of sediments in the deep sea would no doubt influence certain bulk physical properties (e.g., shear strength) as was shown by Morgenstein (1967) in the Pacific.

Hamilton (1970) has shown that mineral composition has a considerable influence on certain bulk physical properties. This influence is especially true in the eastern Indian Ocean where amounts of clay minerals differ greatly from area to area (Fig. 5). Mineral composition would also throw light on the importance of oceanic circulation and climatological influence on sedimentary processes in the ocean basins. In non-biogenic fractions of deep-sea sediments, clay size material is often the dominant component. Griffin et al. (1968), Goldberg and Griffin (1970), and Rateev et al. (1969) have studied clay mineral distributions in the Indian Ocean. These studies, though useful, were based on too few samples and perhaps many details on sedimentation processes were not brought out. Recently detailed mineralogical studies of Indian Ocean sediments have begun at Lamont-Doherty. The following summary is taken from Venkatarathnam and Biscaye (1973b) to show the usefulness of mineralogical studies in understanding the sedimentary phenomena on the sea floor.

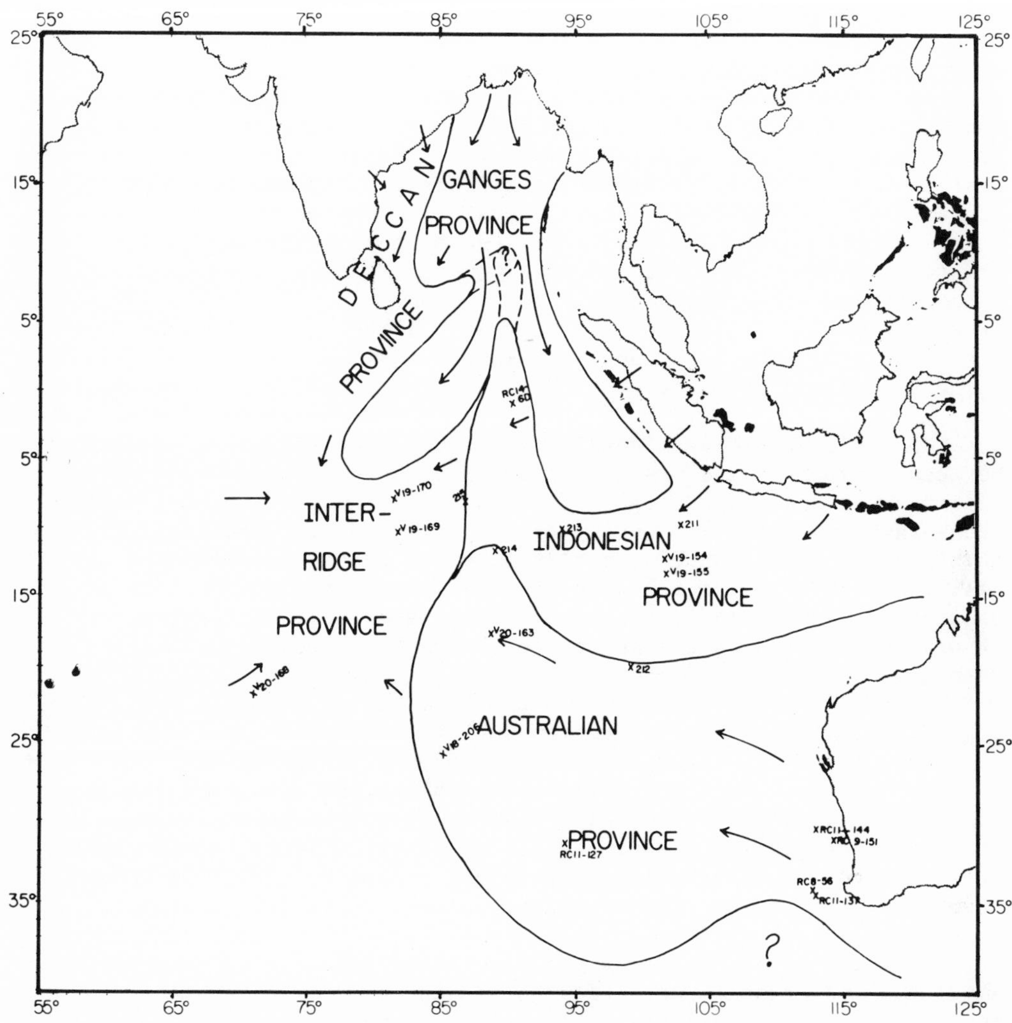

Figure 5. Mineral provinces and dispersal routes of solids in the eastern Indian Ocean (Venkatarathnam, 1973). (The numbers underlined are some of the DSDP 22 sites; other numbers are pre-Quaternary piston cores.)

Large zones around the periphery of the eastern Indian Ocean (Indonesian, Inter Ridge and Deccan provinces) are characterized by abundant smectite (Fig. 5). These are largely volcanogenic sediments and are derived from: the Indonesian Archipelago in the northeast, submarine volcanics of Mid-Indian Ocean Ridge and nearby areas in the southwest, and from the Deccan Traps of India along the western side of the Bay of Bengal. The effect of Indonesian volcanism is also observed in the Cocos-Roo Rise area and on the Ninety East Ridge between 05° north and 15° south latitude. The chief transport agents of Deccan smectite are turbidity and bottom currents and counter-clockwise surface water circulation, and, in the case of Indonesian volcanics, wind. The Ganges province, occupying the Ganges Cone and the Java Trench, is characterized by relatively more abundant illite and chlorite dispersed primarily by turbidity currents from the Ganges and Brahmaputra Rivers. This sediment source extends well south of the equator on both sides of the Ninety East Ridge. The southwestern part of the Ganges Cone is characterized by smectite-rich sediments derived from the Deccan Traps. A zone extending from west of Australia to the Ninety East Ridge (the Australian province) contains relatively abundant kaolinite which is probably windborne from that continent.

Comparison of clay mineralogy in several of the above regions between Quaternary and pre-Quaternary sediments indicates changes in sediment sources and/or transport mechanisms in the geological past (Venkatarathnam, 1973). During the pre-Miocene, the abundance of authigenic minerals such as zeolites, smectite and palygorskite suggest that basalt volcanics were the main source of sediments in the equatorial eastern Indian Ocean. During later periods, sediments of diverse origins were supplied to the eastern equatorial Indian Ocean.

Changes in sedimentary facies during geological time are not uncommon as shown in Figure 6 for some DSDP cores in the eastern Indian Ocean. For example, siliceous skeletal remains began appearing in sediments of equatorial Indian Ocean in the Late Miocene. The supply of turbidite sediments seems to have fluctuated in the area of site 211. Such variations could certainly result in changes in physical and acoustic properties. In addition, the consolidation of sediments does not always increase with depth. In fact, consolidation is likely to be influenced more by cementation by carbonates, amorphous and crystalline silica (e.g., chert), silicates (e.g., zeolites) and Fe-Mn oxides. The variations in sedimentary facies and consolidation may be correlated with distinctive seismic reflectors.

LIMITATIONS OF PREVIOUS WORK

Much work has been done on the physical and engineering properties in the Pacific and Atlantic Oceans. Even in these areas,

---

Figure 6 (opposite). Changes in sedimentary facies in certain JOIDES cores from the eastern equatorial Indian Ocean (Scientific Staff DSDP, 1972).

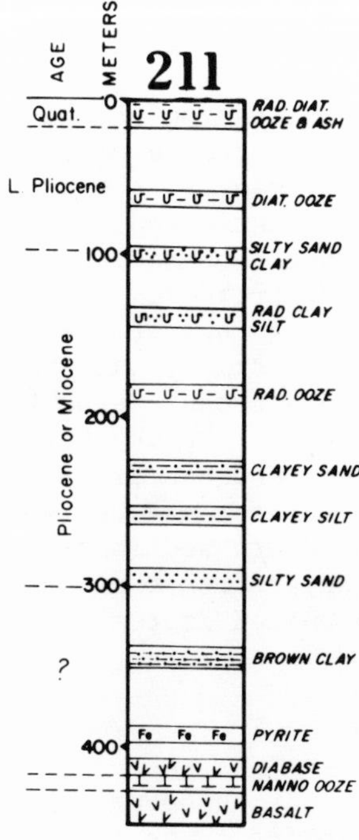
AGE
METERS
211
Quat.
L. Pliocene
Pliocene or Miocene
?
0
100
200
300
400
RAD. DIAT. OOZE & ASH
DIAT. OOZE
SILTY SAND CLAY
RAD CLAY SILT
RAD. OOZE
CLAYEY SAND
CLAYEY SILT
SILTY SAND
BROWN CLAY
PYRITE
DIABASE
NANNO OOZE
BASALT

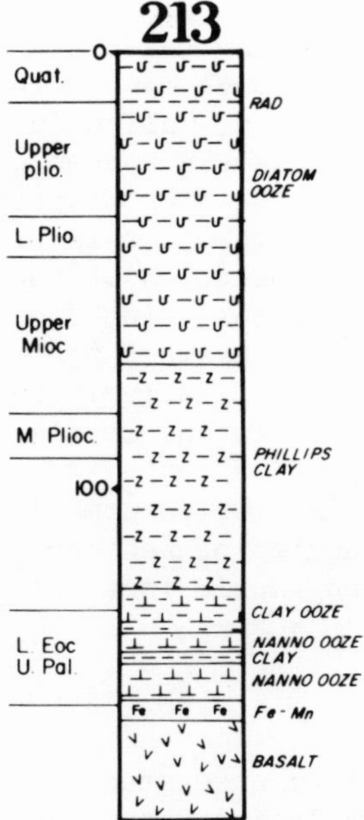
213
Quat.
Upper plio.
L. Plio
Upper Mioc
M. Plioc.
L. Eoc
U. Pal.
0
100
RAD
DIATOM OOZE
PHILLIPS CLAY
CLAY OOZE
NANNO OOZE
CLAY
NANNO OOZE
Fe-Mn
BASALT

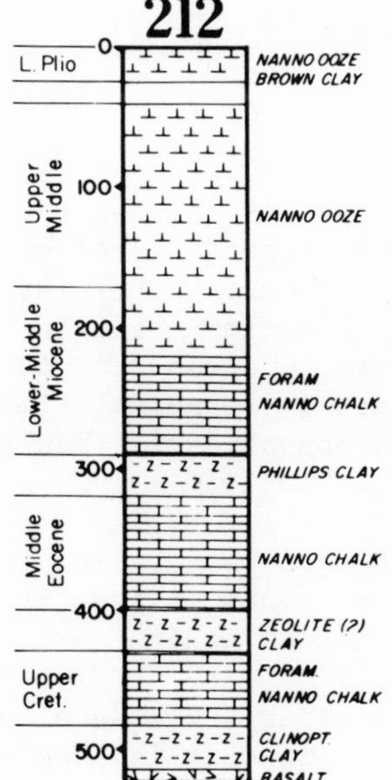
212
L. Plio
Upper Middle
Lower-Middle Miocene
Middle Eocene
Upper Cret.
0
100
200
300
400
500
NANNO OOZE
BROWN CLAY
NANNO OOZE
FORAM
NANNO CHALK
PHILLIPS CLAY
NANNO CHALK
ZEOLITE (?)
CLAY
FORAM.
NANNO CHALK
CLINOPT.
CLAY
BASALT

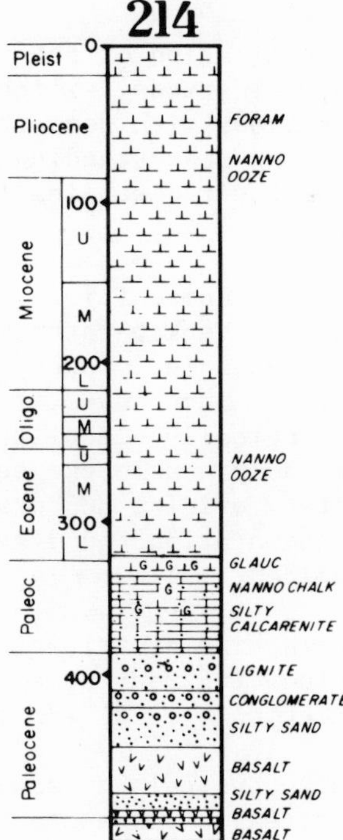
214
Pleist
Pliocene
Miocene
Oligo
Eocene
Paleoc
Paleocene
U
M
L
0
100
200
300
400
FORAM
NANNO OOZE
NANNO OOZE
GLAUC
NANNO CHALK
SILTY CALCARENITE
LIGNITE
CONGLOMERATE
SILTY SAND
BASALT
SILTY SAND
BASALT
BASALT

one of the major sediment components, biogenic silica, has not been even considered in terms of its effects on the physical properties of the material, let alone its quantitative determination. The effects of cementation and diagenesis on consolidation of the deep-sea sediments, though considerable, have been studied very little. In the Indian Ocean, unlike in other oceans, many of the basic sediment components have not been quantitatively estimated. No bulk physical and engineering properties have been measured. Speculations have been made in this paper on the distribution of certain bulk sediment properties of the Indian Ocean. However, as Keller (in this volume) has pointed out, one can not simply extrapolate the knowledge of physical properties from one oceanic area to another without quantitatively determining them. The physiography of the Indian Ocean is too complex and sediment sources are too varied to take any physical property of the sediment for granted. Contrasted with lack of information on Indian Ocean sediments, there are a large number of sediment cores collected from this ocean by various oceanographic institutions in the United States.

RECOMMENDATIONS

In order to understand the processes and phenomena of sedimentation on the Indian Ocean floor, the following types of research are recommended.

1. Determination of the basic sediment components (properties): terrigenous, calcareous, siliceous and Fe-Mn oxide components. Mapping the sea floor with respect to these basic properties is a prerequisite to the understanding of bulk physical and acoustic properties, because their determinations can be easily made on wet or dry sediment cores.

2. Mineral composition of sediments: the type of detailed work reported by Venkatarathnam and Biscaye (1973b) for the eastern Indian Ocean and briefly described in this paper, has to be extended to other areas. Such investigations should also be carried out on DSDP and geologically old piston cores. These mineralogical studies of Holocene and pre-Holocene sediments are important to understand the sedimentary origins and processes during geological time. Influence of cementation and diagenesis on consolidation has to be closely examined.

3. Grain size distribution: Grain size is a basic property greatly influencing many bulk physical and acoustic properties (e.g., Horn et al., 1968) and should be determined.

4. Bulk physical and acoustic properties (bulk density, porosity, shear strength and sound velocity): Since it is likely that some of the cores from the Indian Ocean presently available at var-

ious institutions are dry, it many not be possible to determine bulk properties on all the cores. These properties should be determined for at least a few typical areas of the Indian Ocean selected on the basis of sedimentary facies maps prepared from the basic properties mentioned earlier. Multiple correlations of these bulk properties with the basic properties have to be made to understand their inter-relationships. Extrapolation can then be made to areas where only basic sediment properties are known.

5. Miscellaneous aspects of sediments: Examination of sedimentary structures, echogram and seismic profile records, and bottom photographs along with the direct measurement of bottom currents and water turbidity should be carried out to help understanding the sedimentary dynamics of the sea floor.

6. Mapping of sedimentary facies and seismic reflectors: Based on the above studies, the various sedimentary facies on the sea floor of the Indian Ocean can be quantitatively mapped for the Holocene time. With the help of sedimentary information available on DSDP and piston cores, the prominent seismic reflectors can be correlated with a particular subsurface sedimentary facies. If the various prominent seismic reflectors are mapped areally, the sedimentary facies distributions in the different geological epochs, corresponding to each reflector, could be delineated by extrapolation. The mineralogical and diagenetic studies of sediments should provide information on the processes affecting the nature of sediments and seismic reflectors.

ACKNOWLEDGMENTS

The work reported here has been primarily carried out under Contract N00014-67-A-0108-0004 of the Office of Naval Research. The core collection and curating at Lamont-Doherty are supported by NSF grant GA 35454 and the ONR contract N00014-67-A-0108-0004. Mrs. Dorothy Cooke, acting core curator, was generous in the facilities she made available. We thank Drs. Stephen Eittriem and Allan Be for the critical reading of the manuscript and helpful suggestions.

REFERENCES

Burckle, L. H., W. H. Abbott, and J. Maloney, Sediment transport by Antarctic bottom water in the southeast Indian Ocean (abstract), Trans. Am. Geophys. Union, 54 (4), 336, 1973.

Curray, J. R., and D. C. Moore, Growth of the Bengal deep-sea fan and denudation in the Himalayas, Bull. Geol. Soc. Am., 82, 563-572, 1971.

Ewing, M., T. Aitken, and S. Eittreim, Giant ripples in the Madagascar basin (abstract), Trans. Am. Geophys. Union, 49, 218, 1968.

Ewing, M., S. Eittreim, M. Truchan, and J. I. Ewing, Sediment distribution in the Indian Ocean, Deep-Sea Res., 16, 231-248, 1969.

Goldberg, E. D., and J. J. Griffin, The sediments of the northern Indian Ocean, Deep-Sea Res., 17, 513-537, 1970.

Griffin, J. J., H. Windom, and E. D. Goldberg, The distribution of clay minerals in the world ocean, Deep-Sea Res., 15, 433-459, 1968.

Hamilton, E. L., Sound velocity and related properties of marine sediments, North Pacific, J. Geophys. Res., 75, 4423-4446, 1970.

Heezen, B. C., and C. Hollister, Deep-sea current evidence from abyssal sediments, Mar. Geol., 1, 141-174, 1966.

Heezen, B. C., and M. Tharp, Physiographic diagram of the Indian Ocean, the Red Sea, the South China Sea, the Sulu Sea and Celebes Sea, Geol. Soc. Am., 1964.

Holeman, J. N., The sediment yield of major rivers of the world, Water Resources Res., 4, 787-797, 1968.

Hollister, C., Sediment distribution and deep circulation in the western Atlantic, Ph.D. thesis, Columbia Univ., N. Y., 368 p. (unpublished).

Horn, D. R., B. M. Horn, and M. N. Delach, Correlation between acoustical and other physical properties of deep-sea cores, J. Geophys. Res., 73, 1939-1957, 1968.

Lisitzin, A. P., Distribution of carbonate microfossils in suspension and in bottom sediments, in The Micropaleontology of the Oceans, edited by B. M. Funnel and W. R. Riedel, pp. 197-218, Cambridge Univ. Press, London, 1971.

Lisitzin, A. P., Sedimentation in the World Ocean, Soc. of Econ. Paleontol. Mineral. Spec. Publ. 17, 218 p., 1972.

McKenzie, D., J. G. Sclator, The evolution of the Indian Ocean since the late Cretaceous, Geophys. J. Roy. Astron. Soc., 25, 437-528, 1971.

Morgenstein, M., Authigentic cementation of scoriaceous deep-sea sediments west of the Society Ridge, South Pacific, Sedimentology, 9, 105-118, 1967.

Rateev, M. A., Z. N. Gorbunova, A. P. Lisitzin, and G. L. Nosov, The distribution of clay minerals in the ocean, Sedimentology, 13, 21-43, 1969.

Scientific Staff, Deep-Sea Drilling Project, Leg 22, Geotimes, 17 (6), 15-17, 1972.

Venkatarathnam, K., Mineralogical data from sites 211, 212, 213, 214, and 215, of Deep-Sea Drilling Project Leg 22, and origin of non-carbonate sediments in the equatorial Indian Ocean, Initial Volumes of the Deep-Sea Drilling Project, Leg 22, 1973 (in press).

Venkatarathnam, K., and P. E. Biscaye, Deep-sea zeolites: variations in space and time in the sediments of the Indian Ocean, Mar. Geol., 15, M11-M17, 1973a.

Venkatarathnam, K., and P. E. Biscaye, Clay mineralogy and sedimentation in the eastern Indian Ocean, Deep-Sea Res., 20, 727-738, 1973b.

PHYSICAL PROPERTIES OF SEDIMENTARY PROVINCES, NORTH PACIFIC AND NORTH ATLANTIC OCEANS

DAVID R. HORN, MARILYN N. DELACH, AND BARBARA M. HORN

Lamont-Doherty Geological Observatory

ABSTRACT

The North Pacific and North Atlantic Oceans are sites of very slow and relatively rapid deposition respectively. In this brief account we describe the lithologic, textural and physical properties of sediment cores and define geographic limits of provinces of similar sediment. Values of mean grain size, wet density, moisture content, and porosity are presented.

Red clays of the North Pacific are slightly coarser than their Atlantic equivalents but have similar bulk properties. Radiolarian oozes of the North Pacific are the most porous, water-laden materials of any ocean floor. Diatomaceous sediments have bulk properties between those characteristic of red clay and radiolarian ooze. Carbonate oozes of the North Atlantic have a wider range of properties than their North Pacific counterparts. This is explained by the location of the former along the Mid-Atlantic Ridge where there is ample evidence in cores of winnowing of fines by currents. Deposits of the equatorial Pacific accumulate in a quiet deep-water environment.

Turbidites of the northeast Pacific grade from clay (Mz = 1μ) to sand (Mz - 128μ). In the North Atlantic they are far more widespread and coarser-grained. They grade from clay (Mz - 1μ) to sand (Mz - 136μ) and rarely to gravel (Mz - 1526μ). Ash layers of the North Pacific average 6 cm in thickness and have a Mz = 26μ. Hemipelagites have properties which are the same for both oceans, but ranges of textural parameters of these terrigenous sediments are greater in the North Atlantic.

Interest in physical properties of the ocean floor is experiencing something of a revival today. Mining companies, faced with problems related to exploitation of ferromanganese nodules, are studying the seabed in order to understand its nature and to assure proper design of dredge hardware. Environmentalists demand careful evaluation of all aspects of ocean mining in order to avoid possible damage to organisms living in the water column or on the sea floor. Settling properties of abyssal clays are not well understood. Additional research is needed to determine the effects of large-scale resuspension of these tiny grains on the physical and biological framework of the oceans.

The United States Navy (Naval Ship Systems Command) has shown considerable foresight by supporting investigations related to use of the sea floor as an acoustic interface. Systems analysts in military oceanography need information on the distribution of layered and uniform marine deposits along with their physical characteristics. These data are fundamental to accurate prediction of levels of sound reflection at the water-sediment interface.

This report briefly summarizes some of the results of a five-year analytical program directed at definition of physical and textural properties of ocean sediments in the Northern Hemisphere.

TECHNIQUE

Cores were obtained by personnel aboard research vessels VEMA, ROBERT D. CONRAD, and ATLANTIC SEAL using piston corers with barrels of 6.35 cm (2 1/2 in) ID. The average length of cores is 7.92 m (26 ft) for the North Pacific and 6.70 m (22 ft) for the North Atlantic. A complete description of the coring procedure and method of storage at Lamont-Doherty was given by Ericson et al. (1961). Bulk property samples were removed from cores immediately after their extrusion on the ship's deck. Samples were run in duplicate on air-comparison pycnometers and an average determined for each sample. All were saturated with water and a correction was made for salt content. Porosities were obtained using the formula of Sutton et al. (1957). Samples selected for textural analysis were taken from cores upon their arrival at the laboratory. Grain size analyses followed the combined sieve-pipette technique of Folk (1968). Mz is the best graphic measure for determining overall particle size and was calculated by using the formula $Mz = (\phi 16 + \phi 50 + \phi 84)/3$ (Folk and Ward, 1957). Sound velocities were determined using a sediment velocimeter with four measurements recorded every 5 cm down the core (at each 5-cm interval a reading was taken every 90 degrees around the circumference of the linered core). The procedure and results are described in more detail by Horn et al. (1968).

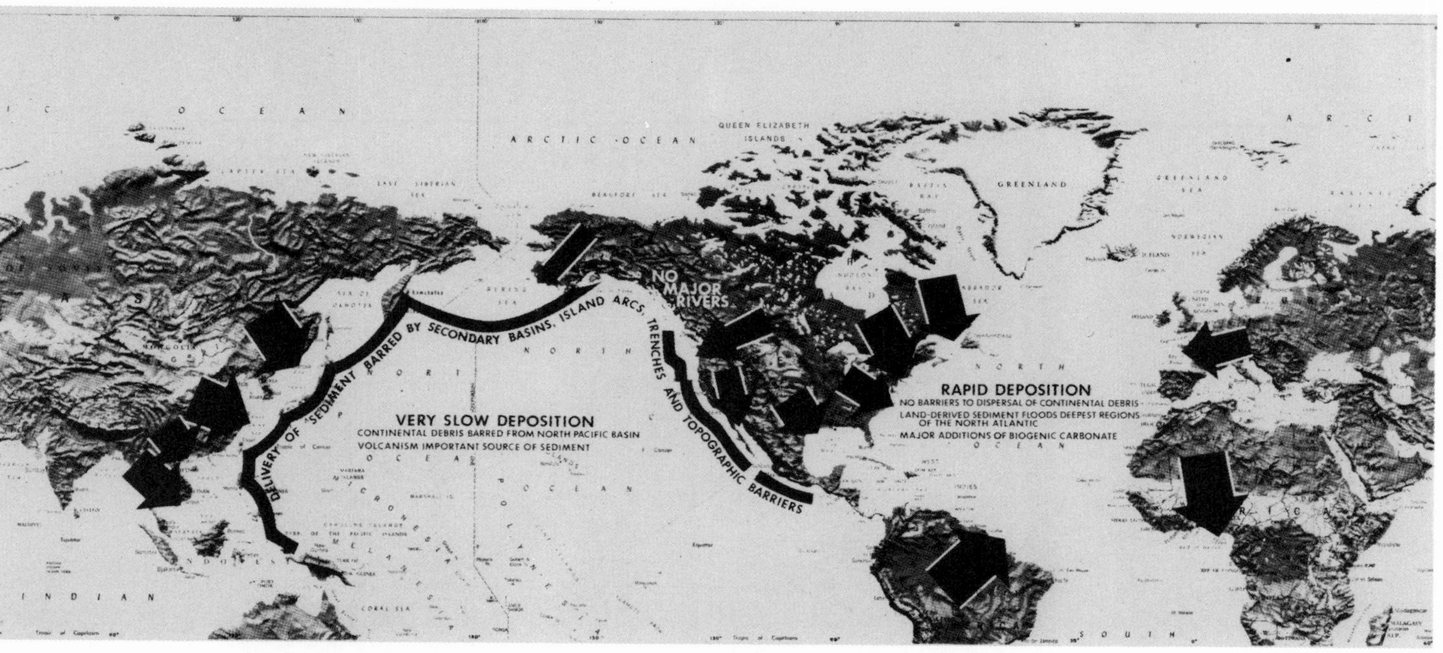

Figure 1. Coarse continental debris does not reach the North Pacific Ocean because of topographic obstacles to sediment dispersal. Only off Oregon and British Columbia is there a break in the circum-Pacific barriers. Here turbidity currents actively transport material from the Columbia River to the deep-ocean floor. The North Atlantic is different in that there are no obstacles to impede sediment dispersal. Coarse-grained terrigenous material is freely transported by turbidity currents. Resulting deposits have leveled major areas of the North Atlantic and produced the great abyssal plains characteristic of the ocean.

TABLE 1

DIFFERENCES BETWEEN SEDIMENTARY FRAMEWORK OF THE NORTH PACIFIC AND NORTH ATLANTIC (Modified after Revelle et al., 1955; Keller and Bennett, 1968; with additions by the writers)

NORTH PACIFIC	NORTH ATLANTIC
Large areas of great depth.	Major mid-ocean ridge system reduces overall depth.
Small inflow of river water.	Large inflow of river water.
24 rivers greater than 400 km empty into the Pacific.	More than 100 rivers empty into the Atlantic.
Drainage area is 1/10th of surface area.	Drainage area is 1/2 of surface area.
Turbidity current activity restricted.	Turbidity current activity widespread, no barriers.
Width approximately 4800 miles.	Width approximately 3000 miles.
Area is approximately twice that of the North Atlantic.	
Pyroclastics common and often in the form of widespread ash layers.	Pyroclastics uncommon.
High frequency of seamounts, volcanism, seismic activity, trenches and ridges.	
Percent of ocean floor covered by clays is twice that of Atlantic.	
Large areas of siliceous ooze.	No siliceous ooze.
2 large areas of red clay.	2 small areas of red clay.
Dominated by red clay due to great distances to land and low input of sediment.	Dominated by carbonate associated with mid-ocean ridge.
Clays higher in iron, manganese and authigenic zeolite.	

FRAMEWORK OF SEDIMENTATION: NORTH PACIFIC AND NORTH ATLANTIC

Results indicate that vast areas of ocean floors are covered by sediments with common lithologies and properties. Designated as provinces, they are products of the global framework of marine sedimentation. Materials within each province vary only slightly in sediment type and physical properties. Continuous deposition has produced uniform sections of biogenic ooze and unfossiliferous red clay. On the other hand, catastrophic events, either in the form of volcanic eruption or earthquake, have built up sediment consisting of a variety of different deposits. Each event is recorded by a texturally graded volcanic ash or turbidite layer. These rapidly deposited materials are generally intercalated with pelagic or hemipelagic muds and clays. They stand out in the sediment column because of their coarser texture.

North Pacific

In Figure 1 and Table 1 are given reasons why sediments of the North Pacific are quite different from those of the North Atlantic. Numerous large rivers drain toward both oceans; however, those of Asia and western North America do not deliver sediment to the open sea. Material transported toward the Pacific is trapped in secondary settling basins (e.g., Japan Sea, Bering Sea), behind island arcs (e.g., Kuril Islands), in deep trenches (e.g., Aleutian Trench), or other topographic barriers (e.g., Santa Rosa Mountains, California). The obstructions to dispersal of sediment combine to form a barrier which encircles the North Pacific. The Columbia is the only major river whose sediment is not blocked from entering the North Pacific Basin. Turbidity currents transport material from near the river's mouth down Cascadia Channel (Fig. 2) and across the southern end of the Tufts Abyssal Plain, a distance of 1770 km or 1100 miles (Horn et al., 1969, 1970, 1972a). Resulting deposits constitute a province of graded sands and silts interlayered with clay in the northeast corner of the Pacific Ocean.

Circum-Pacific volcanoes have delivered much coarse-grained silt to the northern margin of the ocean (Fig. 2). Ash layers occur within an arc-shaped zone 1200 to 1600 km (800 to 1000 miles) wide. They are the dominant coarse-grained sediment derived from adjacent land areas (Horn et al., 1969). Both turbidites and ash are potential reflectors of sound. The coarse layers produce abrupt impedance contrasts within the sediment section. Their areal distribution, thickness, and position in cores can be used to predict performance levels of bottom-bounce sonars.

South of the regions influenced by turbidity-current and volcanic activity is an immense region of abyssal hills. Three sedi-

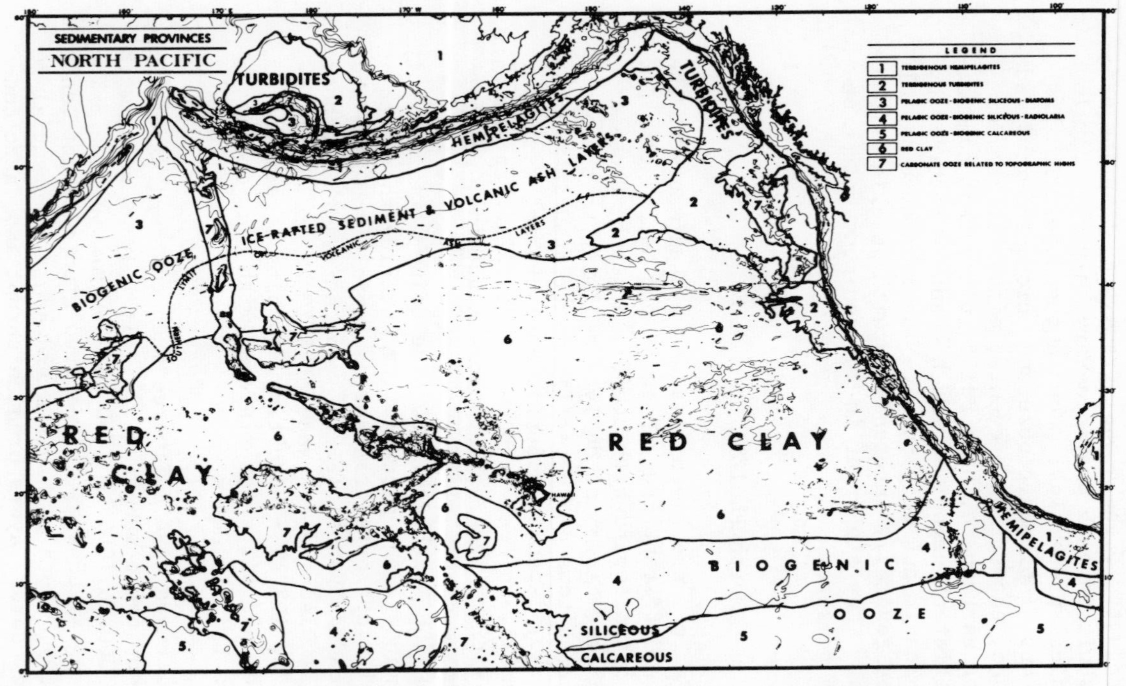

Figure 2. The North Pacific is dominated by pelagic sedimentation. Sedimentary provinces run east-west and parallel major water masses. The central area of the ocean is a site of red clay deposition with biogenic zones lying to the north and south. Siliceous deposits between the equator and 15°N contain skeletons of Radiolaria. A province of carbonate ooze (principally Foraminifera) lies along the equator. At the northern limit of the basin is a wide zone of mixed biogenic ooze (diatomaceous clay), ice-rafted sediment and volcanic silt. Deep-sea sands and silts occur off the coast of Oregon, on the floor of the Bering Sea, and along narrow floors of trenches. They are emplaced by turbidity currents. The latter process is also responsible for transfer and deposition of sediment near major topographic features such as the Hawaiian Ridge and Mid-Pacific Mountains.

mentary provinces occupy the area and reflect an interaction of biologic productivity controlled by distribution of major water masses and a trend of increasing water depth to the north. Sediments accumulating at the equator are carbonate oozes. Farther north are siliceous oozes (radiolarian) and red clay (Fig. 2). Axes of the provinces are east-west and match the limits of principal water masses. Sediments within these provinces are uniform throughout the section penetrated by a coring tool (Fig. 3). The monotonous uniformity of red clays of the central North Pacific suggests it has been a site of extremely slow deposition of finest-grained materials for millions of years. Although the uniform sediments absorb sound rather than reflect it, and therefore are not of interest to military acousticians, it is the radiolarian sediments just north of the equator which are receiving close study by ocean industry. Nodular ferromanganese lying on radiolarian ooze and radiolarian clay are rich in copper and nickel (Horn et al., 1972b, 1973a, b). Mining sites selected by major companies lie along the margins of the province of radiolarian deposits (Fig. 2, province 4).

North Atlantic

The situation is very different in the North Atlantic. There are no ash layers or siliceous radiolarian deposits. Turbidite deposition, which was confined to a small area of the North Pacific, is the dominant process of deposition in deep waters of the Atlantic. Major sedimentary provinces lie in broad north-south bands. This is due to dominance of the framework of sedimentation by the Mid-Atlantic Ridge. It occupies over one-third of the North Atlantic, lies above the compensation depth of calcium carbonate, and is a site of rapid accumulation of carbonate ooze. Two provinces of red clay lie east and west of the flanks of the ridge. They are beyond the influence of material swept from the ridge and are protected by rugged topography from incursion of turbidity currents loaded with terrigenous sediment. Deep-sea sands and silts delivered by turbidity currents have progressively levelled large areas of the North Atlantic. A combination of correct geometry (flat) and highly reflective layers (turbidite sands and silts) within abyssal plains of the North Atlantic indicates that these parts of the sea floor will serve as highly reflective, acoustic interfaces.

PHYSICAL PROPERTIES OF OCEAN SEDIMENTS

Selected determinations of mean grain size and wet density are given in Figures 3, 4, 6, and 7. Mean grain size profiles of the North Pacific are based on 1043 analyses; whereas 711 determinations of texture were made on North Atlantic samples. Wet density profiles include measurements on 1694 samples from the North Pacific and 1479 from the North Atlantic. A list of average mean grain sizes and

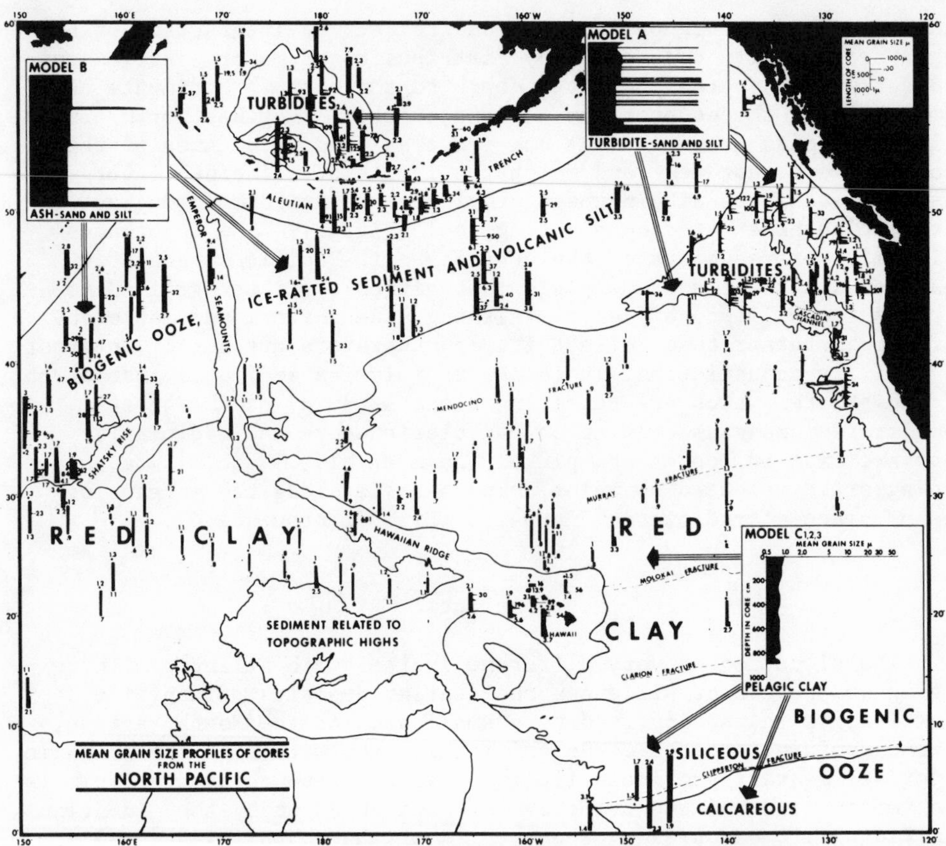

Figure 3. Textural profiles of mean grain size of cores indicate the predominant size of red clay is 1μ. There is only a slight decrease of grain size with depth in cores suggesting conditions have remained much as they are today for several million years. The biogenic zones are coarser because of additions of biological material. However, the relative fine grain of the matrix supports the idea that the North Pacific is an area of pure pelagic deposition of only finest land-derived materials. Coarse ash layers stand out from the uniform profiles and are shown as horizontal lines. The mean grain size of the coarsest ash in each core is labeled. Turbidite sands and silts also stand out from the profiles and the coarsest of these deposits at each site has also been identified and labeled. It is apparent that turbidites occur in the Bering Sea, Aleutian Trench, off Oregon and at the base of the Hawaiian Ridge.

Profiles of mean grain size of representative cores of turbidites, ash layers interlayered with pelagic sediment, and pelagic sediment are given as inserts in the diagram. Scales of the models are similar to those shown for Model C. They illustrate the differences between catastrophic deposition from turbidity currents and ash falls and the uniform accumulation of pelagic clay.

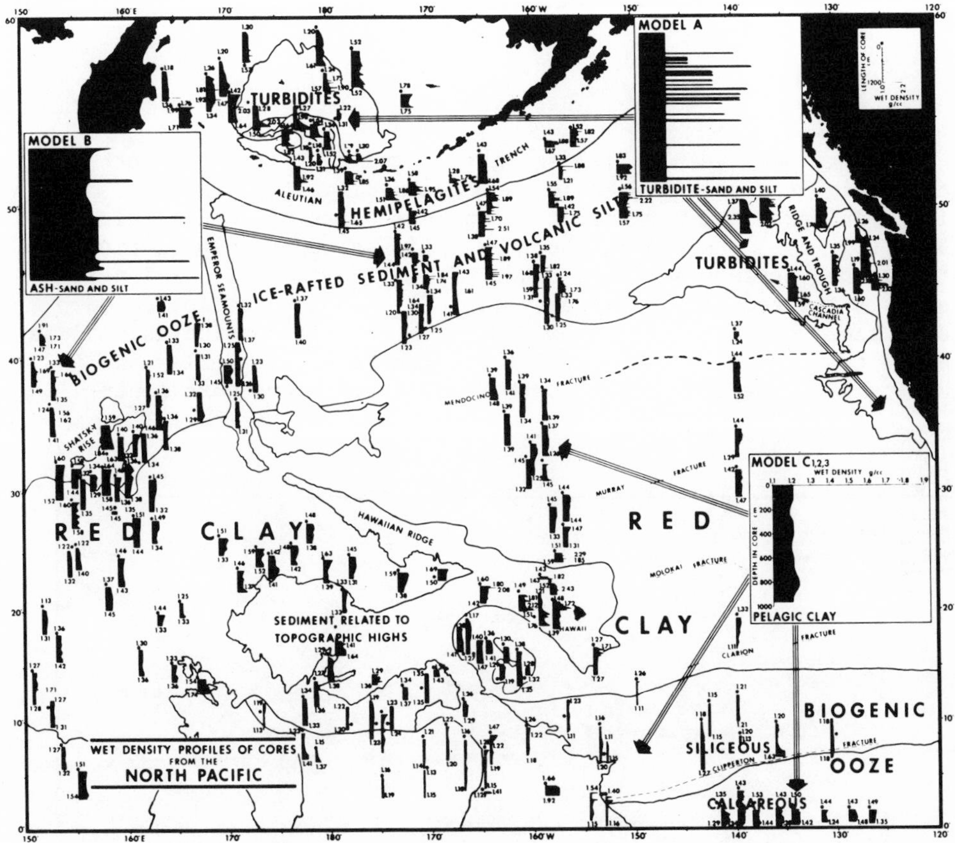

Figure 4. Wet densities are high for cores composed of abundant terrigenous sediment (turbidites and hemipelagites); calcareous deposits such as those of the equatorial Pacific and Shatsky Rise are also quite high; whereas those of the northern siliceous zone and the red clay region are slightly lower. Wet densities of the southern Radiolaria-rich province are unusually low and often approach 1 gm/cm^3. These very low densities reflect the very high porosity and moisture content due to high interstitial pore space and hollow framework grains. The radiolarian sediments are the substrate of Cu-Ni rich ferromanganese nodules.

Profiles of wet density of representative turbidites, ash layers intercalated with pelagic sediment and pelagic deposits are given. Scales are the same as those shown for Model C. Again, as in Figure 3, there is a clear-cut difference between catastrophic sediments and uniform pelagic deposits.

bulk properties is given in Table 2.

Red Clay

Textural analyses of 121 samples of red clay from the North Pacific give an overall average mean grain size of 0.97μ (Fig. 3). This is in agreement with earlier published data of Horn et al. (1970, Fig. 5). The central North Pacific is apparently covered by fine-grained sediments of the order of 1μ. In the North Atlantic there are two red clay provinces (Fig. 5). In the smaller western basin, clays have an average Mz = 0.66μ. This is considerably finer-grained than equivalent clays of the North Pacific. Keller and Bennett (1968) reported that North Pacific red clays were finer-

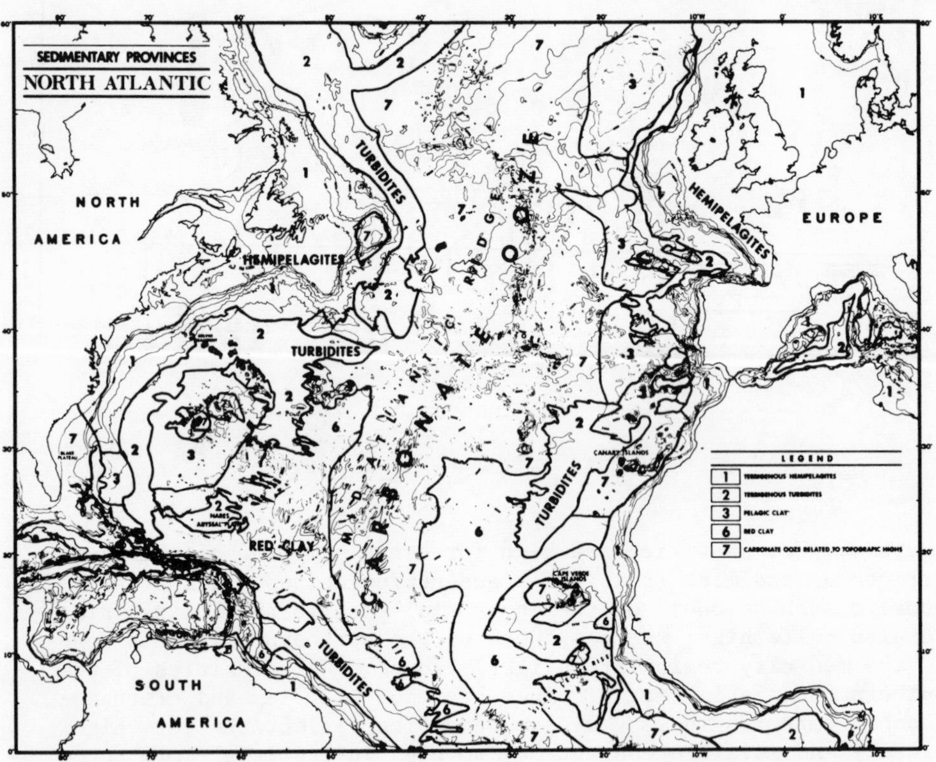

Figure 5. Sedimentary provinces of the North Atlantic are dominated by carbonate ooze associated with the Mid-Atlantic Ridge. Graded sands and silts emplaced by turbidity currents have filled linear troughs between the Ridge and continental margins. Red clays occur in two small provinces on each side of the Ridge. Massive, gray-green muds and clays blanket the hemipelagite provinces.

grained than those of the Atlantic Ocean. It may be that they included as red clay impure forms with admixtures of fine-grained carbonate debris. These medium brown clays are given separate province status by the writers (see Fig. 5, province 3). Possibly the relative number of ferromanganese micronodules has a bearing on measurement of textures.

Red clays of the North Pacific and North Atlantic have similar values of wet density. The average for samples from the North Pacific (where n equals numbers of samples analyzed) is 1.41 g/cc (n = 192). These results are similar and suggest red clays of the world have similar bulk properties (Figs. 2 and 5). Moisture content shows a similar relation: the average for the North Pacific is 126% and that for the North Atlantic, 114%. These data indicate red clays of the North Pacific include more water than their Atlantic equivalents. Average porosity of North Pacific samples is 75%, whereas for Atlantic samples it is 73%.

In summary, red clays from the two oceans have closely similar bulk properties, but material from the North Pacific is slightly coarser grained.

Siliceous Ooze

Radiolarian ooze and radiolarian clay constitute a 885 km (550 mile) wide, east-west province of siliceous deposits (Fig. 2, province 4) unique to the North Pacific. They consist of varying numbers of spherical, opaline skeletons of Radiolaria in a clay matrix. The latter is dominated by montmorillonite (Jacobs and Hays, 1972). Textural analyses suggest a range of mean grain size from 1μ to 3μ with an average of 2μ.

The average wet density of radiolarian sediments is 1.17 g/cc, moisture content 341%, and porosity 88%. Purest radiolarian oozes have lowest wet density - 1.14 g/cc, highest porosity - 89%, and highest moisture content - 389%, of any ocean sediment (refer to Tables 1 and 2). Microscopic inspection of samples of this unusual sediment reveals not only high interstitial porosity, but framework grains are also porous and hollow.

It has been established that there is a coincidence of the distribution of radiolarian sediments in the North Pacific and ferromanganese nodules rich in copper and nickel (Horn et al., 1972b). Nowhere else in the world ocean are metal values as high as they are within this province of fine-grained siliceous deposits (Horn et al., 1973a, b). The correlation between valuable nodules and a specific substrate has led to the theory that upward migration of metal-bearing solutions may be facilitated by the highly porous framework characteristic of the sediments (see Raab, 1972).

TABLE 2

TEXTURAL AND OTHER PHYSICAL PROPERTIES OF DEEP-SEA SEDIMENTS
from the North Pacific and North Atlantic

NORTH PACIFIC									
	Average				Range				
Sediment Type	Mean Size	Wet Density gm/cm^3	Moisture Content % dry wt.	Porosity %	Mean Size μ	Mean Size φ	Wet Density gm/cm^3	Moisture Content % dry wt.	Porosity %
Red Clay	0.97μ 10.01φ	1.41	126	75	0.50-2.45 49 cores n=121	8.67-10.97	1.21-1.69 51 cores n=349	48-242	55-87
Siliceous Ooze (Radiolarian)	1.99μ 8.97φ	1.17	341	88	1.07-3.44 3 cores n=27	8.18-9.86	1.07-1.29 29 cores n=230	190-673	75-94
Pure Rad. Ooze		1.14	389	89			1.07-1.24	214-673	80-94
Carbonate Ooze	1.78μ 9.13φ	1.41	120	74	1.51-2.27 2 cores n=6	8.78-9.37	1.21-1.64 26 cores n=171	62-248	62-85
Diatomaceous Clay	2.42μ 8.69φ	1.34	152	77	0.60-6.25 56 cores n=134	7.32-10.70	1.16-1.61 43 cores n=332	63-321	51-89
Hemipelagites	3.39μ 8.20φ	1.51	90	67	1.13-21.29 11 cores n=32	5.55-9.79	1.19-1.82 15 cores n=151	32-256	47-85
Ashes	26.34μ 5.25φ	1.75	40	48	3.85-122.4 60 cores n=139	3.00-8.02	1.55-2.33 29 cores n=58	5-70	10-63
Turbidites Clay	1.06μ 9.88φ	1.46	104	71	0.78-1.36 12 cores n=31	9.52-10.32	1.24-1.59 3 cores n=8	71-182	66-78
Mud	2.41μ 8.70φ	1.55	86	68	1.14-7.40 19 cores n=49	7.08-9.78	1.37-1.71 6 cores n=30	42-118	52-82
Silt	21.38μ 5.55φ	1.82	47	54	4.06-60.70 28 cores n=111	4.04-7.94	1.60-2.12 11 cores n=48	23-97	40-66
Sand	128.33μ 2.96φ	2.12	20	34	58.7-242.6 12 cores n=30	2.04-4.09	1.97-2.43 6 cores n=9	7-30	15-46

TABLE 2

(CONTINUED)

NORTH ATLANTIC									
	Average				Range				
Sediment Type	Mean Size	Wet Density gm/cm^3	Moisture Content % dry wt.	Porosity %	Mean Size μ	Mean Size φ	Wet Density gm/cm^3	Moisture Content % dry wt.	Porosity %
Red Clay	0.64μ 10.61φ	1.45	114	73	0.43-0.90 15 cores n=15	10.12-11.18	1.28-1.61 32 cores n=192	71-223	61-87
Carbonate Ooze Mid-Atl. Ridge	1.72μ 9.18φ	1.52	87	68	0.81-3.97 8 cores n=74	7.98-10.27	1.31-1.99 33 cores n=402	31-210	43-82
Pelagic Clay	1.72μ 9.18φ	1.51	89	70	0.60-5.13 12 cores n=30	7.67-10.70	1.34-1.70 12 cores n=146	51-144	57-78
Hemipelagites	3.24μ 8.27φ	1.58	80	65	0.87-14.2 n=77	6.14-10.17	1.28-2.05 26 cores n=215	25-168	38-81
Turbidites									
Clay	1.13μ 9.78φ	1.51	88	69	0.63-1.86 37 cores n=45	9.07-10.63	1.40-1.68 36 cores n=132	53-150	61-79
Mud	3.98μ 7.97φ	1.57	72	65	1.32-16.25 17 cores n=19	5.94-9.57	1.46-1.69 11 cores n=24	51- 94	55-73
Silt	28.72μ 5.12φ	1.78	43	52	5.28-72.96 91 cores n=181	3.78-7.57	1.51-2.00 52 cores n=154	24- 74	38-67
Sand	136.17μ 2.88φ	2.03	25	38	10.02-1042. 74 cores n=168	-.06-6.64	1.73-2.28 20 cores n=54	13- 49	26-57

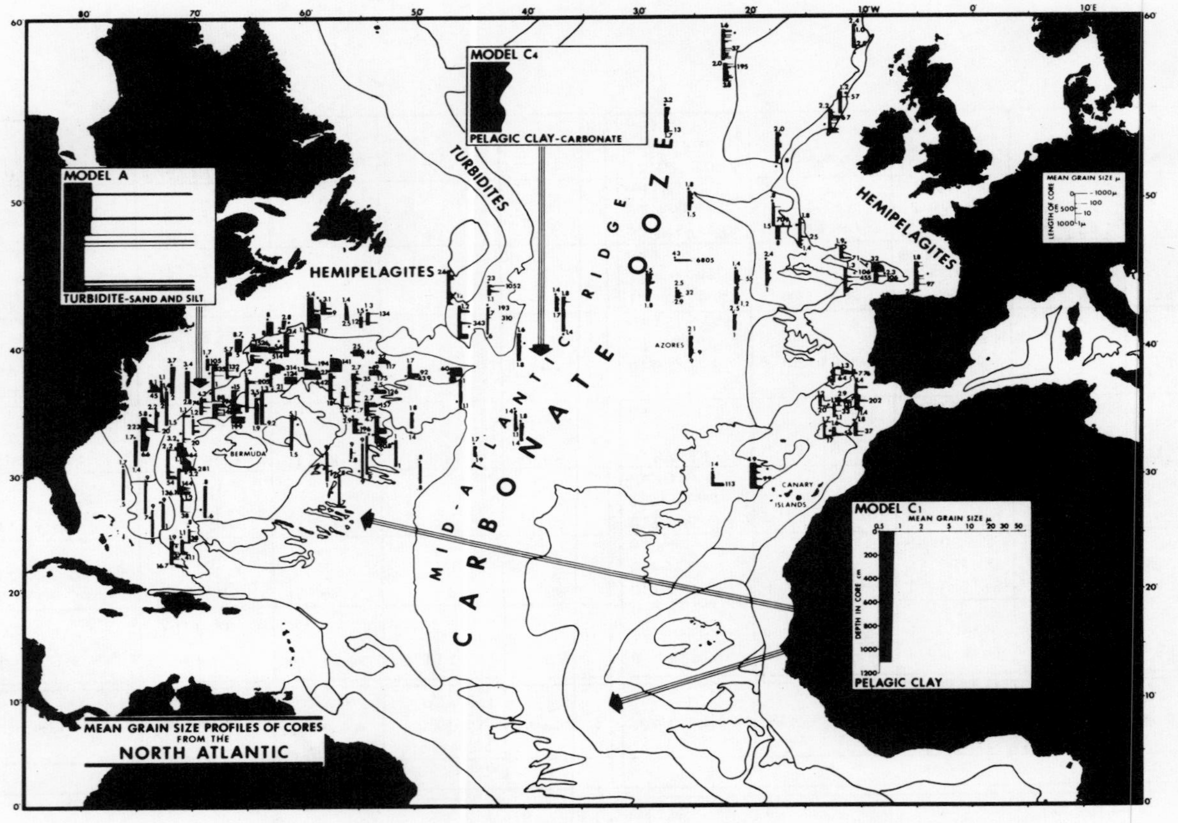

Figure 6. Mean grain sizes of the top and bottom of cores are shown as well as that of the coarsest layer. Abrupt increases of grain size of cores obtained from abyssal plains mark additions by turbidity currents. Turbidites are the coarsest sediment in the North Atlantic. Red clays are the finest (Mz = 1μ). Carbonate oozes of the Mid-Atlantic Ridge are only slightly coarser (Mz = 1 to 2μ) than the clays. Profiles of mean grain size of representative cores of turbidites and pelagic sediments are given. Scales of models are similar to those shown for Model C_1.

Siliceous Deposits - Diatomaceous Clay

There is a broad province at the northern limit of the North Pacific Basin which consists of pelagic sediments containing mixtures of diatom frustules, ice-rafted debris, volcanic silt and finest terrigenous fractions from neighboring land areas (Fig. 2, province 3). Characteristic of many of the cores from the province is extensive mottling. Clear-cut beds of white and brown ash are often intercalated with the diatomaceous sediment (Fig. 3). The ash layers offer good sub-bottom reflectors where beds are thick and topographic relief is subdued.

The average mean grain size of diatomaceous clays is 2.42μ. They are the coarsest pelagic sediment of the North Pacific. Volcanic glass and relatively coarse-grained, ice-rafted material constitute the silt and sand fractions. The average wet density of 332 samples is 1.34 g/cc. This value is intermediate between measurements obtained on red clay and radiolarian deposits. Average moisture content and porosity is 152% and 77%, respectively.

Carbonate Deposits

Only two cores from carbonate oozes of the equatorial North Pacific were analyzed for texture. Based on this very limited data, the average mean grain size of the samples is 1.78μ. Considerably more data are available for similar deposits of the Mid-Atlantic Ridge. Here the average mean size is 1.72μ for uniform clay-sized carbonate on flanks of the Ridge. At its crest there is evidence of winnowing of fines by currents and, consequently, concentrations of coarse size fractions. In such instances the silt and sand are concentrated in irregular layers within the carbonate sections. Much of the coarse detritus is skeletal material of Foraminifera in all states of preservation. The ill-defined coarse layers of the ridge crest have an average and range of particle size different from carbonates accumulating on its lower flanks. A typical sample from the crest has a mean size of 2.50μ and a range of values from 1.30μ to 21.39μ.

A large number of samples of carbonate ooze have been analyzed for bulk properties (Table 2). Results indicate carbonates of the equatorial North Pacific have a wet density of 1.41 g/cc, moisture content 120%, and porosity 74%. These properties are similar to those of red clays of the North Pacific.

Carbonate oozes of the Mid-Atlantic Ridge have average wet density, moisture content, and porosity of 1.52 g/cc, 87%, and 68%, respectively. The results suggest the oozes contain less water and have lower porosities than their counterparts in the Pacific.

Pelagic Clays

The provinces of "pelagic clays" of the North Atlantic (Fig. 5, province 3) are, in fact, intimate mixtures of two end members: red clay and fine-grained carbonate. The latter is swept from the Mid-Atlantic Ridge and settles on adjacent regions dominated by deposition of red clay. The product of mixing is either calcareous red clay or alternating layers of pure red clay and calcareous red clay with a wide spectrum of values of carbonate content. The average wet density, moisture content, and porosity is 1.51 g/cc, 89%, and 70%, respectively. As might be expected, the values lie between those for red clay and carbonate ooze.

Turbidites

Turbidites of the Northeast Pacific have been described and analyzed by Horn et al. (1971, 1972a). They consist of graded layers of very fine-grained sand and medium-grained silt which grade upward to clay. Representing sediment deposited during catastrophic events, they were emplaced during a late phase of the Pleistocene. Modern turbidity currents are limited to channels and upper reaches of submarine fans (Griggs and Kulm, 1970). Thickness and textural properties of turbidite layers are a function of the position of a sampling point relative to the routes of major avenues of turbidity flows across the ocean floor (Horn et al., 1972a).

Upper clay units of graded layers in the North Pacific have an average mean grain size of 1.06μ and range from 0.78 to 1.36μ. Silt units often constitute the base of turbidite layers and have an average mean size of 21μ and range from 4.06 to 60.70μ. Deep-sea sands which comprise the base of some graded layers have an average mean size of 128μ and range from 58.7 to 242.6μ.

Graded layers of sand and silt similar to those of the North Pacific are very extensive in the North Atlantic (Fig. 5, province 2). Material eroded from North America is spread over the ocean floor by turbidity currents. In time, this levelling process has buried topography through constructive sedimentation and has continued to the point where we now have the vast Sohm, Hatteras, and Nares Abyssal Plains. Turbidity flows of the eastern North Atlantic, operating in a similar manner, have filled the area between the east flank of the Mid-Atlantic Ridge and the continental slopes off Africa and Europe. Here sedimentation has produced the Biscay, Iberian, Tagus, Horseshoe, Madeira and Cape Verde Abyssal Plains (Horn et al., 1972a).

Turbidites of the North Atlantic are slightly coarser than those of the North Pacific (Fig. 6). This is as expected with the availability of large amounts of glacial debris along many northern shores, unobstructed transfer of terrigenous sediment, and large-scale transfer of material by turbidity currents right up to modern times (Heezen and Ewing, 1952).

The upper clay units of resulting graded layers have an average mean grain size of 1.13μ, silts 28.72μ, sands 136.17μ, and gravel 1526μ (only one sample of gravel was available as such coarse units are rare in the North Atlantic). In Figure 6 are shown profiles of mean grain size of cores from the North Atlantic along with the grain size of the coarsest layer in each core. The turbidites are generally restricted to areas of little or no relief. The perfect geometry and multiple sub-bottom reflectors of most abyssal plains of the North Atlantic offer excellent interfaces for reflection of sound.

Hemipelagites

Hemipelagites occur on the slope and upper rise of all continental margins. They represent rapid deposition of debris weathered and transported from continents. The average mean particle size of hemipelagic clays and muds of the North Pacific is 3.39μ, wet density - 1.51 g/cc, moisture content - 90%, void ratio - 2.31, and porosity - 67%. Samples from the North Atlantic have an average mean size of 3.24μ, wet density - 1.58 g/cc, moisture content - 80%, and porosity - 65%. Although the values for the two oceans are somewhat similar, the range for North Atlantic samples is considerably greater (Table 2).

VOLCANIC ASH LAYERS

Layers of volcanic ash are widespread in the North Pacific (Fig. 2). White ashes (163 layers) range in thickness from 1 to 29 cm and have an average thickness of 6.5 cm (Horn et al., 1969). They blanket one-fourth of the floor of the North Pacific. Texturally they are medium-grained silts with an average mean grain size of 26μ. They are products of explosive subaerial volcanism of highly siliceous magma. Brown ashes occur east of 180° and are often intercalated with white layers. They range in thickness from 1 to 13 cm and the average thickness of 82 layers is 3.9 cm. The brown ashes are also medium-grained silts (average Mz = 26μ) with textural properties almost identical to those of their white counterparts. There is no evidence of ash layers in cores from the turbidite province of the northeast corner of the Pacific.

Bulk properties of the ashes contrast markedly with those of the pelagic clays with which they are associated. The average properties of 139 samples include wet density - 1.75 g/cc, moisture content - 40%, and porosity - 48%. In many instances, ash layers have properties which will result in the reflection of sound. They are recorded by 3.5 kHz precision depth recorders and in regions of moderate to

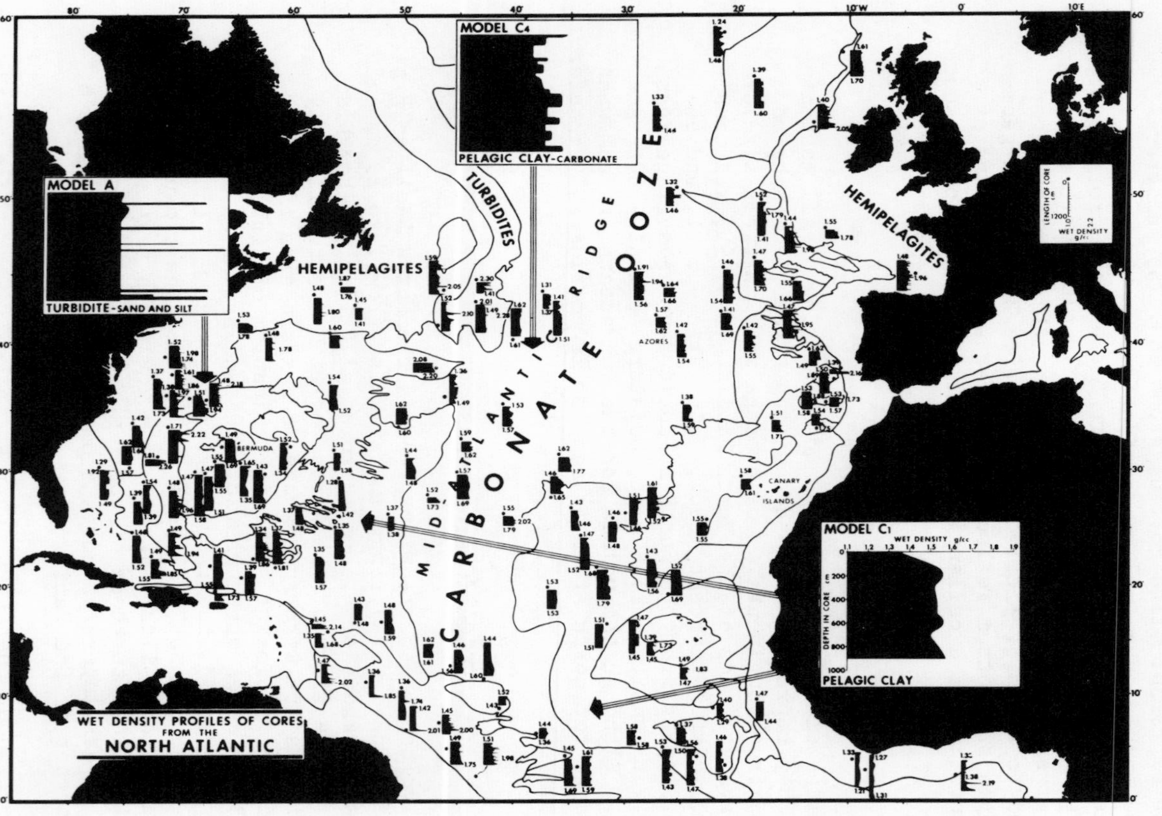

Figure 7. Highest wet densities belong to the deep-sea sands emplaced by turbidity currents. They have values in excess of 1.80 gm/cm^3. Intermediate values were obtained on samples from the Mid-Atlantic Ridge where fluctuating carbonate content and texture often give a sawtooth-like profile. Lowest values indicate sites of red clay and pelagic clay from lower portions of the flanks of the Mid-Atlantic Ridge. Representative profiles of wet density of cores of turbidites and pelagic sediments are given. Scales of models are similar to those shown on Model C_1.

negligible relief they can serve as reflectors of sound.

CURRENT APPLICATIONS OF MAPS OF DEEP-OCEAN SEDIMENTS AND THEIR PHYSICAL PROPERTIES

Ocean mining companies, university researchers, and branches of the United States Navy are the primary users of data on the distribution of abyssal sediments and their physical properties. United States, West German, French, Japanese and Canadian firms with an interest in exploitation of ferromanganese nodules have an immediate application for such information. They need sediment maps as aids in the selection and characterization of mining sites. Information on the physical nature of the seabed is employed in the design of complex dredging devices. It is anticipated that at least three major mining programs will be in production shortly, each raising 10,000 tons of nodules per day, with an estimated annual value of metals recovered set at $2,000,000,000. They will mine nodular ferromanganese deposits within province 4 shown in Figure 2. This is a province dominated by radiolarian ooze and radiolarian clay. It is these deposits which have the lowest wet density and highest porosity and moisture content of any ocean-floor material. Dredge heads then must be designed to hover over or roll lightly across the sea floor without ploughing into the substrate. Because of the complexity of problems related to obtaining large numbers of nodules without raising appreciable amounts of clay, industry is investing millions of dollars to gain an understanding of the properties of both nodules and substrate.

Recent public concern for protection of the environment has been a topic of research of companies planning to exploit ferromanganese nodules as a source of copper, nickel and cobalt. Large-scale dredging operations which involve raking the sea floor for nodules have been equated by some environmentalists to strip mining on land. However, it was the mining companies who took the lead and initiative in this area. For example, at the outset of their research and development program, one organization permitted marine biologists and physical oceanographers from U.S. oceanographic institutions to monitor the effects of dredging at test sites in the Atlantic and Pacific. Preliminary studies have shown little evidence of damage to the seabed or the overlying waters. The tests have been on a very small scale and the results may not represent the situation when mining systems raise 10,000 tons of nodules per day. The National Oceanic and Atmospheric Administration will begin support of further monitoring of mining tests during fiscal year 1975. Included are studies of the substrate and nodular material. The short- and long-term effects of raising the nodules, throwing fine-grained sediment back into suspension in the water column, and scraping the ocean floor with large dredge heads are not known. The relative amount of potential damage to the ocean environment by each of the major mining systems is also

It would seem that this is the time to investigate potential hazards of future mining, yet words of most scientists have fallen on deaf ears. In order to avoid possible errors which could lead to international disputes and permanent damage to the ocean environment, it appears that government should sponsor studies which include data acquisition at sea followed by laboratory analysis and research. Efforts in this direction will lead to a better understanding of ocean floor materials. Without a basic amount of fundamental data on the properties of the nodules and their clay substrate, it is not possible to predict the effects of ocean mining on the marine environment and its inhabitants.

As the same time as commercial activities are moving forward, there is a small research program at universities in the United States directed toward definition of the distribution and genesis of ferromanganese nodules. It is sponsored by the International Decade of Ocean Exploration of the National Science Foundation. Ships from the University of Hawaii, Scripps Institution of Oceanography, and West Germany's R/V VALDIVIA will carry scientists into the area of radiolarian sediments of the equatorial North Pacific. They will collect new data on the physical properties of the sea floor using latest oceanographic technology. As mentioned earlier, the sediments are unusual. Their distribution coincides with the richest Cu-Ni deposits on the sea floor (Horn et al., 1973b). Discovery of this relationship has led to speculation that extremely high porosities of these siliceous materials play an important role in facilitating upward flushing of metal-bearing solutions (Raab, 1972). Some workers feel the metals are added to nodules at the sediment-water interface. The authors suggest that the age of nodules is important. The older the nodule and its substrate, the higher are the values of copper and nickel. The mechanics and chemistry of nodule growth require extensive knowledge of the physical properties of both nodules and substrate. Because the age, physical properties and other parameters correlate with regions of deposits of economic value, far more analytical work on the substrate is needed to fill the present gap in our understanding of both economic and academic oceanology involved in nodule studies. Physical property data may lead to an understanding of the origin of the greatest metal resource of the oceans.

Acousticians of the United States Navy have for many years studied the correlation between properties of ocean sediments and successful operation of bottom-bounce sonars. Their aim has been to determine regions of the ocean floor which will serve as an acoustic interface and reflect sound at high levels. In Figure 8 we give an example of a core from the Mediterranean Sea (Messina Abyssal Plain) analyzed by the writers (Horn et al., 1968). It shows that there is a strong correlation of sound velocity with several other physical properties. Because this important relation exists, it has been reasoned that sound directed at the seabed should be affected to varying degrees within provinces of contrasting sediments. The provinces of

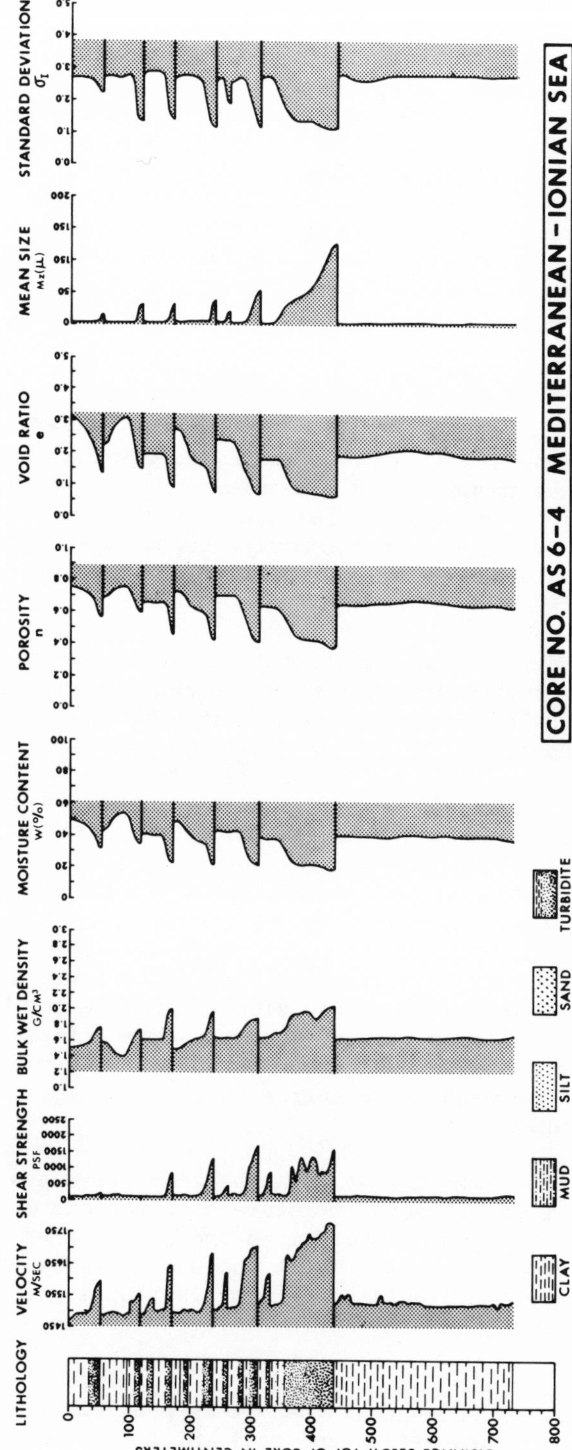

Figure 8. Correlation of lithology, sound velocity, shear strength, physical and textural properties of a turbidite core (modified after Horn et al., 1968). This is an excellent example of a turbidite sequence and illustrates the strong tie between acoustic and other physical properties of cores. There is a positive correlation between velocity and shear strength, wet density, and mean grain size; and a negative correlation between sound velocity and moisture content, porosity, and void ratio. The correlation between sound velocity and physical and textural properties reveals the need for regional maps similar to those shown in Figures 2, 3, 6, and 7.

pelagic clay intercalated with discrete ash and turbidite layers do reflect sound where topography is not too rough. The thickness of the reflective horizons and their frequency within the sediment section influences the level of sound reflection. Models of sedimentary provinces of the North Pacific and North Atlantic are given in Figures 3, 4, 6, and 7. It is mainly texture of layers which affects wet density and, in turn, wet density which controls the acoustic impedance of strata. In short then, regional mapping of sediments of the abyss is fundamental to programs of research into levels of sound reflection at the sea floor.

CONCLUSIONS

This survey shows that it is possible to divide the seabed into vast areas of similar sediment. Each has been called a province named after the dominant sediment. With a knowledge of the distribution of sediment provinces in hand, it has been possible to determine the physical properties of each major marine deposit. In only one instance are the differences in bulk properties slight (i.e., pelagic clay and carbonate ooze).

Radiolarian ooze is quite different from other sediments because it has an extremely low wet density, high moisture content and high porosity. Red clays and calcareous oozes have nearly the same intermediate values of bulk properties. Northern diatomaceous clays of the Pacific contain more water than either red clay or carbonate ooze. Hemipelagites are twice as coarse-grained as pelagic deposits and have lower porosities and moisture contents.

Both textural and physical properties of turbidites show grading. Wet density values decrease toward the top of a turbidite whereas the moisture content, void ratio and porosity increase in this direction. The progressive change of bulk properties within a turbidite is a response to decreasing grain size toward the top of each bed. Turbidite sand and silt is more than a hundred times coarser than the gray clay laid down at the close of turbidity flows. A similar relation exists between these coarse basal units and surrounding or intervening pelagites.

Ashes are generally 26 times coarser than pelagic clays associated with them. Their porosities and moisture contents are low, their wet densities high. Ashes are not given province status because they are products of airborne transport and not a result of the regime of ocean sedimentation. Although distribution is generally restricted to one province in the North Pacific, this is not always true.

The results of this program suggest that it is possible to identify provinces by their bulk properties and sediment type. There is a continuing need of sediment maps and physical property data on abyssal deposits. The information is direct input into current exploration programs of mining companies, university research on genesis of the most important metal resource of the oceans, studies of the environmental impact of ocean mining, and problems included in systems analysis of the United States Navy.

ACKNOWLEDGMENTS

Cores were taken by scientists and crews of the research vessels VEMA and ROBERT D. CONRAD of Lamont-Doherty Geological Observatory and ATLANTIC SEAL of Texas Instruments, Inc. Research and technical assistance were provided by Mary Parsons, Lillian Sussilleaux and Ivana Buric. Illustrations were drawn by Virginia Rippon. This paper is Lamont-Doherty Geological Observatory Contribution No. 2019.

Preparation of the manuscript was supported by the U.S. Naval Ship Systems Command (N00024-72-C-1152) and a grant from the National Science Foundation (GX-33616). Maintenance of the Deep-Sea Core Library at Lamont-Doherty is supported by the Office of Naval Research (Contract N00014-67-A-0108-0004) and the National Science Foundation (GA-35454).

REFERENCES

Ericson, D. B., M. Ewing, G. Wollin, and B. C. Heezen, Atlantic deep-sea sediment cores, Bull. Geol. Soc. Amer., 70, 193-286, 1961.

Folk. R. L., Petrology of Sedimentary Rocks, Hemphill's, Austin, Texas, 1968.

Folk, R. L., and W. C. Ward, Brazos River bar: a study in the significance of grain size parameters, J. Sediment. Petrol., 27, 3-26, 1957.

Griggs, G. B. and L. D. Kulm, Sedimentation in Cascadia Deep-Sea Channel, Bull. Geol. Soc. Am., 81, 1361-1384, 1970.

Heezen, B. C., and M. Ewing, Turbidity currents and submarine slumps, and the 1929 Grand Banks earthquake, Am. J. Sci., 250, 849-873, 1952.

Horn, D. R., M. N. Delach, and B. M. Horn, Distribution of volcanic ash layers and turbidites in the North Pacific, Bull. Geol. Soc. Am., 80, 1715-1724, 1969.

Horn, D. R., M. N. Delach, and B. M. Horn, Metal Content of Ferromanganese Deposits of the Oceans, NSF/IDOE-GX33616, Lamont-Doherty Geological Observatory Tech. Rept. 3, 1973a.

Horn, D. R., J. I. Ewing, and M. Ewing, Graded-bed sequences emplaced by turbidity currents north of 20°N in the Pacific, Atlantic and Mediterranean, Sedimentology, 18, 247-275, 1972a.

Horn, D. R., B. M. Horn, and M. N. Delach, Correlation between acoustical and other physical properties of deep-sea cores, J. Geophys. Res., 73, 1939-1957, 1968.

Horn, D. R., B. M. Horn, and M. N. Delach, Sedimentary provinces of the North Pacific, Geol. Soc. Am. Mem. 126, 1-21, 1970.

Horn, D. R., B. M. Horn, and M. N. Delach, Ferromanganese deposits of the North Pacific, NSF/IDOE-GX33616, Lamont-Doherty Geological Observatory Tech. Rept. 1, 1972b.

Horn, D. R., B. M. Horn, and M. N. Delach, Ocean Manganese Nodules: Metal Values and Mining Sites, NSF/IDOE-GX33616, Lamont-Doherty Geological Observatory Tech. Rept. 4, 1973b.

Horn, D. R., M. Ewing, M. N. Delach, and B. M. Horn, Turbidites of the Northeast Pacific, Sedimentology, 16, 55-69, 1971.

Jacobs, M. B., and J. D. Hays, Paleo-climatic events indicated by mineralogical changes in deep-sea sediments, J. Sediment. Petrol., 42, 889-898, 1972.

Keller, G. H., and R. H. Bennett, Mass physical properties of submarine sediments in the Atlantic and Pacific Basins, XXIII Intern. Geol. Cong., 8, 33-50, 1968.

Keller, G. H., and R. H. Bennett, Variations in the mass physical properties of selected submarine sediments, Mar. Geol. 9, 215-223, 1970.

Raab, Werner, Physical and chemical features of Pacific deep-sea manganese nodules and their implications to the genesis of nodules, in Papers from a Conf. on Ferromanganese Deposits on the Ocean Floor, edited by D. R. Horn, pp. 31-49, IDOE/ Nat. Sci. Fdn., Washington, D. C., 1972.

Revelle, R., M. Bramlette, G. Arrhenius, and E. D. Goldberg, Pelagic sediments of the Pacific, Geol. Soc. Am. Spec. Paper 62, 221-236, 1955.

Sutton, G. H., H. Berckhemer, and J. E. Nafe, Physical analysis of deep-sea sediments, Geophysics, 22, 779, 1957.

THE APPLICATION OF STUDIES OF MARINE SEDIMENT DYNAMICS

JEAN-PIERRE MIZIKOS

Elf RE

ABSTRACT

Possible applications of acoustical properties to present investigations of soil and sea-bottom sediment dynamics are presented. Conclusions are rather pessimistic concerning application of relationships between the acoustics and mechanics of marine sediments. Engineering progress will probably come from an improvement of soil concepts and testing methods.

THE RELATIONSHIP BETWEEN ACOUSTICS AND SOIL MECHANICS

Much work on unconsolidated marine sediment acoustics has been done by persons such as Breslau (1965), Hamilton et al. (1970), and Hastrup (1966, 1968, 1969). With the exception of Hastrup's work on reflectivity, this is empirically based research which has resulted in correlation between velocity or reflectivity of P waves and geotechnical or mechanical parameters. Hamilton indicates that such correlations can be accurate only on a local basis. Since empirical regression curves seem unable to provide a universal law for the relationships, the author undertook a theoretical study of acoustic wave propagation in marine sediments. The study indicates that (Mizikos, 1971):

1. For the acoustical pressure amplitudes usually transmitted in geophysics, measured wave velocities and formulas for predicting wave velocities are identical or similar to those for elastic media. Only the attenuation coefficient is sensitive to the difference between ideal elasticity and actual sediment behavior. In other words, under usual geophysical conditions, sea-bottom sediments exhibit linear dynamic response.

2. Only for muds, i.e. clayey sediments of 60 to 80% porosity, can compressional wave velocity be rigorously related to porosity. Wodd's formula (1941) applies only to the so-called flocculates, and not to other types of marine sediments. Flocculates are very porous muds (porosity exceeding 80%) in a non-rigid (liquid) state.

3. The decrease in wave velocity-porosity correlation with decreasing porosity (i.e. with increasing proportion of sand) results mostly from Biot's dynamic mass coupling (1956) between vibrating grains and pore water. This coupling modifies the porosity value (a static property) during vibration.

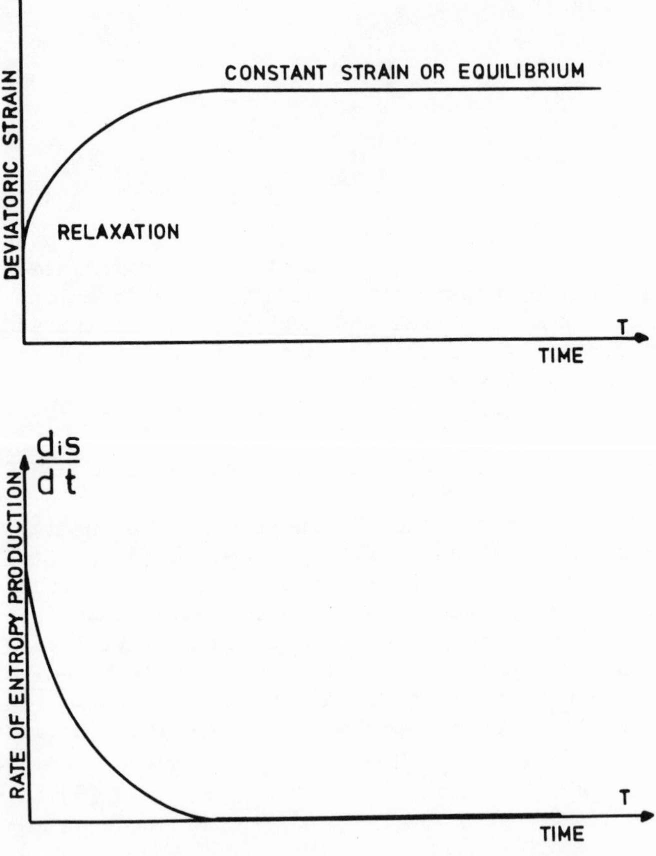

Figure 1. Upper part: Example of linear visco-elastic creep curve corresponding to a linear stress-strain relationship

Lower part: Rate of entropy production curve deduced from the upper curve. Note that final rest state corresponds to $d_i S/dt = 0$.

EXAMINATION OF THE FUNDAMENTAL PHYSICS OF SOIL MECHANICS

Theory

Evolution of particle constraints appears to be the main starting point for a fundamental consideration of soil mechanics. Existing concepts of cohesion and friction angle are based on this property. A theoretical study of this point of view could use non-equilibrium statistical mechanics, for example correlation dynamics could be considered in a generalized space (Prigogine, 1966). This microscopic and fundamental point of view remains beyond the engineer's macroscopic purview. The engineer is obliged to work only in terms of the macroscopic properties which he can measure. Nevertheless, macroscopic effects are statistically determined by microscopic dynamics and bonding structures.

In beginning this new means of investigation, the best macroscopic synthetic concept seems to be entropy production. See, for example, Prigogine's discussion of non-equilibrium thermodynamic methods (1968) or Thompson's paper in this volume. The rate of entropy production can be obtained from a creep test by the following relationship.

$$\frac{d_i S}{dt} = \frac{\sigma}{\tau} \cdot \frac{d\varepsilon}{dt}$$

where S = entropy

t = time

τ = absolute temperature

σ = constant axial stress

ε = axial strain.

The two kinds of commonly observed creep curves are illustrated in Figures 1 and 2. Also shown are curves of the rate of entropy production. Notice that, in agreement with the fundamental theorem of minimum entropy production, the rate of entropy production decreases with time for both curves.

The initial high rate of entropy production results from an internal movement of the grains and of the pore water as a whole. This is confirmed by the fact that the beginning of the creep curve is approximately linear with a slope equal to $1/\eta_2$ (η_2 is the bulk viscosity coefficient). The soil exhibits only a viscous bulk resistance.

The curved part of both types of creep curve indicates a relaxation process before the equilibrium state is reached ($d_iS/dt = 0$) or before creeping takes place. Creep is also a viscous flow but with a larger coefficient, $\eta_1 >> \eta_2$. This inequality shows that soil resistance is higher at the end of relaxation that at the beginning.

Experiments on two-dimensional cylinder packing can explain this conclusion: relaxation corresponds to the build-up of a new internal honeycomb structure of cylinders (or a strength hardening in the general case); such a structure represents a so-called "dissipative structure" (e.g. Benard's honeycomb cell in hydrodynamics--see Prigogine, 1968). If this structure can bear the applied pressure, the soil system reaches a new state of equilibrium rium. If not, the new structure creeps but presents more resistance to applied stress than the initially less-ordered arrangement of grains or bondings. Consequently, creep is resisted more than the initial short flow relaxation and $\eta_1 >> \eta_2$. When the dissipative structure can no longer exist, failure occurs.

These results reveal that creep limit is the exact failure criterion. It corresponds to the end of the linear part of stress-strain curves or shear curves. Although its numerical value varies according to boundary conditions, the variation law is still unknown. Creep limit also corresponds to the slope of the linear part of σ versus log ε curves. Primary results of this theory are the following (Mizikos, 1972):

a) for a linear stress-strain relationship (Fig. 1), the creep equation is

$$\frac{\varepsilon(t)}{\sigma} = \frac{1}{G_2}\left[1 - \exp(G_2 \cdot t/\eta_2)\right].$$

The stress-strain equation, if relaxation is finished, is

$$\sigma = G_2 \cdot \varepsilon$$

The soil behaves like a Kelvin-Voigt model. It is therefore incorrect to compare linearity of the curves with those for elastic models because of the existence of relaxation and permanent residual strain (or settlement) in the loading-unloading cycles.

b) when linearity ceases, which occurs for the so-called critical stress or creep pressure or creep limit σ^*, the previously constant modulus G_2 becomes: (equation $\longrightarrow$)

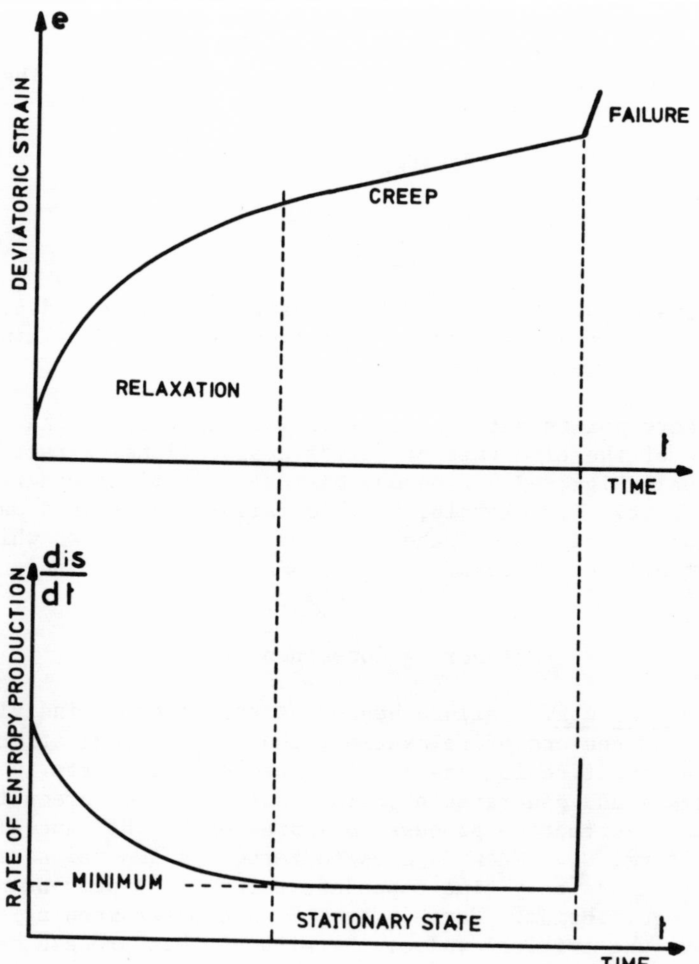

Figure 2. Upper part: Example of viscoliquid creep curve corresponding to an exponential stress-strain relationship

Lower part: Rate of entropy production curve deduced from the upper one. Note that creep corresponds to a stationary process: d_iS/dt is constant.

$$G_1 = G_2 \exp\left[- (\sigma - \sigma^*)/\sigma^*\right],$$

while η_2 remains constant. The creep equation is then (Fig. 2):

$$\frac{\varepsilon(t)}{\sigma} = \frac{t}{\eta_1} + \frac{1}{G_1}\left[1 - \exp\left(- G_1 t/\eta_2\right)\right].$$

The term t/η_1 corresponds to pure creep. For an initial period, the right hand side of the equation is very similar to the curves noted by many authors (e.g. Murayama and Shibata, 1961). The stress-strain curves become exponential:

$$\varepsilon = \varepsilon^* \cdot \exp\left[(\sigma - \sigma^*)/\sigma^*\right]$$

ε^* = critical strain

σ^* = critical stress.

Two successive exponentials are often observed, the second corresponding to development of cracks (variation of η_1). This kind of behavior is called viscoliquid.

The theory points out the decisive influence on engineering measurements of the time rate of stress applications, a factor which is usually ignored. Numerous misinterpretations can result from its neglect. For example, visible failure represents mainly an artificial retardation of the creep or failure process which is initiated at critical stress.

Engineering Consequences

Failure strength. Failure begins at the level of individual grains when, at the end of relaxation, the new internal structure or hardening cannot resist the applied stress. It starts in some isolated points and generates a chain reaction up to "creep time". This internal destruction process is expressed by the cancelling of the modulus term, G_2. When G_2 goes to zero, the general stress-strain equation has only time-dependent terms, thus preventing attainment of equilibrium. Furthermore, G_1 decreases when stress increases above the critical value. Details of this process depend upon the boundary conditions.

Exact failure stress appears then as the critical stress, Menard's creep stress or the usual creep limit. Macroscopic and noticeable failure occurs at a higher value than the critical stress because stress increases above the critical value faster than the rate of the microscopic failure process. Safety coefficients usually correct this discrepancy.

Relationship between acoustics and soil mechanics. It may be possible to determine critical stress from a comparison of frequency spectra of acoustical signals reflected from the top and the bottom of soft sea-bottom layers.

When any sediment is traversed by an acoustic wave and behaves as a quasi-elastic medium, the attenuation coefficient is proportional to frequency.

Above the elastic limit, the sediment behaves as a linear visco-elastic medium. Dynamic visco-elastic creep curves are illustrated in Figure 3. If the acoustic signal duration is short compared to the relaxation time, the attenuation coefficient will be proportional to the square of frequency:

$$\alpha = \frac{\omega^2}{2VG_2}$$

with $\omega = 2\pi f$

and V = acoustic wave velocity.

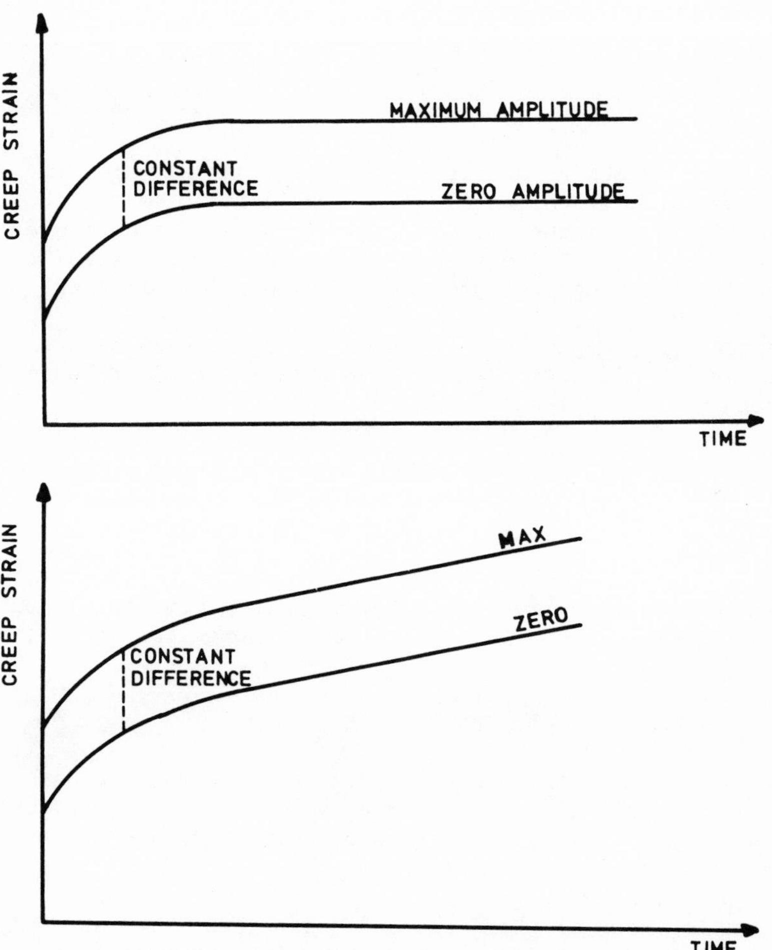

Figure 3. Examples of viscoelastic and viscoliquid dynamic creep curves. Signal is sinusoidal. Note the existence of a constant strain difference between minimum and maximum amplitudes.

When the pulse duration is longer than the relaxation time, a new structure has been established by the grains and propagation is quasi-elastic (but with a modulus G_2 different from the primary elastic one) so that:

$$\alpha = k.f$$

In the general case: $\alpha = k.f^n$ with $1 \geq n \geq 2$.

When the acoustic amplitude is higher than the critical value, the dynamic creep curve, as illustrated in Figure 3, is obtained. Based on G_1 being constant, attenuation depends on frequency only through V. That is:

$$\alpha = \frac{G_1}{2V\eta_1}$$

Generally speaking, an attenuation law such as: $\alpha = k.f^n$ with $n < 1$ is possible. Consequently $n < 1$ would indicate that acoustic stress is higher than the creep value. This point required experimental verification. To carry out such a measurement would require a broadband, high frequency projector or special explosives.

The frequency dependence of attenuation may provide a survey tool for sediment identification. For example it might be determined that in a given locality:

$$\alpha = k.f^2 \text{ for muds}$$

$$\alpha = k.f^n \text{ with } 1 \leq n \leq 2 \text{ for intermediate materials}$$

$$\alpha = k.f \text{ for sands.}$$

Referring to this local identification, seismic profiling could provide a remote sediment identification in some like vicinities.

ACKNOWLEDGMENTS

The author expresses his gratitude to the Direction des Recherches et Moyens d'Essais - Paris, who partly supported this work. The help of Ms. A. M. Mizikos for computation and granulometric analysis is greatly appreciated.

Mr. A. Anderson of the Applied Research Laboratories, University of Texas kindly assisted the author with the translation of his ideas into proper English. His help is gratefully acknowledged.

REFERENCES

Biot, M. A., Theory of elastic waves in a fluid saturated porous solid, J. Acoust. Soc. Am., 28, 168-191, 1956.

Breslau, L., Classification of sea-floor sediments with a shipborne acoustical system, Le Petrole et la Mer, 132, 1965.

Hamilton, E. L., Sound velocity and related properties of marine sediments, North Pacific, J. Geophys. Res., 75, 4423-4446, 1970.

Hastrup, D. F., Reflection of plane waves from a solid multilayered damping bottom, SACLANT ASW Research Centre Tech. Rept. TR50, La Spezia, Italy, 1966.

Hastrup, D. F., The reflectivity of the top layer in the Naples and Ajaccio abyssal plains, SACLANT ASW Research Centre Tech. Rept. TR118, La Spezia, Italy, 1968.

Hastrup, D. F., The effect of periodic bottom layering on acoustic reflectivity, SACLANT ASW Research Centre Tech. Rept. TR149, La Spezia, Italy, 1969.

Mizikos, J. P., Practical application of relations between mechanical and acoustical properties of marine sediments, in Intern. Symp. Eng. Properties of Sea-Floor Soils and their Geophys. Ident., pp. 279-287, Seattle, 1971.

Mizikos, J. P., Mechanisms of soils evolution from laboratory testing equipment, submitted for publication to Géotechnique, Part of a State Doctor degree thesis, Nice University, (AO.6425), 1972.

Murayama, S., and T. Shibata, Rheological properties of clays, Proc. 4th Intern. Conf. Soils Mech. and Fdn. Engrs., I, 99-129, 1961.

Prigogine, I., Non-equilibrium Statistical Mechanics, Wiley Interscience Publ., N. Y., 1966.

Prigogine, I., Introduction to Thermodynamics of Irreversible Processes, J. Wiley & Sons, Inc., N. Y., 1968.

Wood, A. B., A Textbook of Sound, G. Bell and Sons, London, 1941.

DEEP-SEA CARBONATES

RICHARD REZAK

Texas A&M University

ABSTRACT

Calcium carbonate in sediments can greatly effect the engineering properties of those sediments. Factors that may effect engineering properties are: 1) great variation in primary void ratios, (2) inherent strength of carbonate grains, (3) degree of sorting, (4) ratio between grains and matrix, (5) nature and amount of cement, and (6) depth of burial. Deep-sea carbonates appear to be anomalous but do exist in rather large quantities. Their existence must depend upon a combination of physico-chemical and organic-chemical factors that need to be investigated in much more detail.

INTRODUCTION

The presence of calcium carbonate in deep-sea sediments can greatly effect the physical properties of the sediments in two very important aspects. Carbonate sediments do not compact to as low a void ratio as noncarbonate sediments and they are very susceptible to post depositional alteration, i. e. dissolution, cementation, compaction, comminution by biological activity on the sea floor (bioerosion), and chemical alteration due to bacterial activity.

More than a third of the present day deep-sea floor is covered by sediments containing over 30% calcium carbonate. This carbonate is primarily biogenic although in some areas physico-chemical precipitates of calcium carbonate do exist. The biogenic fraction is composed primarily of the skeletons of pelagic organisms living in the upper parts of the water column. In order of abundance, these organisms are, foraminifers, coccolithophorids, and pteropods. The foraminifers and coccolithophorids are composed of low-Mg calcite with minor amounts of high-Mg calcite contributed by some foraminifers. Pteropods are aragonitic and because of their greater susceptibility to dissolution are not very abundant at great depths. In addition to pelagic sedimentation, minor, although locally significant amounts of benthic shallow-water skeletal material are transported into the deep-sea environments either by slumping, density flow, turbidity currents, or simple gravity transport.

The accumulation and cementation of calcium carbonate at oceanic depths seems anomalous in view of the evidence that the solubility of $CaCO_3$ increases with depth. The purpose of this paper is to discuss some deep-sea occurrences and to suggest possible lines of future research.

The term consolidation as used by engineers means something quite different from the geologist's definition of the term. To the geologist, consolidation is a synonym for lithification and includes both physical and/or chemical processes. Therefore, rather than use the term consolidation in the present paper, I use the term compaction which denotes a reduction of pore space and total volume due to overburden pressure.

PHYSICAL PROPERTIES OF CARBONATE SEDIMENTS

Carbonate sediments become lithified either by cementation or compaction. Cementation may occur in the subaerial environment either in the vadose or phreatic ground water zones. Cementation may also occur in the marine environment either in the intertidal zone (beachrock) or the subtidal environments (submarine cements). In all of these environments the effects of compaction are either negligible or absent. The cementation process is simply a growth of crystals on the surfaces of sediment grains. On the other hand, carbonate sediments subjected to high overburden pressures are lithified by a combination of compaction and pressure solution.

Cementation at or near the sediment-water interface is ordinarily not pervasive, but occurs in layers or lenses of irregular distribution. This is amply demonstrated by the occurrence of hard grounds in Cretaceous chalks, cores of carbonate sediments taken by the GLOMAR CHALLENGER, and studies of Recent carbonate cements. Cemented layers may have a sharp upper boundary and a vague lower boundary or both upper and lower boundaries may be vague. The areal extent of the cemented layers varies considerably. Well cemented layers alternate with layers that contain no cement.

The primary effects of cementation, regardless of where or how it occurs, are to decrease porosity by the filling of void spaces and to increase rigidity by welding adjacent grains.

Compaction may reduce void ratios in one of two ways. Calcite has well developed cleavage in two directions and aragonite has it in one direction. These cleavage planes present a multitude of gliding surfaces that can readily relieve stresses acting upon the sediment. Under certain conditions of pressure, temperature, and stress, carbonate dissolves at points of grain contact and is redeposited on grain surfaces facing adjacent voids. The resulting rock has a texture of irregular, interlocking grains that gives the rock great strength.

Factors that complicate the interpretation of the engineering properties of carbonate sediments are: (1) extreme variation in primary void ratios caused by variations in the sizes and shapes of carbonate grains, (2) the strength of carbonate grains as compared with clay grains and quartz sands, (3) degree of sorting (matrix is composed of particles less than 10μ in diameter and grains over 10μ in diamter), (4) ratio between grains and matrix (is the sediment grain-supported or mud-supported?), (5) the nature and amount of the cement (fibrous, prismatic, or microcrystalline?), and (6) depth of burial.

This discussion has considered only the cementation of pure carbonate sediments. However, other kinds of sediments can be cemented by carbonate and consequently small quantities of carbonate cement in terrigenous sediments may create a considerable effect on the engineering properties of these sediments.

$CaCO_3$ STABILITY

The saturation of surface waters with respect to $CaCO_3$, in tropical and subtropical regions, has been known for a long time. It also has been recognized that the concentration of $CaCO_3$ in open ocean sediments decreases with depth of water. Lisitzin (1972) defines the critical or compensation depth as that depth at which the amount of calcium carbonate in the bottom sediment decreases to 10%. The critical depth varies from ocean to ocean and also within ocean basins due to a number of variables such as: (1) the rate of carbonate supply, (2) dilution with terrigenous and siliceous material, and (3) dissolution at great depths. Smith et al. (1968) go to great lengths to prove that there is no world-wide compensation depth. Their correlation analysis of the data on 1,350 pelagic sediment samples once again demonstrates that the computer can be used to prove either side of a question depending upon the information that it is fed. In their conclusions, however, Smith et al. state that, "solution of carbonate does appear to be related to depth in the ocean. The low correlation between $CaCO_3$ percentage and depth . . . suggests that important factors not considered in this statistical analysis are also involved."

Experiments by Peterson (1966) and Berger (1967, 1970) indicate sharp increases in the rate of dissolution at about 3,700 m and 3,000 m. Peterson (1966) states that the carbonate compensation depth is due to a sharply increased rate of dissolution rather than to change from supersaturation to undersaturation.

The works of Pytkowicz (1970) and Hawley and Pytkowicz (1969) give theoretical determinations of the depths of supersaturation, saturation, and undersaturation of oceanic water with respect to calcium carbonate. Surface waters are generally supersaturated between 50°N and 50°S Lat. The saturation depth varies with latitude from about 200 m at 50°N Lat. to about 3,500 m at 10°S Lat.

The relative stability of the important $CaCO_3$ minerals in the deep-sea environment is quite different from that in shallow marine waters or in subaerial conditions. Friedman (1964) demonstrated that when aragonite, high-Mg calcite and low-Mg calcite (all equally stable in the shallow marine environment) are exposed to fresh water, then high-Mg calcite is the least stable, aragonite is more stable, and low-Mg calcite is most stable. In deep-sea conditions, Friedman (1965) found that aragonite is the least stable, high-Mg calcite more stable, and low-Mg calcite most stable. This is supported by Li et al. (1970) who found that in Antarctic water south of the Antarctic Convergence the saturation depth for calcite is 2,000 to 3,000 m while for aragonite it is 200 m.

The reason for this increase in solubility of aragonite with depth may be due to the increase in the partial pressure of CO_2 with increasing depth. Weyl (1959) found in laboratory experiments that as the partial pressure of CO_2 increases from 10^{-3} molal initial concentration to 10^{-2}, the solubility of aragonite exceeds that of calcite by 11%. Park (1966) measured approximately three thousand pH values for sea water in the northeastern Pacific Ocean. He found that there exist two maxima and one minimum. The first maximum (8.2-8.3) exists intermittently near the surface within the first 100 m. The second maximum (about 7.9) exists near 4,000 m. The pH minimum (7.5-7.7) ranges in depth from 200-1,200 m and generally exists at the depth of the oxygen minimum. Park concluded that the deep maximum is the boundary between the upper oceanic layer that is influenced biochemically and the lower layer which is physicochemically influenced. The effect of hydrostatic pressure on the dissociation constants of carbonic acid appears to be important in the formation of the boundary.

Carbonates may be preserved in bottom sediments at depths far greater than the compensation depth. Lisitzin (1972) states that carbonates may escape solution and exist 1,000 m or even as much as

2,000 to 2,300 m below the critical depth. He concludes that the critical depth of carbonate accumulation does not depend mainly on the physico-chemical environment but rather on dynamic factors such as: carbonate supply and dissolution rates, and the rates of supply of diluting components and of total sediment.

ORGANIC INFLUENCES

Bramlette (1961) claimed that coccoliths (low-Mg calcite), are not affected by dissolution while settling. Coccoliths, about 5μ in diameter, will settle to 5,000 m in about 10 years but they accumulate on the bottom with foraminifers which settle to the bottom in a few days. This is a contradiction to the evidence produced by Peterson (1966). Some other mechanism must be operating to prevent dissolution. Shrader (1971) cites the role of faecal pellets in the sedimentation of diatoms in deep ocean basins. The pellets being larger than individual diatoms settle at a greater rate and in addition the diatoms are protected from dissolution by an organic membrane. A similar mechanism could be active in coccolith sedimentation.

Another and probably more important mechanism involves organic films that are adsorbed onto the skeletal surfaces. Smith et al. (1968) offer the hypothesis that organic molecules adsorb to pelagic carbonate particles and prevent or inhibit carbonate-sea water reactions. Kennedy and Hall (1967) suggest a mechanism by which amino acids could prevent dissolution of carbonate. The amino and carboxyl groups are polar with positive and negative charges respectively. In water they are hydrophilic and attract water dipoles, while the other two side chains are hydrophobic. At the surface of carbonate, the charged groups in the amino acid molecules would be attached to unsatisfied CO_3^{--} and Ca^{++} ions and the resulting surface layer of amino acid molecules would present an outer surface of hydrophobic groups which would prevent access of water to the crystals. Pytkowicz (1970) credits the rates of decay of organic films or a layer of $CaCO_3$ saturated water on the sediment for the preservation of $CaCO_3$ at depth. Berger (1967) postulates a semi-saturated layer of bottom water to protect carbonates from dissolution.

Mitterer (1972) reports that the proteinaceous matrix is an important aspect of skeletal calcification. He has found that protein is also associated with non-skeletal carbonates such as oolites and "carbonate mud". This protein has a composition "strikingly similar to that in many skeletal carbonates." Mitterer's work lends support to a hypothesis that I have been pushing for the past several years. My hypothesis recognizes organic molecules adsorbed on mineral surfaces as primitive organic matrices (templates) that influence the nature of the minerals deposited on these surfaces. The organic molecules that have been called upon by several authors to protect carbonate from solution will also, under proper pH conditions, act as templates for the precipitation of calcium carbonate polymorphs. During the past few years we have created carbonate cements in the laboratory and influenced the precipitated cements by addition of various organic acids to the solutions.

DEEP-SEA CARBONATE CEMENTS

Fisher and Garrison (1967) reviewed many occurrences of submarine cementation and concluded that contrary to widespread opinion, carbonate sediments can be lithified on the sea floor. Cifelli et al. (1966) reported friable aggregates of foraminifers from the mid-Atlantic Ridge. They suggested that the cement could be produced by raising the ridge to a depth 2,000 m less than that at which the interstitial waters had equilibrated with the sediment. This may be possible but it seems to be a difficult way to produce a cement.

Milliman (1966) found that pelagic sediments from the tops of guyots and seamounts are in isotopic equiligrium with present ambient waters. Certain shallow water deposits that had undergone subaerial diagenesis retained their isotopic composition and showed no tendency to re-equilibrate with new conditions.

Thompson et al. (1968) reported a wide variety of lithified carbonates in the deep sea of the equatorial Atlantic. These include partially lithified foraminiferal oozes, foraminiferal limestones, tuffaceous limestones, recrystallized limestones, and dolostones, ranging in age from Late Miocene through Pleistocene. They found no correlation between paleontological age and lithification.

Cementation of carbonate sediments is not a regular, orderly process. Submarine cemented carbonates are patchy in their distribution and there is not a progressive increase in cementation with depth below the sediment surface. Indeed, layers of well cemented carbonate may alternate with layers that contain absolutely no cement. The hard grounds that are seen in Cretaceous chalks are examples of this kind of cementation. The upper surfaces of some cemented layers are smooth and exhibit boring by organisms. This type probably formed at the sediment-water interface. However, other layers are quite irregular both on their upper and lower surfaces. These layers probably represent cementation at some depth within the sediment.

The irregular nature of these submarine cements is difficult to explain on a strictly physico-chemical basis. I suspect that organic compounds, trapped within the sediments, exert a major influence upon this process. It is entirely possible that calcium organo-complexes are responsible for the development of these cements. Berner (1968) conducted experiments using dead fish in carbonate solutions. The results of his experiments showed the development of calcium soaps, complexes that are very similar to those found in Recent alewife concretions. Berner states that with the passage of time the calcium soaps would degrade to $CaCO_3$ and hydrocarbons.

SUGGESTIONS FOR FUTURE RESEARCH

A considerable amount of experimental work has been reported by tectonophysicists on the deformation of carbonate rocks. This literature should be digested and applied to engineering studies.

Consolidation experiments using different artificial mixtures of carbonates and of carbonate and terrigenous sediments should be conducted. These experiments could be conducted in laboratory produced cementing and non-cementing environments in order to assess the effects of cementation upon the shear strength and consolidation characteristics of the sediments.

Much of our knowledge concerning chemical reactions in the oceanic water column is based upon laboratory experiments and extrapolation from sea-surface reactions. Very little experimental work has been conducted in the natural environment. More in situ measurements of pH are needed both in the water column and in the bottom sediments. Interstitial waters should be analyzed for alkalinity, Ca/Mg ratios, trace elements, and organic content. The solid phase, both in the water column and in the bottom sediments should be subjected to petrographic and geochemical analyses and compared with the water chemistry to determine what changes take place and the reasons for these changes from the time a sediment particle begins to settle towards the bottom until it is buried under several centimeters of bottom sediment. Sites for such studies should be carefully selected in order to realistically assess the influence of organic compounds on carbonate alteration and cementation processes. One such site might be in the vicinity of an active oceanic hydrocarbon seep. Another site might be in an area of known high concentration of biologically derived organic matter. Analyses should include not only the standard geochemical analyses but also mass spectroscopy and X-ray diffractometry to determine the possible presence of calcium organo-complexes.

In conclusion I should like to quote D. H. Welte (1969): "Organic matter in sediments is not just a component left over from former organic life; it plays an active role in the formation and diagenesis of sediments. Unlike the inorganic components, only small amounts, about 1-2% of organic matter, may strongly influence the post depositional behavior of a sediment."

REFERENCES

Berger, W. H., Foraminiferal ooze, Solution at depths, Science, 156, 383-385, 1967.

Berger, W. H., Planktonic foraminifera: selective solution and the lysocline, Mar. Geol., 8, 111-138, 1970.

Berner, R. A., Calcium carbonate concretions formed by the decomposition of organic matter, Science, 159, 195-197, 1968.

Bramlette, M. N., Pelagic sediments, in Oceanography, edited by M. Sears, pp. 345-390, Am. Assoc. Adv. Sci. Publ. 67, Washington, D. C., 1961.

Cifelli, R., V. T. Bowen, and R. Siever, Cemented foraminiferal oozes from the mid-Atlantic Ridge, Nature, 209, 32-34, 1966.

Fisher, A. G., and R. E. Garrison, Carbonate lithification on the sea floor, J. Geol., 75, 488-496, 1967.

Friedman, G. M., Early diagenesis and lithification in carbonate sediments, J. Sediment. Petrol., 34, 777-813, 1964.

Friedman, G. M., On the origin of aragonite in the Dead Sea, Israel J. Earth Sci., 14, 79-85, 1965.

Hawley, J., and R. M. Pytkowicz, Solubility of calcium carbonate in seawater at high pressures and 2°C, Geochim. Cosmochim. Acta, 33, 1557-1561, 1969.

Kennedy, W. J., and A. Hall, The influences of organic matter on the preservation of aragonite in fossils, Proc. Geol. Soc. London, 1643, 253-255, 1967.

Li, Y. H., T. L. Ku, and G. G. Mathieu, The CO_2-$CaCO_3$ system in the Antarctic as related to the Atlantic and Pacific Oceans, Antarctic J., 5 (5), 1970.

Lisitzin, A. P., Sedimentation in the world ocean, Soc. Econ. Paleontol. and Mineral. Special Publ. No. 17, Norman, Okla., 1972.

Milliman, J. D., Submarine lithification of carbonate sediments, Science, 153, 994-997, 1966.

Mitterer, Richard M., Calcified proteins in the sedimentary environment, in Adv. in Org. Geochem., pp. 441-451, Pergamon Press, Oxford, 1972.

Park, P. K., Deep-sea pH, Science, 154, 1540-1542, 1966.

Peterson, M. N. A., Calcite: rates of dissolution in a vertical profile in the central Pacific, Science, 154, 1542-1544, 1966.

Pytkowicz, R. M., On the carbonate compensation depth in the Pacific, Geochim. Cosmochim. Acta, 34, 836-839, 1970.

Schrader, H.-J., Fecal pellets: role in sedimentation of pelagic diatoms, Science, 174, 55-57, 1971.

Smith, S. V., J. A. Dygas, and K. E. Chave, Distribution of calcium carbonate in pelagic sediments, Mar. Geol., 6, 391-400, 1968.

Thompson, G., V. T. Bowen, W. G. Melson, and R. Cifelli, Lithified carbonates from the deep-sea of the equatorial Atlantic, J. Sediment. Petrol., 38, 1305-1312, 1968.

Welte, D. H., Organic matter in sediments, in Organic Geochem., edited by G. Eglinton and M. T. J. Murphy, 262-264, Springer-Verlag, N. Y., 1969.

Weyl, P. K., The change in solubility of calcium carbonate with temperature and carbon dioxide content, Geochim. Cosmochim. Acta, 17, 214-255, 1959.

PRECIPITATION AND CEMENTATION OF DEEP-SEA CARBONATE SEDIMENTS

JOHN D. MILLIMAN

Woods Hole Oceanographic Institution

ABSTRACT

Inorganic precipitation of calcium carbonate, both as limestone cement and unconsolidated lutite, occurs in several types of deep-sea environments. While such carbonates account for only a small part of the total calcium carbonate budget, they appear to be more important quantitatively than shallow-water precipitates. Recognizing and understanding the processes of precipitation and cementation of carbonates can give added insights into deep-sea carbonate sedimentation prior to the evolution of calcareous nannoplankton and zooplankton.

Carbonate cementation can affect the general physical (and engineering) properties of marine sediments. Sudden variations in such properties as shear strength may reflect cementation of that sediment. Techniques for recognizing small (but perhaps critically important) quantities of carbonate cement in deep-sea sediments should be developed.

INTRODUCTION

Until recently most geologists believed that carbonate sediments could be lithified only upon exposure to meteoric water. The fact that most marine waters are undersaturated with respect to calcium carbonate supposedly would prevent precipitation of carbonate cements. Although intertidal cementation of beachrocks was well documented (for example, see Ginsburg, 1957), the role and importance of subtidal cementation was minimized. As recently as 1964, popular opinion ran as follows: "Sediments that remain in a marine environment almost always resist lithification, whereas those exposed subaerially to fresh water become lithified" (Friedman, 1964, p. 809). This notion of non-lithification of marine sediments changed recently with the "discovery" of many examples of deep-sea limestones (Friedman, 1965; Gevirtz and Friedman, 1966; Milliman, 1966; Fischer and Garrison, 1967).

The concept of submarine cementation, however, is not new. This process was documented by European oceanographers well before the turn of the 20th century. Murray and Hjort (1912) summarized their knowledge about submarine cementation as follows:

> A limited amount of purely inorganic precipitation does, indeed, take place (in the deep sea)...In the Mediterranean, for instance, stonelike crusts are plentiful, consisting of clay cemented by calcium carbonate, which is produced by ammonium carbonate arising from the decay of organic matter in the mud below bottom level meeting with fresh water from above. We have further lime-concretions of the Pourtales, Argus and Seine Banks, the "Challenger" casts of shells from the Great Barrier Reef, and so on." (Murray and Hjort, 1912, p. 178).

Why this well-accepted concept was widely rejected (or forgotten) in subsequent years is not known. It was retained in part, however, by many European geologists working in ancient limestones (for example, see Schmidt, 1965).

Two distinct types of modern marine limestones can be identified: shallow-water and deep-water limestones. Shallow-water limestones usually occur in water depths less than 100 m and contain predominantly shallow-water carbonate constituents. Cement can be either aragonite or high-magnesian calcite (generally 15 to 21 mole percent $MgCO_3$ in solid solution with the calcite). These limestones generally are in isotopic disequilibrium with the ambient environment, suggesting that cementation may occur through some organic or biochemical agent, possibly the activity of blue-green algae (Milliman, 1973). Because the mineralogy, isotopic composition, and mode of formation are uniquely different from deep-sea limestones (Table 1), I will exclude discussion of shallow-water limestones from the following paragraphs.

DEEP-WATER LIMESTONES

Although deep-sea sediments can be cemented by pyrite (Field and Pilkey, 1970), rhodochrosite and siderite (JOIDES, 1970a, 1970b) and altered marine clays, most deep-sea sedimentary rocks reported in the scientific literature are cemented by calcium carbonate or dolomite. Many types of limestones have been dredged from the deep-sea. More than 50 reports of such dredgings have been published and DSDP (Deep-Sea Drilling Program - JOIDES) drillings have penetrated numerous other cemented layers.

Five major types of submarine limestones can be recognized from the deep sea: (1) those occuring in non-depositional environments, (2) those associated with the diagenesis of volcanic material, (3) those related to thermal metamorphism, (4) those related to overburden pressure, and (5) those forming in oceanic basins with restricted circulation (Fig. 1). Each type of limestone will be discussed in the following paragraphs.

Limestones Associated with Non-Depositional Areas

The vast majority of limestones dredged from the deep sea have been recovered from the sides of sea mounts or islands, from the tops of guyots, and plateaus, and from continental slopes. Water depths generally range from 300 to 1000 m although shallower and deeper occurrences are not unusual. These limestones often consist of irregular nodules or heavily bored slabs. Component grains are dominated by planktonic and benthonic foraminifera and mollusks, although displaced shallow-water organisms can be important in limestones dredged from island and continental slopes. In such limestones the early stages of cementation involves formation of a

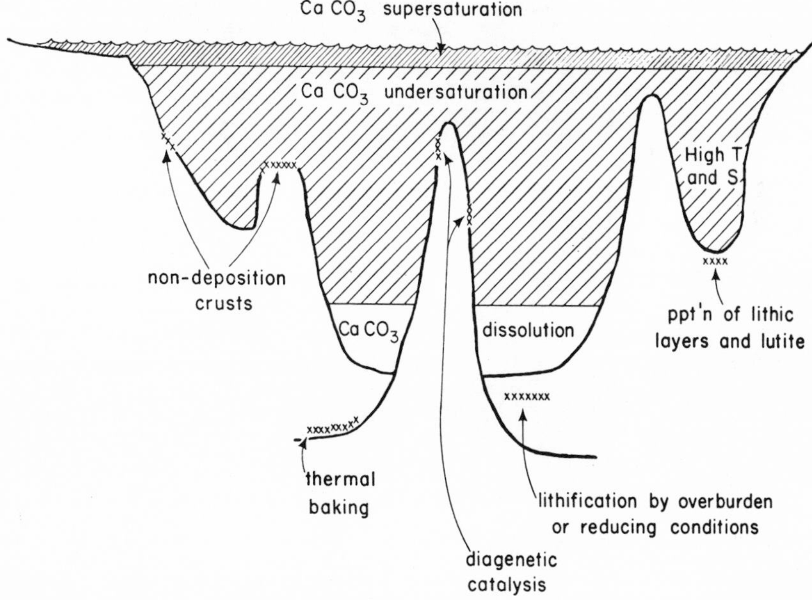

Figure 1. Schematic illustration of the locations of submarine cementation and precipitation in the deep sea

TABLE 1

SUMMARY OF VARIOUS TYPES OF DEEP-SEA LIMESTONES

Rock Type	Range of Water Depths (m)	Cement	Mineralogy	Range of Stable Isotope Ratios $2O^{18}$	$2C^{13}$
Non-depositional	100-3000	Rim cement and/or matrix	Mg-calcite calcite	+1.47 to 3.08	+0.40 to 2.20
Volcanic	1700-4300	Distinct crystals; loose matrix	Calcite	-4.2 to 1.9	-1.29 to 5.3
Thermal metamorphism	>3000 (+sediment cover)	Anhedral to Euhedral matrix	Calcite (mostly)	-2.2 to -7.1	+0.2 to +3.2
Overburden	>3000 (+sediment cover)		Calcite		
Semi-enclosed basins	800-4000	Matrix cement + some rim cement	Mg-calcite aragonite	+3.5 to +7.0	+2.9 to +4.3

(after Milliman, 1974)

fibrous to cryptocrystalline rim cement of magnesian calcite ($MgCO_3$ contents ranging from 8 to 13 mole percent, with a mean of 11 to 12 mole percent). Some limestones, however, do contain prominent amounts of magnesian calcite matrix cement. Stable isotope contents of these limestones are in equilibrium with the ambient waters, thus verifying the process of in situ cementation (Milliman, 1966). McFarlin (1967) found that the manganese nodules on the Blake Plateau contain aragonite-filled veins. The anomalous stable isotope composition of these aragonites suggests that precipitation occurred within some as yet undefined micro-environment.

Although pre-Pliocene limestones also tend to be in isotopic equilibrium with ambient waters, these limestones are composed almost exclusively of low magnesium calcite. Presumably these limestones were cemented originally with magnesian calcite and later inverted to calcite (Milliman, 1966; Gomberg and Bonatti, 1970). Subsequent diagenesis may involve phosphatization as well as the deposition of manganese coatings.

The mode of lithification of non-depositional limestones is not known, although such limestones can lithify in less that 15 thousand years (Friedman, 1964; Marlowe, 1971). Judging from the distribution of these limestones, it seems reasonable to suspect that the low rates of sedimentation on slopes and seamount tops are critical for cementation. On the Blake Plateau, for instance, submarine lithified crusts commonly form on Miocene outcrops (Stetson et al., 1969; Milliman, 1972) and locally can form "lithoherms" which provide suitable substrata upon which various epibenthic organisms can attach (Neumann et al., 1972).

Limestones Associated with Volcanics

Dredgings from submarine volcanoes have recovered a significant number of limestones and rocks with carbonate fillings of vesicles and cracks. Published data indicate that the cements are dominantly low magnesian calcite (Table 1), although some occurrences of aragonite and dolomite have been reported (Thompson, 1972). Cifelli et al. (1966) suggested that such limestones may precipitate during tectonic uplift, thus allowing the sediments to pass from undersaturated to supersaturated waters. A second possibility involves thermal baking (Saito et al., 1966), a process which will be discussed in the following section. Thirdly, many limestones may form during the diagenesis of the volcanic rocks. The fact that some cements occur in the veins and vesicles of altered volcanic tuffs suggests that at least some of the limestones are related to diagenesis. Stable isotope data support this conclusion (Thompson, 1972).

Limestones Associated with Thermal Metamorphism

Recent holes drilled by DSDP have penetrated deeply buried cemented carbonate horizons directly overlying igneous basement. Brecciated limestones recovered at Site 53 in the northwestern Pacific, for example, consist of nannofossils cemented with an anhedral to euhedral calcitic matrix (Pimm et al., 1971). These limestones, which are extensively recrystallized and brecciated, lie close to the basalt basement. Similarly, isotope data from deep limestones drilled in the central Caribbean and dredged from the Mid-Atlantic Ridge indicate lithification at elevated temperatures (Anderson and Schneidermann, 1972; Thompson et al., 1968).

Limestones Associated with Overburden and Reduction

Other limestone layers cored during DSDP operations have been recovered within unconsolidated calcareous oozes (Wise and Kelts, 1971). Available data indicate that the limestones are predominantly calcitic, although some aragonite has been detected (Pimm et al., 1971). Obviously this cementation is not related to thermal metamorphism, since the overlying and underlying sediments do not appear altered. Stable isotope data suggest that lithification occurred at ambient sea temperatures, but no specific origin has been suggested. Perhaps overburden pressure (solution-welding) may cause cementation (Davies and Supko, 1973), although limestones associated with dark-colored layers may have been lithified under reducing conditions (P. R. Supko, personal communication, 1971).

Limestones Associated with Semi-Enclosed Basins

Many of the early reports of deep-sea limestones came from the dredgings made in the eastern Mediterranean Sea and the Red Sea (Natterer, 1894, 1898; DeWindt and Berwerth, 1904; Boggild, 1912). Quaternary sediments within these two basins contain far greater concentrations of lithic crusts and layers than do normal deep-sea sediments. Rapid lithification of the carbonate sediments from these two areas undoubtedly is related to the high salinities and temperatures which characterize the deeper waters (Table 2).

The upper few meters of sediment in the Red Sea contain several prominent lithic layers. Most lithic fragments contain grains of planktonic foraminifera cemented with a dense magnesian calcite matrix (11 to 12 mole percent $MgCO_3$). The youngest layer has been dated at 6000 years B.P. and probably was cemented in oceanographic conditions quite similar to those at present (Milliman et al., 1969). The most prominent lithic layer, however, is cemented by aragonite and its primary biogenic components are planktonic pteropods. It is not known if the aragonite cement is related to the presence of

TABLE 2

		Temperature Range (°C)	Salinity Range (°/oo)
Normal Ocean Water	Surface	15-25	34-36
	2000 m	1.5-5	34-35
	Bottom	0.5-2.5	34-36
Mediterranean Sea	Surface	15-25	37-38.8
	2000 m	13-14	38.5-38.7
	Bottom	13.7	38.4-38.6
Red Sea	Surface	24-30	38-40
	2000 m	22	40.5-41
	Bottom	22	40.5-41

Notes

Temperature and salinity ranges in normal ocean, Mediterranean Sea and Red Sea waters at the surface, 2000 m and near the bottom of the deep basins. Normal ocean values refer to subtropical and temperate climates; Red Sea bottom depths generally range between 1500 and 2000 m, thus the 2000 m and bottom values are the same. (from Milliman and Müller, 1973)

the aragonitic pteropod shells (Glover and Pray, 1971).

The aragonitic layers formed 11 to 20 thousand years ago under supersaline conditions caused during the isolation of the Red Sea from the Indian Ocean by glacially lowered sea level (Milliman et al., 1969; Herman, 1965; Berggren and Boersma, 1969; Deuser and Degens, 1969). When post-glacial sea level rose, salinity decreased, aragonite precipitation ceased and magnesian calcite cementation again became dominant. Prior epochs of aragonite precipitation may have occurred, but they probably have been obscured by the dissolution of aragonite and alteration to magnesian calcite (Friedman, 1965; Milliman et al., 1969).

Cementation in the eastern Mediterranean is more restricted, both horizontally and vertically, than it is in the Red Sea. Lithic layers occur within individual cores, but rarely can be correlated with layers in other cores. Cements in the lithic crusts and fragments are similar in morphology and composition within those in the Red Sea (Milliman and Müller, 1973).

DEEP-WATER PRECIPITATION OF CARBONATE LUTITES

Until recently it was assumed that all modern deep-sea carbonate sediments were biologic in origin, mostly derived from the tests of planktonic foraminifera and coccolithophorids. Even the very fine-grained carbonate sediments from the Black Sea, which previously had been thought to be inorganically precipitated (Caspers, 1957), have been shown to be composed of coccoliths (Bukry et al., 1970). As a result, it was believed that most deep-sea carbonate sediments were composed almost exclusively of low-magnesium calcite (Pilkey and Blackwelder, 1968). Recent studies, however, have shown that in certain deep-sea environments calcium carbonate can precipitate inorganically. Specifically, the Red Sea and the eastern Mediterranean Sea, by virtue of their high salinities and temperatures (Table 2), contain large amounts of magnesian calcite within their lutite fractions. These lutites have mineralogies and crystal structures similar to the cements within co-existing limestone fragments and crusts (11 to 12 mole percent $MgCO_3$). The lack of any biogenic source plus the great similarity with limestone fragments infer an inorganic precipitation of these lutites. As mentioned in a previous section, precipitation was altered during lower stands of sea level when aragonite was the dominant carbonate phase. The eastern Mediterranean, being somewhat less saline and warm than the Red Sea, has experienced a somewhat different sequence of sedimentation. Magnesian calcite apparently precipitated during both interglacial and glacial epochs, but during periods of glacial melting, when circulation of the Mediterranean supposedly was restricted (Olausson, 1965), precipitation of magnesian calcite ceased and carbonate deposition was restricted to the accumulation of coccoliths (Milliman and Müller, 1973).

THE IMPORTANCE OF PRECIPITATION AND CEMENTATION OF DEEP-SEA CARBONATES

Recognition of inorganically precipitated and cemented deep-sea carbonates can have several significant implications upon both geology and marine geology. Elucidation of these implications may aid in interpretation of the geologic record and deep-sea processes evidenced in that record.

1. In terms of the total calcium carbonate budget within the world oceans, precipitation and cementation of modern deep-sea carbonates probably are not major processes. Except for deeply buried limestone layers, carbonate cementation in the deep sea generally is restricted to rather unique areas. Some of these areas are: areas that have high salinities and temperatures, areas exposed to submarine volcanism, and areas that have exceedingly low rates of sediment accumulation.

While modern inorganic precipitation of deep-sea lutite composed of calcium carbonate appears to be restricted to the eastern Mediterranean and Red Seas, a preliminary calculation shows that these two areas account for more inorganically precipitated calcium carbonate than all the shallow-water areas in the world (Milliman, 1974). Not only does this mean that the bulk of inorganic precipitation occurs in the deep sea (not in shallow tropical waters, as previously thought), but also that the primary mineral phase is magnesian calcite, not aragonite.

2. Although modern submarine cementation and lutite precipitation are minor occurrences, their impact upon the understanding and interpreting of older limestones may be critically important. During Tertiary and Quaternary times more than 90% of the biogenic calcium carbonate deposited in the oceans has accumulated in the deep sea (Milliman, 1974). Prior to the evolution and rapid explosion of calcareous nanno- and zooplankton (during the Mesozoic and early Tertiary), however, biogenic deposition of calcium carbonate must have been restricted primarily to shallow water (Kuenen, 1950). As a result, prior to the Tertiary the oceans must have been significantly more supersaturated with respect to calcium carbonate than at present. If one also takes into account the general lowering of bottom temperatures since the Eocene (Emiliani, 1954), it is suspected that inorganic precipitation of deep-water carbonates may have been a more common and important sedimentary process in the past than it is in the modern oceans.

How can one recognize inorganic precipitation in ancient deep-water limestones? Mineralogical data are practically useless, since magnesian calcite (the primary precipitate and cement) is metastable and quickly "inverts" to calcite when exposed to fresh water (Bathurst, 1971). If the exchange of $CO_3^{=}$ is involved in this transformation, stable isotope data also are meaningless. Petrographic techniques may be more helpful. Of course, the presence of planktonic or deep-sea benthic fauna can define the environment of deposition. Similarly, the nodular appearance of lutites and cemented fragments in some Alpine limestones suggest a deep-sea origin (R. E. Garrison, written communication, 1973). Perhaps more important is the fact that many of the cements in modern deep-sea limestones appear to have unique crystal morphologies, the presence of which may help to define ancient counterparts. For example, preliminary analyses of several possible "deep-water" limestones from Mesozoic formations in the southeastern U. S. and Mexico, utilizing a scanning electron microscope, indicate calcitic components and cements with similar morphologies to those occurring in modern deep-sea limestones (Milliman, 1974; J. Koch and C. H. Moore, Jr., personal communication, 1972).

3. Present documentation of precipitation and cementation in

deep-sea carbonates has been limited to obvious examples. In the case of cementation, this means limestones or limestone fragments or crusts. In the case of precipitated lutites it means carbonates containing distinctly unique mineralogic or petrographic properties. But how can these processes be recognized if the material is loosely cemented but not lithified, or if the precipitated carbonate is mineralogically similar to biogenic carbonates? Such occurrences of small quantities of cement or indistinguishable lutites may be more common than presently suspected and may be critically important in understanding deep-sea sedimentary processes.

One example of this phenomenon would be the effect of low concentrations of carbonate cement upon the physical properties of a deep-sea sediment. Several papers within this volume refer to sudden changes in such properties as shear strength or bulk density with core depth, but what sedimentologic factors actually cause these variations? One distinct possibility would be the presence of loosely cemented grains; grains not cemented enough to lithify the sediment, and maybe not enough to affect the composition of the sediment, but enough to alter strongly other physical properties. At present there is no standard method to prove the presence or absence of this cement. Perhaps further studies of such sediments with the scanning electron microscope and microprobe analyses will supply criteria. Certainly the detection of such cements and precipitates will be an important factor in understanding the processes controlling the physical properties of deep-sea sediments.

ACKNOWLEDGMENTS

This review article was written with contract support from the Office of Naval Research, Contract Number N00014-66-C-0241. I thank C. P. Summerhayes and F. T. Manheim for constructive comments on the manuscript. This paper is Woods Hole Oceanographic Institution Contribution No. 3151.

REFERENCES

Anderson, T. F., and N. Schneidermann, Isotope relationships in pelagic limestones from central Caribbean, Leg 15, Deep-Sea Drilling Project (abstract), Trans. Am. Geophys. Union, 53, 555, 1972.

Bathurst, R. G. C., Carbonate Sediments and their Diagenesis, Elsevier Publ. Co., Amsterdam, 1971.

Berggren, W. A., and A. Boersma, Late Pleistocene and Holocene planktonic foraminifera from the Red Sea, in Hot Brines and Recent Heavy Metal Deposits in the Red Sea, edited by E. T. Degens and D. A. Ross, pp. 282-298, Springer-Verlag, N. Y., 1969.

Boggild, O. B., The deposits of the sea bottom, Danish Oceanographic Expedition, 1908-1910, Rept. 1, 255-269, 1912.

Bukry, D., S. A. Kling, M. K. Horn, and F. T. Manheim, Geological significance of coccoliths in fine-grained carbonate bands of post-glacial Black Sea sediments, Nature, 226, 156-158, 1970.

Caspers, H., Black Sea and Sea of Azov, in Treatise on Mar. Ecol., edited by J. W. Hedgpeth, pp. 801-890, Geol. Soc. Am. Mem. 67, Boulder, Colo., 1957.

Cifelli, R., V. T. Bowen, and R. Siever, Cemented foraminiferal oozes from the Mid-Atlantic Ridge, J. Mar. Res., 26, 105-109, 1966.

Davies, T. A., and P. R. Supko, Oceanic sediments and their diagenesis: some examples from deep-sea drilling, J. Sediment. Petrol., 43, 381-390, 1973.

Deuser, W. G., and E. T. Degens, O^{18}/O^{16} and C^{13}/C^{12} ratios of fossils from the Hot Brine deep area of the central Red Sea, in Hot Brines and Recent Heavy Metal Deposits in the Red Sea, edited by E. T. Degens and D. A. Ross, pp. 336-347, Springer-Verlag, N. Y., 1969.

DeWindt, J., and F. Berwerth, Untersuchung von Grundproben der I., II, and IV Reise von S.M. POLA in den Jahren 1890, 1892 and 1893, Denkschrift Akademie Wissenschaft Wien, Mathematik und Naturwissenschaften, 74, 285-294, 1904.

Emiliani, C., Temperatures of Pacific bottom waters and polar superficial waters during the Tertiary, Science, 119, 853-855, 1954.

Field, M. F., and O. H. Pilkey, Lithification of deep-sea sediments by pyrite, Nature, 226, 836-837, 1970.

Fischer, A. G., and R. E. Garrison, Carbonate lithification on the sea floor, J. Geol., 75, 488-497, 1967.

Freidman, G. M., Early diagenesis and lithification in carbonate sediments, J. Sediment. Petrol., 34, 777-813, 1964.

Friedman, G. M., Occurrence and stability relationships of aragonite high-magnesian calcite, and low-magnesian calcite under deep-sea conditions, Bull. Geol. Soc. Am., 76, 1191-1196, 1965.

Gevirtz, J. L., and G. M. Friedman, Deep-sea carbonate sediments of the Red Sea and their implications on marine lithification, J. Sediment. Petrol., 36, 143-151, 1966.

Ginsburg, R. N., Early diagenesis and lithification of shallow-water carbonate sediments in South Florida, in Regional Aspects of Carbonate Sedimentation, edited by R. J. LeBlanc and J. G. Breeding, pp. 80-100, Soc. Econ. Paleontol. and Mineral. Spec. Publ. 5, Norman, Okla., 1957.

Glover, E. D., and L. C. Pray, High-magnesium calcite and aragonite cementation within modern subtidal carbonate sediment grains, in Carbonate Cements, edited by O. P. Bricker, pp. 80-87, Johns Hopkins Univ. Studies in Geol., No. 19, Baltimore, 1971.

Gomberg, D. N., and E. Bonatti, High-magnesian calcite: leaching of magnesium in the deep sea, Science, 168, 1451-1453, 1970.

Herman, Y. R., Etudes des sediments Quaternaires de la Mer Rouge, Ph.D. thesis, Univ. of Paris, pp. 341-415, Masson & Cie Editeurs, Paris, 1965.

JOIDES, Initial Reports of the Deep-Sea Drilling Project, Hoboken, N. J. to Dakar, Senegal, 2, U. S. Govt. Printing Office, Washington, D. C., 1970a.

JOIDES, Initial Reports of the Deep-Sea Drilling Project, Rio de Janeiro, Brazil to San Cristobal, Panama, 4, U. S. Govt. Printing Office, Washington, D. C., 1970b.

Kuenen, P. H., Marine Geology, John Wiley and Sons, Inc., N. Y., 1950.

Marlowe, J. I., Dolomite, phosphorite, and carbonate diagenesis on a Caribbean Seamount, J. Sediment. Petrol., 41, 809-827, 1971.

McFarlin, P. F., Aragonite vein fillings in marine manganese nodules, J. Sediment. Petrol., 37, 68-72, 1967.

Milliman, J. D., Submarine lithification of carbonate sediments, Science, 153, 994-997, 1966.

Milliman, J. D., Atlantic continental shelf and slope of the United States, petrology of the sand fraction-northern New Jersey to southern Florida, U. S. Geol. Survey Prof. Paper 529-J, 1972.

Milliman, J. D., Marine Carbonates, Springer-Verlag, Heidelberg, 375 p., 1974.

Milliman, J. D., D. A. Ross, and T. H. Ku, Precipitation and lithification of deep-sea carbonates in the Red Sea, J. Sediment. Petrol., 39, 724-736, 1969.

Milliman, J. D., and J. Müller, Precipitation and lithification of magnesian calcite in the deep-sea sediments of the eastern Mediterranean Sea, Sedimentology, 20, 29-46, 1973.

Murray, J., and J. Hjort, The Depths of the Ocean, MacMillan and Co., London, 1912.

Natterer, K., Chemische unterscuhungen im oestlichen Mittalmeer, i. Reise S.M. Schiffes POLA in Jahre 1890, Denkschrift Akademie Wissenschaft Wien, Mathematik und Naturwissenschaften, 61, 23-64, 1894.

Natterer, K., Expedition S.M. Schiff POLA in das Rote Meer, Nördliche Hälffe (October 1895-March 1896), Denkschrift Akademie Wissenschaft Wien, Mathematik und Naturwissenschaften, 65, 445-572, 1898.

Neumann, A. C., G. H. Keller, and J. W. Kofoed, "Lithoherms" in the Straits of Florida (abstract), Geol. Soc. Am., Abstr. with Prog., 4, 611, 1972.

Olausson, E., Evidence of climatic changes in North Atlantic deep-sea cores, with remarks on isotopic paleotemperature analysis, in Progress in Oceanography, 3, edited by M. Sears, pp. 221-252, Pergamon Press, Oxford, 1965.

Pilkey, O. H., and B. W. Blackwelder, Mineralogy of the sand size carbonate fraction of some recent marine terrigenous and carbonate sediments, J. Sediment. Petrol., 38, 799-810, 1968.

Pimm, A. C., R. E. Garrison, and R. E. Boyce, Sedimentology synthesis: lithology, chemistry and physical properties of sediments in the northwestern Pacific Ocean, in Initial Reports, Deep-Sea Drilling Project, 6, pp. 1131-1252, U. S. Govt. Printing Office, Washington, D. C., 1971.

Saito, T., M. Ewing, and L. H. Burckle, Tertiary sediment from the Mid-Atlantic Ridge, Science, 151, 1075-1079, 1966.

Schmidt, V., Facies, diagenesis, and related reservoir properties in the Gigas Beds (Upper Jurassic), Northwestern Germany, in Dolomitization and Limestone Diagenesis, edited by L. C. Pray and R. C. Maurray, pp. 124-168, Soc. Econ. Paleontol. and Mineral. Spec. Publ. 13, Norman, Okla., 1965.

Stetson, T. R., E. Uchupi, and J. D. Milliman, Surface and subsurface morphology of two small areas of the Blake Plateau, Trans. Gulf. Coast Assoc. Geol. Soc., 19, 131-142, 1969.

Thompson, G., A geochemical study of some lithified carbonate sediments from the deep sea, Geochim. Cosmochim. Acta, 36, 1237-1254, 1972.

Thompson, G., V. T. Bowen, W. G. Melson, and R. Cifelli, Lithified carbonates from the deep sea off the Equatorial Atlantic, J. Sediment. Petrol., 38, 1305-1312, 1968.

Wise, S. W., and K. R. Kelts, Submarine lithification of middle Tertiary chalks in the South Pacific Ocean basin (abstract), in Prog. 8th Intern. Sediment. Cong., p. 110, Heidelberg, 1971.

WORKSHOP I PHENOMENA

The purpose of this workshop was to discuss our knowledge (or lack of it) of the basic phenomena governing the mass physical and engineering characteristics of marine sediments and to focus upon areas of research which could improve our understanding of these characteristics. It became obvious (in the initial discussion of various problem areas) that at least two approaches are required to understand the mechanical behavior of marine sediments and to develop accurate predictive models. The first could be termed a basic approach. The second approach is an applied approach to focus on immediate problems which must be solved to obtain greater reliability in empirical relationships developed for terrestrial soils. Attention should be directed toward accurate equations based upon more precise techniques for sampling and testing.

Both approaches are necessary to eventually establish accurate predictive models. Both approaches require scientific and engineering support.

In order to determine what problems should be attacked first, the question was asked: "What is unique about the marine environment that will affect sediments?" The following list (not intended to be inclusive of all the unique characteristics) summarizes the discussion.

1. The sea floor is primarily a depositional environment.
2. The sediments are 100% saturated.
3. The sediments are saturated with salt water rather than fresh water.
4. Sea-floor sediments usually have very high void ratios.
5. Carbonate sands occur only in the marine environment and their behavior is different from other sand-sized materials.
6. Biogenic opaline is an important sediment component in the deep sea.
7. Diagenesis and lithification of the sediments: overburden pressure may not be as significant as secondary cementing in the reactions of sea-floor sediments to induced stresses.

Some of the basic problems which should be studied include (these are not in order of priority):

1. The relationships between erodability of sediments and shear strength, grain size, bulk density, water content and animal activity.
2. Which physical, chemical, and mechanical properties of a marine sediment change as a result of diagenesis and lithification?
3. What is the relationship between diagenesis and shear strength?
4. How do depositional processes affect the mechanical properties of sediments?
   (a) How are different materials deposited?
   (b) How does rate of deposition influence the change of shear strength with time?
   (c) How does the degree of aggregation or disaggregation at time of deposition influence the development of shear strength?
5. What is the nature of interparticle bonding in marine sediments and what affect does it have on the mechanical properties?
6. What is the effect of benthic animal activity on the physical and mechanical properties of sediments?
7. What is the time dependency of the physical and mechanical properties of marine sediments? What is the cause of that dependency?

Some of the more immediate problems which require attention include:

1. The development of better sampling and in situ testing techniques, especially in sand.
2. The determination of the behavior of marine sediments under low confining stresses.
3. The determination of the controlling factors for slope failures.
4. The determination of the most important aspects of sediment strength for marine sediments.
5. The determination of a method to evaluate changes in properties resulting from disturbance.
6. The definition of mechanical and mass physical properties of carbonate sediments.

A discussion was held on the shear strength of sediments. Highlights of these discussions were as follows:

1. There is no universal theory describing the development of shear strength. Shear strength values depend upon the rate of testing and drainage (i.e. the effective stress) and water content. The measured value is also dependent upon the type of test.

2. Materials should be tested under conditions as close as possible to the natural state.
3. The vane shear test is only an indicator of shear strength. It is possible to develop a pseudo-strength envelope by using vanes in conjunction with consolidation tests but this is not usually practical.
4. The "constants" of the Coloumb equation, c and ϕ, are not constants. Cohesion (c) is a function of the normal stress and stress history. Phi (ϕ) will vary in tests run at normal stresses below the preconsolidation stress from tests run with normal stress above the preconsolidation stress. The phi (ϕ) value for drained tests on remolded material will however remain constant.
5. Particle boundary effects must be considered (in studying shear strength). Kinetic energy is stored at boundaries and failure begins when the boundaries can no longer absorb energy. Failure is also dependent on rate of energy increase.

It was pointed out that the c/p ratio (c = cohesive, p = overburden pressure) has been used with two different connotations by many authors. One connotation and use is as the slope of the strength profile measured with increasing depth. When used in this manner, it is the slope of the line and not a single point value. It is a rate of change of shear strength over a depth range and not a discreet point.

The ratio has also been used in the literature as a discreet point measurement of the relationship between undrained shear strength and overburden pressure at a given depth in the sediment. It is important for the author to define his use of the term so that the reader can understand the difference.

Other facts brought out during the general discussion include:

(a) In areas of slow sedimentation there does not appear to be much disturbance to the sample when coring. Some investigators report their best results have been with gravity corers.
(b) Vibrocorers appear to greatly disturb the core sample.
(c) The importance of the disturbance to a core sample from the hydrostatic pressure decrease during retrieval from the sea floor is difficult to judge at this time. More controlled test data is necessary to evaluate its relative importance.

Highlights of a discussion on the relationships between acoustic and mechanical properties of sediments were:

(a) Every acoustical parameter of sediment depends upon several physical and some chemical parameters so there is no direct correlation between any two properties. Yet, through linear regression analyses some very general relationships may be established between certain acoustic and mechanical properties of the material.

(b) The basic problem in studying acoustic properties of sediments is to get dependable, repeatable, measurements. The gross characteristics of the material must be known to interpret the acoustical data.

(c) Acoustics can be used to determine in situ values for the elastic properties of sediment.

(d) Geophysical parameters, other than radiological measurements (such as x-ray back-scattering), that appear to be related to changes in the mechanical properties of a marine sediments are:

1. electrical conductivity
2. acoustic velocity
3. acoustic attenuation
4. acoustic impedance (derived directly or from reflectivity)
5. acoustic back-scattering.

Most of the above geophysical parameters relate in a multivariate way to the mechanical properties. Thus velocity is more clearly defined by grain size and porosity together than by either grain size or porosity by itself; and electrical conductivity, while being particularly dependent on porosity, also depends on the permeability of the sediment.

It follows that a single geophysical measurement can only give a gross approximation of mechanical properties. However, if several geophysical measurements are made, it may be possible to define some of the mechanical properties. Those mechanical properties that are subject to this type of determination are:

Porosity/Density - conductivity, velocity, impedance

Mean grain size - velocity, attenuation, and backscattering (impedance can also be included although the relationship is not as clear)

Liquid limit - velocity, impedance, and velocity x impedance

Plasticity index - velocity, and velocity x impedance.

In addition, certain consolidation parameters can be obtained from the acoustic properties.

WORKSHOP II INSTRUMENTATION

The purpose of this workshop was (1) to discuss ways of improving existing instrumentation and measuring techniques, and (2) to develop new concepts to answer the question: How can we get deeper into the sea-floor sediment to monitor physical and mechanical properties?

It is highly inaccurate and misleading to extrapolate changes in physical and mechanical properties of the sediments with depth based upon surficial values. Large errors can be made due to stratification and layering of the marine sediment. It is necessary to develop techniques to actually measure these properties in situ with depth.

Presently, the only satisfactory way to sample to depth in the sea floor is drilling and/or coring. Drilling is expensive, time consuming, and requires elaborate ships and technical systems. It is now possible to take piston cores up to 22 m in length in the deep ocean but care must be taken in interpretation of their physical properties. Frequently, the cohesive layers of sediment will be recovered intact whereas the softer interbedded layers may be missing or greatly disturbed.

Usually, the primary object of remote measurements is to enhance a sampling program. It will probably always be necessary to obtain some samples for correlation with the results obtained by remote methods, but a satisfactory remote-sensing system can greatly minimize the number of samples necessary for accurate correlation over large areas of the ocean basins.

Presently acoustic methods are used to enhance a sampling program, but it is often difficult to correlate acoustic reflectors with the cored samples. The reflectors appear to be a function of many physical properties including mineralogy, density, shear strength, and microstructure. Additional understanding of the acoustic propagation parameters is necessary for improved acoustical predictions of the physical and mechanical properties of sea-floor sediments. Acoustics, however, remains a useful tool for the indication of gross sea-floor stratigraphy. The geophysical methods are valuable as an interpolating technique in that they provide a convenient means of providing average values of the variability of properties over a large area. Spot values can be obtained by placing probes in the bottom.

New techniques are needed for estimates or monitoring of the properties of deep-sea sediments, especially to depths over 10 m. One problem of sending any probe deep into the bottom is that a constant position must be maintained on the sea surface for some length of time. One of the objects of any sampling or monitoring technique should be to minimize that time.

A possible monitoring technique for future development is a cone penetrometer combined with a high energy projectile for deep penetration. The static cone penetrometer has become almost a standard instrument in Europe with data available on the correlation between resistance to penetration and shear strength. By using two sensors, one on the point of the cone and one on a sleeve just behind the cone, a ratio of penetration resistances is obtained. By use of this ratio, the material type can be distinguished and more accurate estimates of shear strength made. At present these devices are used to depths of 22 m and can penetrate materials with a maximum resistance of 88,960 N (10 tons).

Some experiments utilizing deceleration curves of projectiles dropped or shot into the bottom have been done. This method has great potential, especially if the features of a cone penetrometer are included. Some of the problem areas for this technique are:

(a) A better understanding of the deceleration trace in terms of sediment properties must be developed.
(b) A good device to trigger the projectile must be developed which can be placed on the sea floor and be capable of being moved from site to site easily and quickly.
(c) Initial velocities of the projectile should be 76 to 91 m/s for deep penetration (up to 153 m in soft mud).
(d) A means of retrieving the projectile is necessary.
(e) A method of measuring the plotting both cone and sleeve penetration on a projectile must be developed.

Other possible methods for monitoring deep properties include a mole-like device that bores its way into the bottom, and small explosive anchors. The mole-like device could ingest, analyze and discard sediment samples as it moved through the sediment layers. The flukes on small explosive anchors can be used to measure dynamic properties on the way into the sediment and measure static properties during a very slow withdrawal out of the sediment.

Presently, a few instrumented systems exist which measure the physical and mechanical properties of the bottom sediments in situ. By combinations of probes on these systems much information can be gained to enhance a sampling program. It is now possible to combine probes to measure density (by nuclear means), electrical conductivity and shear strength (by vane or static cone penetrometer). Information gained from this suite of instruments includes bulk density, specific gravity, porosity, shear strength, water content and an inference of the mineralogy of the material. Disadvantages of such systems are:

(a) They are large, heavy and require relatively sophisticated systems of deployment.
(b) They require prolonged station-keeping ability.
(c) Presently they do not penetrate more than 4 m into the bottom sediment.

WORKSHOP III STANDARDIZATION

The purpose of this last workshop was to attempt to obtain some degree of agreement on standardization of units, symbols, nomenclature and test techniques. All workers within the field of marine geotechnique or geomechanics realize the diversity of units, symbols and terms used in publications pertaining to this field. By standardizing the nomenclature it is hoped that some confusion and ambiguity can be eliminated.

Table I is the list of SI units and Table II is the list of symbols voted and agreed upon by the seminar group. All members of the workshop agreed to the use of these units and symbols in their publications. Standardization can only be accomplished through general usage of one set of units and symbols. It is the hope of the workshop group that the units and symbols presented in Tables I and II are accepted by the community through universal usage in order to eliminate the ambiguity and confusion which now exists in the technical literature.

A discussion was held on the standardization of vane rotation rate. It was generally agreed that since 6 degrees per minute is the standard rate in terrestrial soil mechanics, it should be the standard for marine measurements for the time being. More research is needed on the effect of rotation rate on vane shear strength. The vane shear test is only valid in cohesive sediments and should not be used in silts or sands. When vane shear results are presented in the literature, the author is requested to also present:

(a) the dimensions of the vane used,
(b) the rate of vane rotation (in rad/s), and
(c) a list of the index properties of the material tested.

A motion was passed that in testing marine sediments for water content, a correction must be made for salt content on the basis of an average salinity of 35 parts per thousand (3 1/2% salt content) or the exact value if measured.

A second motion was passed to define water content as:

$$\frac{\text{wt. of fluid}}{\text{dry wt. of solids}} = \frac{\text{wt. of salt water}}{\text{dry wt. of solids - wt. of salt}}$$

TABLE 1

COMMON SI UNITS IN MARINE SOIL MECHANICS AND MASS PHYSICAL PROPERTIES

Quantity	SI Unit	SI Symbol or Formula	Conventional U. S.	From	Conversion to multiply by
length	meter	m	ft	foot	m: 3.048 000 E-01
mass	kilogram	kg	lbm (avoir-dupois)	pound mass	kg: 4.535 924 E-01
time	second	s	min (mean solar)	minute	s: 6.000 000 E+01
plane angle	radian	rad	deg	degree	rad: 1.745 329 E-02
angular velocity	radian/second	rad/s	deg/min	deg/min	rad/s: 2.909 E-04
area	square meter	m^2	ft^2	$foot^2$	m^2::9.290 304 E-02
density	kilogram/ cubic meter	kg/m^3	lb/ft^3 or g/cm^3	pcf	kg/m^3: 1.601 846 E+01
				g/cm^3	kg/m^3: 1.000 000 E+03
force	newton	N	lb-force	lbf avoir.	N: 4.448 222 E+00
				kip	N: 4.448 222 E+03
stress or pressure	pascal	Pa	psi, psf	psi	kPa: 6.894 757 E+00
				psf	kPa: 4.788 026 E-02
				g/cm^2	kPa: 9.806 650 E-02
				kg/m^2	kPa: 9.806 650 E+01
frequency	hertz	Hz	cps	cps = Hz	
power	watt	W	W	W = W	
voltage	volt	V	V	V = V	
temperature*	Celsius	°C	Centigrade scale	°C = °C	

* Not SI, but permissible instead of the SI kelvin (K).

The dry weight of the solids shall be determined by drying in an oven at 105°C.

When running Atterberg limit tests on marine sediments, it is necessary to use salt water and not fresh water. Normally, the material is not oven dried or sieved for the tests. Samples are used at their natural water content or air dried for the plastic limit test.

The question of calibrating corers and comparing test results from samples obtained by various corers was discussed. Several people expressed the opinion that in stiff, relatively strong sediments, samples obtained with various corers appear to yield similar test results for index properties and shear strength. This observation does not hold true for softer, weaker sediments. In the latter case, samples obtained with different corers yield varying test results. The investigator must realize that no core sample is truly "undisturbed" and recognize the limitations of the tool and his test results.

It appears that cores with lengths greater than 20 times their diameter may be obtained without any apparent significant disturbance. However each core must be closely examined to determine the true extent of disturbance and how valid test results on that material will be.

It is known that corers may rotate as they travel through the sediment which adds a torsional stress to the material. It is suggested that an orientation device be added to each corer in order to determine if rotation took place during each coring operation.

<u>Key references for Table 1 (opposite)</u>:

ASCE, Guidelines for the use of SI units of measurement in the publications of the Am. Soc. of Civil Engrs., ASCE, N. Y., 1971.

ASTM, Standard Metric Practice Guide, Standard E380-72, Am. Soc. Test. Mat., Phila., 1972.

Page, C. H. and P. Vigoreux, Editors, The International System of Units (SI), U. S. National Bureau of Standards Spec. Publ. 330, 1972.

TABLE 2

SELECTED SYMBOLS

GENERAL	
t	time
g	acceleration due to gravity
V	volume
W	weight
F	factor of safety
STRESS AND STRAIN	
u	pore pressure
u_w	pore water pressure
σ	normal stress
$\bar{\sigma}$	effective normal stress
τ	shear stress
ε	linear strain ("strain" ASTM)
SOIL PROPERTIES	
γ	unit weight of soil
γ_w	unit weight of water
γ_{sw}	unit weight of sea water
γ_{sat}	unit weight of water-saturated soil
γ'	unit weight of submerged soil
G_s	specific gravity of solid particles
e	void ratio
n	porosity
w	water content

TABLE 2

(CONTINUED)

w_L	liquid limit (3rd choice ASTM; "LL" 1st)
w_p	plastic limit (1st choice ASTM)
I_p	plasticity index (1st choice ASTM)
I_L	liquidity index (3rd choice ASTM; "B" 1st)
C_c	compression index
c_v	coefficient of consolidation
τ_f	shear strength ("s" ASTM)
$\bar{c}$	effective cohesion intercept
$\bar{\phi}$	effective angle of internal friction
c_u	apparent cohesion intercept ("c" ASTM)
ϕ_u	apparent angle of internal friction ("ϕ" ASTM)
S_t	sensitivity
s_u	vane shear strength
N	bearing capacity factor
OTHER (from various sources)	
OCR	over consolidation ratio (p_c/p_o)
$\bar{\sigma}_v$	effective overburden stress
p_c	preconsolidation stress (greatest effective pressure)

Note: For further information concerning these symbols, see Table 1 in Richards' article on standardization of symbols.

LIST OF PARTICIPANTS AND CONTRIBUTORS

Aubrey L. Anderson
Applied Research Laboratories
The University of Texas
at Austin
Austin, Texas 78712

Victor C. Anderson
Marine Physical Laboratory
Scripps Institution of
Oceanography
University of California
San Diego, California 92132

Burt B. Barnes
Geo-Testing, Inc.
P. O. Box 4339
San Rafael, California 94903

William R. Bryant
Department of Oceanography
Texas A&M University
College Station, Texas 77843

Jon W. Carlmark
Physical Oceaography Program
Ocean Science & Technology
Division
Office of Naval Research
Arlington, Virginia 22217

Bobb Carson
Department of Geological
Sciences
Lehigh University
Bethlehem, Pennsylvania 18015

Barbaros Celikkol
Department of Mechanical
Engineering
University of New Hampshire
Durham, New Hampshire 03824

Andre P. Deflache
Department of Civil Engineering
Lamar Technical University
Beaumont, Texas 77700

Marilyn N. Delach
Old Core Laboratory
Lamont-Doherty Geological
Observatory
Columbia University
Palisades, New York 10964

Kenneth R. Demars
College of Engineering
University of Rhode Island
Kingston, Rhode Island 02881

John T. Fuller
Ocean Operations Division
Kennecott Exploration, Inc.
10306 Roselle Street
San Diego, California 92121

Daniel K. Gibson
Marine Physical Laboratory
Scripps Institution of
Oceanography
University of California
San Diego, California 92132

Royal M. Hagerty
Deepsea Ventures, Inc.
Gloucester Point, Virginia 23062

John E. Halkyard
Ocean Operations Division
Kennecott Exploration, Inc.
10306 Roselle Street
San Diego, California 92121

Edwin L. Hamilton
Code 5033
Naval Undersea Center
San Diego, California 92132

Loyd D. Hampton
Applied Research Laboratories
The University of Texas
at Austin
Austin, Texas 78712

James D. Hayes
Deep-Sea Sediments Laboratory
Lamont-Doherty Geological
Observatory
Columbia University
Palisades, New York 10964

Donald F. Heinrichs
Marine Geology & Geophysics
Programs
Ocean Science & Technology
Division
Office of Naval Research
Arlington, Virginia 22217

Herbert G. Herrmann
Foundation Engineering Division
Naval Civil Engineering
Laboratory
Port Hueneme, California 93043

Terence J. Hirst
Marine Geotechnical Laboratory
Lehigh University
Bethlehem, Pennsylvania 18015

Barbara M. Horn
Old Core Laboratory
Lamont-Doherty Geological
Observatory
Columbia University
Palisades, New York 10964

David R. Horn
Old Core Laboratory
Lamont-Doherty Geological
Observatory
Columbia University
Palisades, New York 10964

Anton L. Inderbitzen
College of Marine Studies
University of Delaware
P. O. Box 286
Lewes, Delaware 19958

George H. Keller
Marine Geology and Geophysics
Laboratory
Atlantic Oceanographic &
Meteorological Laboratories
National Oceanic and Atmospheric
Administration
15 Rickenbacker Causeway,
Virginia Key
Miami, Florida 33149

William E. Kelly
College of Engineering
University of Rhode Island
Kingston, Rhode Island 02881

Homa J. Lee
Foundation Engineering Division
Naval Civil Engineering Laboratory
Port Hueneme, California 93043

Shun C. Ling
Chesapeake Division
Naval Facilities Engineering
Command
Building 57, Washington Navy Yard
Washington, D.C. 20390

Bramlette McClelland
McClelland Engineers, Inc.
6100 Hillcroft
Houston, Texas 77036

Alex Malahoff
Marine Geology & Geophysics
Programs
Ocean Science & Technology Division
Office of Naval Research
Arlington, Virginia 22217

John D. Milliman
Geology & Geophysics Department
Woods Hole Oceanographic Institution
Woods Hole, Massachusetts 02543

Jean-Pierre Mizikos
Elf RE
7, rue Nélaton
Paris 75015
France

Neil T. Monney
Naval Systems Engineering Department
U. S. Naval Academy
Annapolis, Maryland 21402

Vito A. Nacci
College of Engineering
University of Rhode Island
Kingston, Rhode Island 02881

Iraj Noorany
Department of Civil Engineering
California State University, San Diego
5402 College Avenue
San Diego, California 92115

Richard Rezak
Department of Oceanography
Texas A&M University
College Station, Texas 77843

Adrian F. Richards
Marine Geotechnical Laboratory
Lehigh University
Bethlehem, Pennsylvania 18015

Gilbert T. Rowe
Department of Biology
Woods Hole Oceanographic Institution
Woods Hole, Massachusetts 02543

G. Edward Shank
Ocean Technology Program
Ocean Science & Technology Division
Office of Naval Research
Arlington, Virginia 22217

Armand J. Silva
Civil Engineering Department
Worcester Polytechnic Institute
Worcester, Massachusetts 01609

Frank Simpson
Lockheed Ocean Laboratory
Lockheed Missiles & Space Company
3380 N. Harbor Drive
San Diego, California 92101

Atwar Singh
Lockwood-Singh & Associates
Los Angeles, California 90230

Denzil T. Smith
Marine Sciences Laboratories
University College of North Wales
Menai Bridge
Anglesey, Wales

John B. Southard
Department of Earth and Planetary Sciences
Massachusetts Institute of Technology
Cambridge, Massachusetts 02139

Richard W. Sternberg
Department of Oceanography
University of Washington
Seattle, Washington 98195

Robert E. Stevenson
Office of Naval Research
Scripps Institution of Oceanography
La Jolla, California 92037

Louis J. Thompson
Civil Engineering Department
Texas A&M University
College Station, Texas 77843

Peter K. Trabant
Department of Oceanography
Texas A&M University
College Station, Texas 77843

Philip J. Valent
Civil Engineering Laboratory
Naval Construction Battalion
Center
Port Hueneme, California 93043

Kolla Venkatarathnam
Deep-Sea Sediments Laboratory
Lamont-Doherty Geological
Observatory
Columbia University
Palisades, New York 10964

Mian C. Wang
College of Engineering
University of Rhode Island
Kingston, Rhode Island 02881

Christian A. Wethe
College of Marine Studies
University of Delaware
Lewis, Delaware 19958

INDEX